Ken Ring's

IRELAND WEATHER ALMANAC

2025

BY THE SAME AUTHOR

New Zealand Weather Almanac (1999-Present)
Australia Weather Almanac (2006-Present)
Ireland Weather Almanac (2010-Present)
How to Get Your Kid to Like Maths
Predicting Weather by The Moon
New Meteorological Techniques
We Used to Call It Summer
The Romany's Apprentice
Moon and WeatherLore
Secrets of the Moon
Notes From the Pit
The Lunar Code
Fingermaths
Supertramp
A Fiery Tale

Paperback ISBN: 978-1-73862-306-8
Hardback ISBN: 978-1-73862-307-5
eBook ISBN: 978-1-73862-308-2

Email: enquiries@predictweather.com
Web: www.predictweather.com

CONTENTS

INTRODUCTION

DISCLAIMER

The forecasts in this work are best-endeavour opinions of the authors and associates. No guarantees of 100% accuracy is claimed. Neither can the author and/or publisher be held accountable for decisions made on the basis of information in these pages.

ABBREVIATIONS

MOON SYMBOLS

1st Q	Moon in first-quarter phase.
3rd Q	Moon in third-quarter phase.
^	Northern declination Moon at furthest point north for month.
A	Apogee Moon furthest from earth for the month.
F	Full moon.
N	New moon.
P	Perigee Moon at closest point to the earth out of the 27 day cycle.
P(n)	Priority closest e.g. P3 means third closest for year.
V	Southern declination Moon at furthest point south for the month.
XhN	Moon crossing the equator heading north.
XhS	Moon crossing the equator heading south.

RAIN SUMMARY TABLES

d	Dry most of the day.
h	Heavy rain.
l	Light, intermittent showers.
n	Fog, mist, dew, or non-recordable showers.
p	Passing Showers
r	Rain.
s	Showery.

SUNSHINE SUMMARY TABLES

F	Fine, sunny (8+ hours sunshine)
pc	Partly cloudy (4-7 hrs sunshine)
c	Cloudy (1-3 hrs sunshine)
o	Overcast (0 hrs sunshine)

TEMPERATURE SUMMARY TABLES

clr	Cooler than average temperatures.
a-c	Average to cooler temperatures.
av	Average temperatures.
a-w	Average to warmer temperatures.
wrm	Warmer than average temperatures.

THE NATURE OF WEATHER

Weather is determined by astronomical and physical reasons which are under relatively fixed control of cycles. It is not determined by any activities by Man. Climate scientists claim that the heavier-than-air "trace" gas, carbon dioxide, which is absent in 99.96% of the atmosphere; and methane, which is absent in 99.9999%, have a controlling effect on the climate. This has not been confirmed scientifically.

Approximately 80-85% of the weather, in this author's opinion, is caused by the moon, changing and repeating its position above the earth. Planning for major droughts or floods is therefore possible, by comparing the position of past lunar cycles, but not by measuring any emissions of a negligible gas.

Here is the weighting system. There is space weather, climatic weather, and everyday weather.

1. **CHANGES TO THE WHOLE PLANET:**
   (Irrespective of geography)
   • Deviation of Earth's rotational axis to the vertical.
   • Variation of Earth's orbit.
   • Magnetosphere
   • Heliosphere

2. **CLIMATE AT LOCATION:**
   (long-term trends in weather patterns)
   • Distance from the equator (angle of the Sun's rays = latitude)
   • Direction of prevailing winds
   • Topographic factors – the shape of the land, proximity to other land masses, elevation, and distance from the nearest coast.

3. **WEATHER AT LOCATION:**
   (trends and daily occurrences, observable and recordable)
   • Influence of Moon on tides (including air tides), cycle of rain from ocean, and back to ocean. (80-85%), also oceanic cycles due to lunar influence, e.g. El Nino.
   • Level and intensity of cloud cover (retains local heat, determines what, where, when, of rain (which in turn is subject to ocean and air tides)
   • Solar winds, cycles of Sunspots (contributing to temperature of seasons), hemispherical and seasonal factors.

4. **NOT IN THE MIX AT ALL, EITHER FOR CLIMATE OR WEATHER**
   • Land use changes, deforestation and greenification through agriculture.
   • Volcanic activity ongoing, which does not alter climate.
   • Ice-reflectivity feedback, constant.
   • Industrial and commercial pollution.
   • other activities of Man e.g. diet.

Immediate weather is modified by cycles of solar activity (position and number of occurring Sunspots), and the cyclic position of the Moon relative to the earth. Solar influences can determine floods and drought periods.

The weather has no brain, so it does not know that there is a small percentage of earth that is not covered in the ocean, called the Land. To the moon, it is all oceans. From the height rain falls from, up to 12km, it is immaterial whether there is land below or ocean. That's why rainforests make no difference to rainfall amounts. Forests grow where it rains often, not vice versa.

About 85% of all the weather is controlled, directly or indirectly by the Moon. If you live by the ocean, then the influence is more obvious and direct. If you don't live on or near the coast, then weather systems take a little longer to reach you. But you can see, in a broad sense, the influence of the moon. For example, on almost every full moon night, you will see the full moon in all its glory. And that's because, on the night of the full moon, the moon clears the sky of clouds close to midnight, by dragging more air underneath it. This pushes away the cold air of space and prevents condensation of water vapor which makes the clouds visible. The moon controls the weather every night, however, it is just not as obvious.

When the moon is in the sky i.e. above the horizon, it is less likely to rain. This is a Rule. The main times of rain (if rain is about), are when the moon is underfoot (IC), overhead (MH), or on the horizon (ASC=rising or DES=setting). The in-between positions i.e. every three hours, also have rain potential. Sir Isaac Newton identified that when the moon reached a 45° angle to the horizon, a gravitational force corridor was established, and gravity was lessened, i.e. the moon effect was enhanced. Just before it rains, there is a noticeable chill in the air. There are also extra wind gusts in the hours of moonrise and moon set. This was first noticed and recorded by the Greek philosopher Theophrastus in 300 BC. It still applies today. When the trees suddenly rustle for no reason, a useful exercise is to check out where the moon is.

Jupiter, Saturn, Uranus and Neptune, in that order, affect the Sun, and the sun and moon together affect the oceans. Venus and Mercury affect the tides, but not as much as the sun and moon. Jupiter and Saturn mainly, are responsible for what we call the sunspot cycle. That is when the sun gets overactive and seems to put out visible blotches which some cultures like the Chinese have been watching and tabulating for hundreds of years, for they coincide with extra radiation levels. This in turn translates into increased thunderstorms, and growth, agriculturally.

The sunspot cycle is about 12 years, which is slightly more than the Jupiter cycle of 11.8 years although this can vary between 9 and 14 years. There are variations in the positions of the other planets, and it's the oceans that get affected first. Because the moon is the tiny player in the big team, it gets affected by everything that the big team does, in the same way that a baby is affected by the moods of the whole family. Because the moon is the closest planet to us and so to the oceans, everything in terms of planet aspects involves the moon. There is the timing of the tides, the height of the tides, the speed of the tides, all factors in how the ocean behaves, and all that gets imparted to the weather because all weather begins in the ocean.

That is not a known viewpoint, or if it is, it hasn't reached the ears of meteorologists yet, who still look to the sky for signs of change. That's why, in my opinion, they're wasting their time talking about cloud-seeding, weather modification, and geoengineering, those are arguably scams that people get away with, using research funds made available by western governments, thinking by mistake that the weather originates there. But by the time a weather event reaches the sky, where we on the land see it, it's already half-over, it's obsolete, and just the final dregs of the system.

ORIGINS OF A STORM

All adverse weather comes from *under the boat*. I have never heard a sailor say, aw, a violent storm was going on above me, whilst we were sailing *on a flat, calm sea*. Usually, the flat sea accompanies a high-pressure system. Higher air pressure pushes the sea flat. It is as if your hand was the size of a bath and pushed down on the bath's surface. But the flatter sea arrives first and allows more stable air to gather and consolidate over it, exerting more pressure.

It has been further observed that the deeper a low-pressure system is in the air, the faster the wind blows, imparting more energy, and the longer the wind blows over the same area, the bigger the swells, which are the strong underwater currents. It begins in the sea. If currents increase in the deep of the ocean, then waves will increase at the surface and disturb the air over long distances, enough to generate those faster winds.

Strong winds will always accompany a swell. That's because the waves, which arise from the swell, cause the wind, The etymology says it all. Wind, weather, and wing are words that come from the word *wave*, not the other way around. Fishermen have noticed that the direction in which the wind is blowing matters. The word *west* comes from wind as well. So does the word weather. When there is an onshore wind coming off the sea, what's happening is that underwater waves are being whipped up, which disturbs marine worms out of their burrows, dislodges shellfish such as mussels, limpets, and cockles from their home on rocks, and forces small fish, crabs, and other forms of marine life out of weed beds. The currents change first. How could a surface wind possibly affect a seabed? Surface currents do not transmit downwards to the depths of the ocean.

To prove this theory, put some sand along the bottom of a bath and then splash along the surface, and see if the sand moves. Now place floating things, like corks, on the surface, and generate underwater currents with your hand, and see if the floating things are affected. Which moves more? There is no question that the energy transmission in the ocean is always upward. It means that winds are more likely to be created by currents and swells than vice versa. It would be reasonable to conclude that a clear day indicates little wind BECAUSE the sea is calm, indicating deeper currents have lessened. Deep sea fish may have a sense of impending stillness in the tides and may plan to be there when small creatures that they prey on emerge from nooks and crannies.

An experienced fisherman can tell when there is a storm coming. The ocean 'speaks' to him and the sound of the swell reaches him first. Deep ocean currents are a fact, but science does not link them to winds. Seismic activity generates ocean currents, and earthquakes are energy waves up to 20,000 mph through solid rock. It is not likely that ocean currents go downwards to generate earthquakes, and besides we already know that some earthquake epicenters originate about 400 or 500kms below the seafloor.

It is more likely that the wave regimes that earthquakes are part of, transmit through rock to vibrate the ocean floor enough to generate the ocean currents, and on into the air via oscillating waves to generate unstable airflows. The fact that larger earthquakes happen around full moons and what we call perigees, together with stronger currents and king tides, attest to that, also cyclones form around these times. The moon and larger gravitational forces are responsible for all three. So, when it comes to what happens at sea and in the air above the surface of the sea, it is more feasible that ocean currents are in charge, rather than the surface winds.

As soon as the temperatures reverse, the winds reverse. Sailors who monitor Sea Surface Temperatures (SSTs) know this to be more reliable than isobaric maps because sea-surface temperatures change before changes in air pressure, which gives warning of storms before they happen. Just by watching the temperature of the water, you can tell what the weather will do, in the next few hours or longer.

SUMMARY

To recap, a typical storm begins with the flat sea, the Sun shining above, heating the surface, accompanied by non-mixing of waters at various depths, relatively warm sailing conditions, and SSTs (sea surface temperatures) starting high. The surface of the sea is warm because the layer of water constituting the surface is not being replaced by cooler water. Layers of water closer to the surface are the warmest, being closer to the warmth of the Sun. Then, as mixing begins below the surface, cool water is brought up from below, causing surface cooling. Further mixing brings more cooler waters up from the depths and there is an interchange with the warm waters close to the surface. Due to the drop in temperature at the surface, clouds may form as cooler air mixes with the remainder of warm air. Because warm air always moves to cold, the air begins to move, and wind "picks up". There is a storm brewing. Warmer air continues to supply the area which has now considerably cooled, and the surface is described as windy, and then as a gale. The clouds increase and darken with the load of water that they gather. Condensation accompanies cooling, supplied by the movement of air. The surface of the sea begins to ripple strongly as the swell increases. Each wave has a convex and concave side. The convex has a higher pressure and the concave has a lower pressure. Waves are created, with higher pressure moving up each crest. The waves are joined together, and big areas of high pressure are juxtaposed with big areas of low pressure, and strong winds are generated which go from high pressure to low pressure. The storm keeps up if it is fed by the waves and currents, which are the dynamics of the moon, interacting with gravitational factors that produce underwater earthquakes and faster swells, e.g. new moon, full moon, perigees, and when the moon crosses the equator, which is called the lunar equinoxes.

The storm dissipates in the reverse order. Winds lessen and the sea surface temperature rises. Mixing below the surface ceases gradually. The Sun comes out as clouds seem to dissipate and the ocean surface flattens. It is all controlled by the size and direction of the swells. Sailors know that to escape a storm, you sail into the incoming wind, which is into the whitecaps. They have known it for thousands of years, and there are ancient accounts in ancient texts, but they may not have realized why.

QUESTIONS FOR SCIENCE

If we ignore the moon then how could a huge celestial body between a third and a quarter the size of Earth, and only ten Earth circumferences away, not have any effect on Earth, sea, and the air? Astronomers may have let us down over the years. They say that the moon only affects the tides, but nothing else. If it affects something like tides on the earth, then why does it not affect other things that are on the earth too, like humans? And how could all three layers (air, sea, and land) not be interconnected - considering they completely and perfectly interface over most of the surface of the planet? How could it be that only one could get away with having a tide, but not the other two, when all three are joined? Aren't joined things part of a joined system that acts together? Science should answer all questions, for that is the essence of scientific pursuit.

HEAVY RAIN

Cyclones are formed in countries when the sun within 6° latitude of the equator warms the waters that must be 26°C -28°C, over a wide ocean area. And that produces massive evaporation. The heat from a full moon day, especially when the moon is close, in certain areas can create a tropical cyclone close to the equator. When the full moon's heating capacity diminishes, then so will the cyclone. The huge accumulation of water in the sky, ready to fall and cause destruction, can only come from the sea.

That is why they immediately begin to fall apart, the minute that full moon moves away. And that's why cyclones don't form south of Fiji or in Europe. The water is just not warm enough That's why you never get cyclones occurring in Ireland. Cyclones travel from either north or south depending on which hemisphere they are formed in. They will very rarely cross the equator.

Cyclones, hurricanes, and typhoons predominate during equinox months, which are mainly when the atmosphere undergoes a twisting action, through the plane of what we call the ecliptic, due to the increase in speed of the Sun relative to the orbit of the earth. If you can picture the midpoint of a pendulum, that happens in both March and September. Furthermore, with spring and autumn full and new moons, king-tides occur when the moon rises at or near the equator. This represents the midpoint of yet another pendulum, the Sun on the equator being the first, and now the Moon on the equator for the tides. Sun and moon are moving faster relative to earth around equinoxes, so atmospheric disturbance increases.

The end of 2025 is expected to reveal El Nino conditions in the Pacific region, which typically suppresses activity during cyclonic seasons. Currents along the equatorial band, between Tahiti and Darwin stall, allowing for warmth to build up, as it did in 1998. That was an El Nino year, along with 1982/3, 1986/7 and 2016. 1997/8 is still regarded as a powerful El Nino-Southern Oscillation event. Further down the track the Gulf Stream will also slow and stall, allowing for colder Arctic conditions to affect the northern areas of the globe, for a couple of years.

There are twice as many weak La Nina years as there are weak El Nino years. But there are about equal moderate to strong La Nina and El Nino years. La Nina can be viewed as the planks on a deck, with the El Nino akin to the gaps between them. La Nina years are 2x wetter than El Nino years.

It is worth noting that the El Niño/La Nina index has stayed close to the neutral range in 2023 and 2024. In 2025 the index edges up into the El Nino episode range.

PRECIPITATION CYCLES

We return now to what makes more rainfall and what makes less, over a long period. First, we must go back to the cycles of the sun. With each passage of Jupiter, when Jupiter and Saturn sit on one side of the sun in conjunction, there is usually a major drought in the area of Earth that they call, say, eastern Australia. Those years were 1901, 1920, 1940, 1961, 1981, 2020, and they are about 20 years apart. Halfway into this decadal period, Jupiter and Saturn face each other on opposite sides of the sun, and we call that an opposition, and those become flood years. Flood years have higher tides than drought years, and that leads to more rainfall which affect the rivers, and everything falls into place accordingly from the oceans to the farmland. We can note that the conjunctions i.e. drought years, were 1862, 1901, and every 12, 24, and about 36 years, so the next worst was 1959, and then around 2020 (but more broadly 2016-2023), we call them the **solar minimum** years, they were years of world-wide droughts in the drier regions because of low rainfall and low agricultural yields.

Solar cycles determine annual temperatures and amounts of annual rain. Moon cycles determine the timing of weather events. Cycle #23 began in 1998, peaked around 2000/1 (solar maximum), and died off (solar minimum) from 2007-2009. After a 2–3-year period of relative inactivity Cycle #24 began around midway through 2010 and peaked in 2014. It "bottomed out" in 2020-21. 2020/2021 saw the end of Sunspot cycle #24. Cycle #25 began in 2022, is expected to peak in 2026, and then bottom out again in 2032.

By combining lunar and solar influences, temperature (solar) may lead to amounts of rainfall (due to evaporation rates), whereas the timing of severe weather events may follow strict lunar cycles.

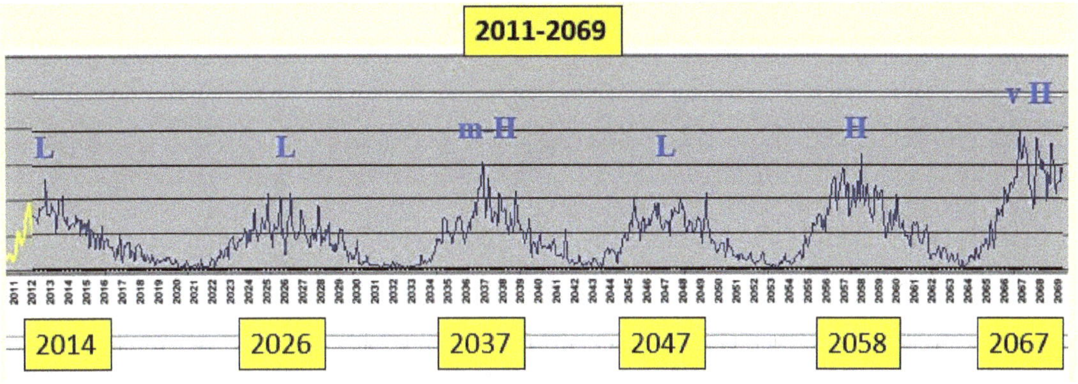

Contrary to what is expected, globally warm periods have fewer and weaker hurricanes while cooler periods have more and stronger hurricanes. 2022 was such a year.
It is the Sun's heat that causes evaporation which brings rain. The amount of heat that the Sun puts out determines the amount of evaporation, which must then fall within about 7 days, near the water source. Coastal locations are generally wetter than places that are some distance inland. Without a source of nearby water, a place quickly dries out. The *amount* of rain falling is due to the amount of solar radiation which causes evaporation. Therefore, rain amounts depend on the sunspot cycle. The amount of rain falling is also dependent on the extent of the air tide, which can be likened to the volume of water in each sea tide. As the solar cycle of radiation is cyclic then so too are rain intensities.

Sea tide timing is predictable, subject to the positions of the Moon and Sun. The volume of water flowing into an estuary depends on wind direction subject to the shape and size of surrounding hills, wind forces increased or confused by surface currents, and less visible factors like the proximity of underwater eruptions, fissures, earthquakes, and time of the day due to temperature gradients. Amounts of rain are influenced by the Sun's warmth for evaporation and then how much cold air comes to the layers of the atmosphere containing clouds.

Sometimes colder air is at a higher or lower altitude than rain-filled clouds, and expected rains must wait for the marrying of the two factors. Until then a cloud may look like rain but heat from the ground is sufficient to buoy up the rain-laden cloud and prevent rain from falling. Analyzing sunspot cycles is at least better than having nothing at all, by looking to see which previous years correspond to the same points of the cycle.

THE ATMOSPHERE

Temperature, volume, and pressure became the focus of climate science a few centuries ago when the Ideal Gas law was discovered. This states that the three factors are dependent, and if given a closed system, and played around with in the laboratory, then by changing one variable you can make others change as well. These principles were applied to the air in the atmosphere regarded as an enclosed gas. It is supposed to be enclosed because it is held here by gravity. This is where science broke down due to a fundamental misunderstanding.

This is not a laboratory, and the earth does not have a lid by which one can change one factor and watch the changes in the other two. In other words, the ideal gas law, which sounds logical, cannot be applied to the atmosphere, which is open to space. Nor can it be shown to work in the laboratory without enclosing the gas in some container. Gravity is still working in the laboratory when you remove the lid of your container. Therefore, we could arguably scrap the ideal gas law, as it applies to our environment, and the idea of temperature (which is a feeling) along with it.

What of volume and pressure? Well, except in isolated and controlled cases, we cannot watch the effects of air pressure on the air, because it is generated too locally close to the ocean surface to be of any use to the atmosphere. Besides, any wind is generated by currents in the sea that work their way to the surface, and create high and low-pressure areas, and wind between high and low pressure, which is only measured on the ocean's surface. Yet the surface of the sea is NOT where the air pressure is generated.

Which leaves us with volume. Science has not come up with a way of accurately quantifying the amount of air in the world, and therefore telling us that there is a danger of it running out. If every tree on the land was instantly destroyed, we would still have 4 million years' worth of oxygen in the atmosphere, because air contains 21% oxygen produced by photosynthesis on the biomass in the ocean. That means 99.9% of our vegetation, which produces oxygen for the survival of every species of life, operates mostly in the sea, and away from the air. That means CO2, or anything else introduced to the air, can have no effect.

The sun warms the earth, but the oceans cool it. Even on the hottest summer's day, a swim in the ocean will be refreshing because the water is always cooler than the air above it. What's more, it can never get hotter than the air above. Everything is in balance and will remain so, and what we call the weather is what we perceive as the outcome of planetary cycles that began thousands of years ago.

WEATHER GLOSSARY

To get the most out of the forecasts, it helps to be able to read a weather map and to understand the language used. A high air pressure – a fair-weather sign, is when the isobaric values increase inwards towards the center. A low-pressure zone or depression, indicates worsening conditions, and when isobaric values decrease towards the center. Closer isobars = stronger winds.

TERM	MEANING	WEATHER
Apogee	Moon furthest from Earth for month	Winds generally light. Calm conditions.
Perigee	Moon closest to Earth for month	Spells trouble.Potential Earthquake risk.
Cold front	Cold air advancing, pushing away warm air	Changeable
Depression	Area of low barometric pressure	Unpredictable, bad weather, strong winds
Ridge of high pressure	Edge of anticyclone	Good weather, cooler nights in winter
Stationary front	Slow-moving weather	No change for 2-3 days
Northern declination	Moon at northernmost point for month. Moon rises in the north.	Slow moving weather system, northerlies, rain in north, colder conditions
Southern declination	Moon at southernmost point for month	Slow moving system, southerlies, generally milder, precipitation in south.
Moon crossing equator	Faster moving weather system, moonrise is due east.	Turbulent, changeable, potential electrical storms, gales or winds expected.

SPEED READING OF CLOUDS

METHOD

For many years, there have been people who have looked at the clouds and determined the coming weather. The techniques are relatively simple to master. Generally, the weather in the early morning does not change much during the day. Whatever reading is gained from looking at the clouds, should usually hold for 24 hours. When we look up and take note of the types of cloud shapes that are forming, and then we look again in an hour, what is developing or fading away is noted. Your guru skills have already noted the sky, at least an hour or two before you announce a prediction.

TYPES OF CLOUD

There are essentially 3 types of clouds, which have Latin names to easily describe their shapes –

1. high (cirrus, literally "mares's tails"),
2. middle layer (cumulus, literally "piles')
3. low cloud (stratus, literally "lines").

HIGH
(20,000-40,000 FEET)

Cirrus; the cloud of change

After a run of clear or bad weather, light streaks of very high cirrus clouds are often the first signs of change. Because of their height, they only move with the rotation of the earth. Bad weather is about 36 hours away. Check in about an hour; if cirrus increases across the sky, it indicates an approaching depression, but if curls and wisps dissolve or vanish, fine weather continues. The higher the cloud, the finer the weather. Rapid cirrus, in bars, or at right angles to bars, mean gale or wind. The bands generally precede a storm by 20-48 hours.

When cirrus become lines (cirrostratus), expect a quick change.

If cirrus streaks point upwards, it means rain. If downwards, wind and dry weather.

If mare's tails are followed by altocumulus (literally "high piles") the next day, it means rain within 12 hours. Cirro-stratus, (literally "spread-out hair"), also means rain. When cirrostratus and stratocumulus gather over the sea, it means rain and probably wind on the second day. Cirrocumulus (literally "piles of hair") is a 24-hour warning. It is also called mackerel sky and means fair weather for that day but rains a day or two after.

Before thunder, cirrocumulus can appear as dense and compact masses, in close contact. If their outline is soft and delicate, if fleeces gently merge, it means a continuance of fine weather. Thunder is usually mainly noise but very little rain. But if the clouds are dense, and associated with cirrus, it means lightning in summer or gales in winter. High cirrocumulus commonly appears a few hours before thunderstorms. The harder and more definite the outline, the more unsettled the coming weather. Cirrocumulus in the early morning, which diminishes means a fine and warm day.

MIDDLE
(6,000-20,000 FEET)

Altocumulus (literally "*high* piles") is a sign of coming thunder. It is also called tower clouds, or Cauliflowers. It consists of turret-like upward growths. A thicker cloud generally arrives some hours later. Altocumulus often occurs together with altostratus, (literally "high, spread out"), and is the cloud that carries rain and snow. It can cover the whole sky, and when it increases, rain and snow over a wide area may be the result. It may also bring drizzle.

Cumulonimbus, (literally "*gathered* piles of rain"), is a sign of heavy rains. They are cumulus clouds which are large enough to produce showers. From a distance, they appear as tall piles, but to stand underneath and look up is to see the dark look of rain. They bring thunderstorms, heavy rain, and occasionally hail. They can be anything from 1000 to 60,000 feet high. Towering thunder clouds that look like mountains, early on a summer's day can become intense storm activity during the heat of the afternoon and often become torrents of rain in the evening.

When the cumulus rain clouds look smaller toward sunset than they were at noon, they are diminishing, and you can expect fair weather. The formation of cumulus clouds to leeward during a strong wind indicates the approach of a calm with rain. But cumulonimbus clouds forming rapidly in the mountains on warm afternoons may mean an intense lightning storm is on its way in minutes.

Cumulus (or ice cream clouds"), are the most common clouds and are caused by moderate, generally westerly winds. All have a well-defined flat base, between 1500 and 2500 feet high. If they appear early in the morning they will develop into bigger shower clouds. If they do not form till late afternoon, they will die away to be followed by a clear sky night. If clouds open and close, rain will continue.

If the clouds are like sheep's wool, deep and dense, the edges being very white, the surrounding sky bright and blue; they are of a frosty coldness, and will speedily fall in hail, snow or rain.

Strato-cumulus, (literally "spread-out gatherings"), are the doors to a cold front. 1500-6500 feet. They cover the sky, sometimes in parallel bands but they usually bring little or no rain. When at sea, the strato-cumulus clouds appear on the horizon, it is a sign that the weather will break up. If over the land in the winter, they may produce a light sprinkling of snow.

If long stripes of cloud drive at a slow rate, high in the air, and gradually become larger, the sky having been previously clear, expect rain.

LOW
(500-2,000 FEET)

Stratus, or hill cloud. Nimbostratus (literally "rain-layer") is a cloud of light rain. Sometimes simply called 'nimbus', these are low, uniform, basic rainclouds, 300-2000 feet. Thick enough to blot out the sun, they can disgorge continuous rain or snow. If a little cloud appears in a clear sky, especially if it comes from the west or somewhere in the south; there is a storm brewing. If there are long points, the look of tails or feathers hanging from rain clouds, close to the horizon, with little wind in the summer, thunder may be expected but the storm will be short. Resembling fog and often covering hills, stratus may bring drizzle.

Harbingers of fine weather are when the day breaks out from the morning fog. Stratus at night is followed by a fine day if the barometer stays high. When mountains extend north and south, if fog or mist comes from the west, expect fair weather. If mist comes down from the top of the mountains, expect rain in summer, and snow in winter.

GENERAL

If two layers of clouds move in opposite directions, then bad weather is on the way.

Clouds do not indicate the presence of water vapour, which is everywhere at once. Clouds represent masses of air of a different temperature. The wind is the movement of an air mass and is not necessarily the movement of clouds. Clouds do not move, that is an illusion. They are disintegrating and reforming at the same time, but we view them at a distance. A fog is a close cloud, and if we were to stand in the middle of a mist we experience the air as still around us.

The wind blows rain in, but also blows it away again. It is more likely to rain when the wind drops. The times when the moon is on the horizon, either rising or setting, and when it is directly overhead or underfoot, are times of greatest rainfall. Atmospheric pressure is highest when the tide is lowest, and vice versa. Atmospheric turbulence, and therefore wind and rain, is more likely at mid-tide going out or coming in.

FINAL WORD

Long range forecasting is different from regular meteorology. It is more trends-based, and it relies on archived records. Even then rain statistics from earlier years may not tell the full story. Statistics taken from e.g., Cork Airport, may describe results that are at variance to what falls in the county because the airport is not always representative.

A possibility is that whereas rain may arrive, winds around the rain gauge, especially at an airport, can push rain quickly away, which then gives a subsequent low readout and a possible misread for the region, even though the winds have already blown the rain in.

A region can easily get confused with its main city, because of population numbers. For that reason, the reader is advised to consider notes about a whole region when planning, and regarding any weather event within a 50-80-mile range as a potential for their area also, even if it is listed for a neighboring district.

Often the amount of rain may be localized, and it is not uncommon for each hill valley to have a different microclimate to the next one over. In a forested and hilly terrain, overall amounts of rain for one small area is practically impossible to determine accurately beyond a general trend. A flat region is much easier as rain is more likely to be widespread due to less quickly changing temperatures.

Quite another problem is that rain gauges often shift locations, to suit the convenience of meteorological staff.

It would seem from what has been said that we should give up before we begin. But it is worth in the exercise because if this book is used correctly, to spot trends rather than look for literal amounts, then good planning decisions may be made.

A 24-hour error is the most obvious because we cannot know from our archived data files of weather 20-50 years away, incorporating at least two lunar cycles, which side of midnight historical data was gathered. Most gathering stations display a rain reading taken in the 24hrs before 9 am, but did the rain on the day happen just before 9 am, at say, 8.55 am and hence on the day in the listing, or, say, 7 pm some 14 hours before, qualifying as previous day's rain?

The scientist Benjamin Franklin (1706 - 1790), the inventor of the lightning conductor, was also a lunar weather forecaster. His regular *Poor Richard's Almanac* ran for 25 years, and he had 10,000 subscribers. Franklin hedged his bets slightly by stating: 'I should ask the indulgence of the reader for a day of grace on either side of the date for a specific weather prediction.'

Similarly, temperatures are gathered as maximums and minimums, one of each per day per town. Suppose reading is taken at 4 pm (often the practice) but let's say on a particularly hot day a very high maximum is reached at 5 pm. The hotter value will miss the cut, and the 4 pm reading will be recorded as the day's maximum. But if the next day brings a cold maximum due to an anomalous and temporary southern wind shift, nevertheless the previous day's 4.30 pm reading will be the relevant maximum figure when taken over the previous 24 hours. The error in temperatures can be large because, over a month, this error in average maximums can be magnified. Before long the "normal" is invalid, and the breaking of records is simply fiction media hype to sell newspapers.

Also, as cold air layers have to match with rain-laden clouds for rain to fall, there can sometimes be up to about a 24-hour error in rain arrival. When such errors for every town every day and every year are collated, a so-called trend can be compromised in favor of whatever bias an operator may have. Weather can be generated between 200ft and 8 miles up, which can incur the potential overshoot of a rain system over a radius of up to 50-60 miles (80-100km). So, if rain is forecast for a coastal location but drops 100 km out to sea, it may arguably still qualify as a successful forecast. It is not uncommon for one forecast report heard on radio or TV to cover half the country in one sentence, yet there will be many microclimatic events within even one small area, especially in hilly terrain.

Most historical data is taken at airports, and these are usually well-suited for wind which makes it easier for planes to take off and land safely, and as windier places dry out more rapidly, it means they are typically drier areas than the surrounding hills. Unless your location is at or near the airport the weather may turn out slightly different to the area forecast in this book. This is not a problem in an unchanging landscape but may be one in a varying terrain whereby a farmer living high on a hill or in a valley can't understand why the airport-derived data he finds in this book doesn't tally with what weather occurred on his farm, yet in the same breath he may be aware of his small microclimates around mounds, hills, valleys, and flats.

The almanac process may be the meteorology of the future. There is a respected place now in our society for differing viewpoints. We already have alternative music and alternative medicine. From the old man on the hill who could tell from the spitting of logs in his fire when snow was coming to the way the sailors hauled in sail believing (correctly) that the appearance of dolphins heralded approaching storms, there has always been room for alternative weather. Because weather is so much a part of our lives, we need all the tools, modern and ancient, and all the past observations about cycles that we can acquire.

THANKS

I am hugely grateful to the support team of Glenys and Victoria, for the many hours of work on project design, research and layout, without which the almanac would not be possible. My heartfelt thanks, also to Vicki in dispatch. We also acknowledge Met Eireann for historical archival records. We hope, as always, that the information provided here will assist in your planning.

GRAPHS

MAX/MIN TEMPERATURES FOR COUNTIES

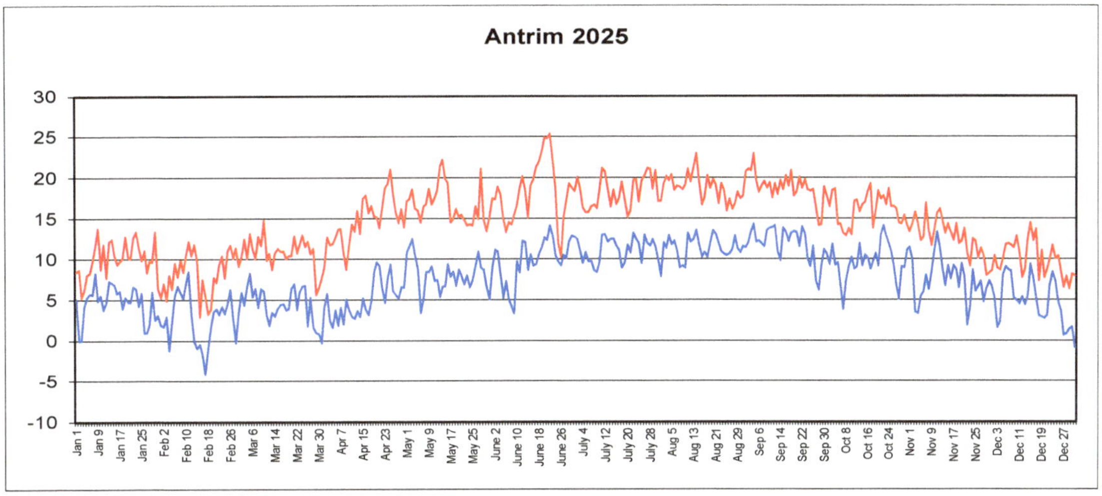

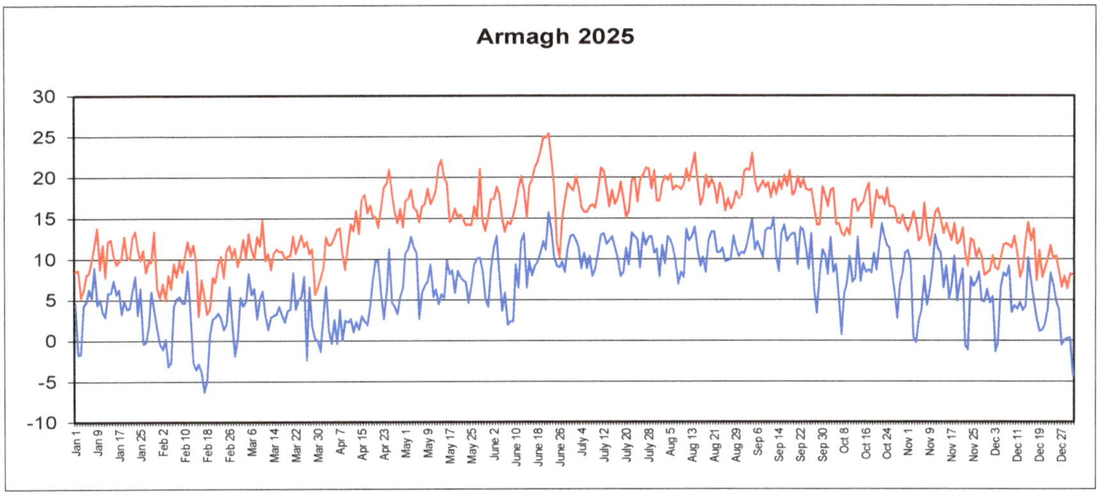

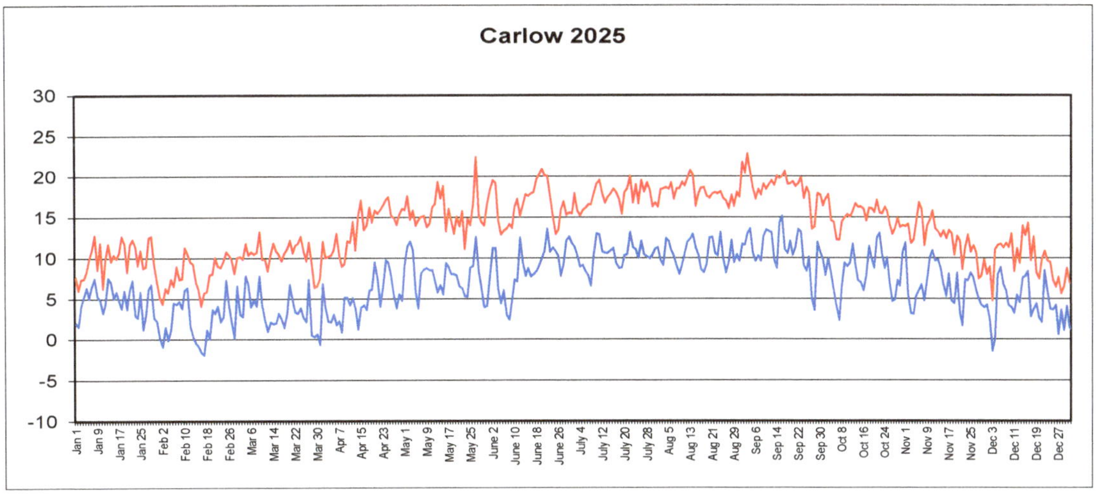

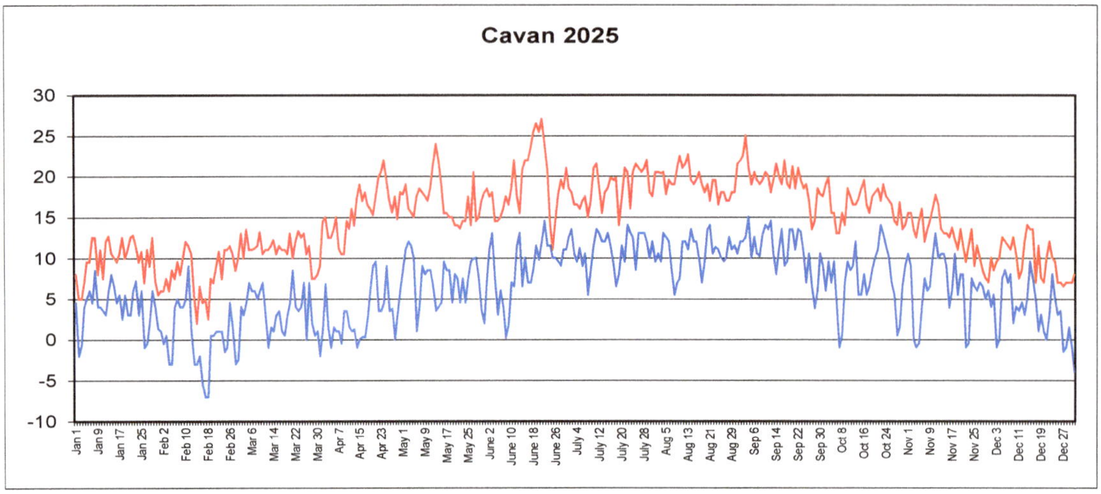
Cavan 2025
30
25
20
15
10
5
0
-5
-10
Jan 1
Jan 9
Jan 17
Jan 25
Feb 2
Feb 10
Feb 18
Feb 26
Mar 6
Mar 14
Mar 22
Mar 30
Apr 7
Apr 15
Apr 23
May 1
May 9
May 17
May 25
June 2
June 10
June 18
June 26
July 4
July 12
July 20
July 28
Aug 5
Aug 13
Aug 21
Aug 29
Sep 6
Sep 14
Sep 22
Sep 30
Oct 8
Oct 16
Oct 24
Nov 1
Nov 9
Nov 17
Nov 25
Dec 3
Dec 11
Dec 19
Dec 27

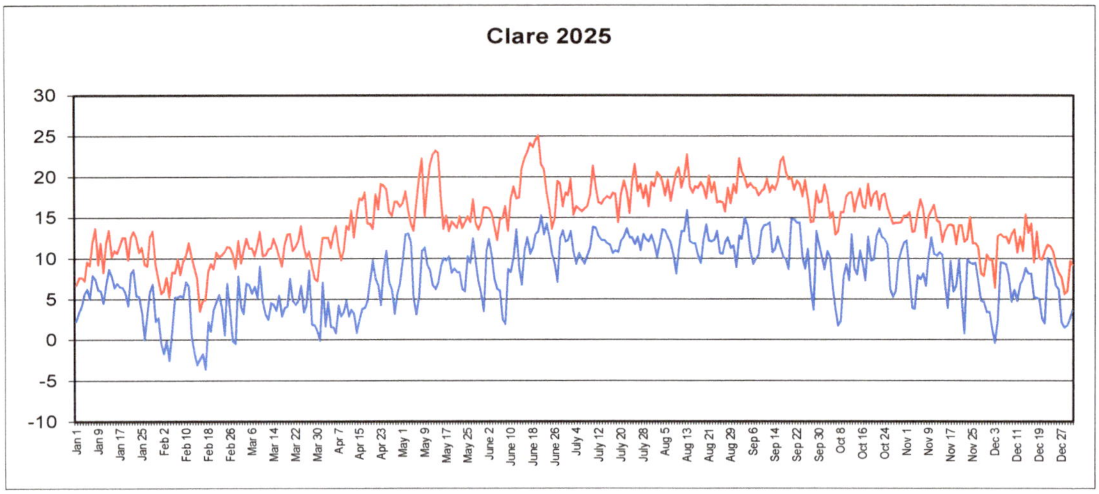
Clare 2025
30
25
20
15
10
5
0
-5
-10
Jan 1
Jan 9
Jan 17
Jan 25
Feb 2
Feb 10
Feb 18
Feb 26
Mar 6
Mar 14
Mar 22
Mar 30
Apr 7
Apr 15
Apr 23
May 1
May 9
May 17
May 25
June 2
June 10
June 18
June 26
July 4
July 12
July 20
July 28
Aug 5
Aug 13
Aug 21
Aug 29
Sep 6
Sep 14
Sep 22
Sep 30
Oct 8
Oct 16
Oct 24
Nov 1
Nov 9
Nov 17
Nov 25
Dec 3
Dec 11
Dec 19
Dec 27

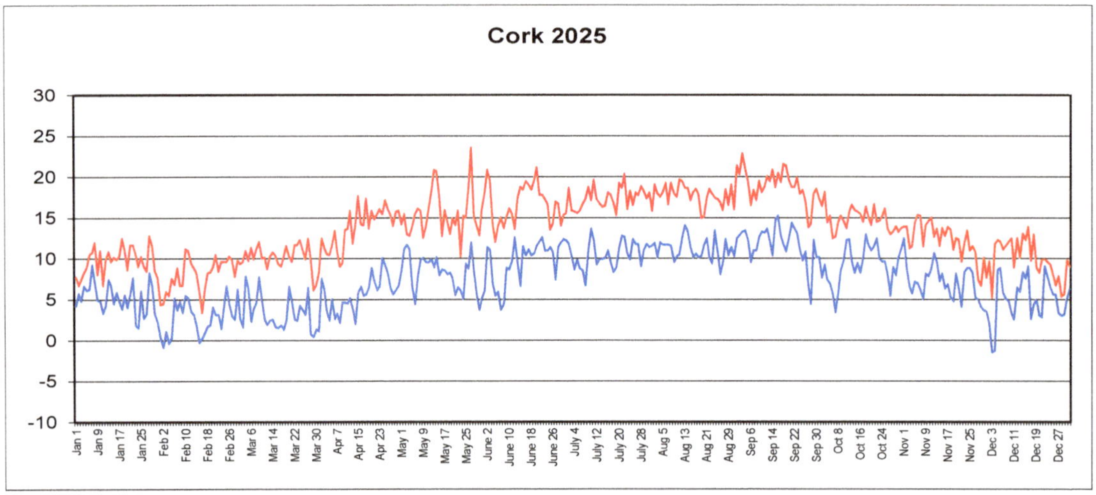
Cork 2025
30
25
20
15
10
5
0
-5
-10
Jan 1
Jan 9
Jan 17
Jan 25
Feb 2
Feb 10
Feb 18
Feb 26
Mar 6
Mar 14
Mar 22
Mar 30
Apr 7
Apr 15
Apr 23
May 1
May 9
May 17
May 25
June 2
June 10
June 18
June 26
July 4
July 12
July 20
July 28
Aug 5
Aug 13
Aug 21
Aug 29
Sep 6
Sep 14
Sep 22
Sep 30
Oct 8
Oct 16
Oct 24
Nov 1
Nov 9
Nov 17
Nov 25
Dec 3
Dec 11
Dec 19
Dec 27

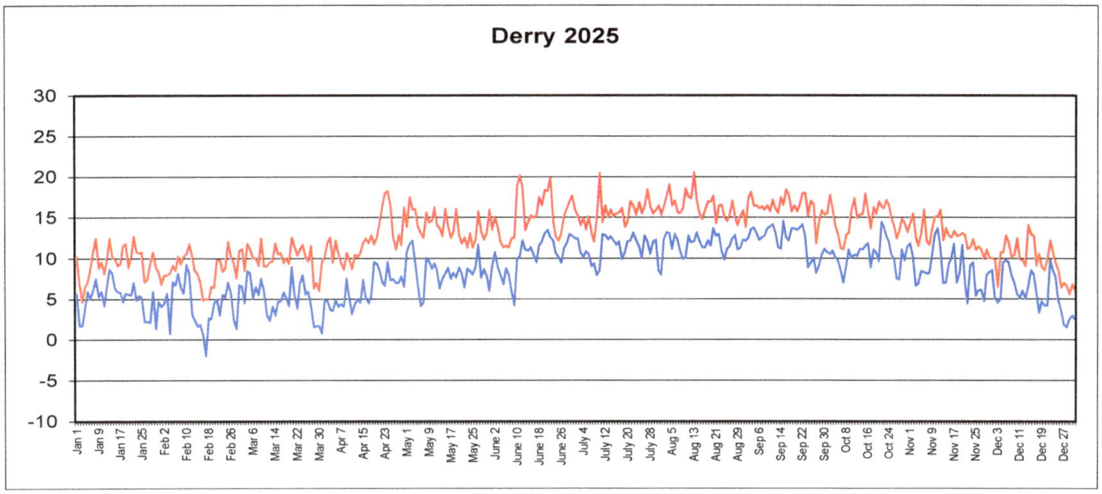
Derry 2025
30
25
20
15
10
5
0
-5
-10
Jan 1 Jan 9 Jan 17 Jan 25 Feb 2 Feb 10 Feb 18 Feb 26 Mar 6 Mar 14 Mar 22 Mar 30 Apr 7 Apr 15 Apr 23 May 1 May 9 May 17 May 25 June 2 June 10 June 18 June 26 July 4 July 12 July 20 July 28 Aug 5 Aug 13 Aug 21 Aug 29 Sep 6 Sep 14 Sep 22 Sep 30 Oct 8 Oct 16 Oct 24 Nov 1 Nov 9 Nov 17 Nov 25 Dec 3 Dec 11 Dec 19 Dec 27

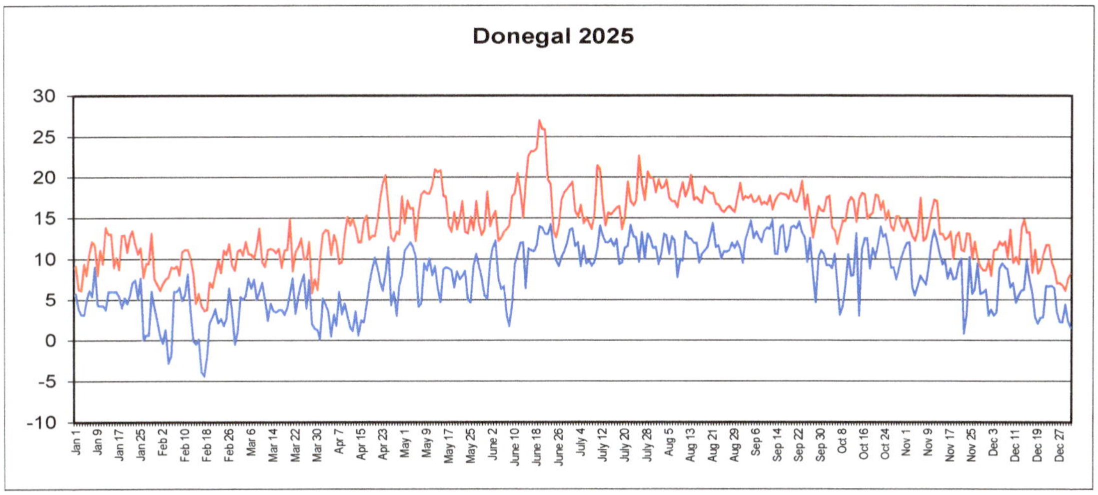
Donegal 2025
30
25
20
15
10
5
0
-5
-10
Jan 1 Jan 9 Jan 17 Jan 25 Feb 2 Feb 10 Feb 18 Feb 26 Mar 6 Mar 14 Mar 22 Mar 30 Apr 7 Apr 15 Apr 23 May 1 May 9 May 17 May 25 June 2 June 10 June 18 June 26 July 4 July 12 July 20 July 28 Aug 5 Aug 13 Aug 21 Aug 29 Sep 6 Sep 14 Sep 22 Sep 30 Oct 8 Oct 16 Oct 24 Nov 1 Nov 9 Nov 17 Nov 25 Dec 3 Dec 11 Dec 19 Dec 27

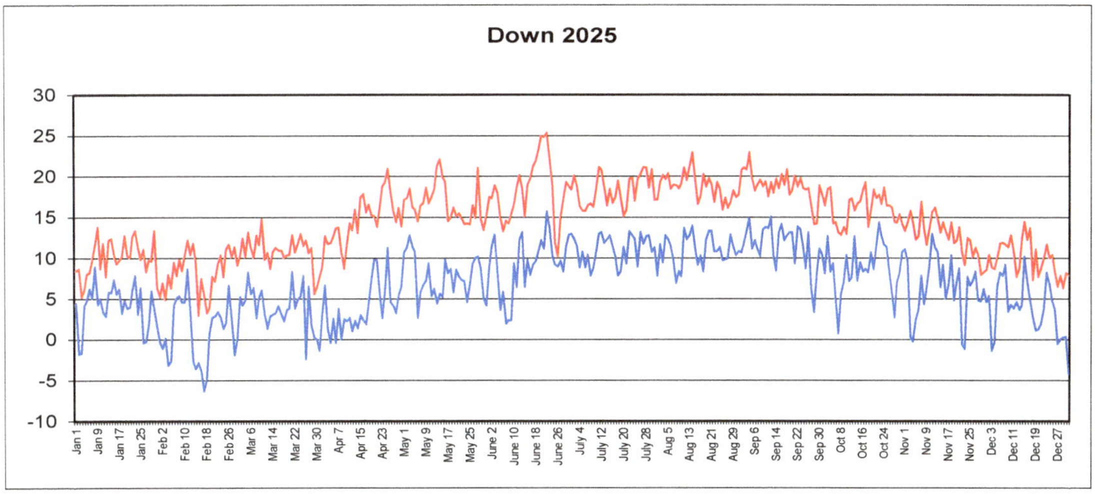
Down 2025
30
25
20
15
10
5
0
-5
-10
Jan 1 Jan 9 Jan 17 Jan 25 Feb 2 Feb 10 Feb 18 Feb 26 Mar 6 Mar 14 Mar 22 Mar 30 Apr 7 Apr 15 Apr 23 May 1 May 9 May 17 May 25 June 2 June 10 June 18 June 26 July 4 July 12 July 20 July 28 Aug 5 Aug 13 Aug 21 Aug 29 Sep 6 Sep 14 Sep 22 Sep 30 Oct 8 Oct 16 Oct 24 Nov 1 Nov 9 Nov 17 Nov 25 Dec 3 Dec 11 Dec 19 Dec 27

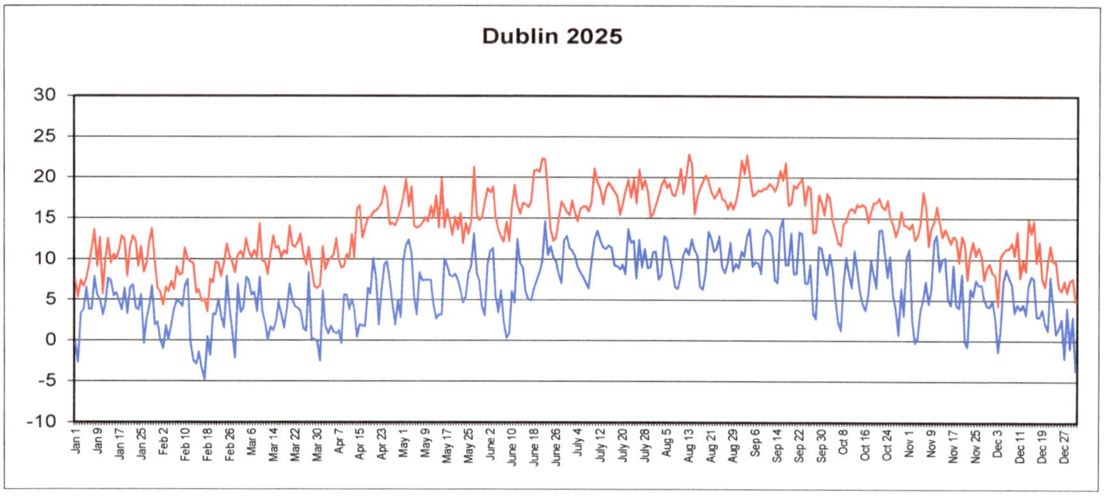
Dublin 2025
30
25
20
15
10
5
0
-5
-10
Jan 1
Jan 9
Jan 17
Jan 25
Feb 2
Feb 10
Feb 18
Feb 26
Mar 6
Mar 14
Mar 22
Mar 30
Apr 7
Apr 15
Apr 23
May 1
May 9
May 17
May 25
June 2
June 10
June 18
June 26
July 4
July 12
July 20
July 28
Aug 5
Aug 13
Aug 21
Aug 29
Sep 6
Sep 14
Sep 22
Sep 30
Oct 8
Oct 16
Oct 24
Nov 1
Nov 9
Nov 17
Nov 25
Dec 3
Dec 11
Dec 19
Dec 27

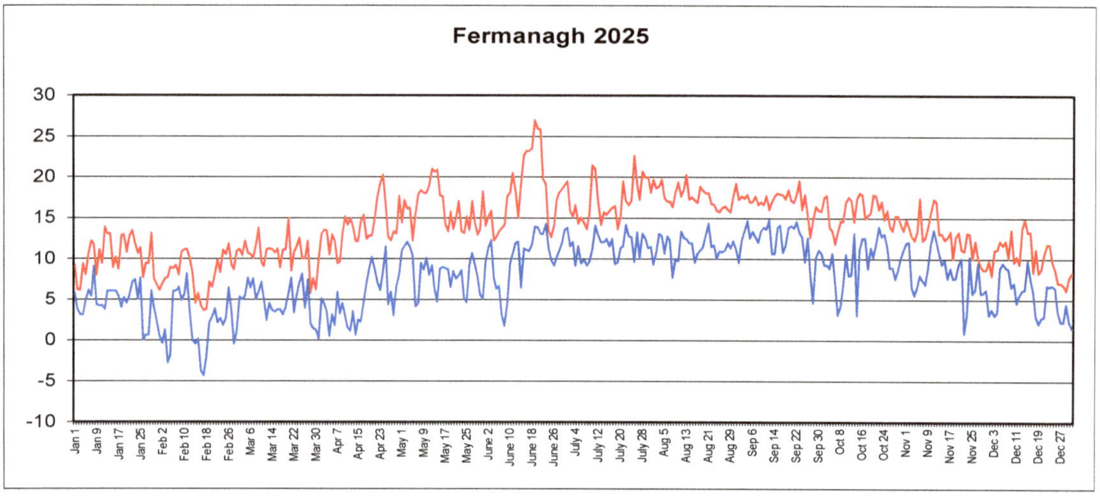
Fermanagh 2025
30
25
20
15
10
5
0
-5
-10
Jan 1
Jan 9
Jan 17
Jan 25
Feb 2
Feb 10
Feb 18
Feb 26
Mar 6
Mar 14
Mar 22
Mar 30
Apr 7
Apr 15
Apr 23
May 1
May 9
May 17
May 25
June 2
June 10
June 18
June 26
July 4
July 12
July 20
July 28
Aug 5
Aug 13
Aug 21
Aug 29
Sep 6
Sep 14
Sep 22
Sep 30
Oct 8
Oct 16
Oct 24
Nov 1
Nov 9
Nov 17
Nov 25
Dec 3
Dec 11
Dec 19
Dec 27

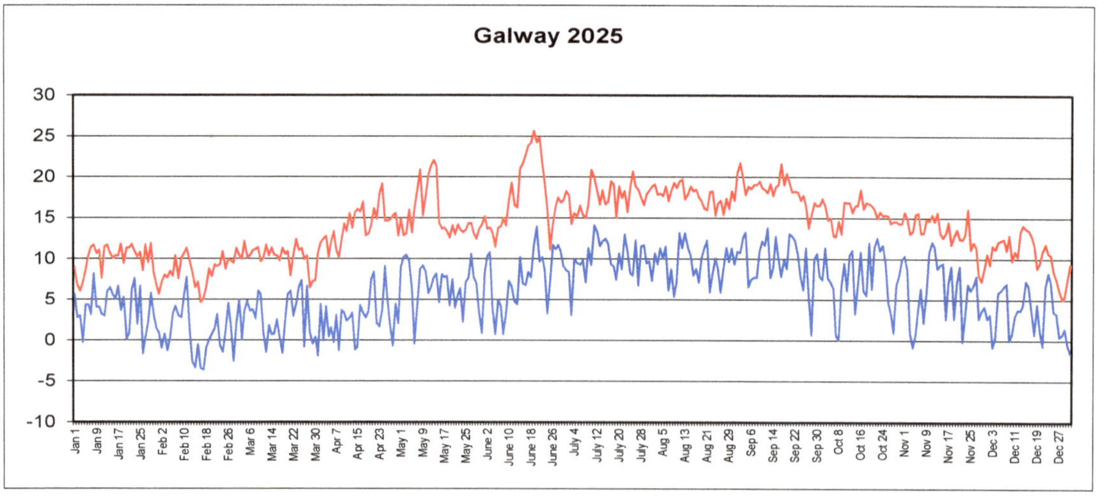
Galway 2025
30
25
20
15
10
5
0
-5
-10
Jan 1
Jan 9
Jan 17
Jan 25
Feb 2
Feb 10
Feb 18
Feb 26
Mar 6
Mar 14
Mar 22
Mar 30
Apr 7
Apr 15
Apr 23
May 1
May 9
May 17
May 25
June 2
June 10
June 18
June 26
July 4
July 12
July 20
July 28
Aug 5
Aug 13
Aug 21
Aug 29
Sep 6
Sep 14
Sep 22
Sep 30
Oct 8
Oct 16
Oct 24
Nov 1
Nov 9
Nov 17
Nov 25
Dec 3
Dec 11
Dec 19
Dec 27

Kerry 2025

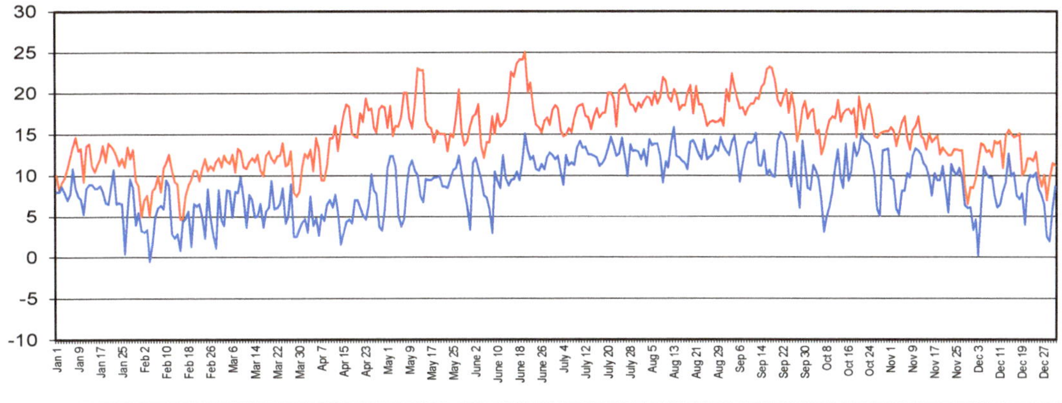

Kildare 2025

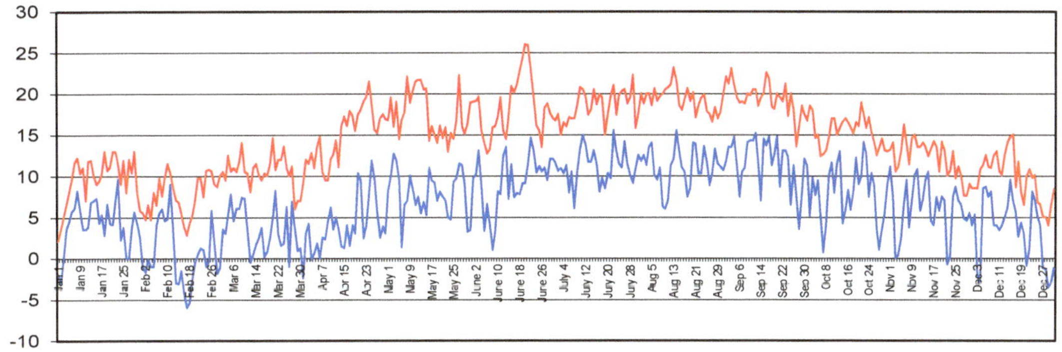

Kilkenny 2025

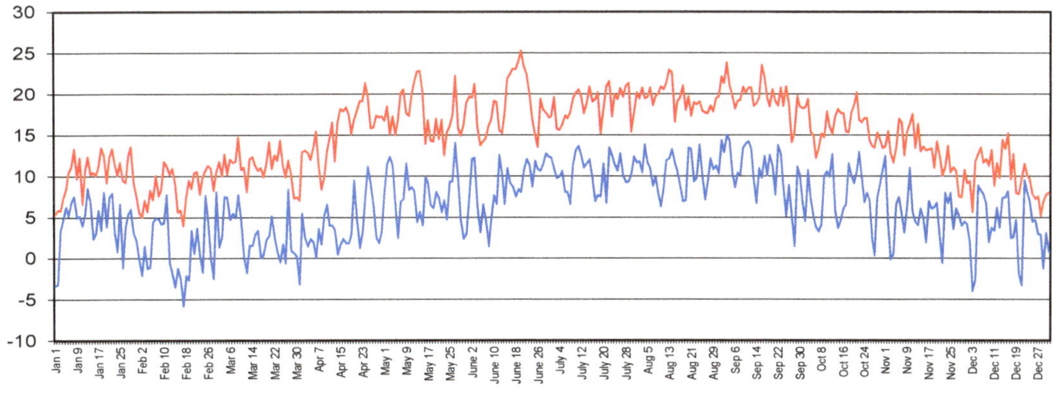

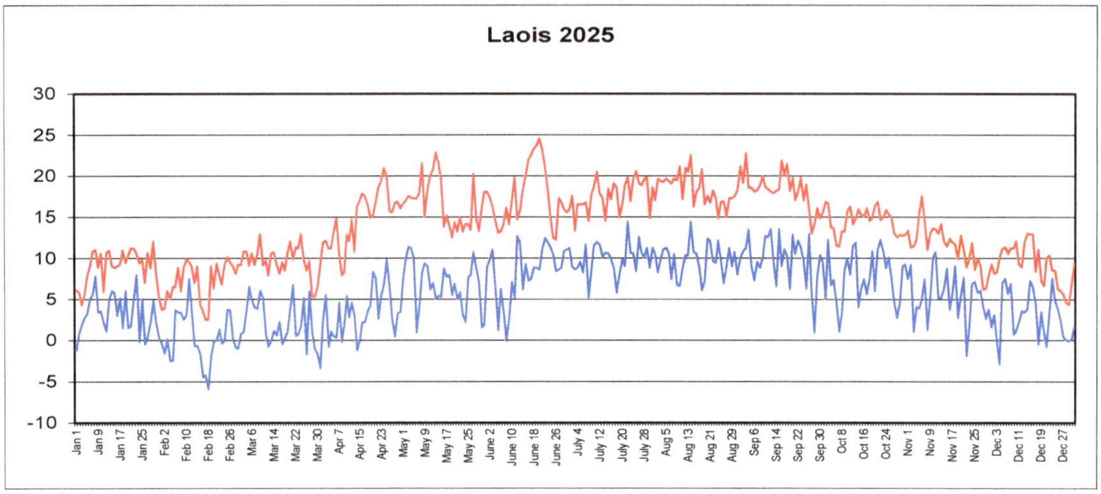
Laois 2025

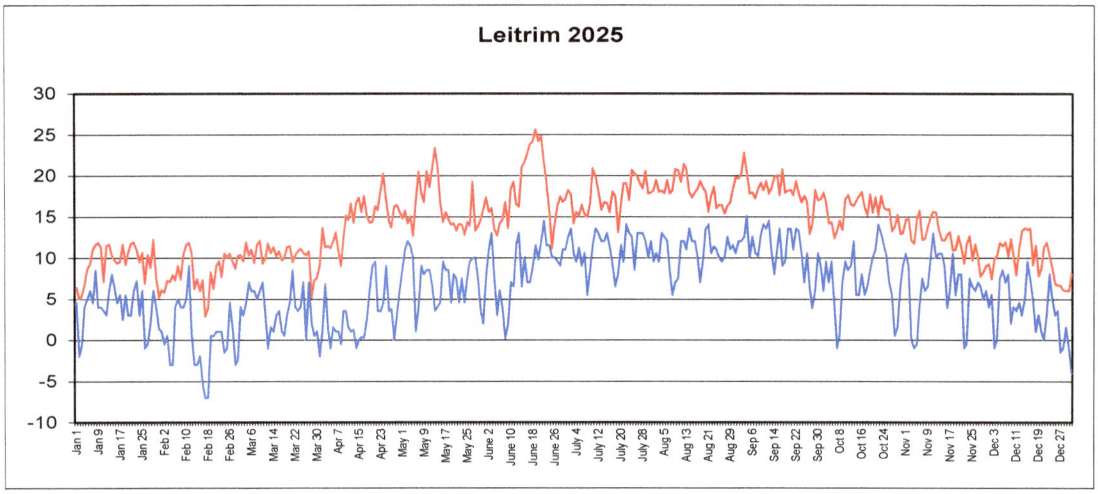
Leitrim 2025

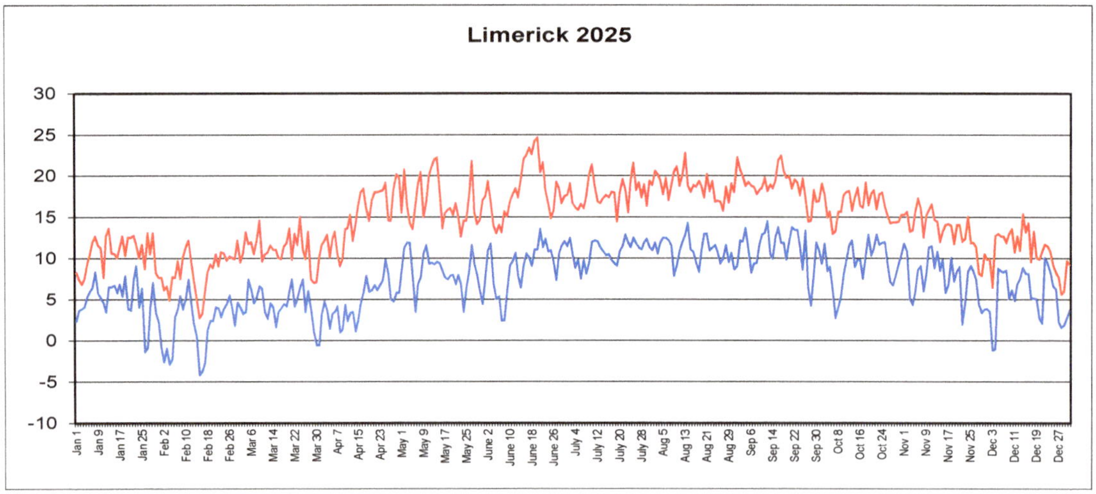
Limerick 2025

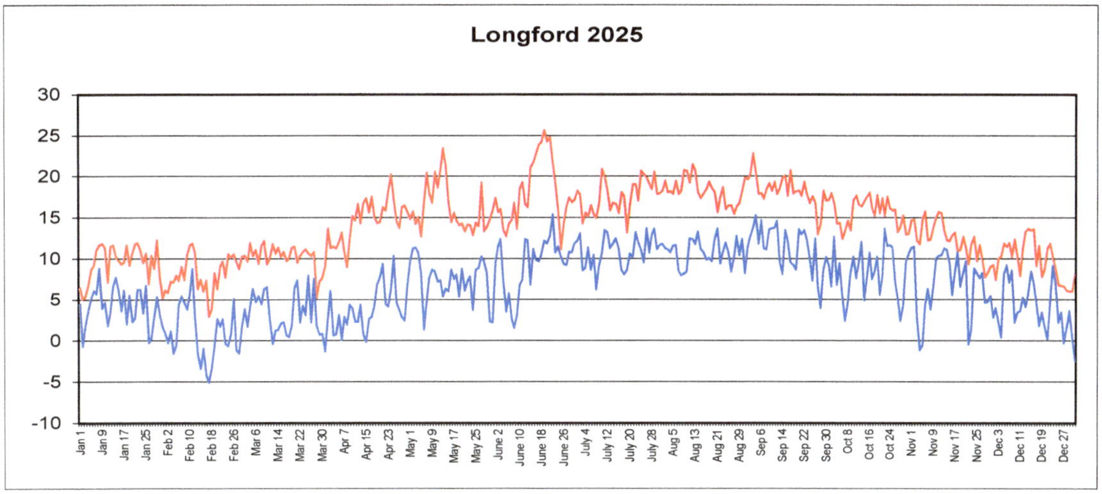
Longford 2025
30
25
20
15
10
5
0
-5
-10
Jan 1
Jan 9
Jan 17
Jan 25
Feb 2
Feb 10
Feb 18
Feb 26
Mar 6
Mar 14
Mar 22
Mar 30
Apr 7
Apr 15
Apr 23
May 1
May 9
May 17
May 25
June 2
June 10
June 18
June 26
July 4
July 12
July 20
July 28
Aug 5
Aug 13
Aug 21
Aug 29
Sep 6
Sep 14
Sep 22
Sep 30
Oct 8
Oct 16
Oct 24
Nov 1
Nov 9
Nov 17
Nov 25
Dec 3
Dec 11
Dec 19
Dec 27

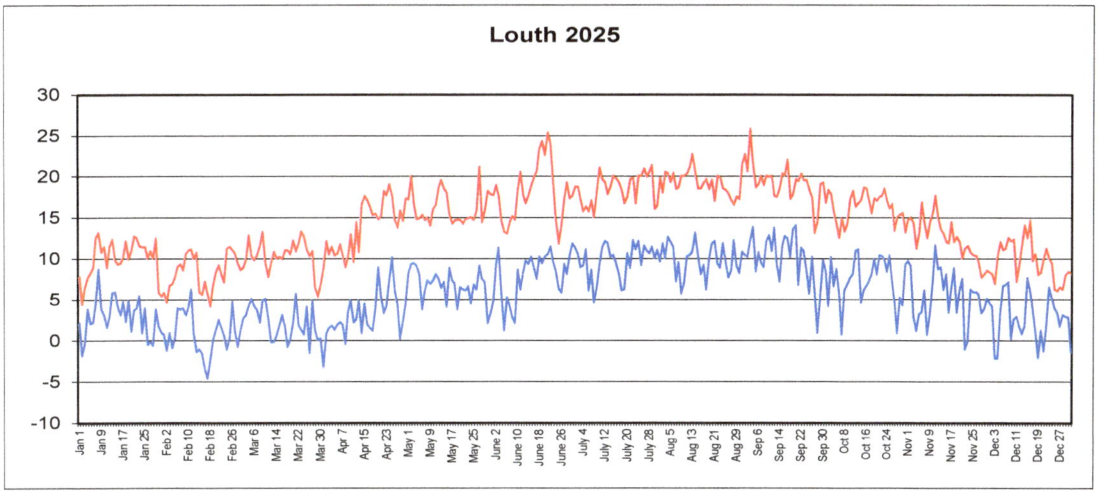
Louth 2025
30
25
20
15
10
5
0
-5
-10
Jan 1
Jan 9
Jan 17
Jan 25
Feb 2
Feb 10
Feb 18
Feb 26
Mar 6
Mar 14
Mar 22
Mar 30
Apr 7
Apr 15
Apr 23
May 1
May 9
May 17
May 25
June 2
June 10
June 18
June 26
July 4
July 12
July 20
July 28
Aug 5
Aug 13
Aug 21
Aug 29
Sep 6
Sep 14
Sep 22
Sep 30
Oct 8
Oct 16
Oct 24
Nov 1
Nov 9
Nov 17
Nov 25
Dec 3
Dec 11
Dec 19
Dec 27

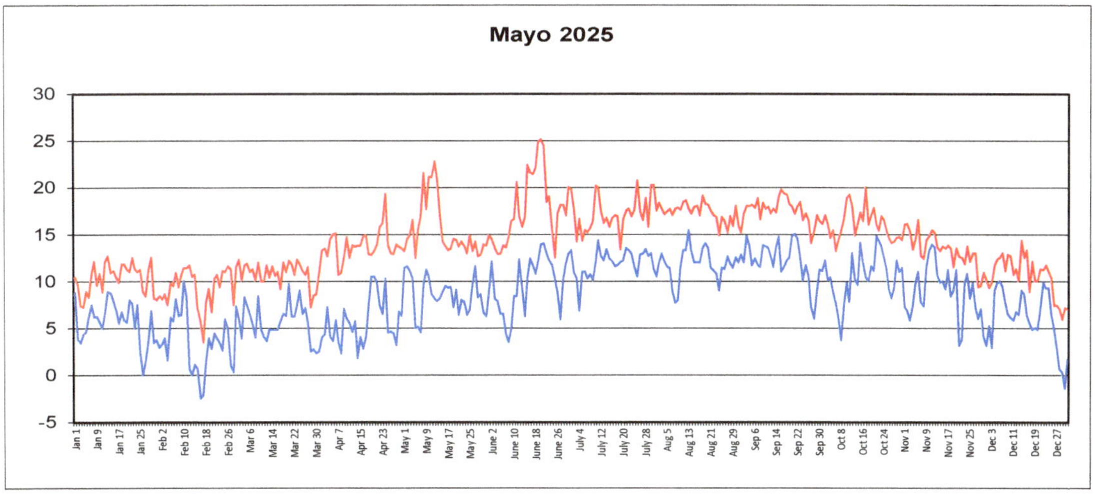
Mayo 2025
30
25
20
15
10
5
0
-5
Jan 1
Jan 9
Jan 17
Jan 25
Feb 2
Feb 10
Feb 18
Feb 26
Mar 6
Mar 14
Mar 22
Mar 30
Apr 7
Apr 15
Apr 23
May 1
May 9
May 17
May 25
June 2
June 10
June 18
June 26
July 4
July 12
July 20
July 28
Aug 5
Aug 13
Aug 21
Aug 29
Sep 6
Sep 14
Sep 22
Sep 30
Oct 8
Oct 16
Oct 24
Nov 1
Nov 9
Nov 17
Nov 25
Dec 3
Dec 11
Dec 19
Dec 27

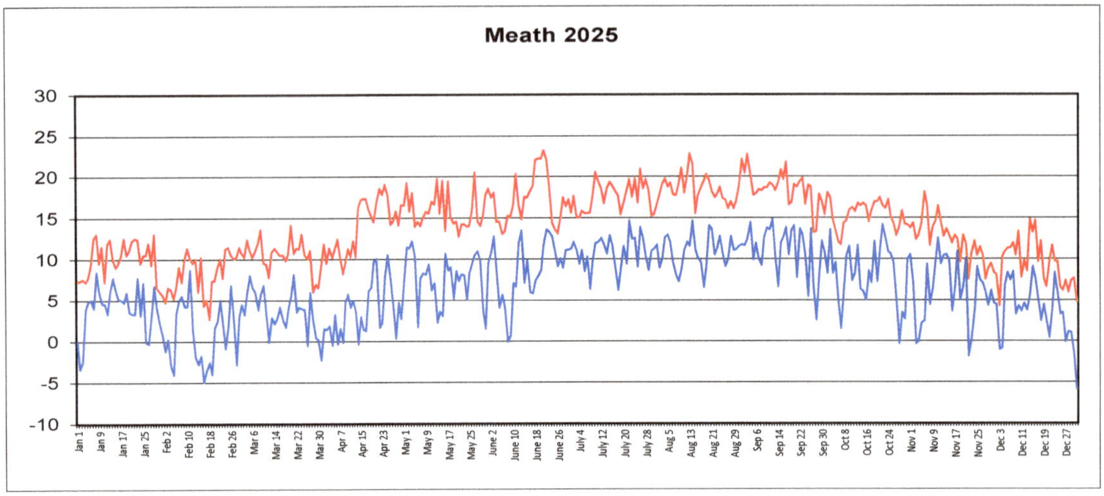
Meath 2025
30
25
20
15
10
5
0
-5
-10
Jan 1
Jan 9
Jan 17
Jan 25
Feb 2
Feb 10
Feb 18
Feb 26
Mar 6
Mar 14
Mar 22
Mar 30
Apr 7
Apr 15
Apr 23
May 1
May 9
May 17
May 25
June 2
June 10
June 18
June 26
July 4
July 12
July 20
July 28
Aug 5
Aug 13
Aug 21
Aug 29
Sep 6
Sep 14
Sep 22
Sep 30
Oct 8
Oct 16
Oct 24
Nov 1
Nov 9
Nov 17
Nov 25
Dec 3
Dec 11
Dec 19
Dec 27

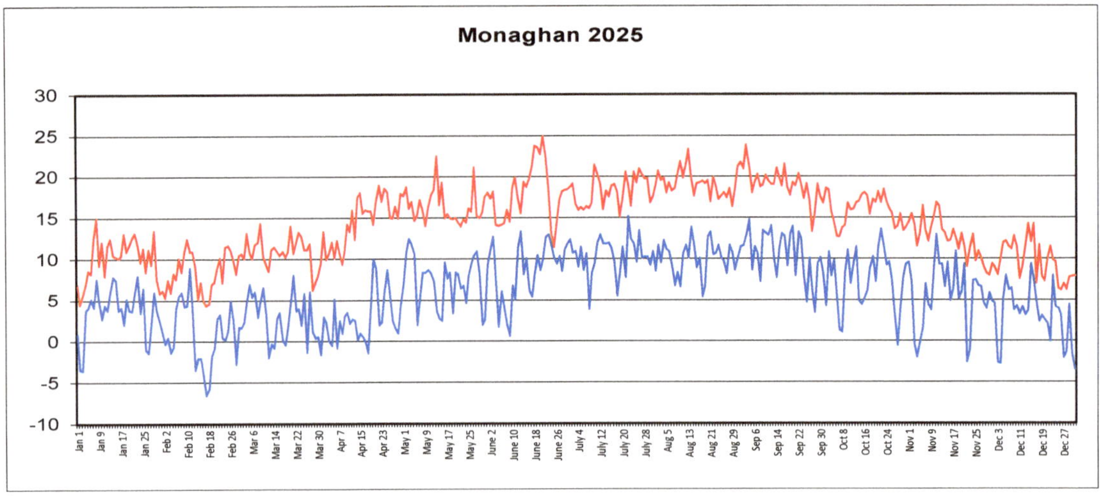
Monaghan 2025
30
25
20
15
10
5
0
-5
-10
Jan 1
Jan 9
Jan 17
Jan 25
Feb 2
Feb 10
Feb 18
Feb 26
Mar 6
Mar 14
Mar 22
Mar 30
Apr 7
Apr 15
Apr 23
May 1
May 9
May 17
May 25
June 2
June 10
June 18
June 26
July 4
July 12
July 20
July 28
Aug 5
Aug 13
Aug 21
Aug 29
Sep 6
Sep 14
Sep 22
Sep 30
Oct 8
Oct 16
Oct 24
Nov 1
Nov 9
Nov 17
Nov 25
Dec 3
Dec 11
Dec 19
Dec 27

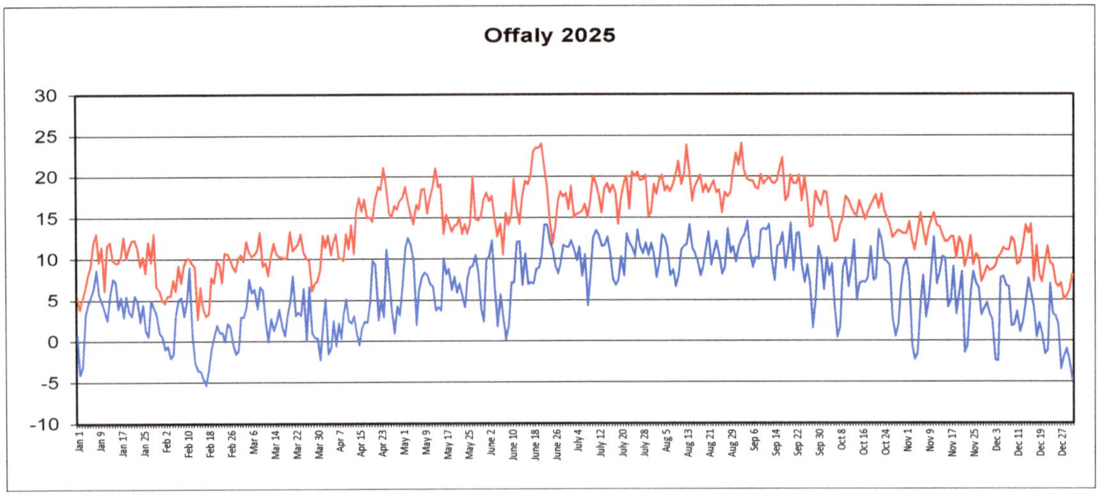
Offaly 2025
30
25
20
15
10
5
0
-5
-10
Jan 1
Jan 9
Jan 17
Jan 25
Feb 2
Feb 10
Feb 18
Feb 26
Mar 6
Mar 14
Mar 22
Mar 30
Apr 7
Apr 15
Apr 23
May 1
May 9
May 17
May 25
June 2
June 10
June 18
June 26
July 4
July 12
July 20
July 28
Aug 5
Aug 13
Aug 21
Aug 29
Sep 6
Sep 14
Sep 22
Sep 30
Oct 8
Oct 16
Oct 24
Nov 1
Nov 9
Nov 17
Nov 25
Dec 3
Dec 11
Dec 19
Dec 27

Roscommon 2025

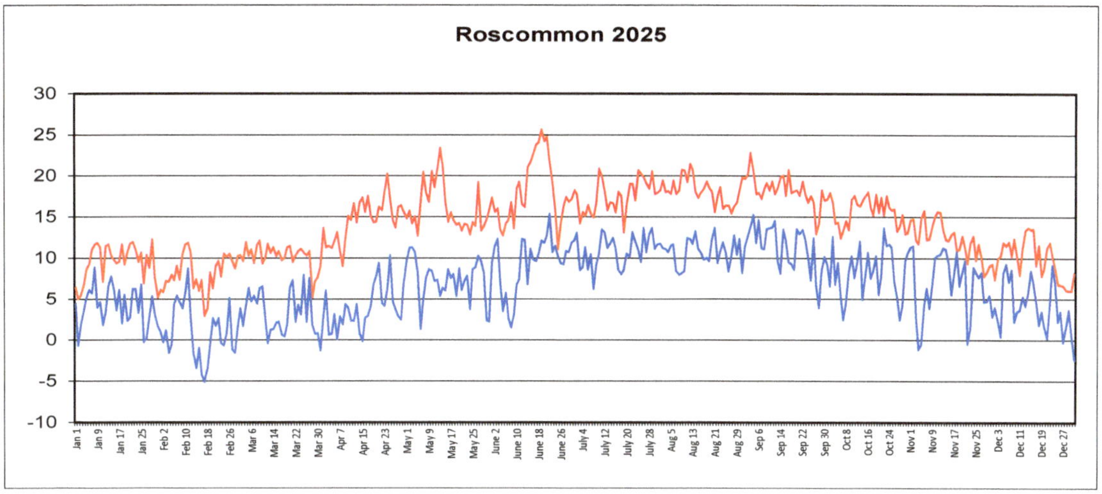

Sligo 2025

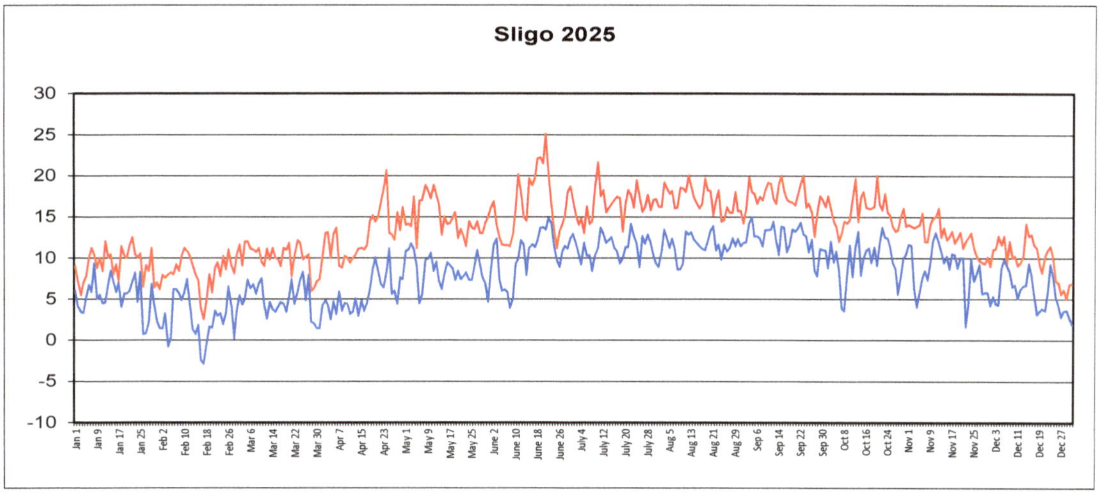

Tipperary 2025

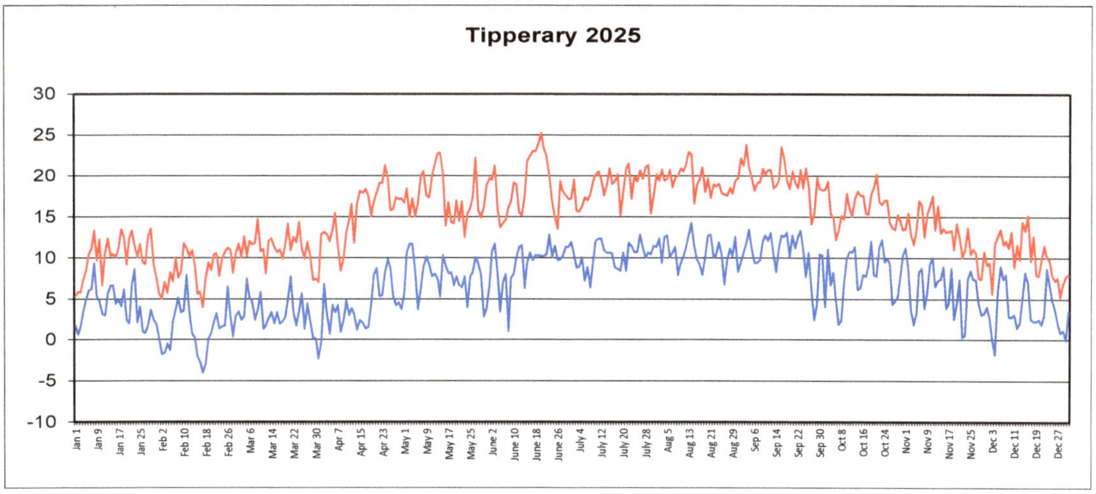

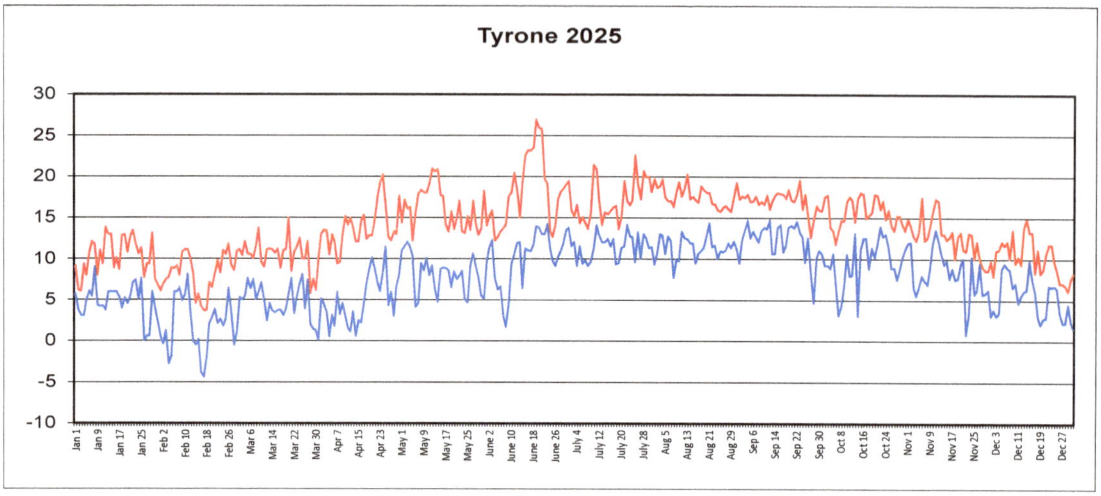
Tyrone 2025

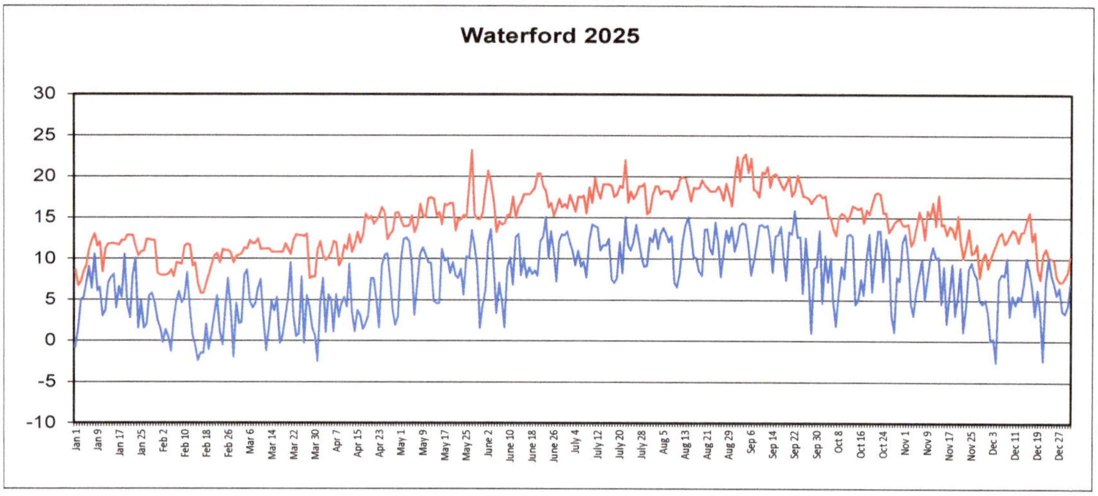
Waterford 2025

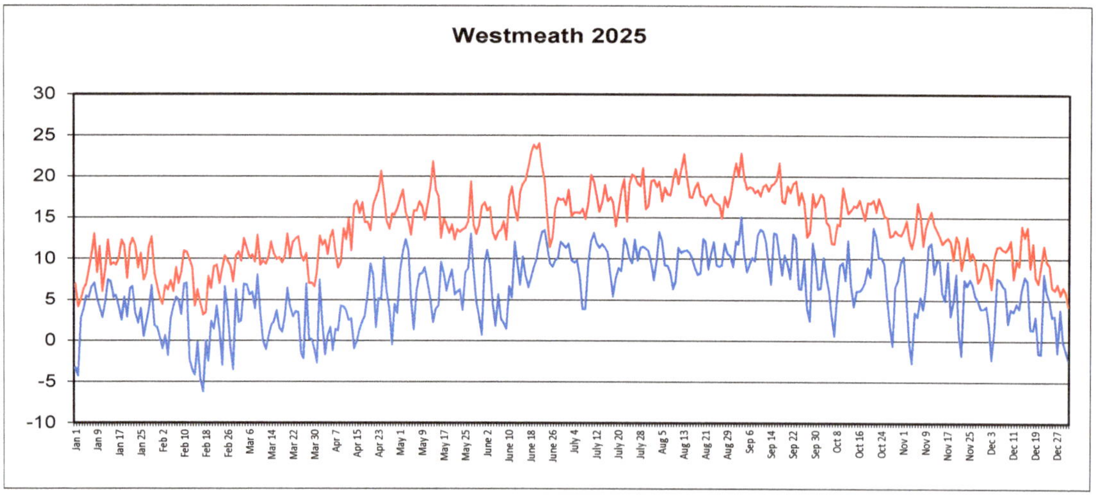
Westmeath 2025

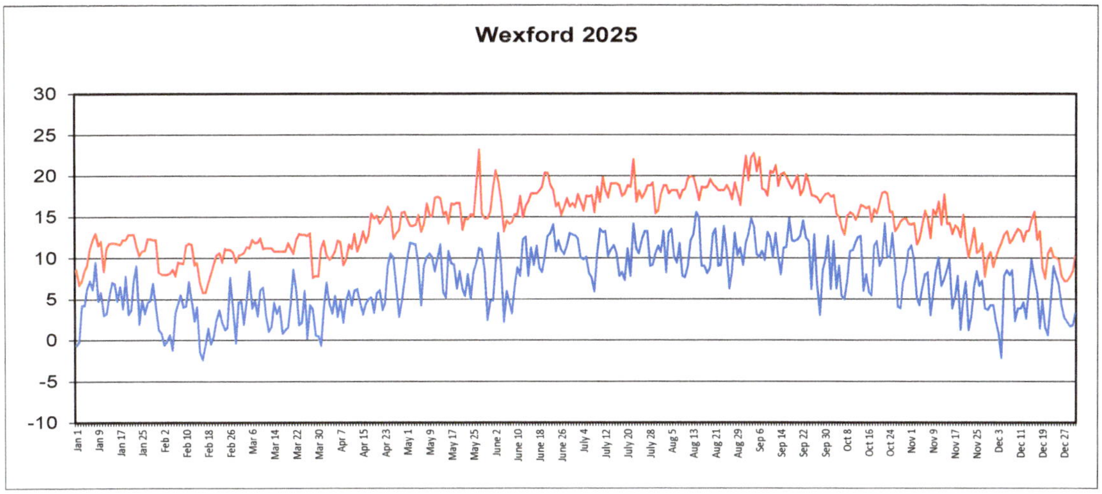

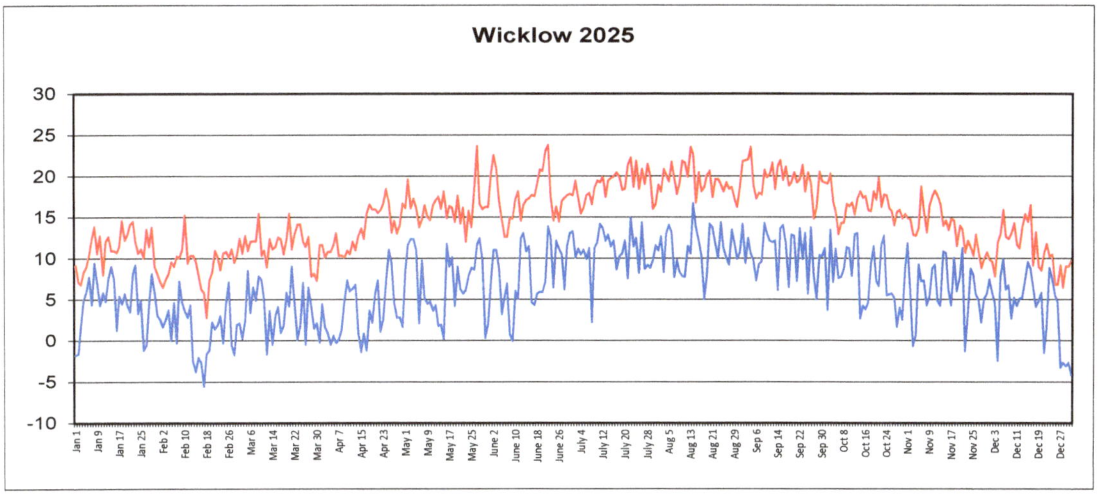

TEMPERATURE TRENDS

AVERAGED ACROSS THE WHOLE COUNTRY

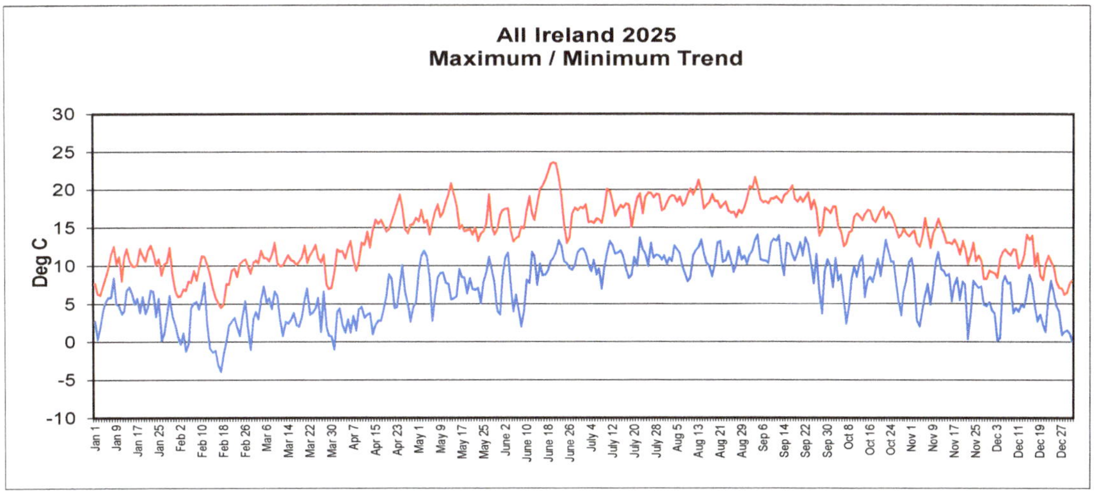

WIND

The graph below shows estimated daily wind force trends, for 2025, as calculated for Knock Airport, County Mayo. This western county has been chosen because of the availability of continuous historical data, and because most wind systems in Ireland come from the west. Other areas may show corresponding trends even if dissimilar force values.

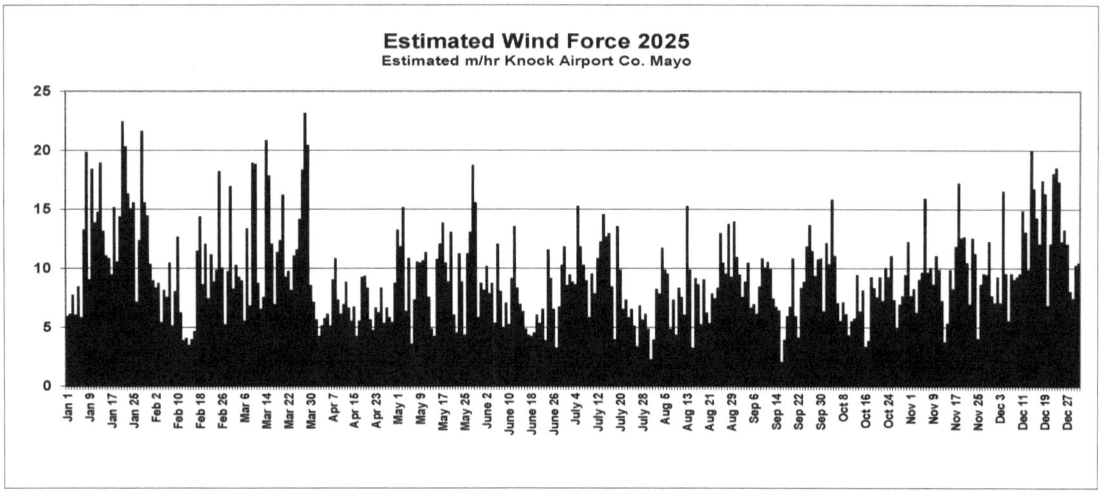

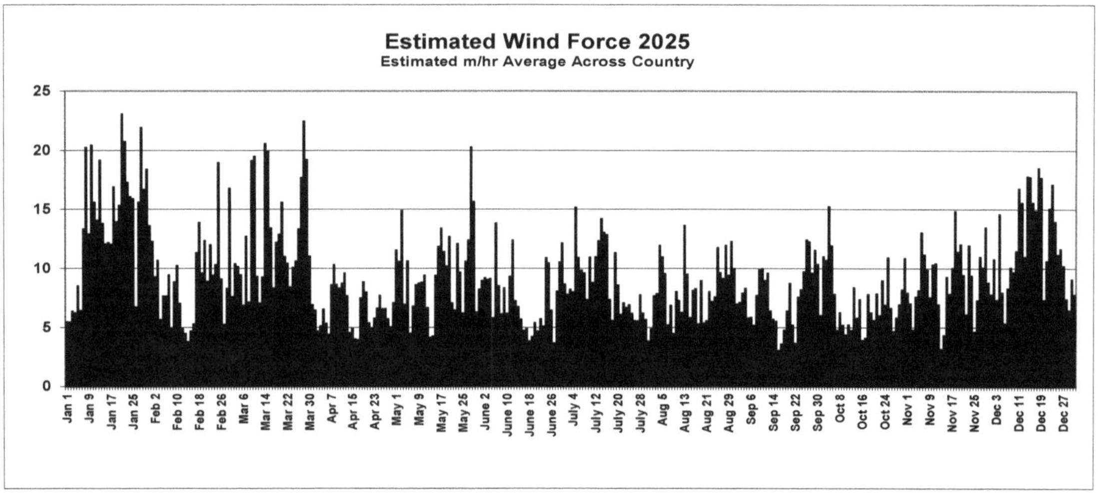

Averaging wind force values across the months reveals that January may be the windiest followed by March and December, with September October and April the calmest. The order of overall average windiest months is estimated to be January, March, December, May, November, February, July, August, June, September, October then April.

This makes the average windiest season, in order of windiest, winter, spring, summer, then autumn.

PRECIPITATION (RAIN/SNOW)

WHEN WILL RAIN AND LOW TEMPERATURES PRODUCE SNOW?

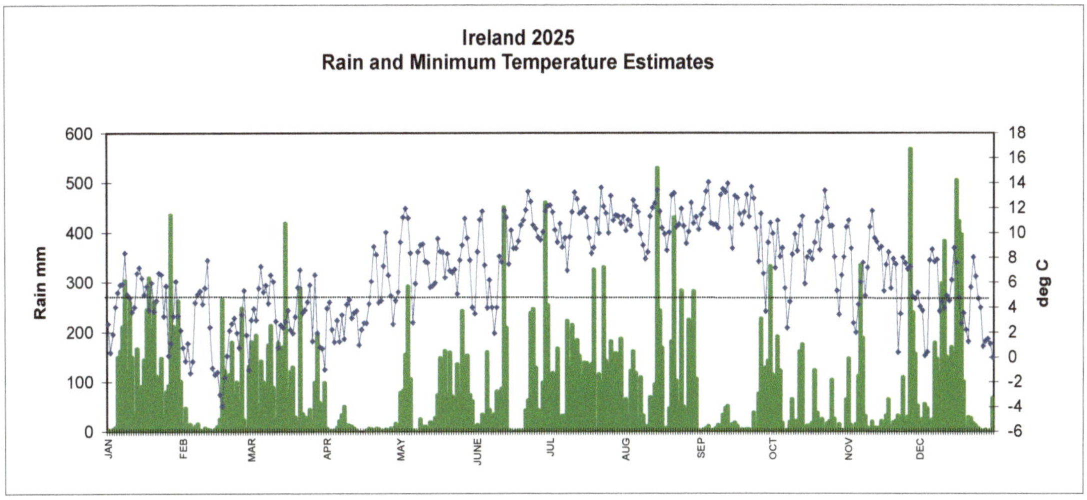

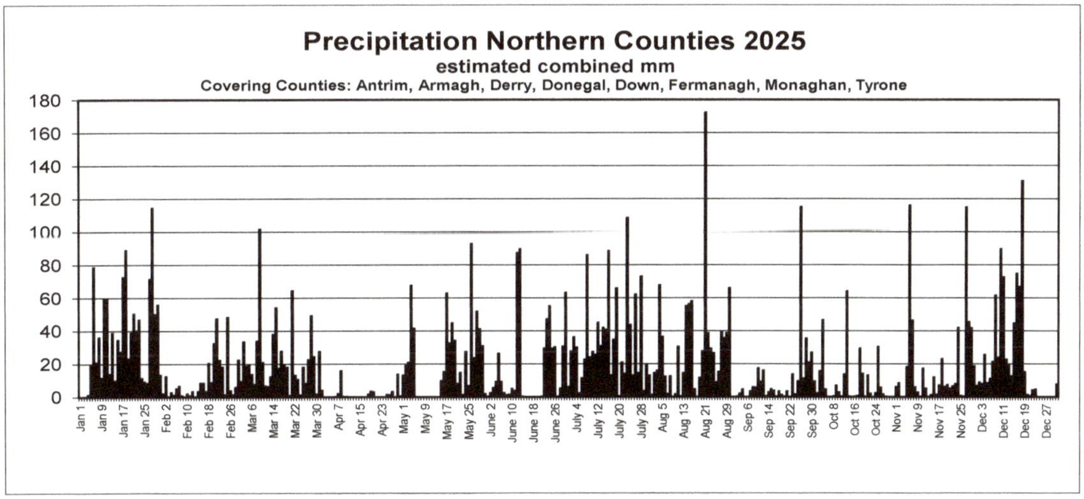

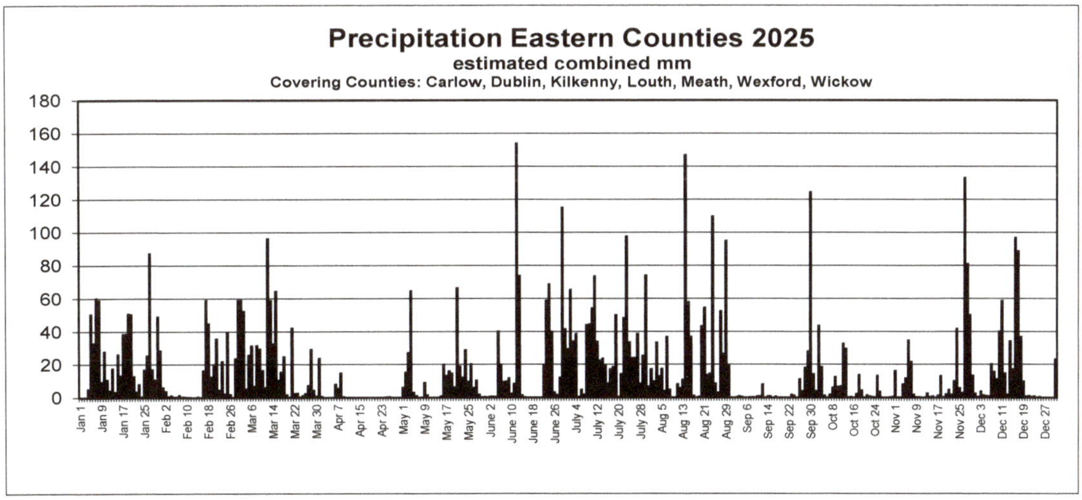

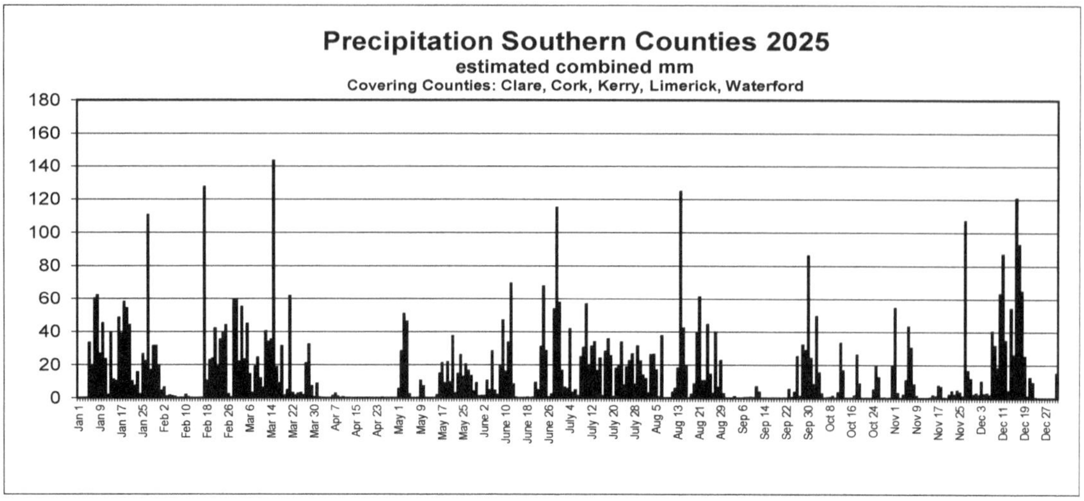
Precipitation Southern Counties 2025
estimated combined mm
Covering Counties: Clare, Cork, Kerry, Limerick, Waterford

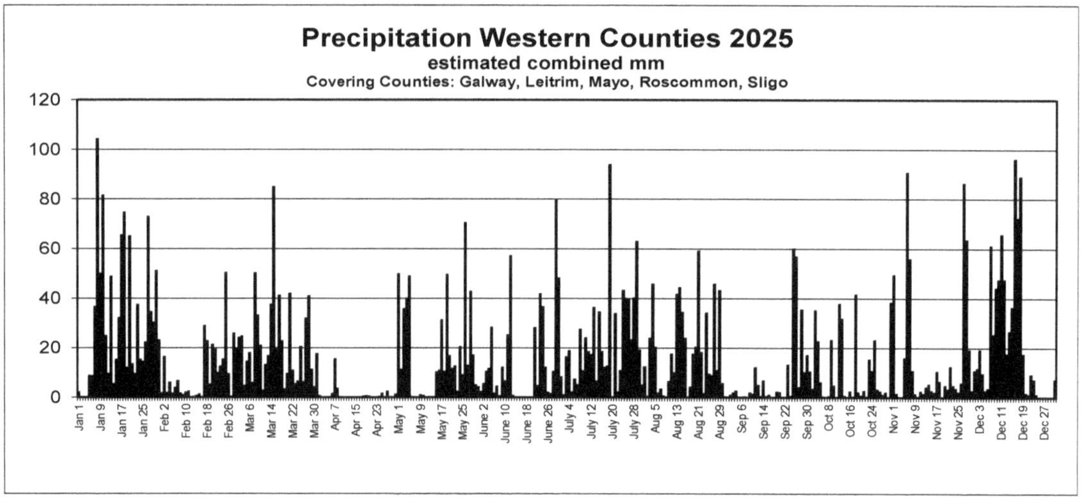
Precipitation Western Counties 2025
estimated combined mm
Covering Counties: Galway, Leitrim, Mayo, Roscommon, Sligo

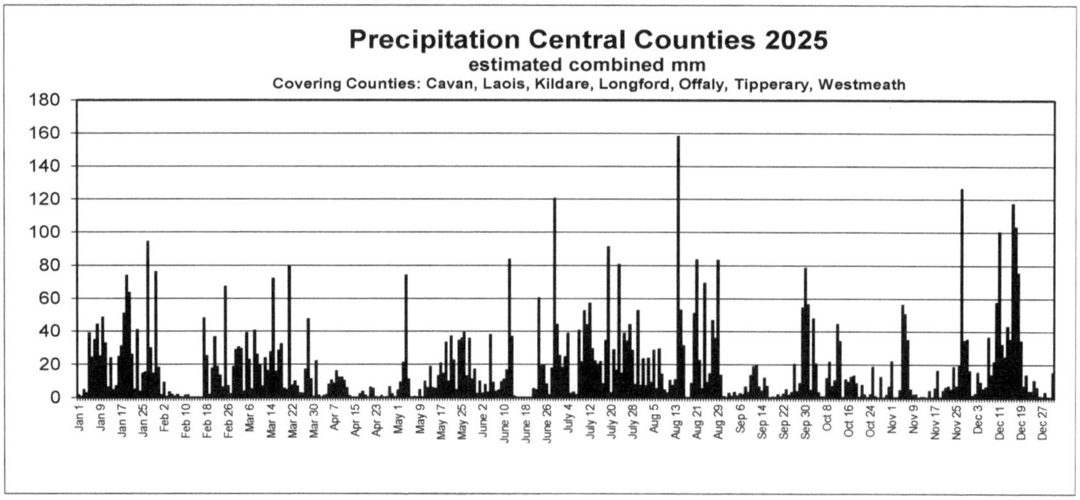
Precipitation Central Counties 2025
estimated combined mm
Covering Counties: Cavan, Laois, Kildare, Longford, Offaly, Tipperary, Westmeath

RAIN TRENDS 2025 -2026

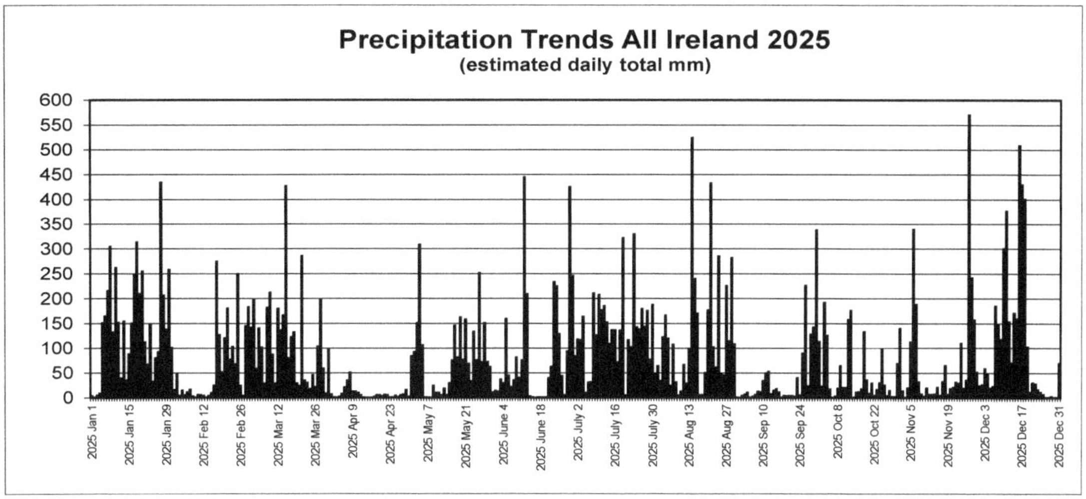

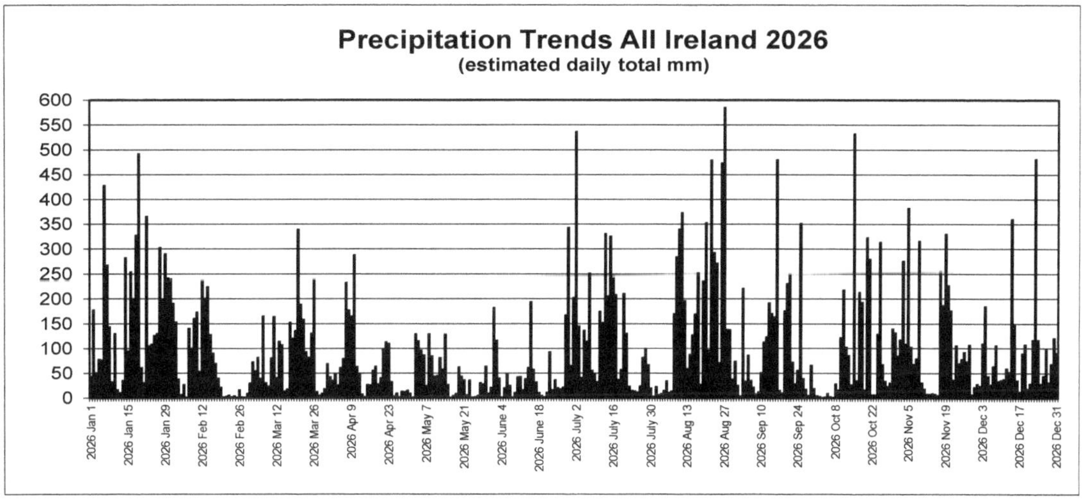

WHEN WILL THE SUN BE OUT?
AVERAGED ACROSS THE WHOLE COUNTRY.

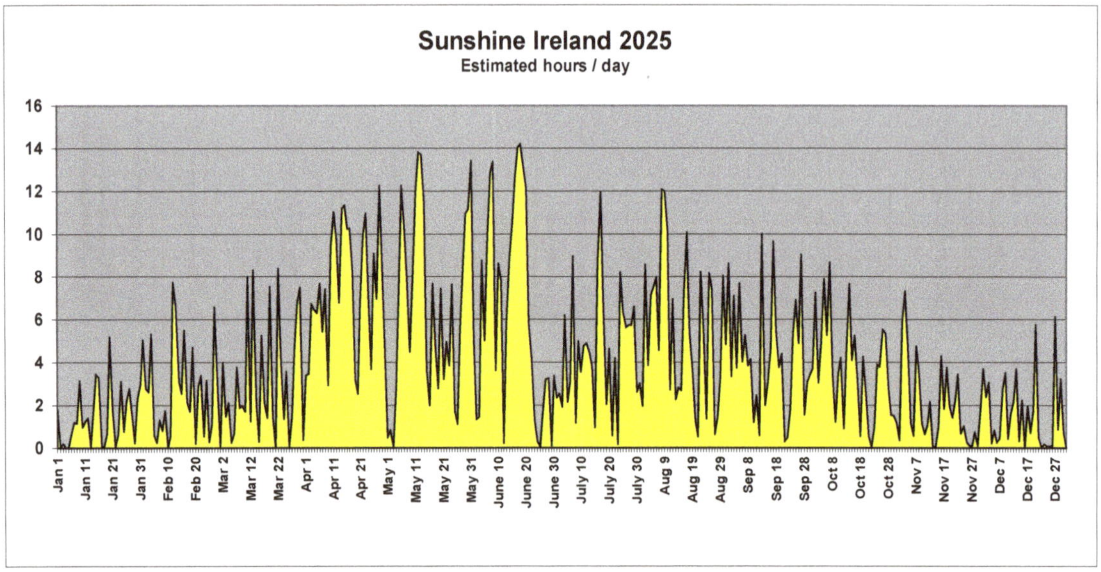

GRAPHS, NOTES, AND TRENDS

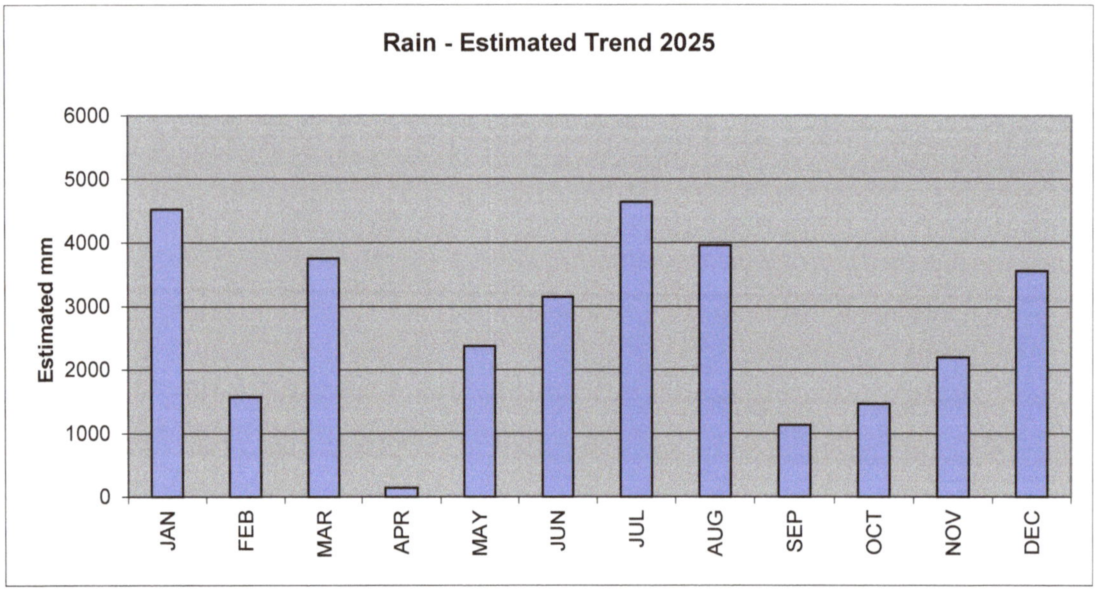

The **most precipitation** for the whole country in 2025 may be **July followed by January and then August, with April, September and October** likely to be the driest. The largest amount of precipitation in one region/month may be in July in the northern and eastern counties. The driest period for the year may be in April everywhere, but especially in the southern and eastern counties.

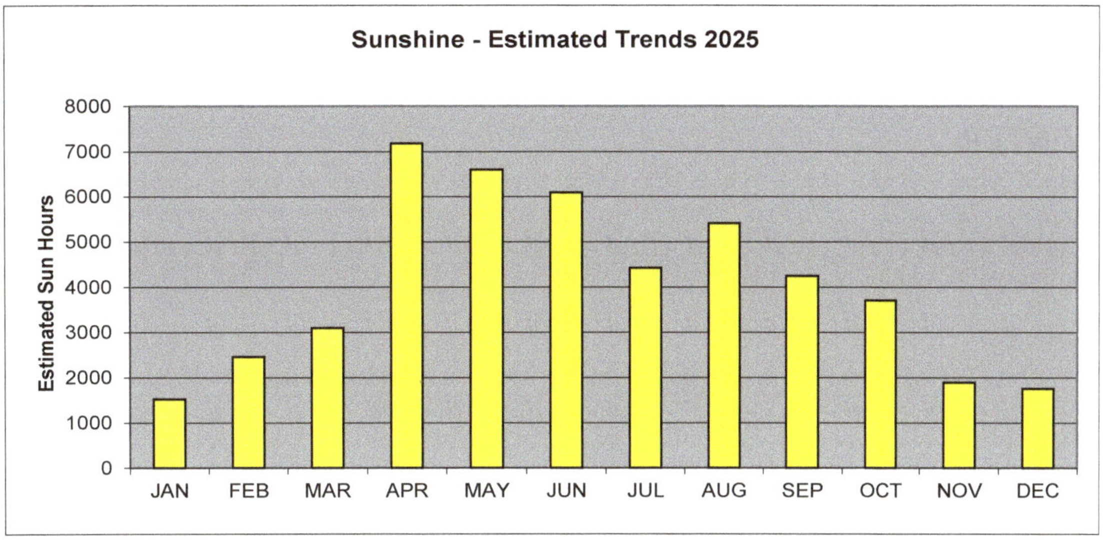
Sunshine - Estimated Trends 2025
Estimated Sun Hours
8000
7000
6000
5000
4000
3000
2000
1000
0
JAN FEB MAR APR MAY JUN JUL AUG SEP OCT NOV DEC

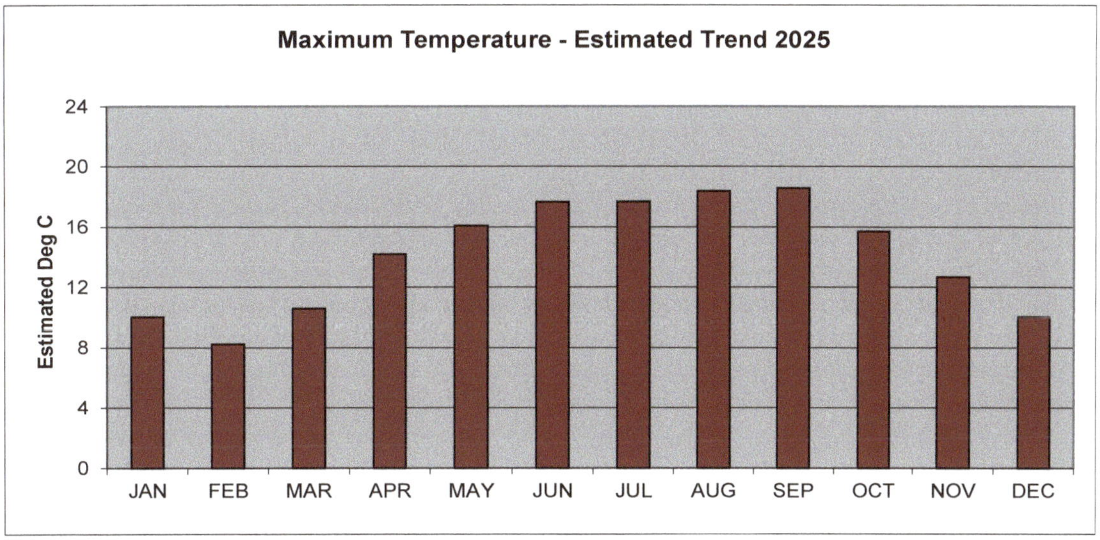
Maximum Temperature - Estimated Trend 2025
Estimated Deg C
24
20
16
12
8
4
0
JAN FEB MAR APR MAY JUN JUL AUG SEP OCT NOV DEC

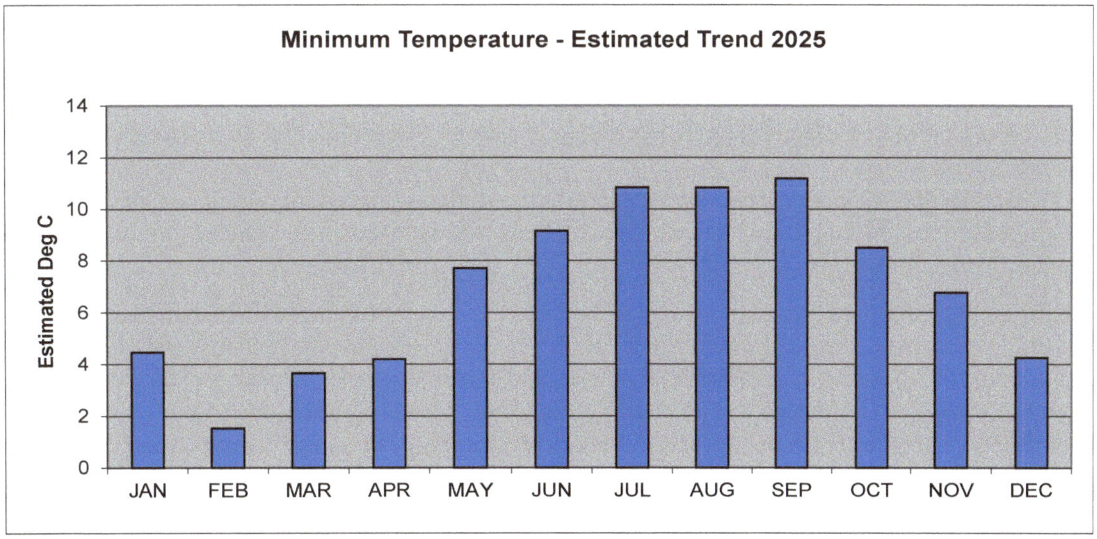
Minimum Temperature - Estimated Trend 2025
Estimated Deg C
14
12
10
8
6
4
2
0
JAN FEB MAR APR MAY JUN JUL AUG SEP OCT NOV DEC

SEASONAL OVERVIEWS

WINTER

Many districts will experience an unusually wetter than normal winter, with many districts receiving around 35-45% more than normal rainfall than an average winter. Despite this, winter will be sunnier than average overall and temperatures will be warmer than normal. The windiest period will be in the fourth week of January.

RAINFALL:

The number of rain days (as measured at 0.2mm or more recorded each day) will be above average almost everywhere especially in the north and northwest, with the country as a whole averaging between 60-80 rain days compared to the norm for a typical winter of between 50-65 rain days. Only the eastern parts of Ulster can expect rainfall amounts just above normal to 10% more, while the remainder of the country can expect to be wetter by more than 15% above the average. Many districts in the west up to 50% wetter, the Killarney area up to 150% wetter and most other parts of the country in the east, south, and northwest between 20-40% wetter. Overall, the Ulster and Leinster Provinces can expect around 15-20% more rainfall than is typical for the season, the Connaught Province around 35% wetter on average, the Munster Province averaging up to 45% wetter than a normal winter. December 2024 and January 2025 will both be much wetter than average, while February brings below average rainfall due to the first half of the month being almost rain free.

SUNSHINE:

By the end of winter, expected sunshine hours will be around 15-20% above average for this season for the country as a whole, with only the western coastal fringes of Connaught cloudier than average but only by around 10% less than the seasonal norm. Overall, only the Connaught district can expect a mix of sunnier and cloudier than average pockets while everywhere else across Ireland will be sunnier. The Ulster Province can expect around 30% more sunshine than normal, the Leinster Province around 20-25% sunnier, the Munster Province around 10-15% sunnier, and even with a mix of sunnier and cloudier areas, the Connaught Province will also receive a regional average of around 5-10% more sunshine than is typical for winter. The winter breakdown by month will see December around 50% sunnier than normal, January around 10% cloudier, and February again sunnier by up to 20% more than normal. The sunniest spells in January can be expected just after the Full Moon early in the third week and at the end of the third week again, just before the Third Quarter. The sunniest spell in February will be at the Full Moon in the last few days of the second week, and again just before the New Moon at the end of the month. The dullest period of the winter season can be expected over the first half of January, and again midway through the third week of January.

TEMPERATURE:

Mean air temperatures will be generally around 1–2°C warmer than normal for the season overall, with maximum temperatures averaging around 1-1.5°C more than average, and minimum temperatures up to 2°C warmer overall. The only exception will be in the southwest, where maximum temperatures will be below average by around a degree and minimum temperatures considerably higher than the seasonal average overall by nearly 5°C.

Maximum Temperatures: The warmest spell in December 2024 will be midway through the second week and at the end of the third week into the first half of the fourth week, and the coolest spell will be in the last few days of the month at the New Moon. The second and third weeks of January will be the warmest, while the first half of the first week and the last couple of days of January will be the coolest. The warmest spell in February will be in the few days up to the Full Moon midway through the second week, and again just before the New Moon in the second half of the last week. Overall, the warmest spell of winter will be in January and the coldest spell at the end of December.

Minimum Temperatures: The mildest spell in December 2024 will be midway through the second week and again in the first half of the fourth week, and the coldest spell will be in the last few days of the month either side of the New Moon. The mildest spell in January will be for a couple of days at the start and end of the second week and in the first half of the third week, while the coldest spells will be early in the first week and again between Southern Declination and the New Moon in the second half of the fourth week. The mildest time in February will be over the second week, while the coldest minimas will be from the end of the second week to near the end of the third week, as well as briefly in the middle of the first week. Overall, the mildest spell of winter will be in the middle of January from the Full Moon to the Third Quarter, while the coldest period of winter will be in the third week of February between the Full Moon and Third Quarter.

Frosts: There are likely to be around 35-40 days with frosts observed during the winter season.

SPRING

Spring will be drier overall this year, except in the west, where average to slightly above average rainfall is expected. Spring will also be much sunnier than average everywhere and generally warmer for both maximum and minimum temperatures by an average of between 0.5-1°C above the seasonal norm. The windiest spells during spring will be in the second and third weeks of March and in the last few days of May.

RAINFALL:

The number of rain days (as measured at 0.2mm or more recorded each day) will be around 55-65 days for the country as a whole, compared to the average range for spring of between 35-40 wet days. Despite a higher than average number of rain days, the rainfall quantities will be less than normal resulting in a drier than average spring across the country as a whole. Overall, the Ulster, Connaught, and Munster Provinces can all expect to be average rainfall to around 10% drier than than normal, while the Leinster Province up to 20% drier overall. Only the far western coastal fringes of the country from Malin Heads to Galway can expect above average rainfall, but still within 10% of the norm, and the Killarney area the only exception of up to 40% wetter than average by the end of the season. The wettest spell in March will be around the Full Moon mid-month, and in May around the First Quarter midway through the first week and again around the New Moon midway through the fourth week. In contrast, April will be considerably drier than average throughout the country, with most days rain free, and the only most significant rainy spell for the month just after the First Quarter around the turn of the first week into the start of the second week.

SUNSHINE:

Overall, sunshine hours will be above average everywhere, almost all districts expecting around 15-20% more sunshine hours than is typical for spring. However, the districts in a line from Armagh through Drumsna and down into the Kilkenny districts may all see up to 30-35% more sunshine than normal, as well as in the Newport area, while the west of Ulster and coastal parts of Connaught only around 10-15% sunnier. The sunniest periods in March will be on a couple of days over the second week just before the Full Moon, and again for a day near the end of the third week. It will also be very sunny at the Third Quarter early in the fourth week, and some good sunny spells can be expected around the New Moon in the last 3-4 days of the month. April will be very sunny almost every day throughout the month, with the only dull couple of days likely between Southern Declination and the Third Quarter towards the end of the third week. Many days during April will see 10 hours or more of sunshine recorded almost everywhere. The sunniest spells in May will be between the First Quarter and Southern Declination, including over the Full Moon period, from midway through the first week to the end of the second week, and again in the last few days of the month after the New Moon and Northern Declination. Overall, March will be the cloudiest, April the sunniest with over twice as much sunshine hours as March, and May still sunnier than March but not as bright as April.

TEMPERATURE:
Following a similar trend to winter, both maximum and minimum temperatures will average around 0.5-1°C warmer than average across the country for the season. All parts of the country will have maximum temperatures of between 1-3°C above the average, with only the southwest close to their seasonal norm and inland parts to the north up to 3°C warmer than normal for springtime. In contrast, minimum temperatures will average around 1°C above the norm in most parts of the country with a mix of warmer and cooler pockets across the country. The northwest quarter of Ireland can expect minimum temperatures of around 1.5-2°C above the norm, while inland parts from Lough Navar to Galway may be cooler by a similar variant. The east of the country, including Dublin as well as in the Shannon area may be slightly cooler, but still within half a degree of the springtime average.

Maximum Temperatures: The warmest spell in March will be at the First Quarter in the second half of the first week and again just after Southern Declination / Third Quarter over most of the fourth week, and the coolest spell in March will be in the last few days either side of the New Moon. In April, the warmest periods will be between the Full Moon and New Moon over the second half of the month, in particular just after the Third Quarter over the second half of the fourth week, while the coldest spell in April will be over the first ten days, either side of the First Quarter. The warmest spell in May will be over the second week, for a few days either side of and including the Full Moon, while the second half of the month from midway through the third week until midway through the fourth week, will be the coolest, as well as in the last few days. Overall, the warmest spell of spring will be in May around the Full Moon in the second week, and the coolest spell of spring will be in March in the last few days of the month either side of the New Moon.

Minimum Temperatures: The mildest spell in March will be around the First Quarter from midway through the first week to midway through the second week, and again briefly just before the Third Quarter at the start of the fourth week and in the last couple of days of the fourth week. The coldest spell in March will be at the New Moon in the second half of the last week of the month. April will be the mildest in the second half of the month at the Third Quarter at the end of the third week and midway for a day or so in the fourth week. The coldest spells in April will be over the second half of the first week around the First Quarter, and just after Apogee at the end of the second week. The mildest spells in May will be in the first half of the first week in the build up to the First Quarter, hen midway through the second week, and finally for a couple of days at the New Moon towards the end of the fourth week. The coldest spell in May will be for a couple of days during the second half of the first week.

Overall, the mildest spell of spring will be in May and the coldest at the end of March and into the first week of April.

Frosts*:* March is likely to see 14 days with frosts, April 10 days and May 4 days with frosts.

SUMMER

This summer will be unusually wetter than normal, with most districts receiving above normal rain, averaging across the country around 40-45% wetter than normal. Sunshine hours will be up to 15% above average in most parts of the country by the end of the season. Both maximum and minimum temperatures will average around the norm for the season overall, with a mix of warmer and cooler districts scattered over the country. The windiest period of summer will be during the second week of July, the first week of August and again during the middle of the fourth week of August.

RAINFALL:
The number of rain days (as measured at 0.2mm or more recorded each day) will be around 75-85 days across the country as a whole, compared to the summer average of between 30-40 days. Overall, the Ulster Province can expect up to 45-50% more rainfall than average, the Connaught Province up to 35% wetter, the Munster Province around 35-40% wetter and the Leinster Province up to 60-65% wetter.

Some parts of the Leinster Province may see up to 90% more rainfall than is typical for summer, as well as in parts of Ulster, while in contrast, the central west of Connaught may only be wetter by nearer to 10% more. Most days throughout the season can expect some rain activity in most districts, with only a brief spell during mid-June over the days from the Full Moon to just after the Third Quarter likely to be dry everywhere.

SUNSHINE:

The above average sunshine conditions established over the first half of this year will continue throughout summer, despite the higher than normal rainfall. Overall, Ireland as a whole can expect around 10-15% more sunshine than is typical for summer, with the Connaught and Munster Provinces around 15-20% sunnier than normal, and the Ulster and Leinster Provinces both around 5-10% sunnier overall. Only the immediate area around Dublin may have a slightly cloudier than average summer but still within 10% of the norm. In June, it will be the sunniest from midway through the first week until the end of the third week, between the First and Third Quarters, with only a dull spell just after the Full Moon around the 9th-12th. Much of the third week will see sunshine hours of up to 12-14 hours per day widespread throughout the country. In contrast, the last 10 days of June will be very dull in comparison to the rest of the month. July will be duller overall, with only a couple of significantly sunny days expected, one at the end of the first week, and another mid-month for a couple of days just before the Third Quarter. August will be the sunniest either side of the Full Moon over the second week, and then just after the Third Quarter mid-month. Overall, June will be the sunniest month of summer and July the dullest with August in the middle. The longest sunny spell will be at the start of summer from midway through the first week to the end of the third week. June will also see the longest spell of dull weather of the season, over the fourth week to the end of the month, for about 4-5 days either side of the New Moon.

TEMPERATURE:

Overall, maximum temperatures will be slightly cooler for the country as a whole, but still within half a degree of the summer average. Maximum temperatures will be up to 1-2°C warmer than average in central parts of the north of the country, as well as in the Killarney county, while the eastern districts, including the Dublin area may be around 1-1.5°C cooler. All other districts are likely to be slightly cooler but within half a degree of the norm. Minimum temperature averages will vary a lot more across the country this season, with the northwest and most western parts, as well as in the east in the Dublin region around 1-2°C cooler than average, while the inland parts, northeast, and south-western parts of the country around 1-2°C warmer than average.

Maximum Temperatures: The warmest spell in June will over the third week either side of the Third Quarter and the coolest in the second half of the first week and again, at the New Moon midway through the fourth week. July will be the warmest at the Full Moon midway through the second week and for a few days after the New Moon over the second half of the fourth week, and in contrast, it will be coolest in July over the first week, and at the Third Quarter midway through the third week. The period between the Full Moon and the Third Quarter during the second week of August will be the warmest, while the coolest spell in August will be after the New Moon over the second half of the fourth week. Overall, June will bring both the warmest and coolest spells of summer, with the third week of June being the warmest, and the second half of the first week of June being the coolest.

Minimum Temperatures: The mildest spell in June will be from the middle of the third week to midway through the fourth week, between the Third Quarter and the New Moon, while the coldest spell in June will be from midway through the first week to midway through the second week between the First Quarter and the Full Moon. July will be the mildest in the first couple of days, then from the Full Moon to when the Moon crosses the Equator in the period midway through the second week to early in the third week, and again at Northern Declination to the New Moon in the couple of days early in the fourth week. The coldest days in July will be from midway through the first week to midway through the second week between Apogee and Southern Declination, and again briefly at the Third Quarter for a couple of days midway through the third week. The mildest spells in August will be between the Full Moon and the

Third Quarter over the second half of the second week, and again in the last couple of days of the third week. The coldest minimum temperatures will be over the Full Moon period early in the second week, again at the Third Quarter for a few days midway through the third week, and again briefly for a couple of days in the middle of the fourth week. Overall, the mildest spell of summer will be between the Third Quarter and the New Moon in the third and fourth weeks of June, and the coldest also in June either side of Apogee from midway through the first week to midway through the second week.

Frosts*:* Only 3 days with frosts can be expected over summer, with these occurring around the end of the first week and through into the first couple of days at the start of the second week of June. After this, no further frosts are expected until October.

AUTUMN

Autumn will be drier throughout the country this year, with most districts likely to only see around half their usual rainfall that is normal for autumn. It will be a much sunnier than normal autumn as well, with many districts likely to see around 20% more sunshine hours than is typical for this season. Subsequently, due to the unusually sunny autumn, temperatures will also be above normal everywhere, averaging around 2-2.5°C above the autumn average. Generally, wind speeds will be below normal everywhere throughout the season, with the windiest spell during autumn will be over the last few days of November.

RAINFALL:

The number of rain days (as measured at 0.2mm or more recorded each day), will be around 55-65 days across the country as a whole, compared to the autumn average of between 35-55 wet days. Again, like in spring, despite the number of rain days being more than average, the overall rainfall amounts will be well below average everywhere. Overall, all Provinces can expect only around half their usual rainfall for the season. The Ulster and Munster Provinces can expect to be around 50% drier, the Connaught Province around 45% drier, and the Leinster Province around 55% drier. The wettest spell in September will be between the New Moon and the end of the month including over the First Quarter, during the latter part of the month. October can expect four waves of shower activity evenly distributed through the month, with the first half of the first week wet, the second half the second week rainy, midway to the end of the third week showery, and midway through the fourth week showery. November will be the wettest month of autumn, with only a few dry days over the first half of the first week and again either side of the Third Quarter over much of the second week. The wettest period in autumn will be right at the end of November in the last few days, from the First Quarter, and into the first half of December.

SUNSHINE:

The above average seasonal conditions continue throughout autumn as well, with the country as a whole expected to see around 15-20% more sunshine hours than is typical for the season. The Ulster, Connaught, and Leinster Provinces can all expect on average about 15% more sunshine overall, while the Munster Province can expect up to 30% more sunshine than normal. Only the northwest of the country can expect sunshine hours to be around 5-10% of the norm, while in contrast, the southwest may see up to 50% more sunshine. The sunniest spells in September can be expected over the first week in the days leading up to the Full Moon, and again over the second half of the fourth week, either side of Apogee, with the sunniest day of the month likely to be just before the Third Quarter at the end of the second week, around the 13th. October may be the sunniest over the first week of the month, especially at the Full Moon, but also for a few days after the Third Quarter early in the third week. November will be the sunniest at the start of the month in the lead up to the Full Moon, just over a few days early in the first week. The dullest period in autumn will be in November from midway through the second week until the end of the month, with only a brief sunnier respite around the 16th-18th. As expected, September will be the sunniest, October slightly sunny, and November the dullest and with less than half the sunshine hours of September.

TEMPERATURE:
Autumn temperatures will return to above average throughout the country by the end of the season, with most parts of the country seeing both maximum and minimum temperatures of up to 2-3°C above the seasonal norm. The only counties that may see average to slightly above average temperatures but still within half a degree of the norm will be the minimum temperatures in the inland western parts from Lough Navar to Galway once more. Everywhere else will record seasonal maximum and minimum temperatures of between 1-3°C above their average, with parts of the northwest up to 3.5-4°C warmer and the southwest and the Dublin area only around 0.5-1°C warmer. All other parts of the country will be around 2-3°C warmer overall.

Maximum Temperatures: The warmest spell in September will be just after Southern Declination early in the first week of the month, and again just after the Third Quarter early in the third week and the coolest spell in September will be at the end of the month between Apogee and the First Quarter in the last few days. The warmest spells in October will be for a couple of days each time, the first time in the first half of the first week, then midway through the second week and finally at the end of the third week. In contrast, the coolest spells in October will be for a couple of days at the end of the first week immediately prior to the Full Moon, and again either side of the First Quarter in the last few days of the month. The warmest spells in November will be at the end of the first week and midway through the second week while the coolest spell will be in the last few days of the month. Overall, the warmest spell of autumn will be at the start of September and the coldest at the end of November.

Minimum Temperatures: The mildest spells in September will be briefly midway through the first week, again midway through the second week, early in the third week, and briefly either side of the New Moon at the end of the third week into the first days of the fourth week. The coldest spells will be in the last couple of days of the second week at the Third Quarter, and over the last week of the month either side of Apogee to Southern Declination. The mildest spells in October will be for a couple of days at the Third Quarter towards the end of the second week and from the New Moon to just after Apogee in the fourth week. The coldest spells in October will be around the turn of the first week into the early days of the second week either side of the Full Moon. This will be when the first frosts make their reappearance in some districts. Then another cold spell can be expected at the end of the month between Southern Declination and the First Quarter in the second half of the fourth week. The mildest spells in November will be immediately at the start of the month briefly, and then again just before the Third Quarter from midway to the end of the second week. The coldest periods in November will be just before the Full Moon in the middle of the first week, again with further frosts appearing in many districts, and again at Southern Declination early in the fourth week. Overall, the mildest spell of autumn will be in September towards the end of the second week and the coldest in the first week of November.

Frosts: Frosts can be expected on 10 days during the season with most in November in the inland parts of the northwest.

2024 OVERALL WEATHER SUMMARY

Winds: The strongest winds of the year will be during the last few days of January, with the year overall seeing lower than average wind levels than is typical for all seasons.

Rain: The wettest spell of the year is likely to occur over the third week of December.

Dry Spells: The longest dry spell of the year will be throughout most of the whole month of April.

Sun: The sunniest month will be in April and the sunniest period will be over the third week of June. The dullest month will be January, and the dullest periods will be the first week of January, the last week of November and the period between the New Moon and First Quarter in December, either side of Christmas in the fourth week.

Maximum Temperatures: The hottest spell in the year will be over the third week of June and the coolest spell will be in the first few days of January and over the first half of the third week of February.

Minimum Temperatures: The mildest spell in the year will be during mid-July and again in the second week of September for a few days, and the coldest period of the year will be between the Full Moon and the Third Quarter in February midway through the month.

Frosts: Frosts are likely on at least 75 days during the year.

RELATIVELY DRIER SPELLS FOR 2025

January	Despite some rain in most districts every day, the driest spell will be over the first half of the first week in the build-up to the First Quarter.
February	From the last couple of days of the first week just after the Moon crosses the Equator, through the First Quarter and Full Moon periods to when the Moon crosses the Equator once again around the 16th.
March	Despite some rain on almost every day during March, a mostly dry spell is expected during the fourth week from the Third Quarter/Southern Declination to the end of the week, with mostly wet conditions returning before the New Moon. Perigee around the 30th may bring a brief dry respite.
April	Most days throughout this month will be dry throughout, with only some showery spells expected around the turn of the first week into the first couple of days of the second week, again spotty showers around Southern Declination, and again at the New Moon at the end of the month.
May	From the end of the first week, through the second week and into the early part of the third week, between the First Quarter and Southern Declination, including over the Full Moon.
June	Throughout the third week, from a couple of days after the Full Moon to a couple of days after the Third Quarter.
July	Some rain can be expected daily almost everywhere during this month, with the least rainfall expected around Apogee during the last few days of the first week. Perigee at the end of the third week may also be the driest day of the month.
August	Again, some rain can be expected on almost every day during this month, with only a couple of mostly dry days either side of the Full Moon midway through the second week, at Northern Declination midway through the third week and in the last day of the month on the First Quarter.
September	Throughout the first week between Southern Declination and the Full Moon, and again over the third week between the Third Quarter to just after the New Moon.
October	Either side of the Full Moon at the turn of the first week into the start of the second week, then for a few days after the Third Quarter over the first half of the third week. Further mostly dry spells expected for a couple of days between the New Moon and Apogee in the fourth week, and finally for a couple of days either side of the First Quarter in the last few days of the month.
November	Early in the first week before the Full Moon, again midway through the second week between Northern Declination to just after the Third Quarter.
December	Over the fourth week to the end of the month, from the New Moon to the First Quarter and for the next few days after the First Quarter to just before New Year's Eve.

NEXT THREE YEARS

Comparing estimated rain over the next three years averaged for the whole country; 2026 should have most rain and 2025 the least, with 2027 in the middle position, but very similar to 2026.

REGIONAL EXPECTATIONS BY REGIONS:

West: Wettest 2026, driest 2025, and middle 2027

South: Wettest 2027, driest 2025, and middle 2026

East: Wettest 2027, driest 2025, and middle 2026

North: Wettest 2026, driest 2025, and middle 2026

Central: Wettest 2026, driest 2025, and middle 2027

2025 may be the best year for sunshine amounts, with 2026 the cloudiest, and 2027 in the middle position.

For maximum temperatures, 2025 may be the warmest overall, with 2026 and 2027 both very similar and milder.

For minimum temperatures, 2026 may be the coldest overall, with 2025 the mildest, and 2027 in the middle position.

JANUARY

PHASES OF THE MOON

6th	First Quarter
7th	Perigee #13
7th	Crossing Equator
12th	Northern Declination
13th	Full Moon
20th	Crossing Equator
21st	Third Quarter
21st	Apogee
26th	Southern Declination
29th	New Moon

1	2	3	4	5	6	7	8	9	10	11	12	13	14	15	16	17	18	19	20	21	22	23	24	25	26	27	28	29	30	31
Mon	Tue	Wed	Thur	Fri	Sat	Sun	Mon	Tue	Wed	Thur	Fri	Sat	Sun	Mon	Tue	Wed	Thur	Fri	Sat	Sun	Mon	Tue	Wed	Thur	Fri	Sat	Sun	Mon	Tue	Wed
					1Q	P13 XhN					^	F							XhS	3Q A					V			N		

MONTHLY SUMMARY

1st-5th Anticyclonic and mostly dry spell. Cooler than normal during the day, especially in areas where fog is slow to clear, and frosts may be widespread in the first couple of days. Some long sunny spells after the fog clears in many parts at first, before cloudier conditions set in. No snow likely.

1st-2nd Very cold overnight with severe frosts in inland parts.

6th-10th Depressions move close to the northwest of the country bringing wet and windy conditions almost everywhere. Rain may be heavy at times in southern and western parts. Strong to gale force southerly winds prevail bringing a milder than normal spell for this time of year.

8th Chance of thunderstorms and hail showers. Unusually mild day.

10th Stormy air with possible thunderstorms, hail showers, and very strong and blustery winds. Rain likely to be widespread and particularly heavy at times in northern parts.

Over the next four weeks a series of deep depressions pass to the northwest of the country, bringing a dominance of milder than normal temperatures, and wet and windy conditions for the first three weeks, followed by a northwest change and high pressure, cooler temperatures and drier conditions for the last week of this outlook period. There may be a wide variation in rainfall across the country, with the most rain affecting northern and western parts. Overall, temperatures almost everywhere are likely to be warmer than is normal. No frosts are anticipated in coastal areas, and inland parts will see less frosts than normal for this time of year. It will be sunniest in the east. Winds to gale force may be frequent during the next four weeks, with strong and blustery southwest winds with the potential to cause localized damage prevailing for the first three weeks before winds ease with the arrival of an anticyclonic and northwest change. During this outlook period, fogs may be infrequent but rain activity almost every day over the first 3 weeks. Thunderstorms are expected on 8 days, hail and snow on 4 days, and frosts on 16-18 days.

11th-31st A succession of intense Atlantic depressions pass near the northwest coast bringing mild and very unsettled weather. Southwest winds generally prevail, often very strong and with gale gusts especially around the coasts. Despite these depressions prevailing, temperatures are likely to be above normal throughout. Rain is expected almost daily in most districts.

11th-12th Scattered thunderstorms likely with passing showers of hail. Possible snow.

12th-19th Unusually mild spell.

17th-20th Very wet spell.

20th-21st Chance of thunderstorms and heavy rain.

21st Very strong and unsettled winds throughout.

26th-27th Chance of thunderstorms and possible hail showers. Very cold with severe frosts and chance of snowfalls in many districts

27th-30th Very wet spell.

28th Strong and blustery southwest winds prevail.

29th Unusually warm day with many districts seeing maximum temperatures in the low teens.

30th Unsettled and blustery southwest winds to gale force in many parts, especially around the coasts.

30th-31st Stormy conditions prevail with gale winds, thunderstorms, hail, and possible snow likely.

IRELAND RAINFALL ESTIMATES

January 2025

d=mainly dry, n= nonrecordable, l=light shrs, s=significant shrs, r=rain, h=hvy falls

		ULSTER PROVINCE					CONNAUGHT PROVINCE						LEINSTER PROVINCE				MUNSTER PROVINCE						
		Alderg	Hillsbo	Armag	Loughl	Malin H	Sligo	Ardtar	Belmul	Drums	Galway	Roscom	Markre	Edende	Kilkenn	Dublin	Shanno	Newpo	Tralee	Killarne	Cork	All	Moon
1st	Wed	d	d	d	d	d	d	d	l	d	d	d	d	d	d	d	d	d	d	d	d	1	
2nd	Thur	d	d	d	d	d	d	d	d	d	d	d	d	d	d	d	d	d	d	d	d	0	
3rd	Fri	d	d	d	d	d	d	d	d	d	d	d	n	d	d	d	d	d	d	d	d	1	
4th	Sat	d	d	d	d	d	d	d	d	d	d	d	n	l	d	d	d	d	d	d	d	2	
5th	Sun	l	s	l	d	d	l	d	d	s	d	s	d	s	r	l	d	d	l	s	s	60	
6th	Mon	s	l	s	d	h	l	d	l	l	l	l	s	s	l	s	l	l	l	l	r	80	1Q
7th	Tues	l	l	l	s	l	s	s	s	s	s	r	l	s	r	l	l	s	r	h	s	131	P13 XhN
8th	Wed	s	s	s	r	s	h	r	h	r	r	r	s	l	r	s	s	h	h	r	h	242	
9th	Thur	l	l	s	r	l	h	r	s	l	s	h	s	l	l	l	s	s	s	r	s	136	
10th	Fri	s	s	l	r	r	r	r	h	h	r	l	s	s	l	s	r	r	r	s	s	197	
11th	Sat	s	l	s	s	r	s	s	s	s	s	s	s	l	l	l	s	r	s	s	s	112	
12th	Sun	s	l	s	l	l	l	l	l	l	n	l	l	l	n	l	n	l	l	l	l	35	^
13th	Mon	l	s	l	s	s	r	s	r	s	h	s	s	l	l	l	l	r	s	h	s	140	F
14th	Tues	l	l	s	l	l	l	l	l	l	l	l	s	l	l	l	l	l	d	s	n	36	
15th	Wed	l	l	l	l	r	l	l	l	l	s	l	n	d	s	n	l	l	l	l	l	48	
16th	Thur	l	l	d	s	s	s	s	s	l	s	s	h	s	l	n	s	r	s	h	s	125	
17th	Fri	l	s	l	r	r	h	r	h	r	s	r	h	s	s	l	l	h	s	r	s	192	
18th	Sat	r	r	s	r	r	r	r	r	r	r	h	h	s	l	s	s	h	s	h	s	231	
19th	Sun	s	l	r	l	l	l	l	l	l	s	l	l	r	s	s	r	l	r	h	s	121	
20th	Mon	l	s	d	r	s	r	r	r	h	h	h	r	s	s	r	s	r	l	h	n	186	XhS
21st	Tues	s	l	s	l	s	l	l	l	l	l	l	l	l	n	s	s	s	s	l	s	71	3Q A
22nd	Wed	l	l	l	d	r	l	d	l	l	l	l	l	l	d	d	d	l	l	l	l	35	
23rd	Thur	l	l	l	s	s	s	s	r	s	l	s	r	l	l	l	l	r	s	s	l	101	
24th	Fri	l	d	l	d	s	n	d	s	l	l	l	l	d	d	d	d	l	d	l	d	24	
25th	Sat	n	l	d	l	l	l	l	s	l	l	l	d	l	l	l	l	l	s	s	s	54	
26th	Sun	l	l	d	s	n	r	s	l	s	s	s	s	l	l	d	d	r	l	s	l	80	V
27th	Mon	r	h	l	r	s	h	r	r	h	h	h	h	r	r	l	r	h	h	h	r	302	
28th	Tues	s	l	r	l	h	l	l	r	l	s	l	l	s	d	r	r	l	s	l	r	133	
29th	Wed	l	l	l	s	r	r	s	s	l	s	s	r	l	l	d	l	s	l	h	l	106	N
30th	Thur	l	s	l	h	s	r	h	s	s	r	s	r	h	s	r	s	h	l	s	l	184	
31st	Fri	s	l	s	l	l	l	l	s	l	l	l	l	l	s	l	s	s	s	s	l	75	
Estimate:		103	107	99	153	200	210	153	220	165	196	175	186	110	103	97	125	242	159	292	145	3239	
Average:		80	91	75	125	120	131	133	134	105	117	130	131	93	80	63	102	167	175	127	131	2309	
Trend:		wtr	wtr	wtr	wtr	wtr	wtr	wtr	wtr	wtr	wtr	wtr	wtr	wtr	wtr	wtr	wtr	wtr	drr	wtr	wtr	wtr	

IRELAND SUNSHINE ESTIMATES

January 2025

F=fine (8-12 hours of sunshine), pc= partly cloudy (4-7 hours), c=cloudy (1-3 hours), o=overcast (0 hours)

		ULSTER PROVINCE					CONNAUGHT PROVINCE						LEINSTER PROVINCE				MUNSTER PROVINCE						
		Alderg	Hillsbo	Armag	Loughl	Malin I	Sligo	Ardtar	Belmul	Drums	Galway	Roscoı	Markre	Edende	Kilkenı	Dublin	Shann	Newpo	Tralee	Killarne	Cork	All	Moon
1st	Wed	c	c	c	c	c	o	c	o	c	c	o	c	c	c	c	c	c	o	o	o	32	
2nd	Thur	o	o	o	o	o	o	o	o	o	o	o	o	o	o	o	o	o	o	o	o	1	
3rd	Fri	o	o	o	o	o	o	o	o	o	o	o	o	o	o	o	o	o	o	o	o	3	
4th	Sat	o	o	o	o	o	o	o	o	o	o	o	o	o	o	o	o	o	o	o	o	0	
5th	Sun	o	o	o	o	o	o	o	o	o	o	o	o	o	o	o	o	o	o	o	o	0	
6th	Mon	o	o	o	o	o	o	o	o	o	o	o	o	o	c	o	o	o	c	c	c	13	1Q
7th	Tues	c	c	c	c	c	o	c	o	c	o	o	c	c	c	c	o	o	o	o	o	20	P13 XhN
8th	Wed	c	c	c	c	c	o	c	o	c	c	o	c	c	c	c	c	c	o	o	o	21	
9th	Thur	c	c	c	c	c	c	c	c	c	c	c	c	c	c	c	c	c	c	c	c	62	
10th	Fri	o	o	o	o	o	o	o	o	c	o	o	o	c	c	c	c	c	c	c	c	19	
11th	Sat	c	c	c	c	c	c	c	c	c	c	c	c	c	c	c	c	c	c	c	c	26	
12th	Sun	c	c	c	c	c	c	c	o	c	o	o	c	c	c	c	o	o	o	o	o	26	^
13th	Mon	o	o	o	o	o	o	o	o	o	o	o	o	o	o	o	o	o	o	o	o	2	F
14th	Tues	c	c	c	c	c	c	c	o	c	o	o	c	c	c	c	o	c	o	o	c	28	
15th	Wed	c	c	c	c	c	c	c	c	c	c	c	c	c	c	pc	c	c	c	c	c	68	
16th	Thur	c	c	c	c	c	c	c	c	c	c	c	c	c	c	c	pc	pc	c	c	pc	62	
17th	Fri	o	o	o	o	o	o	o	o	o	o	o	o	o	o	o	o	o	o	o	o	1	
18th	Sat	o	o	o	o	o	o	o	o	o	o	o	o	o	o	o	o	o	o	o	o	1	
19th	Sun	o	o	o	o	o	o	o	c	o	c	c	o	o	o	o	o	o	o	o	o	12	
20th	Mon	pc	pc	pc	pc	pc	c	pc	c	pc	c	c	pc	pc	pc	pc	pc	pc	c	c	pc	98	XhS
21st	Tues	c	c	c	c	c	c	c	c	c	c	c	c	c	c	c	c	c	o	o	c	35	3Q A
22nd	Wed	o	o	o	o	o	o	o	o	o	o	o	o	o	o	o	o	o	o	o	o	1	
23rd	Thur	o	o	o	o	o	o	o	o	o	o	o	o	o	c	o	o	o	c	c	c	13	
24th	Fri	c	c	c	c	c	o	c	c	c	c	c	c	c	pc	pc	c	c	c	c	c	57	
25th	Sat	o	o	o	o	o	o	o	o	o	o	o	o	o	c	o	c	c	o	o	c	14	
26th	Sun	c	c	c	c	c	c	c	o	c	o	o	c	c	c	c	o	c	c	c	c	42	V
27th	Mon	c	c	c	c	c	c	c	c	c	c	c	c	c	c	c	c	c	o	o	c	53	
28th	Tues	c	c	c	c	c	c	c	c	c	o	c	c	c	c	c	o	o	o	o	o	30	
29th	Wed	o	o	o	o	o	o	o	o	o	o	o	o	o	o	o	o	o	o	o	o	4	N
30th	Thur	c	c	c	c	c	c	c	o	c	o	o	c	c	c	c	c	c	c	c	pc	44	
31st	Fri	c	c	c	c	c	c	c	o	pc	o	o	c	pc	pc	pc	o	c	c	c	pc	55	
Estimate hours:		46	46	46	46	46	29	46	27	50	31	29	46	50	54	64	35	44	31	31	44	844	
Average hours:		46	46	44	41	37	47	47	46	43	51	53	54	51	53	56	49	43	44	44	53	946	
Trend:		av	av	av	av	av	less	av	less	av	less	less	av	av	av	av	less	av	less	less	av	less	

JANUARY **1** WEDNESDAY

A large anticyclonic system sits over Ireland bringing a mostly dry, calm, and sunny day throughout the country, after morning fog clears in most districts and before some patches of evening misty or fog return by evening. Cold and frosty overnight, particularly in the east to central parts, while milder with increased cloud in the west and south. Chance of some isolated showers briefly in the far northwest.

Belfast: Cloudy, dry with passing threats of rain, cool, fog patches likely
Dublin: Mostly calm, dry, cold with frosts, occasional sunny spells
Cork: Calm, dull and gloomy, cool, dry with passing threats of rain
Galway: Cloudy, dry, gentle breezes to mostly calm air

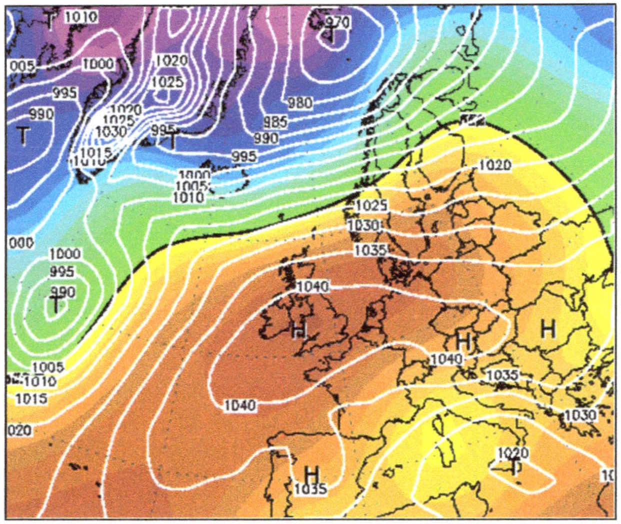

RAIN POTENTIAL

FROST/SNOW

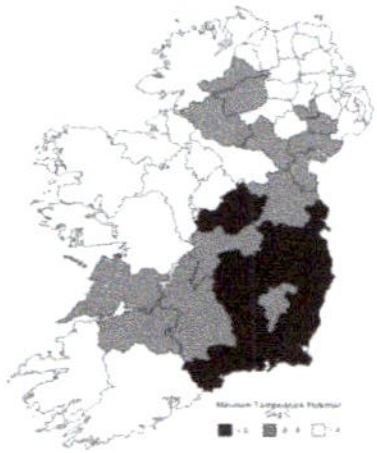

JANUARY **2** THURSDAY

A depression to the northwest of Ireland edges closer to the country pushing the high pressure system slowly away in a gentle south-easterly flow. Cloud increases everywhere, with dull and gloomy conditions in most districts and increasing threats of rain later in the day. Mostly dry with frost and fog patches at first mostly in central places to the north before clearing later to dry but overcast almost everywhere. Chance of some light wintry precipitation overnight, with possible light snow observed in the far north.

Belfast: Dull and gloomy, dry, threats of rain, frost and fog patches, chance of snow
Dublin: Dull and gloomy, dry, cold, frosts, mostly dry with passing threats of rain
Cork: Overcast, cool, dry, calm air
Galway: Overcast and cooler, dry with odd threat of rain, gentle southwest breezes

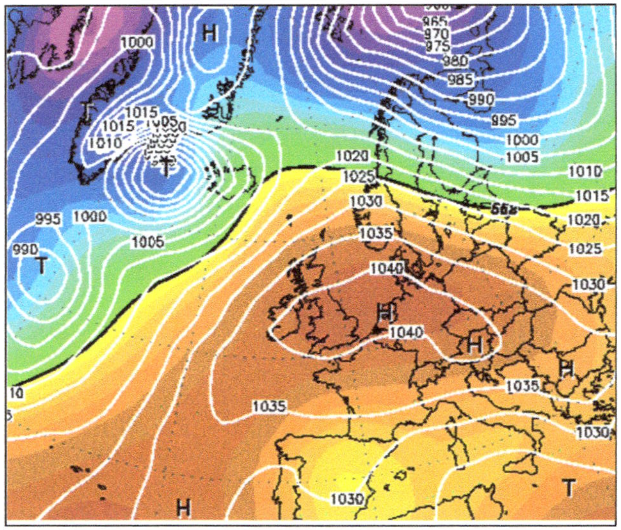

RAIN POTENTIAL

FROST/SNOW

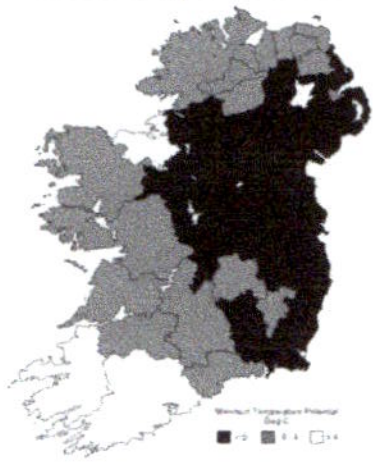

JANUARY 3 FRIDAY

The large high pressure system drifts back over Ireland once more in a very gentle west to southwest flow. Widespread cloud and mostly dry and calm conditions prevail throughout, with patches of fog and mist in the north to central districts. Some brief sunny breakthroughs likely in the east.

Belfast: Calm, overcast, cold, dry, frosts, misty patches
Dublin: Calm, dry, cool, overcast
Cork: Overcast, dry, threats of rain, cool, calm, serene air, chance of misty patches
Galway: Overcast, cool, dry with odd threat of rain, calm atmosphere, mist likely

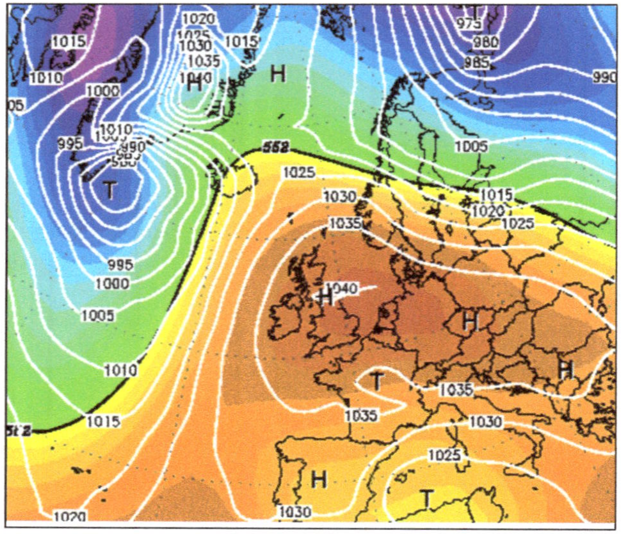

RAIN POTENTIAL

FROST/SNOW

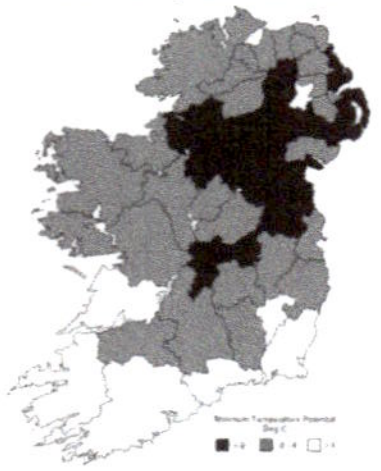

JANUARY 4 SATURDAY

High pressure continues over Ireland with light and variable breezes prevailing. Widespread cloud continues to bring dull and gloomy conditions almost everywhere, with patches of mist, fog in many districts, and pockets of frost mostly in inland regions. While most districts will remain dry, the overcast conditions may result in some outbreaks of light drizzle here and there.

Belfast: Overcast, mostly dry, increasing threats of rain, milder, scattered fog
Dublin: Cold, dry southerly air, dull and gloomy, chance of misty patches
Cork: Overcast, dry with some misty patches, cool, mostly calm air
Galway: Dull and gloomy, dry, cold, frost and fog patches likely, mostly calm air

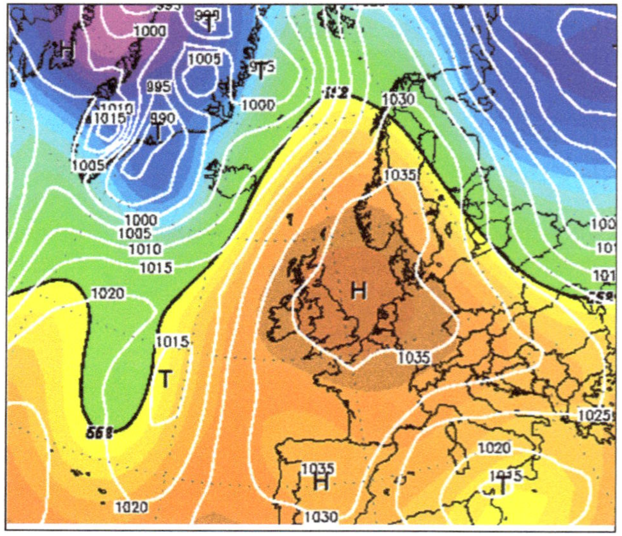

RAIN POTENTIAL

FROST/SNOW

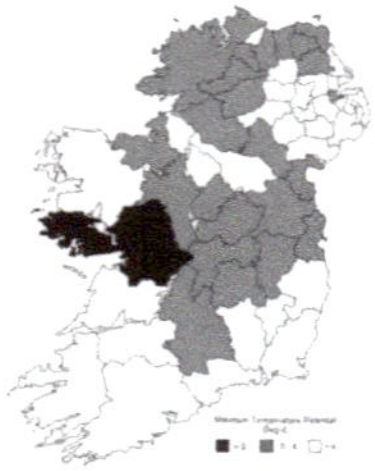

JANUARY **5** SUNDAY

A low pressure system to the north of Ireland pushes the high pressure system to the southeast of the country once more. Generally dull and gloomy throughout the day with outbreaks of light rain and drizzle expected almost everywhere except in the western coastal fringes where some brief sunny spells may be observed. Temperatures likely to be milder than of recent with a light to moderate south to southeasterly flow prevailing, gustier at times along the northern coastal fringes.

Belfast: Overcast, mild, passing light rainy spells, mist and fog likely
Dublin: Overcast, cool, light showers and drizzle patches, mostly calm air
Cork: Mostly calm, dull and gloomy air, some drizzly showers
Galway: Cool, overcast, dry, increasing threats of rain, calm and misty atmosphere

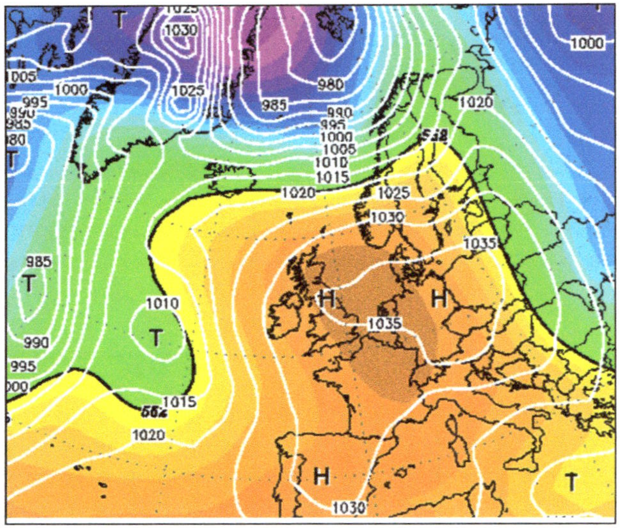

RAIN POTENTIAL

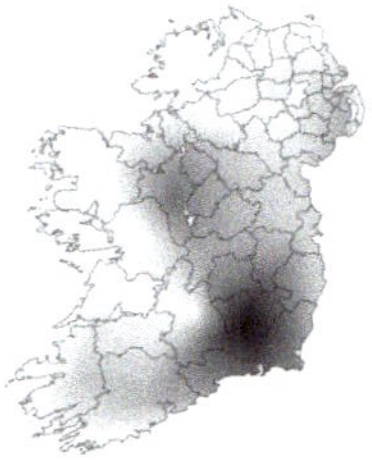

FROST/SNOW

JANUARY **6** MONDAY

Low pressure systems to the north and west of Ireland continue to approach Ireland directing a moderate to light south to southeast flow throughout. Widespread cloud prevails everywhere, with patches of mist and fog overnight, later turning to drizzly showers in most districts.

Belfast: Overcast, brief drizzly showers, patches of shallow fog
Dublin: Overcast, occasional showers, some misty patches, light southerly breezes
Cork: Cloudy, milder, rain and drizzle outbreaks, gentle breezes, chance of fog
Galway: Overcast, milder, light drizzle, gentle air, mist and shallow fog patches likely

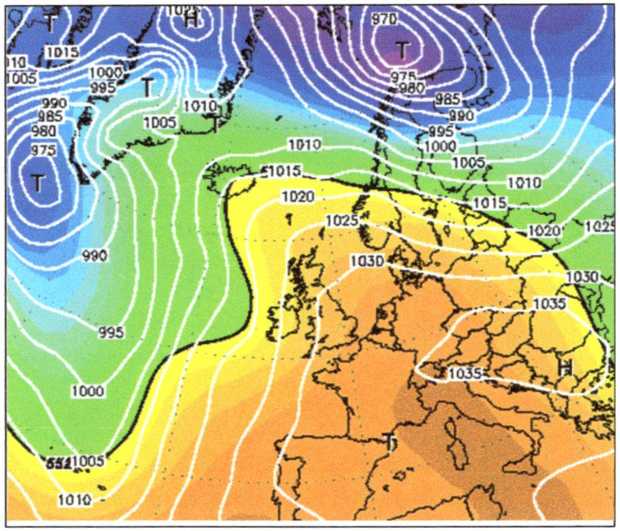

RAIN POTENTIAL

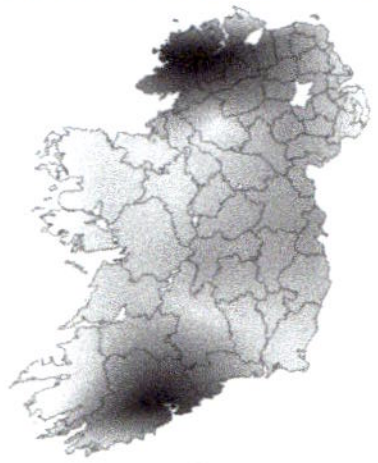

FROST/SNOW

JANUARY **7** TUESDAY

A rain embedded front between the departing high pressure system to the southeast and advancing low pressure systems to the west spreads widespread cloud and outbreaks of showers throughout the country, heavier in the northwest and southeast for a time. Overnight patches of fog in many districts prevail while light southerly winds prevail, later freshening ahead of the front.

Belfast: Mostly cloudy, overnight fog, light showers, freshening winds later in the day
Dublin: Mostly cloudy and calm, light showers, warmer day cool night, fog patches
Cork: Overcast, passing showers, light breezes freshening later, misty patches
Galway: Overcast, occasional showers, light breezes, chance of misty patches

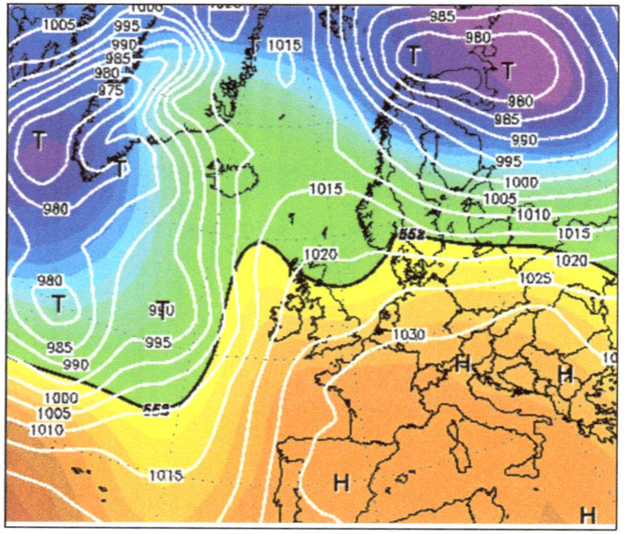

RAIN POTENTIAL

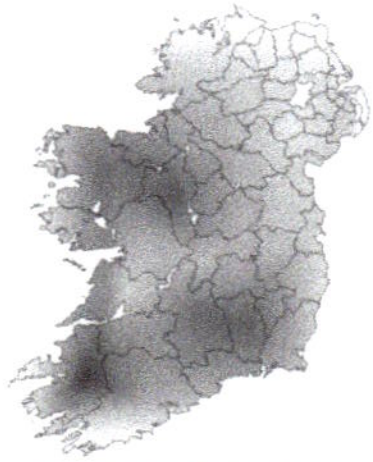

FROST/SNOW

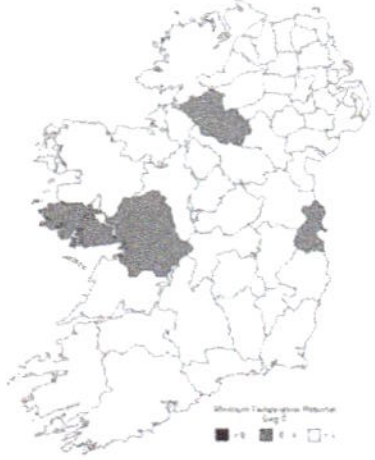

JANUARY **8** WEDNESDAY

A depression centred to the northwest of Ireland extends a stormy air over the country. Strong stormy southerly winds and widespread rain prevails throughout with particularly heavy rain in the western half of the country. Chance of localised flooding for a time in some central districts, and gale winds for a time may cause some distruptions for a time.

Belfast: Stormy air, cloudy, milder, outbreaks of squally rain, very strong winds
Dublin: Milder temperatures, cloudy, occasional showers, blustery at times
Cork: Stormy air, overcast, heavy rain, strong winds with gusts and possible gales
Galway: Unsettled and blustery air, milder temperatures, cloudy, rain

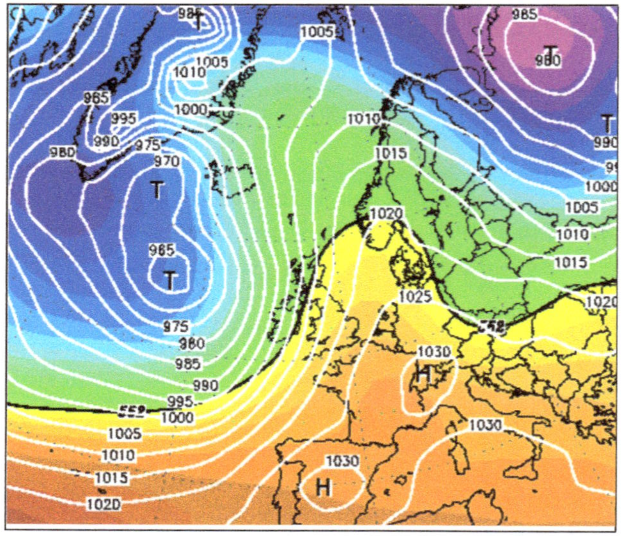

RAIN POTENTIAL

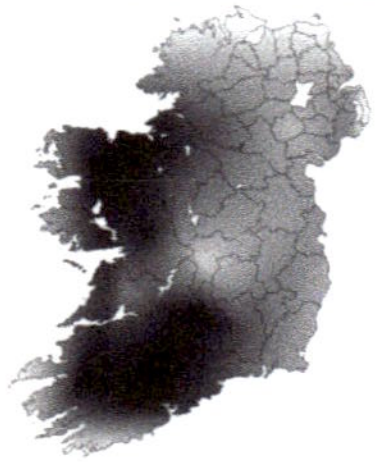

FROST/SNOW

JANUARY **9** THURSDAY

Stormy conditions associated with a depression centred to the northwest of Ireland continues to bring unsettled weather throughout, although clearing for a short time in the east and north. Strong southwest winds prevail, but decrease as the day progresses. Outbreaks of heavy showers continue in the western half of the country, while lighter in the east. Some brief sunny outbreaks likely in the Midlands.

Belfast: Isolated showers, warmer, odd sunny spells, blustery, possible gale gusts
Dublin: Unsettled and blustery winds, colder, some sunny spells, isolated showers
Cork: Stormy winds, cloudy, scattered squally showers, colder
Galway: Strong gusty winds, cloudy, squally showers

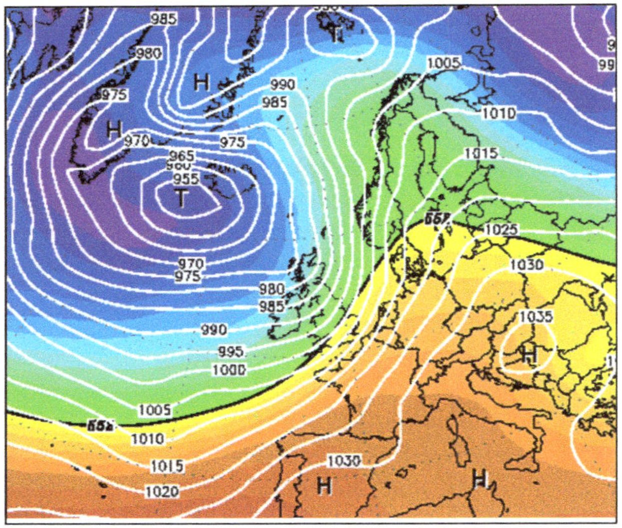

RAIN POTENTIAL

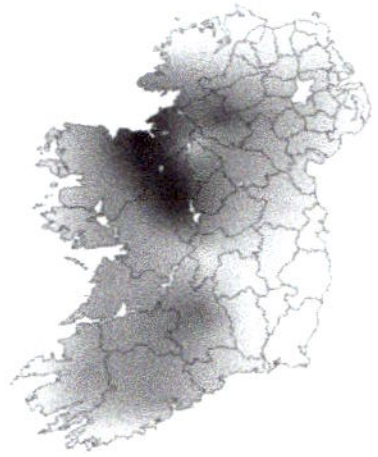

FROST/SNOW

JANUARY **10** FRIDAY

The depression to the northwest of Ireland maintains a very strong and blustery west to southwest flow with winds sometimes to storm force possible as well as bringing the return of widespread rain throughout the country. Rain likely to be particularly heavy in the north and west, possibly accompanying thunderstorm activity and hail showers.

Belfast: Colder, overcast, showers, strong gusty winds, chance of thunderstorms
Dublin: Cloudy, milder, showers of heavy rain, windy
Cork: Cloudy, occasional showers, unsettled and blustery winds
Galway: Overcast, rain, blustery, chance of thunderstorms, hail and gale gusts

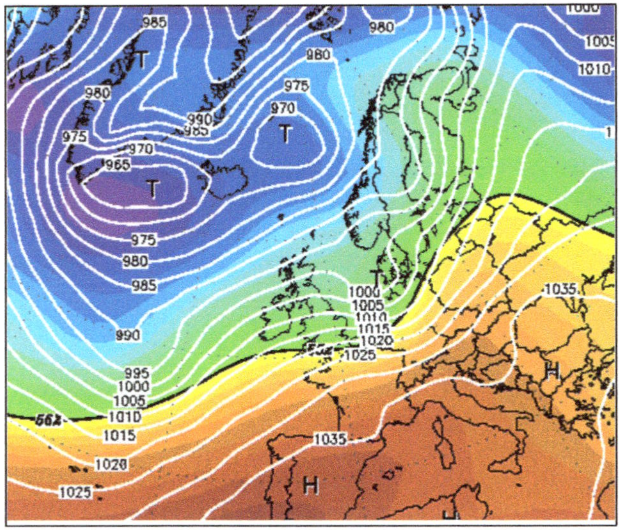

RAIN POTENTIAL

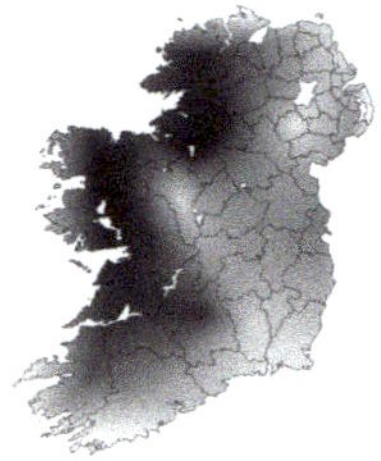

FROST/SNOW

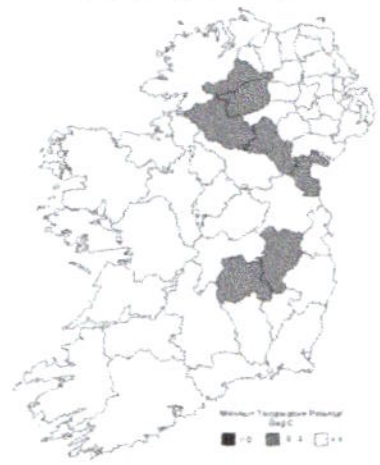

JANUARY **11** SATURDAY

Storm activity intensifies over the country with a west to northwest change. Thunderstorms, hail, and snow likely at times particularly in the north and west, with widespread shower activity elsewhere. Rain activity likely to contract to the far north and west by the end of the day, allowing for a few sunny spells to develop here and there.

Belfast: Stormy air with gale gusts, cloudy, brief squally showers, mild day cold night
Dublin: Blustery and stormy air, cloudy, brief squally showers, colder
Cork: Cloudy, colder, scattered showers, strong gusty winds, gales possible
Galway: Stormy, cloudy, colder, wintry showers of light hail likely, chance of snow

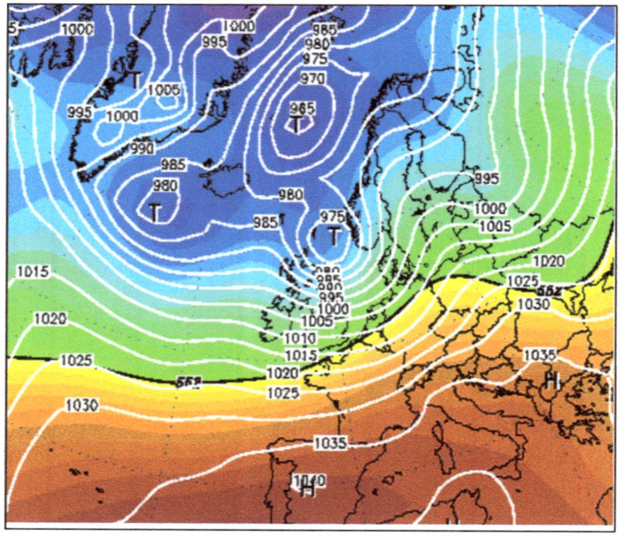

RAIN POTENTIAL

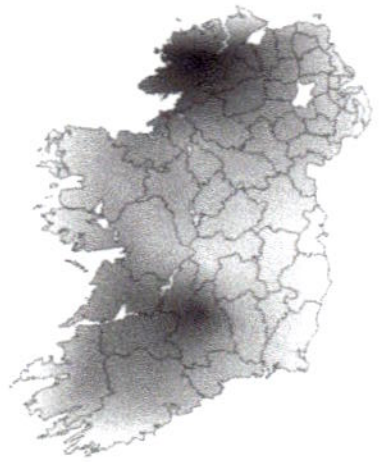

FROST/SNOW

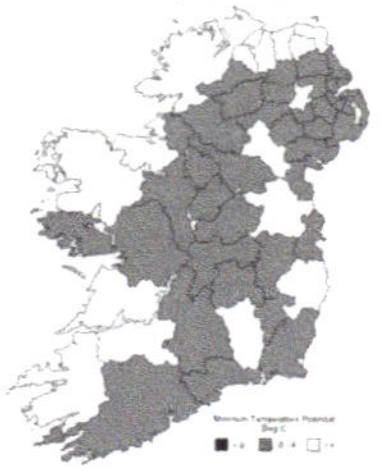

JANUARY **12** SUNDAY

A high pressure system to the south of Ireland extends a ridge up over the southern half of the country, easing storm activity to the occasional passing light shower in most districts, more frequent in the north. Winds however, remain very strong and blustery in a cold northwest flow. Chance of overnight snow flurries in the west, north, and midlands.

Belfast: Cloudy, colder, strong gusty northwest winds, isolated showers
Dublin: Windy, cloudy, brief spotty showers
Cork: Dull and gloomy, isolated showers, winds easing, possible overnight fog
Galway: Stormy air, overcast, isolated squalls, chance of snow flurries

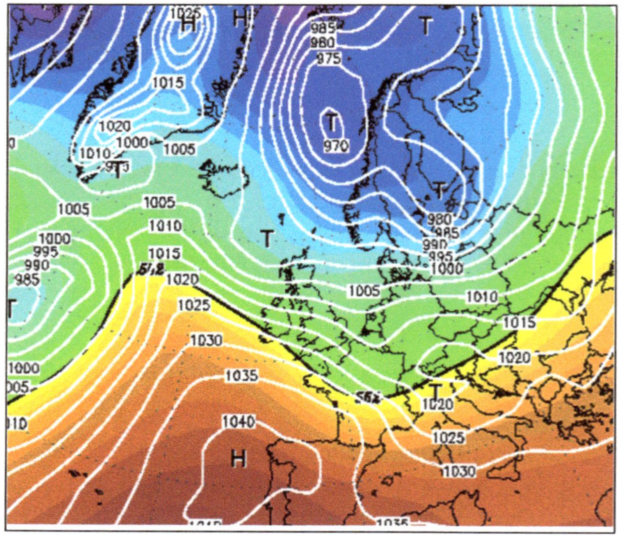

RAIN POTENTIAL

FROST/SNOW

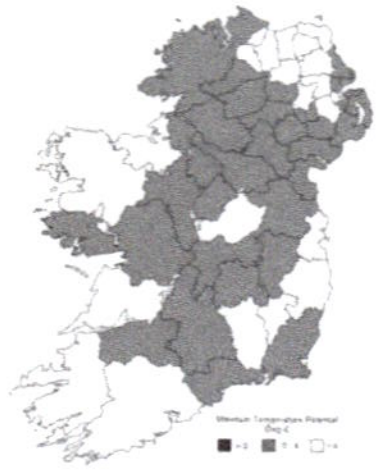

JANUARY 13 MONDAY

The depression to the northwest of Ireland pushes the high to the south further off the country, with a strong and gusty southwest change; winds to gale force at times around the coasts. Outbreaks of rain and drizzle return to most districts, heavier in the west, with rain activity most persistent during the middle of the day, with some clearer spells by evening.

Belfast: Overcast, milder, squally showers, stormy air with gale winds likely
Dublin: Milder, overcast, light showers, fresh to strong winds
Cork: Overcast, light showers, fresh breezes, chance of drizzly fog patches
Galway: Overcast, milder, heavy rain, strong winds

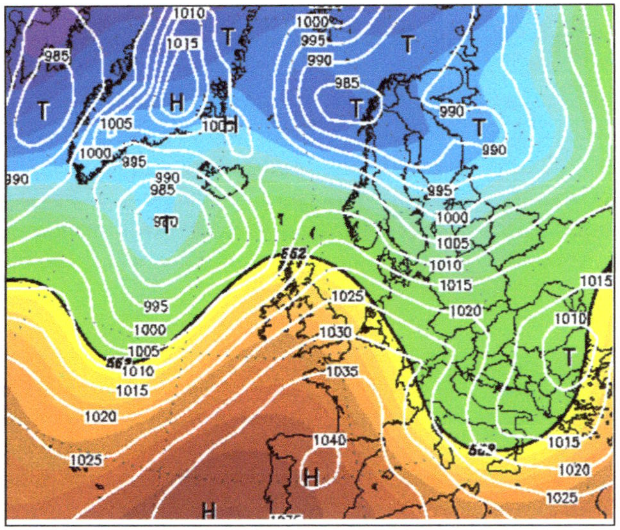

RAIN POTENTIAL

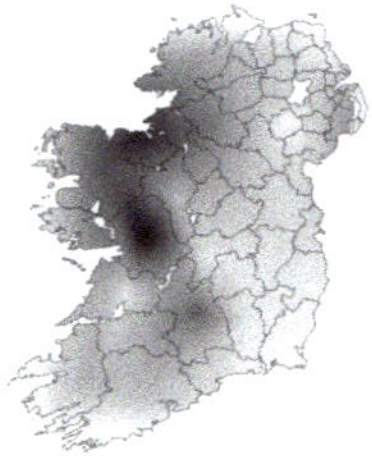

FROST/SNOW

JANUARY 14 TUESDAY

Low pressure to the north of Ireland brings a northwest change across the country with winds fresh to strong in most districts, but possibly up to gale force for a time in the north and around the coastal fringes. Mostly dry in the south and around the coasts of Connaught and Leinster, with isolated to light showers elsewhere, and more frequent in the northeast.

Belfast: Cloudy, brief drizzly showers, windy with occasional gale gusts
Dublin: Cloudy, isolated showers, blustery, chance of gale gusts
Cork: Cloudy, blustery, isolated showers
Galway: Strong and gusty winds, overcast, occasional showers

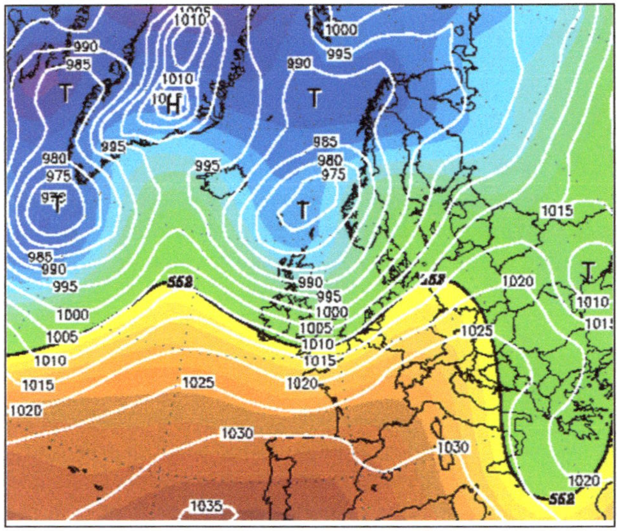

RAIN POTENTIAL

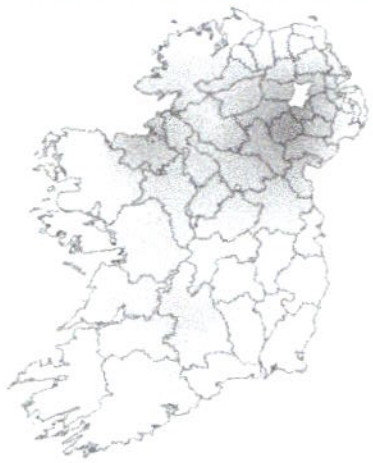

FROST/SNOW

JANUARY 15 WEDNESDAY

Stormy conditions ease over Ireland as the depression north of Ireland drifts to the northwest of the country while maintaining maintains a strong westerly flow, backing southwesterly and easing later in the day. Increasing sunny spells develop between passing showery outbreaks in most districts, with showers more persistent at times in the northwest and southeast. Temperatures generally milder than normal for this time of year.

Belfast: Mostly cloudy, isolated showers, some misty patches, winds easing
Dublin: Passing sunny spells, brief showers, milder day cool night, moderate breezes
Cork: Passing scattered cloud, moderate winds, some light drizzly showers
Galway: Cloudy, scattered showers, winds easing

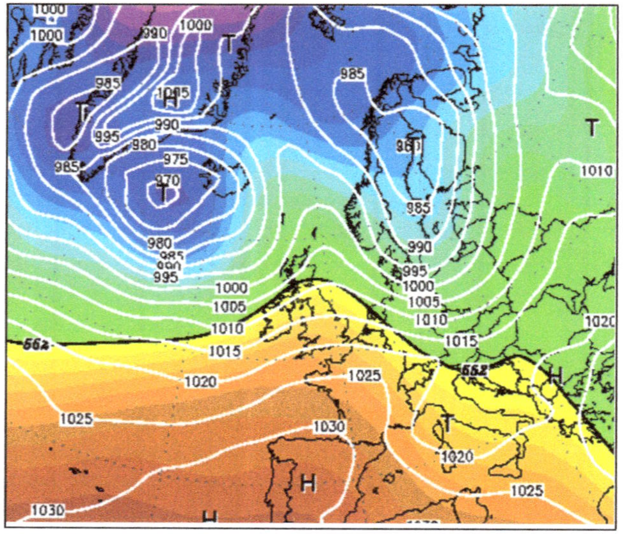

RAIN POTENTIAL

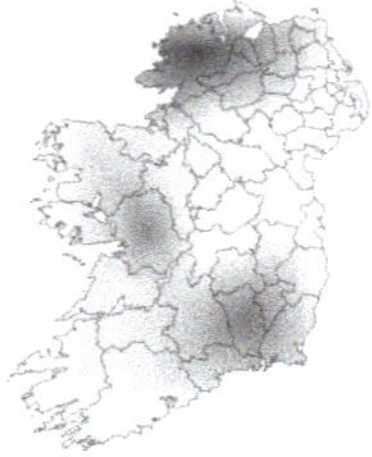

FROST/SNOW

JANUARY 16 THURSDAY

Low pressure continues to bring a strong westerly flow over the country and with a mix of misty drizzle patches and brief showers throughout the day, interspersed with the occasional sunny interlude, more frequent in the east and south and more persistent in the western half of the country. Temperatures will continue to be milder than usual for this time of year.

Belfast: Mostly cloudy, strong winds at times, brief showers, chance of misty patches
Dublin: Cloudy, cool, occasional spells of drizzle, moderate breezes
Cork: Changeable, mix of sunny spells and showers, moderate breezes
Galway: Cloudy, scattered showers, windy spells with odd gust

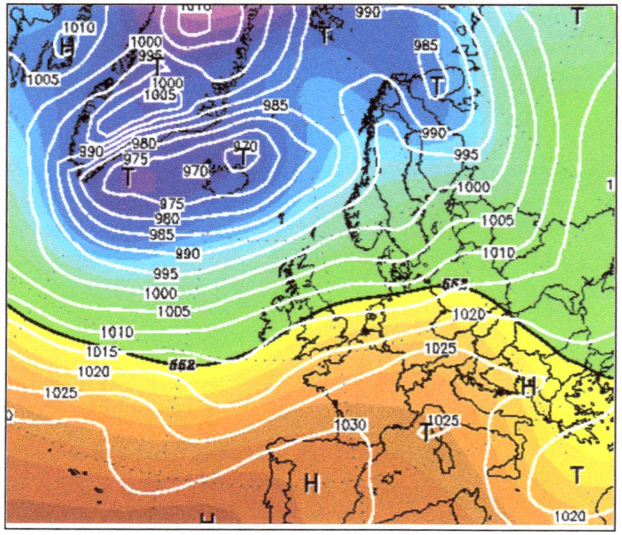

RAIN POTENTIAL

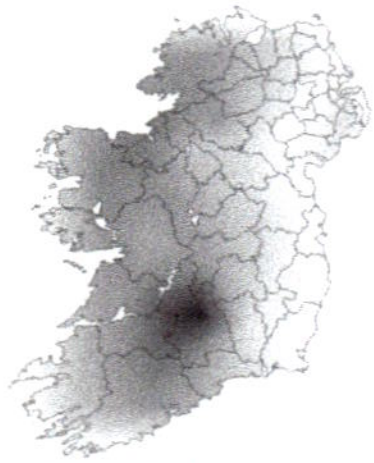

FROST/SNOW

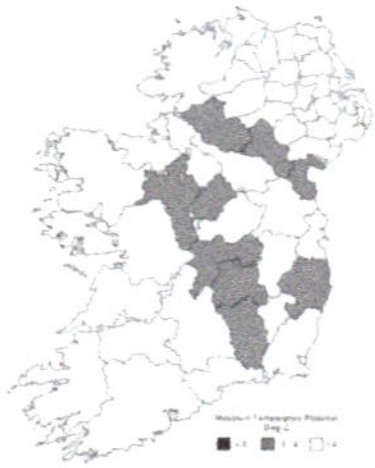

JANUARY **17** FRIDAY

The depression to the north of Ireland edges closer to the north of the country, bringing a stormy air throughout. South to southwest winds will prevail everywhere, very strong and gusty at times, possibly with gale gusts now and then. Despite a few brief sunny spells, widespread rain activity can be expected, including wintry showers of hail in the north, and heavier falls in the north and west, with chance of pockets of brief flooding in some central districts.

Belfast: Dull and gloomy air, scattered showers, windy
Dublin: Overcast, breezy, showers of light rain, mild day cold night
Cork: Overcast, showers, windy spells
Galway: Unsettled breezes, widespread cloud, scattered showers, hail possible

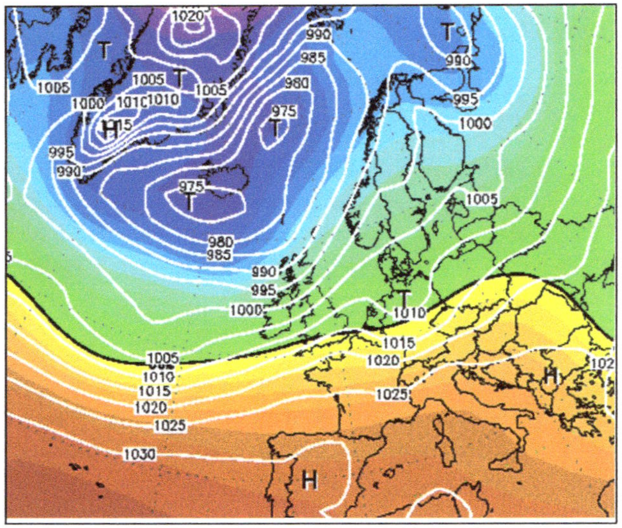

RAIN POTENTIAL

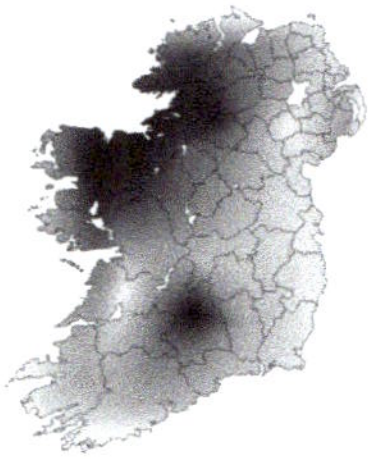

FROST/SNOW

JANUARY **18** SATURDAY

A blustery southwest to westerly flow builds as it crosses the country accompanied by widespread cloud and outbreaks of rain and drizzle everywhere, with some heavy falls in places. Temperatures still milder than is typical for this time of year.

Belfast: Overcast, rain, mild day cold night, fluctuating breezes
Dublin: Overcast, scattered showers, shifting breezy spells
Cork: Fluctuating breezes, dull and gloomy air, light rain and drizzle patches
Galway: Overcast, fluctuating winds, rain and persistent drizzles

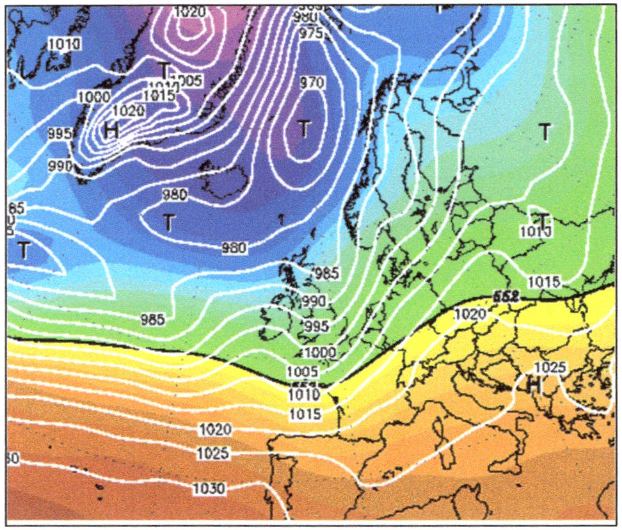

RAIN POTENTIAL

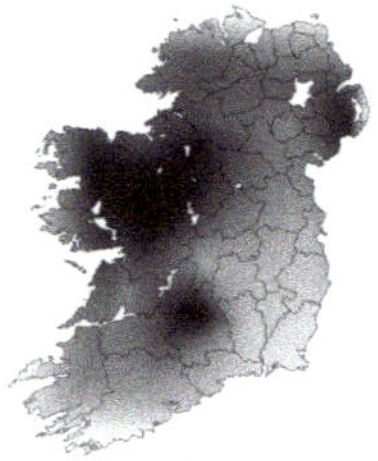

FROST/SNOW

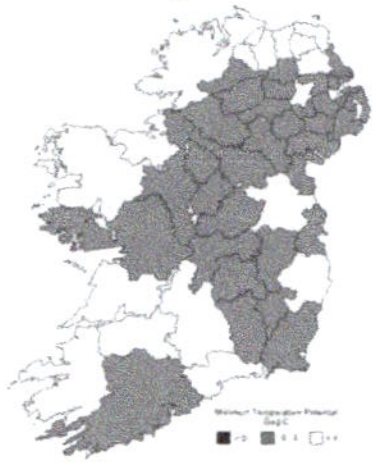

JANUARY **19** SUNDAY

A depression sits immediately to the north of Ireland directing a very strong southwest winds throughout, up to gale force at times in many districts. A stormy air prevails with widespread cloud and outbreaks of rain and squally drizzle in most districts, lighter in the northwest and more frequent and heavier through central to southern and eastern parts. Chance of pockets of flooding once more in some parts of the midlands.

Belfast: Overcast, mild, light showers, unsettled winds, chance of gale gusts at times
Dublin: Overcast, rain, blustery, chance of gales and stormy air
Cork: Overcast, rain and drizzle, strengthening winds with chance of gale gusts
Galway: Cloudy, scattered showers, strong winds, possible gales

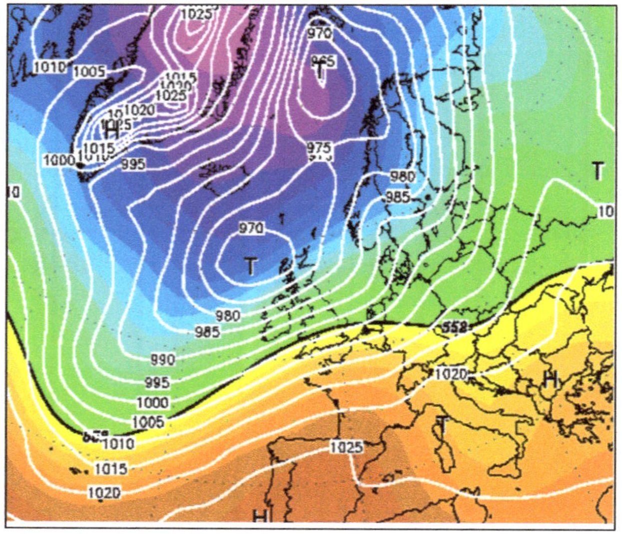

RAIN POTENTIAL

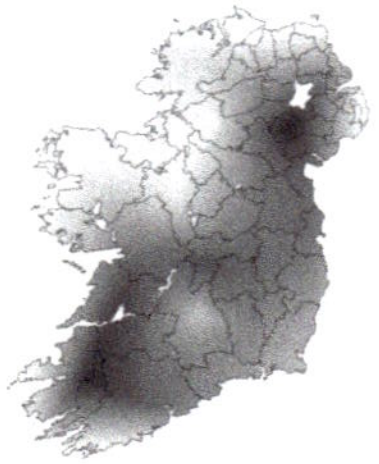

FROST/SNOW

JANUARY **20** MONDAY

A trough between two depressions sits to the north of Ireland while high pressure flows to the south, with a general northwest wind prevailing in the west of the country and a southwest flow to the east, with winds occasionally increasing to gale force for a time, especially in the west. A changeable day with sunny spells interspersed with bands of rain in most districts, heaviest in the midlands to the northeast at first then becoming heavier in the west later in the day. Chance of snow in the north of the country

Belfast: Changeable, mix of sunny spells and showers, winds easing at times
Dublin: Winds easing to moderate, rain, colder, occasional sunny spells
Cork: Mostly sunny, isolated showers, colder, winds easing to light
Galway: Stormy air easing, heavy overnight wintry showers, possible snow

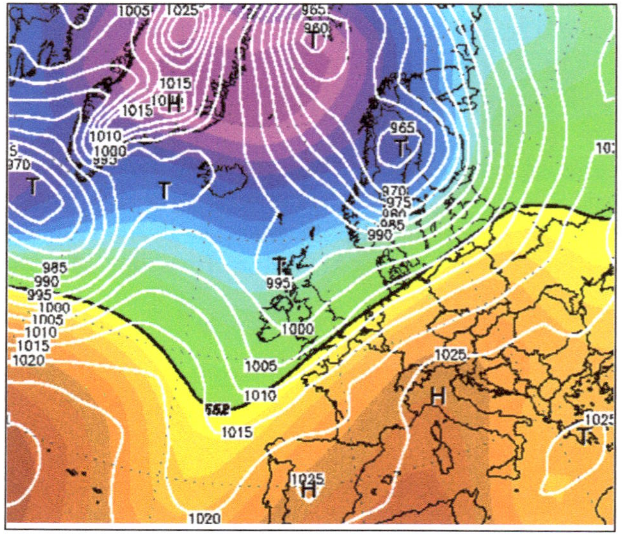

RAIN POTENTIAL

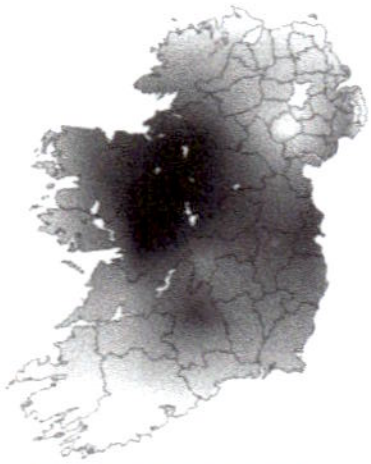

FROST/SNOW

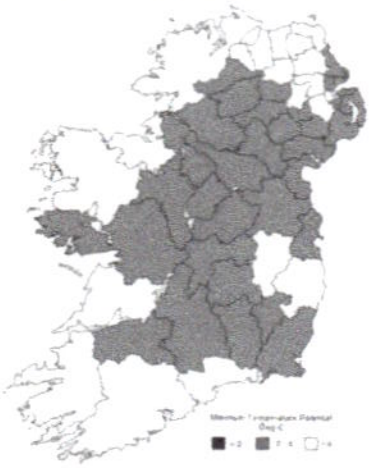

JANUARY **21** TUESDAY

A depression to the north of Ireland brings very strong and unstable west to southwest winds throughout the country, with gale force gusts at times along the coastal fringes. Widespread overnight rain prevails, easing to showers later in the day. A band of thunderstorms may cross the northern half of the country and bring possible wintry showers of hail or snow into some western parts.

Belfast: Stormy air, cloudy, light squally showers, strong winds to gale force at times
Dublin: Stormy air, gale winds likely, cloudy, milder temperatures, heavy showers
Cork: Stormy winds, cloudy, milder temperatures, squally showers
Galway: Stormy air, cloudy, showers, chance of thunderstorms, hail, and snow

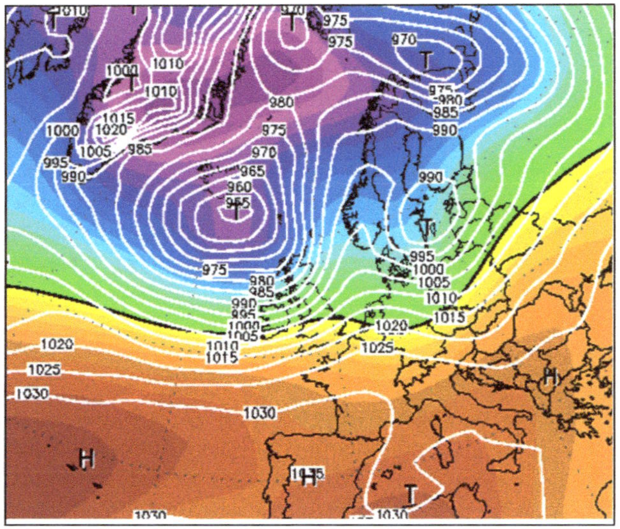

RAIN POTENTIAL

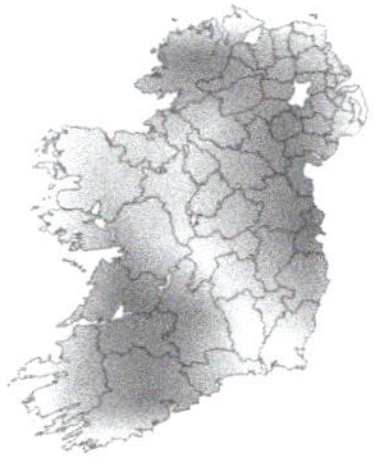

FROST/SNOW

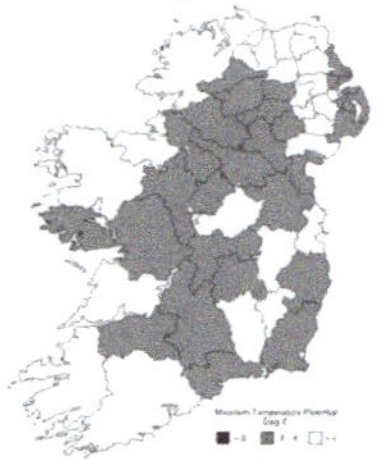

JANUARY **22** WEDNESDAY

High pressure ridges over the southern half of Ireland bringing a mostly dry day along the south and eastern coastal fringes, while a series of depressions build to the north of the country and continue showery activity elsewhere, with some heavy falls at times in the northwest and in the midland districts, where some pockets of localised flooding may be observed. Unstable and very strong westerly winds prevail throughout with gales and gusts to storm force at times likely.

Belfast: Overcast, milder, drizzly showers, unsettled winds with occasional gusts
Dublin: Overcast, mostly dry with odd threats of rain, light drizzle, gusty winds
Cork: Overcast, blustery, light drizzly showers
Galway: Strong and gusty winds, overcast, spells of light rain

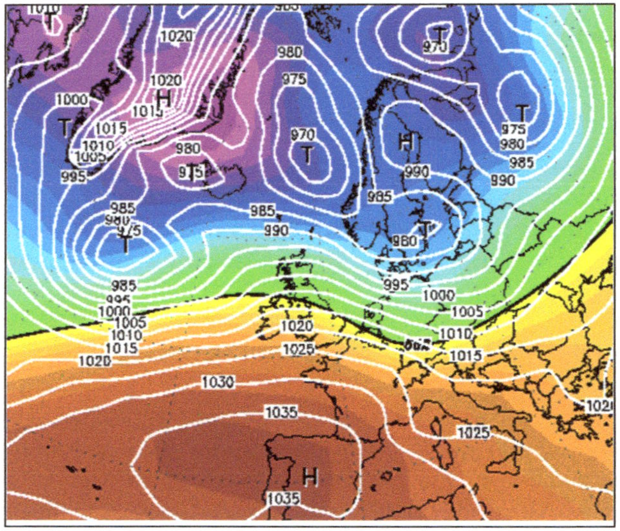

RAIN POTENTIAL

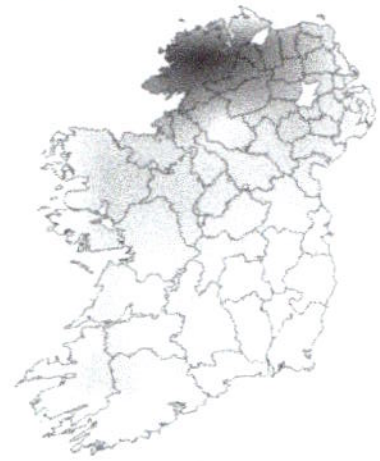

FROST/SNOW

JANUARY **23** THURSDAY

The high pressure ridge slips away from the south of Ireland while bands of low pressure from the depressions to the north expand over the country. Dull and gloomy conditions prevail with outbreaks of showers and rain everywhere, including the chance of some wintry showers possibly falling as snow for a time, particularly in the north and west. West to southwest winds will be strong and blustery, especially in the west.

Belfast: Dull and gloomy, unsettled winds, mild temperatures, occasional showers
Dublin: Overcast, light showers, fresh winds, mild day cool night
Cork: Cloudy, cooler, brief drizzly showers, moderate winds
Galway: Overcast, wintry showers, blustery spells, chance of evening snow

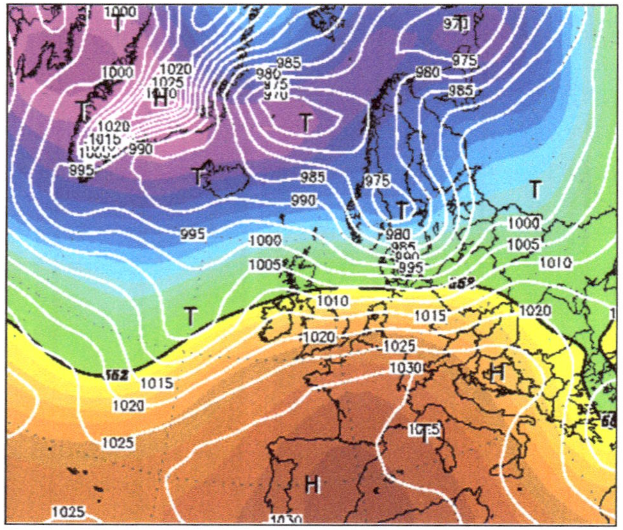

RAIN POTENTIAL

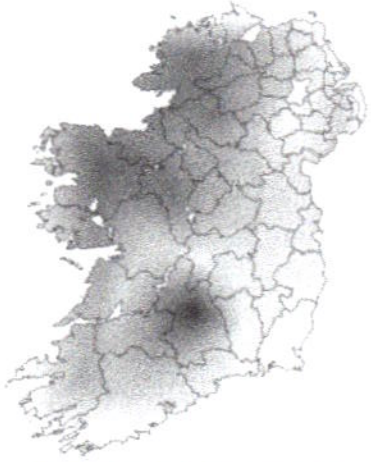

FROST/SNOW

JANUARY **24** FRIDAY

The high pressure ridge moves back up over the southern half of the country bringing a mostly dry day with some good sunny spells in the Leinster and Munster provinces, while depressions north of Ireland maintain more cloud and unsettled showery conditions in the west and north. Winds will remain very strong to gale force at times, in a west to southwest flow.

Belfast: Fresh westerly winds, cloudy, mostly dry, odd threat of rain, cooler overnight
Dublin: Colder, dry, occasional sunny spells, moderate breezes
Cork: Cloudy, mostly dry with odd threat of rain, cold overnight, breezy spells
Galway: Unsettled and breezy air, cloudy, isolated wintry drizzle patches

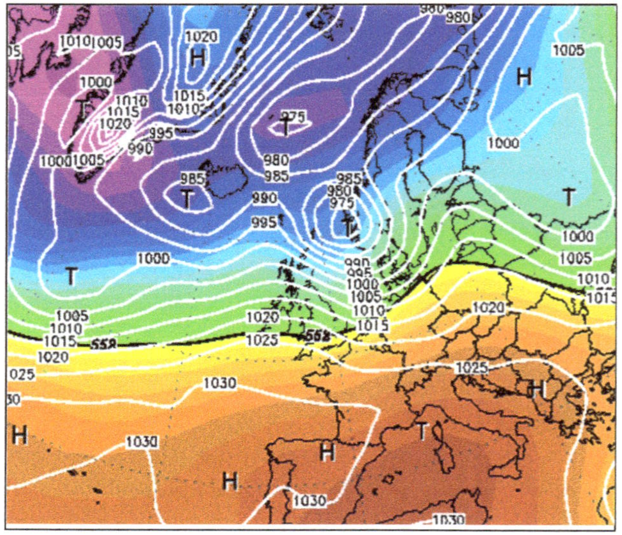

RAIN POTENTIAL

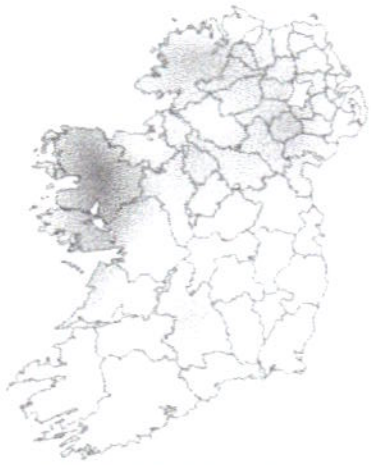

FROST/SNOW

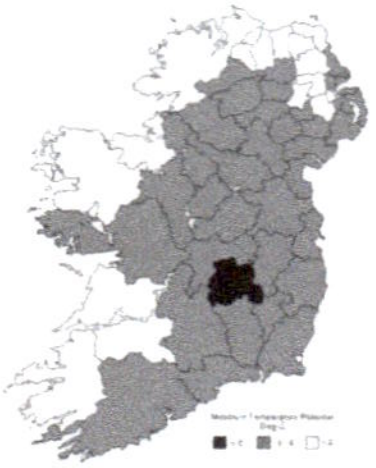

JANUARY 25 SATURDAY

The high pressure ridge over the south of Ireland drifts offshore while the blustery low pressure flow associated with the depressions to the north moves back into the southern districts. Widespread cloud and strong southwest winds prevail throughout with some light showers in most districts, more isolated along the far northern coasts.

Belfast: Overcast, cooler, some brief drizzly patches, fresh southwesterly breezes
Dublin: Moderate winds, dull and gloomy with light showers, milder
Cork: Cloudy, scattered showers and drizzle patches, milder, breezy at times
Galway: Overcast, windy, cool, some light showers

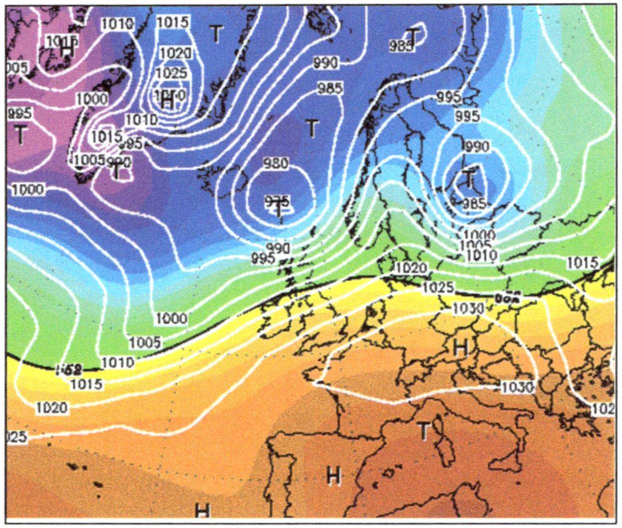

RAIN POTENTIAL

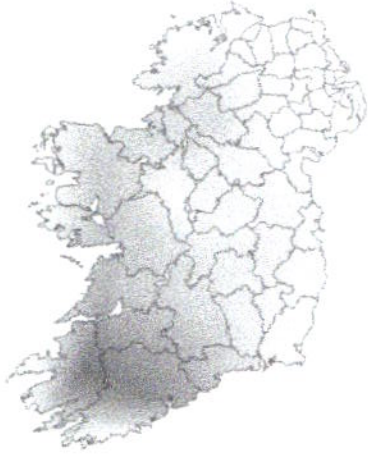

FROST/SNOW

JANUARY 26 SUNDAY

A low pressure trough between high pressure to the south of Ireland and depressions to the north of the country prevails and directs a generally light to mostly calm south to southeast flow throughout. A mixed day of cloudy spells and the odd sunny breakthrough with patches of fog most districts. Mostly dry along the northern and eastern coasts, with some showers in the west, south, and midland districts.

Belfast: Cloudy, light showers, cold overnight with frosts likely, gentle breezes
Dublin: Mostly dry, some sunny spells, frosts likely, spells of drizzle possible
Cork: Mostly calm air with possible misty patches, cloudy, colder, brief showers
Galway: Dull and gloomy air, cold, frosts and fog patches likely, evening showers

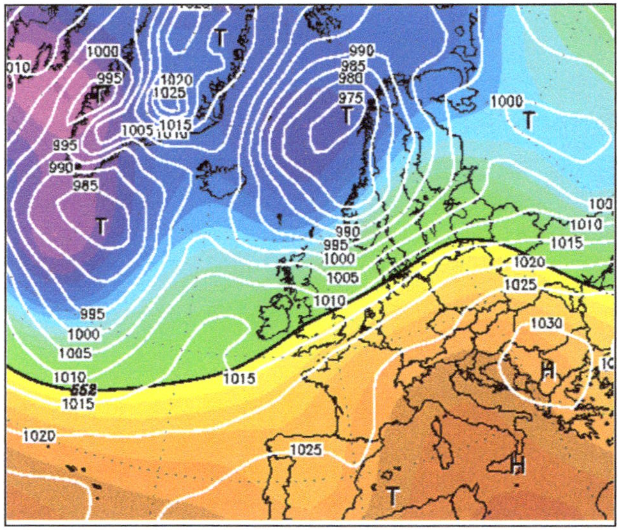

RAIN POTENTIAL

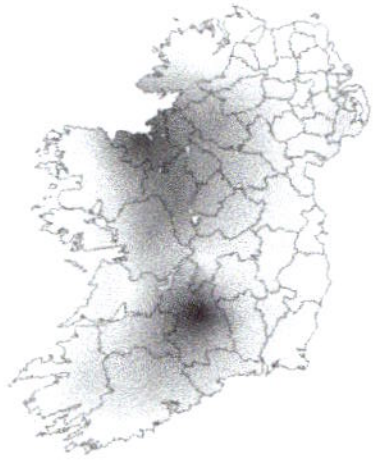

FROST/SNOW

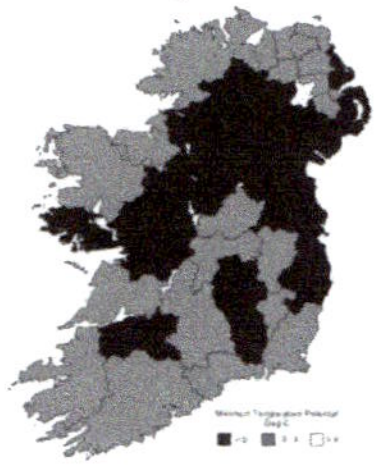

JANUARY **27** MONDAY

Stormy conditions return throughout the country as a large depression centred to the northwest of Ireland extends a fresh to strong south to southwest flow throughout, with winds up to gale force at times and accompanied with stronger gusts. Despite some brief morning sunny spells, generally dull and gloomy conditions prevail, with wintry showers including those of hail and snow likely in many districts.

Belfast: Cold, frosts, some sunny spells, heavy rain, chance of overnight snowfalls
Dublin: Cloudy spells, cool, brief showers, moderate breezes, chance of hail
Cork: Cloudy, rain, freshening winds to strong at times
Galway: Stormy, cloudy, heavy showers of wintry rain, snow, and hail, gales likely

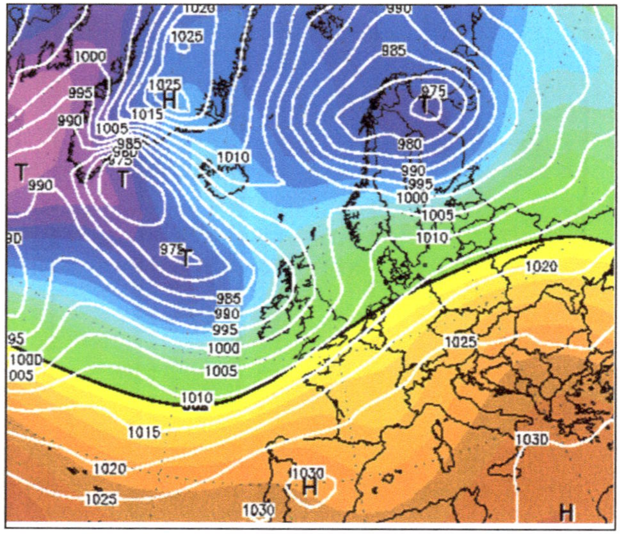

RAIN POTENTIAL

FROST/SNOW

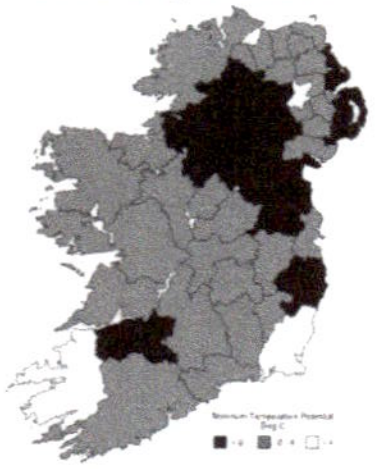

JANUARY **28** TUESDAY

A couple of depressions to the north of Ireland continue to bring strong and gusty southwest winds throughout the country, with storm conditions prevailing in most districts. Overnight snow likely in many places, especially in the north, and widespread wintry rain expected everywhere, including some of hail likely.

Belfast: Cloudy, cold, overnight light snow likely, moderate breezes, occasional gusts
Dublin: Cloudy, milder, rain, windy, chance of overnight hail showers
Cork: Overcast, rain, milder temperatures, blustery at times
Galway: Overcast, cold, wintry rain, snow likely, strong wind gusts at times

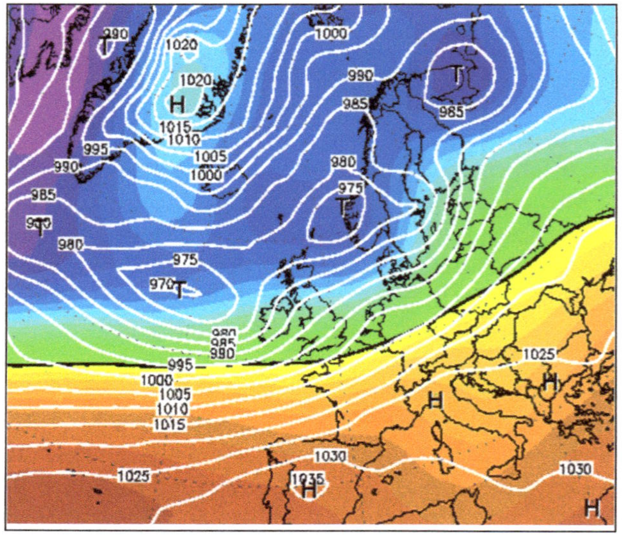

RAIN POTENTIAL

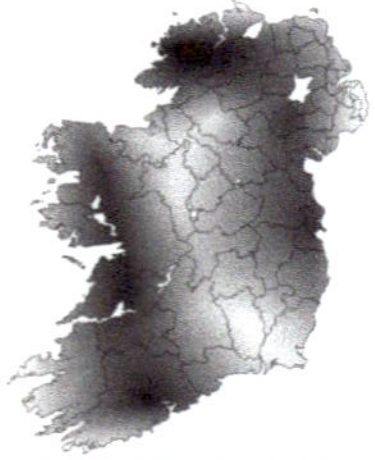

FROST/SNOW

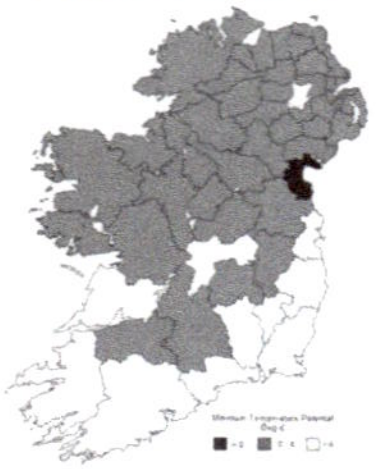

JANUARY **29** WEDNESDAY

Strong to forceful winds prevail in a mostly west to southwest flow as a couple of depressions to the north of Ireland dominate the country. Widespread cloud and outbreaks of showers and persistent drizzles throughout, starting in the south and spreading northward as the day progresses, with heavier falls likely in the northwest later. Chance of fog briefly overnight in the south of the country.

Belfast: Dull and gloomy, moderate breezes, patches of light rain and misty drizzles
Dublin: Overcast, mostly dry with passing threats of rain, windy at times
Cork: Overcast, light showers, windy spells, chance of overnight mist and fog
Galway: Milder, overcast and dull, scattered showers, strong winds

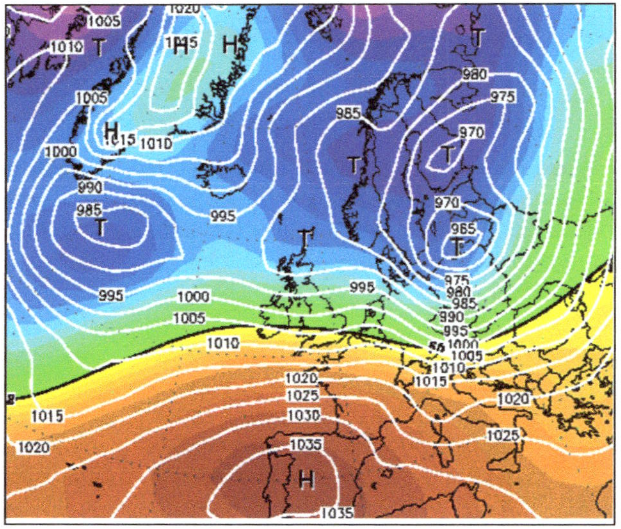

RAIN POTENTIAL

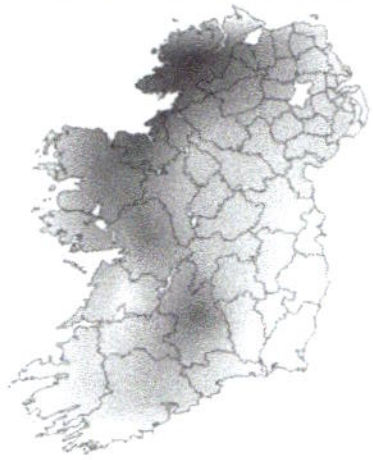

FROST/SNOW

JANUARY **30** THURSDAY

A depression moves southwards onto the north of the country bringing stormy conditions and directing very strong and blustery westerly winds throughout. Scattered thunderstorms and wintry rain in all districts, with the chance of hail and snow at times particularly in the west to central parts. Some brief sunny breakthroughs likey mostly in the south of the country.

Belfast: Unsettled and windy at times, cloudy, mild day cold night, scattered showers
Dublin: Cloudy, colder, rain, windy, thunderstorms possible
Cork: Changeable, mix of sunny spells and passing showers, unsettled winds
Galway: Stormy air, overcast, thunderstorms, rain and hail likely, chance of snow

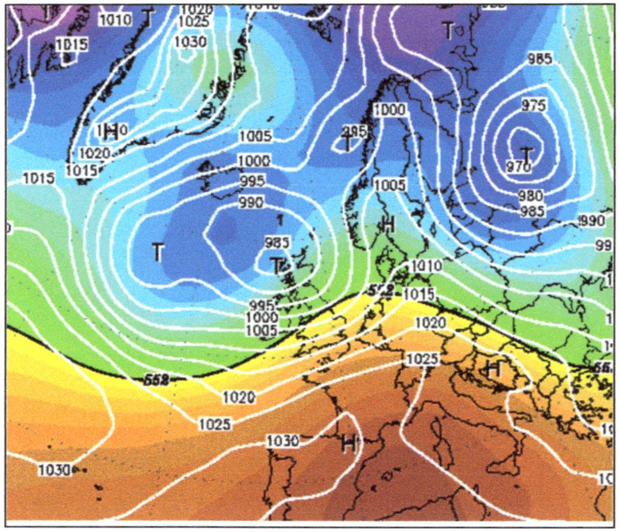

RAIN POTENTIAL

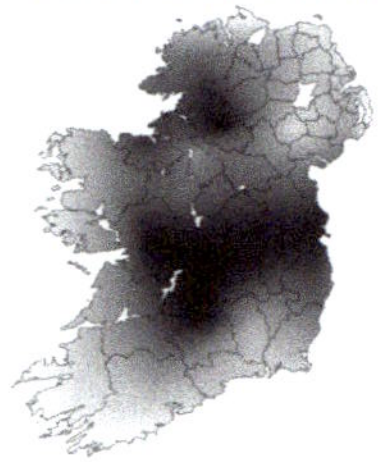

FROST/SNOW

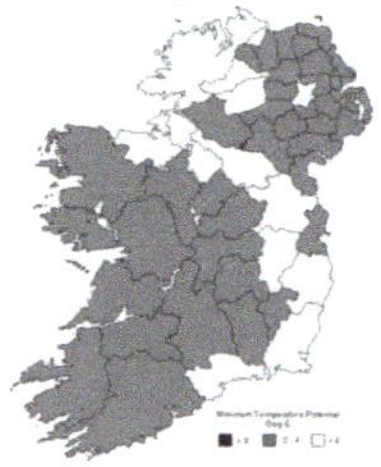

JANUARY **31** FRIDAY

Low pressure continues to dominate the country, with very cold temperatures and strong northwest winds prevailing. Widespread cloud and passing showers expected everywhere, including the chance of hail showers and snow, particularly in the west to the central parts. Some brief sunny spells likely, mostly in the east and south.

Belfast: Mostly cloudy, cold and dry, chance of overnight showers or snow flurries
Dublin: Changeable, mix of sunny spells and isolated showers, colder, windy at times
Cork: Cold, passing scattered cloud, some showers of light rain
Galway: Cold gusts, overcast, wintry showers, chance of snow and hail showers

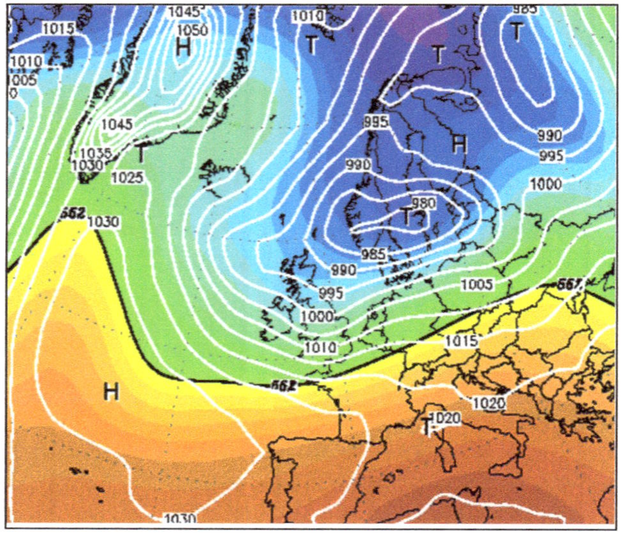

RAIN POTENTIAL

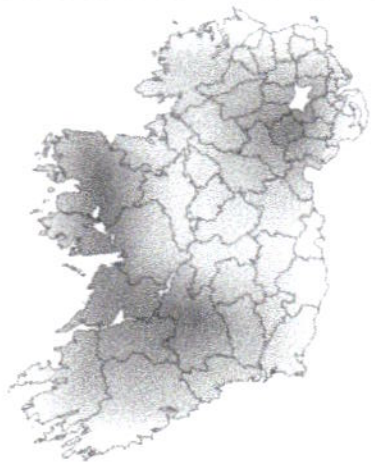

FROST/SNOW

FEBRUARY

PHASES OF THE MOON

2nd	Perigee #10
3rd	Crossing Equator
5th	First Quarter
8th	Northern Declination
12th	Full Moon
16th	Crossing Equator
18th	Apogee
20th	Third Quarter
22nd	Southern Declination
28th	New Moon

1	2	3	4	5	6	7	8	9	10	11	12	13	14	15	16	17	18	19	20	21	22	23	24	25	26	27	28
Thur	Fri	Sat	Sun	Mon	Tue	Wed	Thur	Fri	Sat	Sun	Mon	Tue	Wed	Thur	Fri	Sat	Sun	Mon	Tue	Wed	Thur	Fri	Sat	Sun	Mon	Tue	Wed
	P10	XhN		1Q			^				F				XhS		A		3Q		V						N

MONTHLY SUMMARY

1st-10th A large anticyclone to the west of Ireland slowly crosses the country. Colder with severe frosts especially in inland areas especially in the first few days. Winds generally north to north-easterly at first before bringing milder winds with a northwest change.

1st A cold front moves in a southerly path down through the country bringing stormy conditions throughout with wintry showers and spells of sleet and snow in many parts. Chance of gale winds, thunderstorms and hail showers. Despite this, some good sunshine hours may be recorded in some parts of the country.

1st-3rd Spell of wintry showers and snow likely.

2nd-5th Severe frosts in many districts. Clearing to mostly dry during this period.

4th Unusually sunny day throughout with some places seeing up to 8 hours of recorded sunshine.

6th A high pressure change.

6th-10th Mostly dry and mild spell with only the isolated drizzle patches likely to break through and less than normal ground frosts for this time of year.

Over the next four weeks despite low pressure being frequent and lower than is normal for this time of year, the first week will start with a spell of high pressure and mostly dry weather throughout. After this, the remainder of this outlook period will be wetter overall for the country as a whole, with above normal rainfall in the south of the country, and below average rainfall in the north. Some rain can be expected on most days during the remainder of this outlook period, but most of the rain overall will be recorded in the second week. It will be sunnier than normal in most districts except in some parts of the southwest, with the first week of this outlook period being the sunniest. Temperatures will be milder despite frequent ground frosts, and continue the trend for milder than normal temperatures experienced this winter, with widespread frosts over the first week ahead, but fewer than normal frosts over the remainder of this outlook period. Winds will be lighter than normal for this time of year. During this outlook period there will be around 16-20 days when rain is observed, 3 days with thunderstorms, 6 days with hail, 7 days with snowfalls, 4 days with widespread fogs, 8-11 days with widespread frosts but up to 20-23 frost days in inland parts, and 4 days when winds will be up to gale force at times.

11th-16th Period of settled weather as a high pressure system to the south of the country drifts slowly northwards before decaying. Light winds throughout in a southwest flow, later backing to northerly. Cold overnight with widespread and often severe frosts and patches of freezing fog and mostly sunny after the fogs lift.

12th-17th Very cold spell with widespread frosts and unusually sunnier than normal.

14th-15th Widespread fog likely.

15th-20th Very wintry spell with widespread sleet or snow, especially in the higher reaches, followed by hail or snow showers. Snow falls likely to be particularly widespread around the 18th.

16th-17th Unusually sunny spell.

17th-20th A very cold spell with an active frontal system crossing the country moving into the southwest at first and slowly advancing northwards. This front will bring spells of heavy precipitation, with many falls in the form of sleet or snow and wintry showers with hail likely, with most of the snow is likely to accumulate in northern parts and on the higher reaches. This precipitation will ease to scattered wintry showers in the last couple of days and temperatures will return to near normal values with strong east to southeast winds prevailing particularly around the coasts.

21st-25th An unsettled period with low pressure near the country bringing wet and windy conditions throughout.

21st-28th Milder temperatures with few frosts with south to southwest winds prevailing.

25th Heavy rain and strong winds prevail.

26th-28th A weak ridge moves onto the country bringing a drier spell.

26th Chance of thunderstorms, particularly around the south and south-western parts of the country.

27th Widespread fogs likely, then unusually sunnier than normal after the fog lifts.

IRELAND RAINFALL ESTIMATES

February 2025

d=mainly dry, n= nonrecordable, l=light shrs, s=significant shrs, r=rain, h=hvy falls

		ULSTER PROVINCE					CONNAUGHT PROVINCE						LEINSTER PROVINCE				MUNSTER PROVINCE						
		Alderg	Hillsbo	Armag	Loughl	Malin H	Sligo	Ardtarı	Belmul	Drums	Galway	Roscoı	Markre	Edende	Kilkenı	Dublin	Shannı	Newpo	Tralee	Killarne	Cork	All	Moon
1st	Sat	n	d	l	l	d	l	l	l	d	d	d	d	d	l	l	d	l	d	l	l	12	
2nd	Sun	l	l	d	s	l	l	s	s	l	l	l	s	n	l	d	d	l	l	s	d	43	P1
3rd	Mon	l	d	l	d	d	d	d	l	d	d	d	d	d	d	d	d	l	l	l	d	6	XhN
4th	Tues	d	d	d	l	l	l	l	l	l	d	l	l	n	d	d	d	l	d	l	d	12	
5th	Wed	d	d	d	d	d	l	d	l	n	d	d	l	d	d	d	d	d	n	l	d	7	1Q
6th	Thur	d	d	n	l	n	l	l	l	d	d	d	l	d	d	d	d	l	d	n	d	8	
7th	Fri	d	d	d	s	l	l	s	l	d	d	l	d	d	d	d	d	n	d	d	d	14	
8th	Sat	d	d	d	d	d	d	d	n	d	n	n	d	d	d	d	d	n	d	d	d	2	^
9th	Sun	d	d	d	d	d	d	d	d	n	l	d	d	d	d	d	d	l	d	d	d	3	
10th	Mon	d	d	d	d	l	d	d	d	d	l	d	d	d	d	d	n	l	n	d	d	4	
11th	Tues	d	d	d	d	d	d	d	l	d	l	d	d	d	d	d	l	l	l	n	d	5	
12th	Wed	d	d	d	d	l	d	d	d	d	d	d	d	d	d	d	d	d	l	d	d	2	F
13th	Thur	d	d	d	d	d	d	d	d	d	d	d	d	d	d	d	d	d	d	d	d	0	
14th	Fri	d	d	d	l	l	d	l	d	d	d	d	d	d	d	d	d	d	d	d	d	2	
15th	Sat	d	d	d	d	l	d	d	l	d	d	d	d	d	d	d	d	n	d	d	d	3	
16th	Sun	d	d	d	d	l	d	d	d	d	d	d	d	d	h	d	d	d	d	d	d	19	XhS
17th	Mon	l	l	d	d	d	s	d	l	s	h	s	s	l	l	l	s	r	h	h	h	148	
18th	Tues	s	r	d	l	l	s	l	s	s	s	s	s	s	l	h	r	s	l	l	l	102	A
19th	Wed	l	l	s	d	n	l	d	n	d	s	d	d	n	l	d	l	l	r	r	s	51	
20th	Thur	s	s	l	s	l	s	s	s	s	l	s	s	l	l	s	l	s	s	s	r	92	3Q
21st	Fri	s	s	l	s	r	s	s	s	l	s	l	l	s	l	s	s	r	r	r	s	115	
22nd	Sat	l	l	s	s	l	s	s	l	l	l	l	s	n	d	l	l	s	s	r	s	69	V
23rd	Sun	l	l	l	l	l	l	l	l	l	l	s	l	d	s	d	l	l	r	s	s	57	
24th	Mon	l	d	l	l	d	s	l	l	n	s	l	n	d	l	d	d	s	d	h	d	56	
25th	Tues	l	s	d	l	l	s	l	r	s	s	h	s	r	s	l	s	s	r	s	r	135	
26th	Wed	l	d	s	d	l	d	d	s	d	d	l	d	d	d	d	d	l	s	l	l	23	
27th	Thur	d	d	d	d	n	d	d	d	d	d	d	d	d	d	d	d	d	d	d	d	1	
28th	Fri	l	l	d	s	l	r	s	l	s	s	s	r	l	l	d	l	s	r	h	l	113	N
stimate:		30	36	24	40	36	69	40	51	49	66	58	51	29	42	39	43	74	105	150	73	1104	
Average:		58	68	54	91	87	92	94	97	78	88	98	92	70	57	49	76	127	141	103	98	1717	
	Trend:	drr	drr	drr	drr	drr	drr	drr	drr	drr	drr	drr	drr	drr	drr	av	drr	drr	drr	wtr	drr	drr	

IRELAND SUNSHINE ESTIMATES

February 2025

F=fine (8-12 hours of sunshine), pc= partly cloudy (4-7 hours), c=cloudy (1-3 hours), o=overcast (0 hours)

		ULSTER PROVINCE					CONNAUGHT PROVINCE						LEINSTER PROVINCE				MUNSTER PROVINCE						
		Alderg	Hillsbo	Armag	Loughl	Malin I	Sligo	Ardtarı	Belmul	Drums	Galway	Roscoı	Markre	Edende	Kilkenı	Dublin	Shannı	Newpo	Tralee	Killarne	Cork	All	Moon
1st	Sat	pc	pc	pc	pc	pc	pc	pc	pc	pc	pc	pc	pc	pc	c	pc	pc	pc	c	c	c	105	
2nd	Sun	c	c	c	c	c	c	c	o	c	c	c	c	c	pc	c	c	c	c	c	pc	52	P1
3rd	Mon	c	c	c	c	c	c	c	o	c	c	o	c	c	c	c	c	c	c	c	c	52	XhN
4th	Tues	pc	pc	pc	pc	pc	pc	pc	c	pc	pc	c	pc	pc	pc	pc	pc	pc	pc	pc	F	109	
5th	Wed	o	o	o	o	o	o	o	o	o	o	o	o	o	o	o	o	o	c	c	c	12	1Q
6th	Thur	o	o	o	o	o	o	o	c	o	o	o	o	o	o	o	o	o	o	o	o	6	
7th	Fri	c	c	c	c	c	c	c	c	c	c	c	c	c	c	c	o	c	o	o	o	25	
8th	Sat	c	c	c	c	c	o	c	o	c	o	o	c	c	c	c	o	o	o	o	o	15	^
9th	Sun	c	c	c	c	c	o	c	o	c	o	o	c	c	c	c	c	c	c	c	o	34	
10th	Mon	o	o	o	o	o	o	o	o	o	o	o	o	o	o	o	o	o	o	o	o	0	
11th	Tues	o	o	o	o	o	o	o	o	o	o	o	o	o	o	o	o	o	o	o	c	11	
12th	Wed	pc	pc	pc	pc	pc	pc	pc	pc	pc	pc	pc	pc	pc	F	pc	pc	pc	pc	pc	F	154	F
13th	Thur	pc	pc	pc	pc	pc	pc	pc	pc	pc	pc	pc	pc	pc	pc	F	pc	pc	pc	pc	pc	132	
14th	Fri	pc	pc	pc	pc	pc	pc	c	pc	c	pc	pc	c	c	c	c	pc	c	o	o	c	65	
15th	Sat	c	c	c	c	c	o	c	o	c	o	o	c	c	pc	pc	o	c	o	o	c	43	
16th	Sun	pc	pc	pc	pc	pc	F	pc	F	pc	pc	pc	pc	pc	c	pc	pc	pc	o	o	o	113	XhS
17th	Mon	c	c	c	c	c	o	c	o	c	o	o	c	c	c	pc	o	c	o	o	o	35	
18th	Tues	o	o	o	o	o	o	o	o	c	c	c	o	c	c	o	c	c	pc	pc	pc	33	A
19th	Wed	pc	pc	pc	pc	pc	pc	pc	pc	pc	pc	pc	pc	pc	pc	pc	pc	pc	c	c	pc	95	
20th	Thur	o	o	o	o	o	o	o	c	o	o	o	o	o	o	o	o	o	c	c	o	6	3Q
21st	Fri	c	c	c	c	c	o	c	c	c	c	c	c	c	pc	c	c	c	c	c	pc	49	
22nd	Sat	c	c	c	c	c	c	c	c	c	c	c	c	c	pc	pc	c	c	c	c	pc	61	V
23rd	Sun	o	o	o	o	o	o	o	o	o	o	o	o	o	o	o	c	o	c	c	c	14	
24th	Mon	c	c	c	c	c	c	c	o	c	c	c	c	c	pc	pc	c	c	c	c	pc	56	
25th	Tues	o	o	o	o	o	o	o	c	o	o	c	o	o	o	o	o	o	o	o	o	7	
26th	Wed	o	o	o	o	o	c	o	pc	o	c	pc	o	o	o	o	o	o	o	o	o	25	
27th	Thur	pc	pc	pc	pc	pc	F	pc	pc	pc	pc	pc	pc	pc	pc	pc	pc	pc	pc	pc	pc	137	
28th	Fri	c	c	c	c	c	c	c	c	c	c	c	c	c	c	pc	c	c	o	o	c	50	N
e hours:		78	78	78	78	78	74	78	71	79	70	70	78	79	80	81	69	75	63	63	79	1496	
age hours:		64	64	62	63	62	62	62	67	60	64	64	65	62	64	70	66	61	62	60	64	1268	
	Trend:	more	more	more	more	more	more	more	av	more	av	av	more	more	more	more	av	more	av	av	more	more	

FEBRUARY **1** SATURDAY

A large high pressure system ridges along the western coast of Ireland bringing a much colder and fresh northwest flow over the country, later backing northeasterly. Some overnight snow likely in most districts, clearing to a mostly dry and cloudy day, with more frequent sunny spells in the north and east.

Belfast: Cold northerly air, frosts, mostly sunny and dry, chance of snow
Dublin: Cold, changeable, some sunny spells, some wintry showers, chance of snow
Cork: Cloudy, colder, showers of light snow likely, moderate breezes
Galway: Cold, good sunny spells, dry, chance of frost and light overnight snow

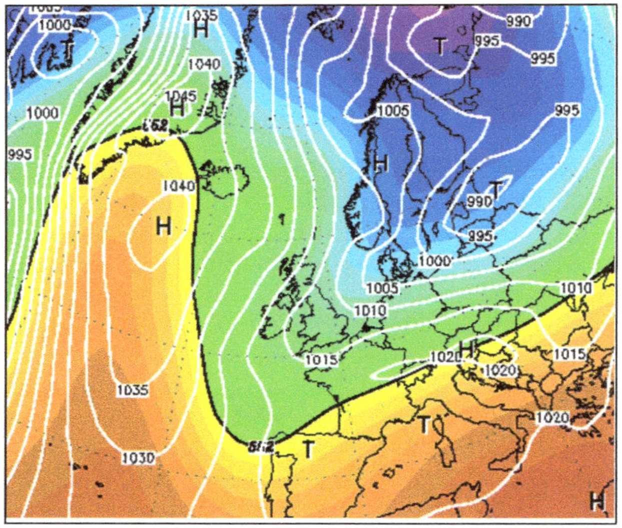

RAIN POTENTIAL

FROST/SNOW

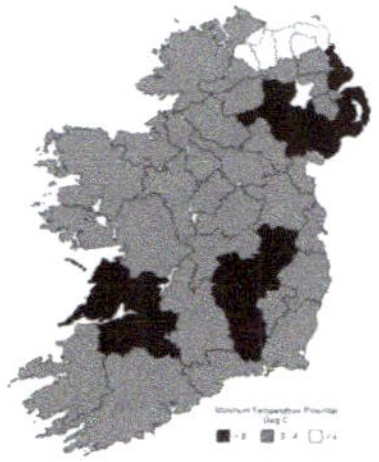

FEBRUARY **2** SUNDAY

A high pressure ridge to the northwest of Ireland extends moderate to fresh northerly winds down through the country, with wintry showers including light snow and snowflurries likely in most districts, lighter around the eastern and southern coasts. Mostly cloudy with a few sunny spells around the middle of the day except inthe east where sunny spells will be longer. Widespread frosts overnight.

Belfast: Cloudy, cold, frosts, light northerly breezes, wintry showers, chance of snow
Dublin: Cloudy, very frosty, moderate breezes, mostly dry, chance of overnight snow
Cork: Cold, frosts, mostly dry and sunny, moderate breezes
Galway: Cloudy, cold, frosts, mostly dry, overnight snow flurries likely

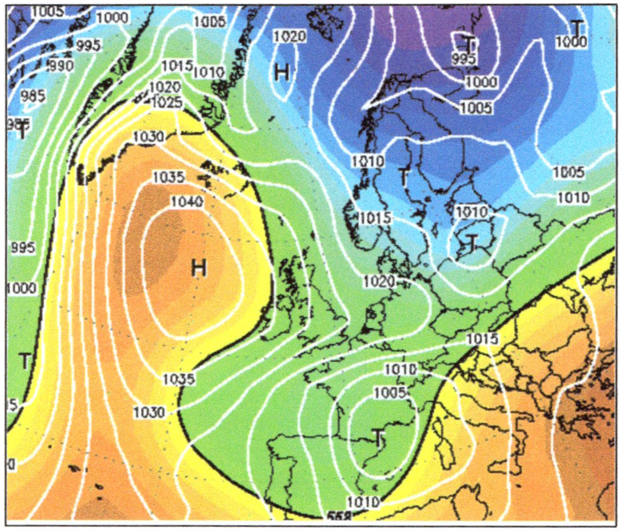

RAIN POTENTIAL

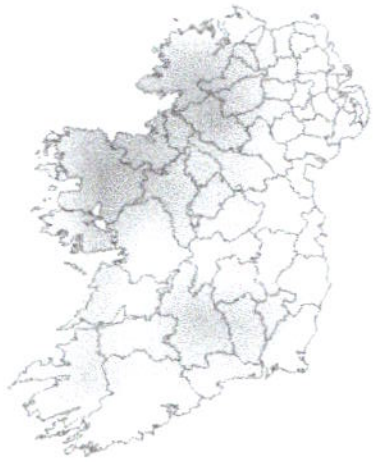

FROST/SNOW

FEBRUARY **3** MONDAY

A large high pressure system to the west of Ireland continues to bring cold and frosty conditions throughout with moderate to fresh northerly winds prevailing, stronger along the western coasts. Mostly dry and with some sunny spells everywhere except for some isolated showers possible here and there. Overnight fog patches possible in the far north, while gusty winds in the west.

Belfast: Cold and cloudy, dry, gentle breezes, fog patches possible
Dublin: Cloudy, cold northerly air, dry
Cork: Cloudy, cold, mostly dry, chance of brief drizzle patches
Galway: Cloudy, dry, cold northerly breezes, chance of gale gusts at times

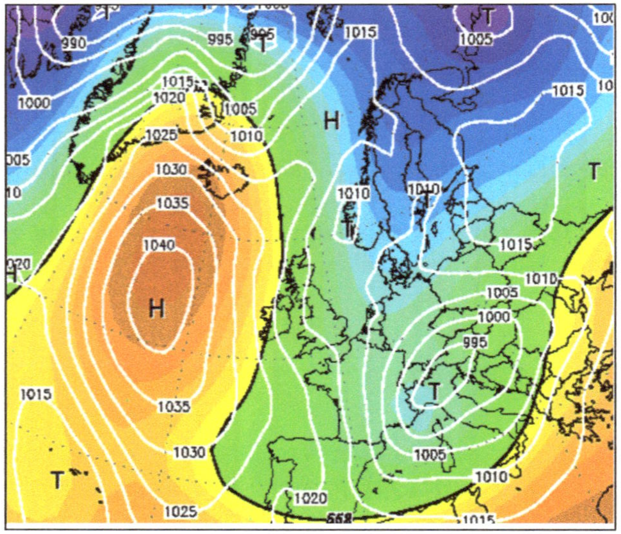

RAIN POTENTIAL

FROST/SNOW

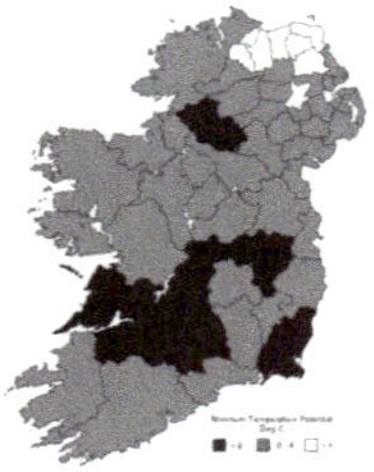

FEBRUARY **4** TUESDAY

Anticyclonic conditions continue to prevail over Ireland in a cool and dry air, with very light and variable breezes, generally north to northeasterly but later backing northwesterly and freshening. Some overnight wintry showers likely including the chance of spotty hail in central parts, before patches of frost and fog form, later clearing to a mostly sunny day throughout.

Belfast: Partly cloudy, mostly dry, cold, mostly calm air, frost and fog patches
Dublin: Mostly calm, dry, some sunny spells, very cold
Cork: Sunny, dry, cold with patches of frost, light breezes
Galway: Cold, frosts, dry, occasional sunny spells, moderate northerly breezes

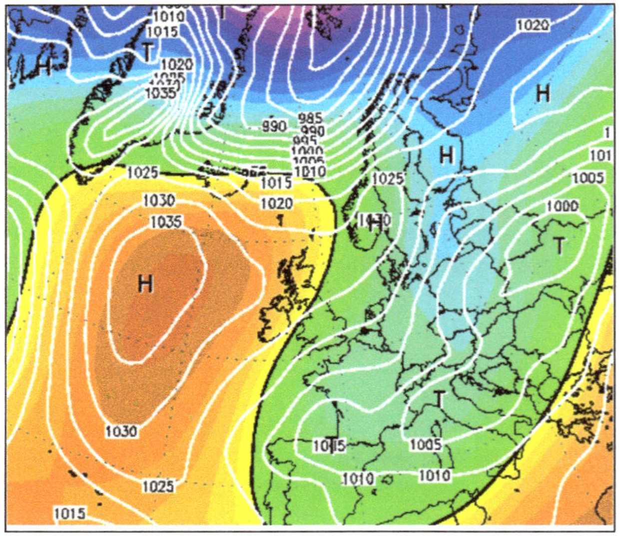

RAIN POTENTIAL

FROST/SNOW

FEBRUARY **5** WEDNESDAY

The large anticyclonic system to the west of Ireland directs a light northwesterly air over the country. Mainly dry with increasing cloud throughout, with brief sunny spells likely in the south of the country while dull and gloomy elsewhere with the occasional passing threat of rain and possible overnight misty drizzle patches here and there. Chance of brief overnight snow flurries noted in some inland parts.

Belfast: Overcast, dry, cold northwesterly air to mostly calm, frost and fog patches
Dublin: Dull and gloomy, dry, cold, odd threat of rain, some misty patches possible
Cork: Cloudy, milder, dry, gentle breezes, possible morning drizzle patches
Galway: Overcast and dry, odd threat of rain, cold northwesterly breezes

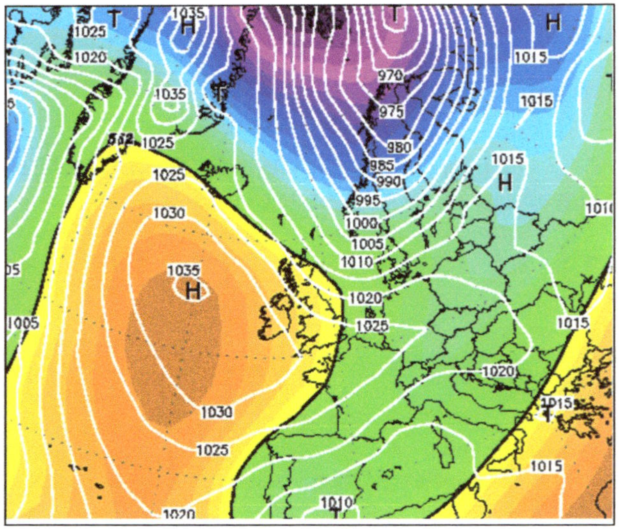

RAIN POTENTIAL

FROST/SNOW

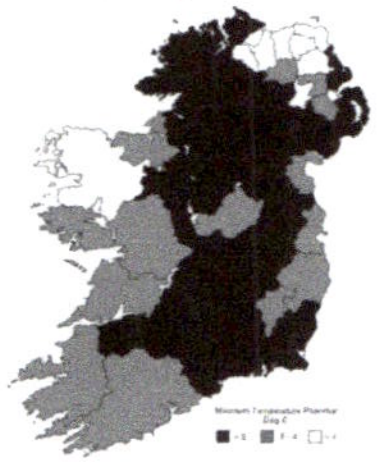

FEBRUARY **6** THURSDAY

A large high pressure system maintains a light west to northwest flow over the country and widespread cloud throughout. Dull and gloomy everywhere and although mostly dry, passing threats of rain and scattered misty patches prevail. Chance of some drizzle patches in the northwest, and possibly light snow for a time in central parts.

Belfast: Dull and gloomy, milder, dry, gentle breezes, fog likely
Dublin: Overcast, mostly dry, odd drizzle outbreak possible, very cold, light breezes
Cork: Overcast, dry, passing threats of rain, milder, mostly calm air
Galway: Overcast, cold, mostly dry, gentle breezes, chance of misty patches

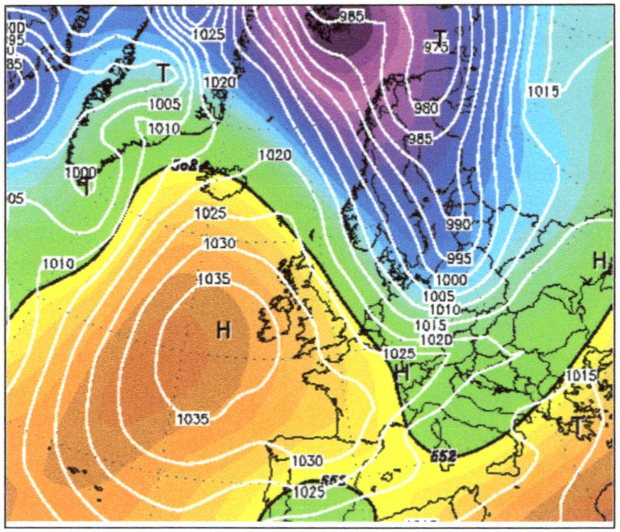

RAIN POTENTIAL

FROST/SNOW

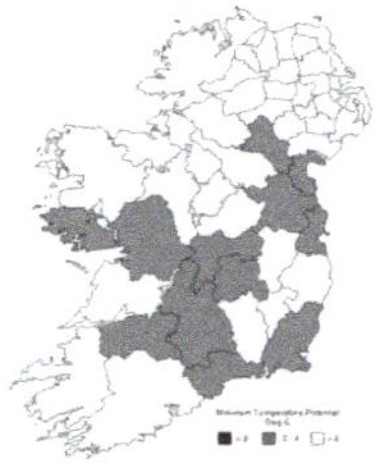

FEBRUARY **7** FRIDAY

A strong high pressure system continues to slowly drift over Ireland bringing further mostly calm conditions throughout at first, but freshening later in the day in a westerly flow, mostly over northern parts. Widespread cloud prevails with some showers in the northwestern quarter of the country, remaining mostly dry elsewhere with patches of fog here and there, especially in central parts.

Belfast: Cloudy, mild, dry, freshening breezy spells, drizzle and fog patches likely
Dublin: Cloudy, milder, dry, moderate breezes
Cork: Overcast, cool, mostly dry with odd threat of rain, mostly serene air
Galway: Cloudy, light breezes, dry and milder

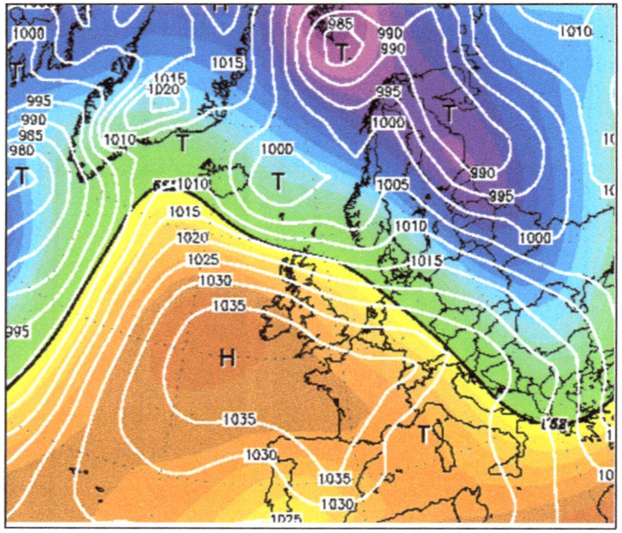

RAIN POTENTIAL

FROST/SNOW

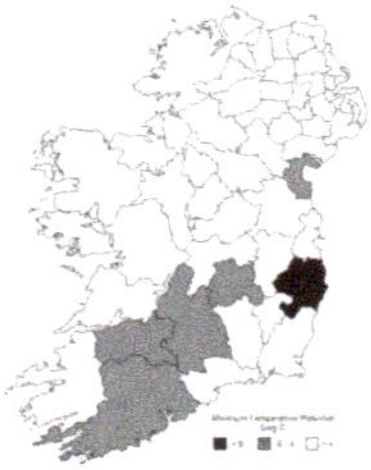

FEBRUARY **8** SATURDAY

High pressure weakens as it lingers over Ireland. A cloudy and light westerly air prevails throughout with pleasant and mostly dry conditions everywhere. Chance of some drizzle patches briefly around the west and northern coastal fringes and patches of overnight frost in sheltered parts.

Belfast: Cloudy, dry, mild temperatures, light breezes to mostly calm air
Dublin: Light northerly air, cold, dry, cloudy, chance of morning drizzle patches
Cork: Overcast and gloomy, calm and serene air, dry
Galway: Calm and dull atmosphere, colder, isolated drizzly showers likely

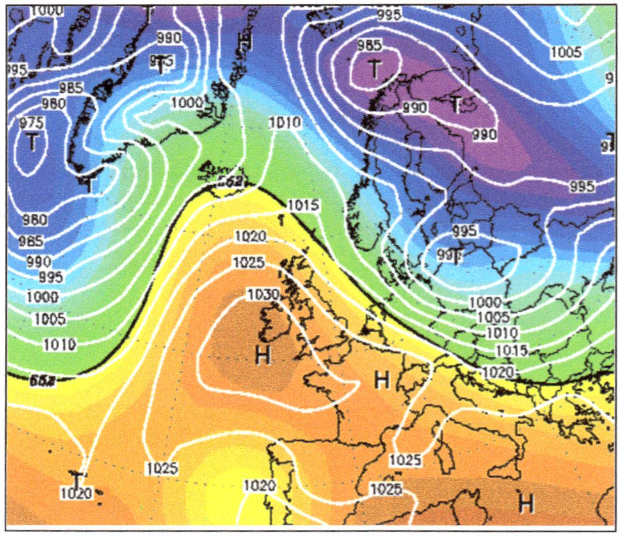

RAIN POTENTIAL

FROST/SNOW

FEBRUARY 9 SUNDAY

A decaying high pressure system continues to linger over Ireland bringing another mostly dry and cloudy to overcast day throughout with patches of fog overnight in many places, but milder temperatures and no frost anticipated. Winds generally light and variable everywhere, but freshening later in the north with a gradual southwest change.

Belfast: Calm, cool, and cloudy air, dry, chance of fog
Dublin: Calm and serene air, cold, dry, occasional sunny outbreaks
Cork: Overcast and dry, odd threat of rain, cold, calm serene atmosphere
Galway: Overcast, milder, isolated showers, mostly calm air freshening later

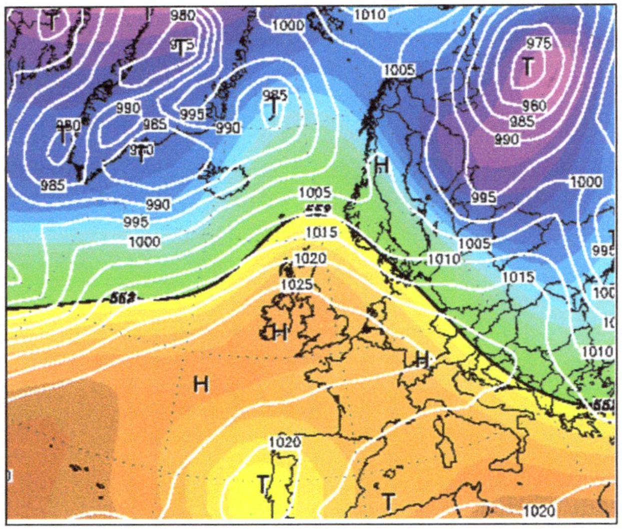

RAIN POTENTIAL

FROST/SNOW

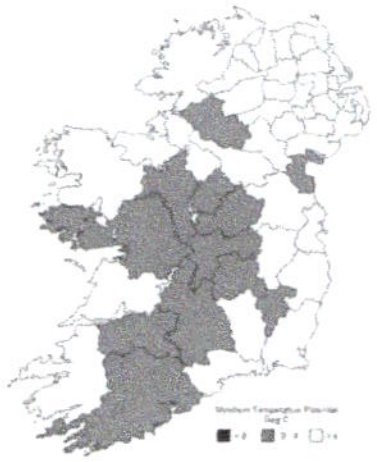

FEBRUARY 10 MONDAY

The high pressure system that has prevailed over the previous week slips to be centred to the south of Ireland allowing for a fresh and blustery west to southwesterly change to develop throughout. Widespread cloud continues everywhere, with some isolated showers and drizzle patches mostly in the northwest at first, but later spreading briefly into central parts for a time.

Belfast: Overcast, dry with passing threats of rain, windy spells, chance of gale gusts
Dublin: Milder, overcast, dry, light breezes with occasional fresh breezy spell
Cork: Dull and gloomy air, warmer, dry, light westerly breezes
Galway: Unsettled windy spells, overcast, milder, brief showers

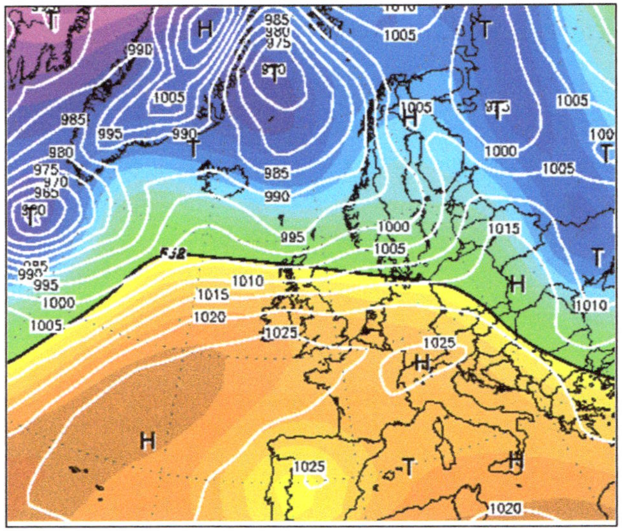

RAIN POTENTIAL

FROST/SNOW

FEBRUARY 11 TUESDAY

The high pressure ridge associated with the anticyclonic system now situated to the south of Ireland brings an easing of breezes to a light southwest flow. Widespread cloud continues to bring dull and gloomy conditions everywhere, except for some brief sunny outbreaks possible in eastern parts. Patches of fog and mist overnight, some turning to drizzle for a time. Warmer than normal for this time of year.

Belfast: Overcast and milder, dry, gentle breezes to mostly calm air, chance of fog
Dublin: Overcast, calm and dull air, milder, mostly dry, chance of light drizzle
Cork: Mostly calm, cloudy, and dry air, milder, chance of brief drizzle patches
Galway: Overcast, milder, light drizzle patches, gentle breezes

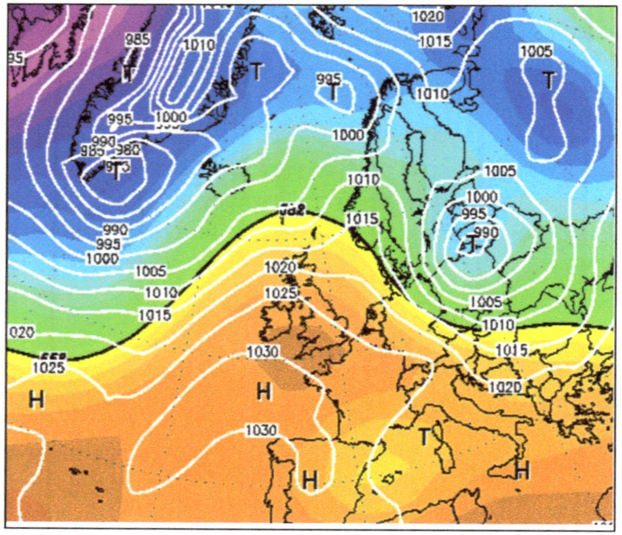

RAIN POTENTIAL

FROST/SNOW

FEBRUARY 12 WEDNESDAY

A northwest change moves over Ireland as the high pressure centre slowly drifts to be southwest of Ireland, with winds generally light throughout, except for fresher northwesterly spells along the northern coasts. Mostly sunny and dry everywhere. Patches of overnight fog likely, particularly in the north and west and frost in the east and central parts.

Belfast: Mostly sunny and dry, gentle northwesterly air, mild day cold night, fog likely
Dublin: Sunny, dry, calm and serene air, overnight frost patches likely
Cork: Sunny and dry, cooler, calm and serene air
Galway: Mostly sunny and dry, cool, light breezes, chance of shallow fog patches

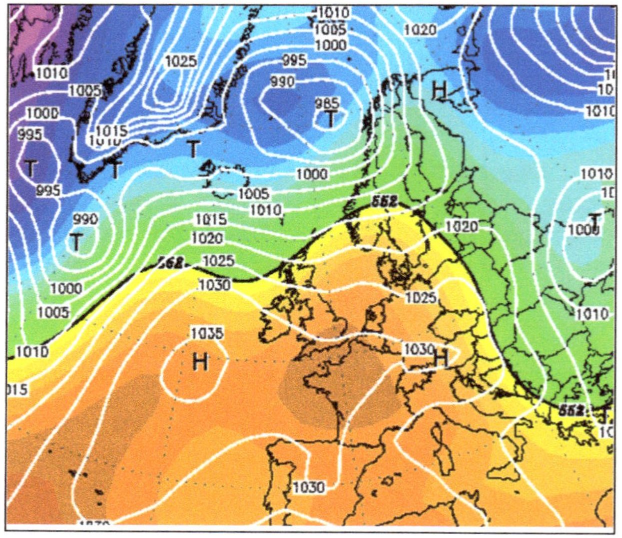

RAIN POTENTIAL

FROST/SNOW

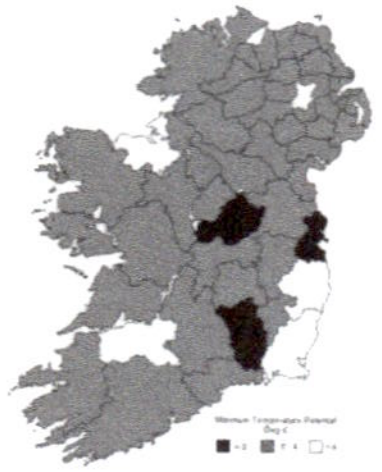

FEBRUARY 13 THURSDAY

The high pressure system strengthens as it moves back over Ireland once more, bringing a calm and serene atmosphere throughout with patches of fog and frost in many districts. Frosts may be severe in numerous eastern and central parts, and fogs slow to clear. Mainly sunny and dry day everywhere. Chance of snow flurries for a time in the north.

Belfast: Calm, mostly sunny, dry, overnight frost and fog, chance of snow flurries
Dublin: Sunny, cold with frosts, dry, calm and serene atmosphere, fog likely
Cork: Calm and serene air, passing sunny spells, dry, chance of misty patches
Galway: Partly cloudy, mostly dry, cold, frosts, calm air, shallow fog patches likely

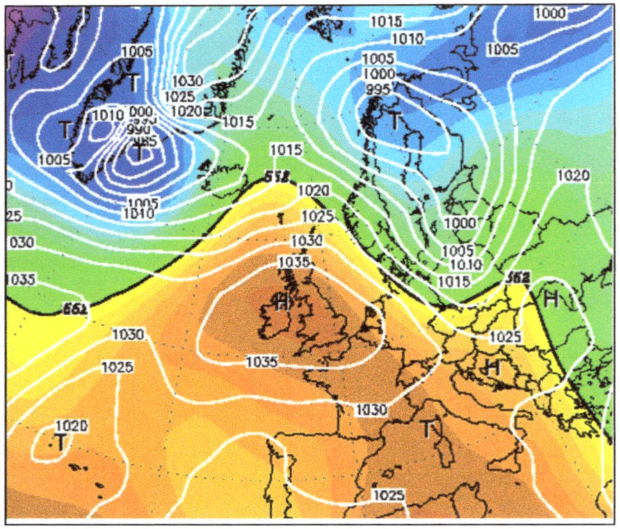

RAIN POTENTIAL

FROST/SNOW

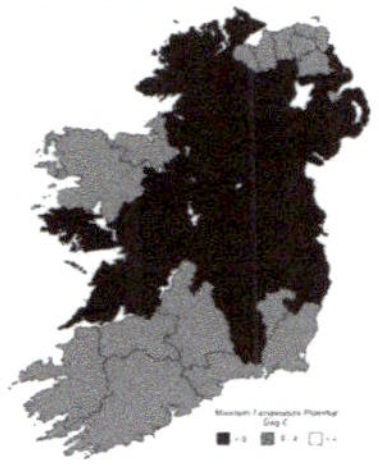

FEBRUARY 14 FRIDAY

High pressure continues to bring dry and cold conditions throughout, with frosts and fog in many districts and the chance of some snow flurries in the north. Cloudy everywhere, with fog possibly lingering for most of the day in some central and eastern dsitricts. Light shifting breezes to a mostly calm air prevails throughout.

Belfast: Cloudy spells, dry, cold, severe frosts likely, patches of fog, chance of snow
Dublin: Cloudy, dry, frosts and freezing fog patches, mostly calm air
Cork: Cloudy, cold, gentle shifting breezes, mostly dry, possible misty patches
Galway: Cold, dry, severe frost pockets and chance of shallow fog, some sunny spells

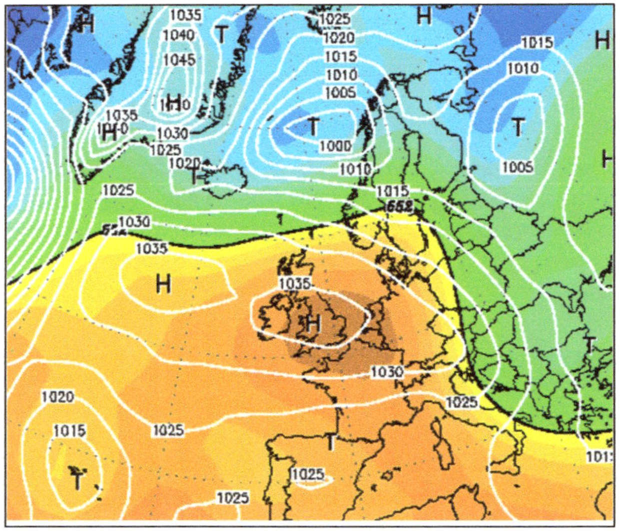

RAIN POTENTIAL

FROST/SNOW

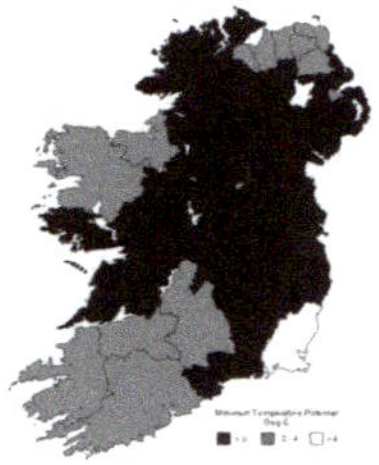

FEBRUARY **15** SATURDAY

The high pressure influence of the past couple of weeks weakens with a low pressure to the northeast of Ireland nearing the country in a light northerly air. Mostly cloudy everywhere, except for some sunnier spells in the east and central parts. Patches of fog in many districts overnight, slow to clear in a number of places. Chance of evening sleet and snow flurries once more in the north.

Belfast: Cloudy, cold, dry, severe frosts and patches of fog, chance of snow flurries
Dublin: Partly cloudy, dry and cool, frosts, mostly calm and serene atmosphere
Cork: Mostly cloudy, calm and dry, cold, frosts, possible misty drizzle patches
Galway: Overcast, calm cold atmosphere with frost, dry, patches of freezing fog likely

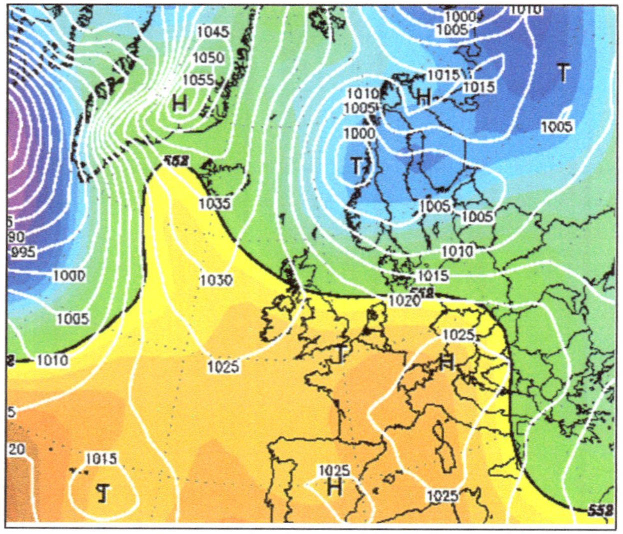

RAIN POTENTIAL

FROST/SNOW

FEBRUARY **16** SUNDAY

Low pressure moves down over Ireland behind the departing high pressure ridge, bringing very cold temperatures throughout and light variable breezes. Widespread overnight frost and fog patches, with fog slow to clear in places. Dull and gloomy in the south of the country with some rain in the south of Leinster to the northeast of Munster, while sunnier and mostly dry elsewhere. Chance of snow in the north and west overnight.

Belfast: Mostly sunny, cold, dry, severe frosts, chance of freezing fog
Dublin: Very cold with severe frosts, dry, occasional sunny spells, mostly calm air
Cork: Overcast, very cold, dry with increasing threats of rain, gentle breezes
Galway: Mostly sunny and dry, very cold, severe frosts, gentle breezes, chance of fog

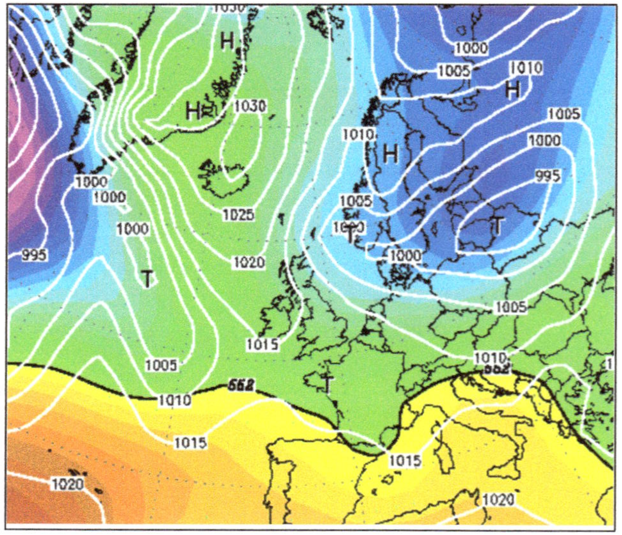

RAIN POTENTIAL

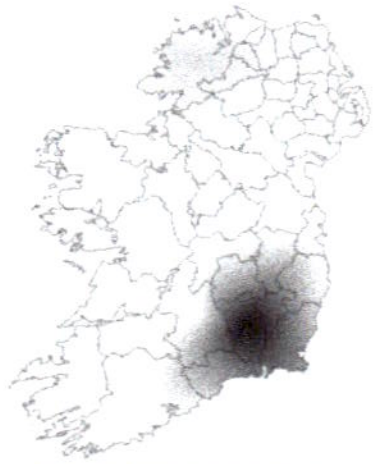

FROST/SNOW

FEBRUARY **17** MONDAY

A low pressure trough between two depressions crosses the country in a fresh to strong east to southeast flow, with possible gale gusts at times in the north. Very cold and cloudy throughout with widespread severe frosts, and wintry rain activity almost everywhere. Some patches of fog possible in the north, while snow falls likely through central parts from the west coast to the east coast.

Belfast: Very cold, cloudy, severe frosts, wintry showers, snow possible, shallow fog
Dublin: Cold with severe frosts, occasional sunny spells, light snow
Cork: Overcast, cold, heavy wintry rain, chance of overnight snow
Galway: Cold, severe frosts, overcasts, heavy wintry rain and snow likely

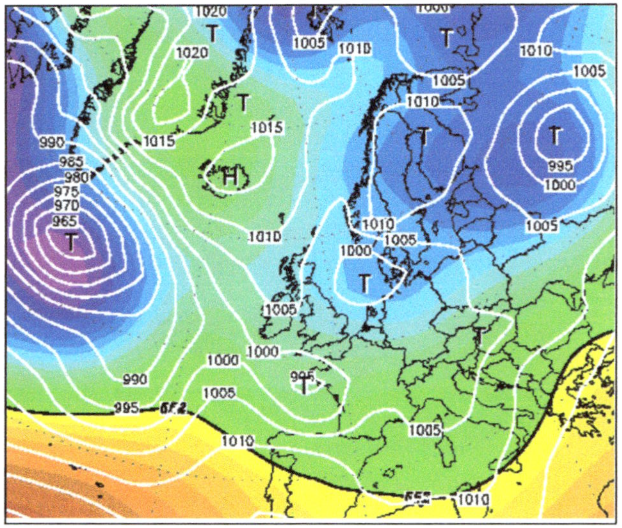

RAIN POTENTIAL

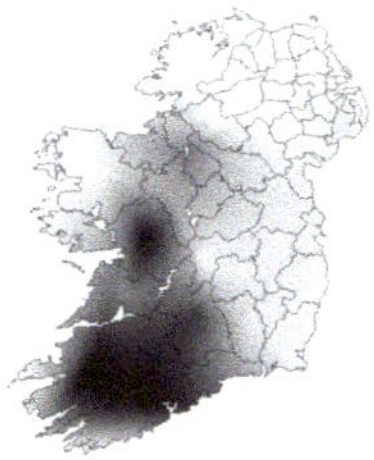

FROST/SNOW

FEBRUARY **18** TUESDAY

A depression to the northwest of Ireland moves over the country bringing widespread wintry storm conditions throughout. Severe frosts almost everywhere, with snow in many districts. Unsettled and blustery winds prevail throughout, generally southeasterly in direction. Dull and gloomy almost everywhere, except for some brief sunny breakthroughs possible in southern parts for a short time.

Belfast: Stormy, overcast, very cold, severe frosts, wintry rain, snow and gales gusts
Dublin: Overcast, heavy wintry precipitation including snow, very cold, windy, gusts
Cork: Unsettled winds, mix of sunny spells and brief drizzly showers, slightly milder
Galway: Unsettled and windy, cloudy, cold, frosts, patches of rain, chance of snow

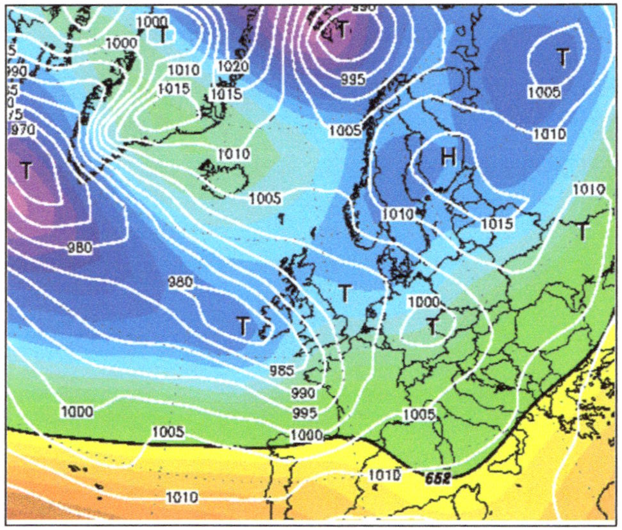

RAIN POTENTIAL

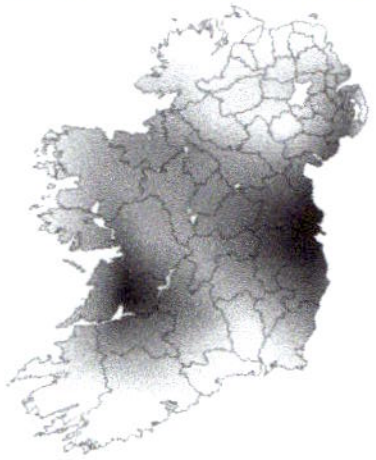

FROST/SNOW

FEBRUARY 19 WEDNESDAY

Low pressure and southerly breezes prevails over Ireland with a westerly change later in the day. Unsettled and changeable throughout with a mix of sunny spells and wintry showers in most districts, but mostly dry along the coastal fringes of the northwest and in Leinster. Chance of heavy rain for a time in the southwest.

Belfast: Changeable, isolated wintry drizzle patches, some sunny spells, very cold
Dublin: Partly cloudy and dry, milder day overnight frosts, shifting breezes
Cork: Mix of sunny spells and wintry showers, fluctuating breezes
Galway: Changeable, mix of sunny spells and showers, some breezy spells

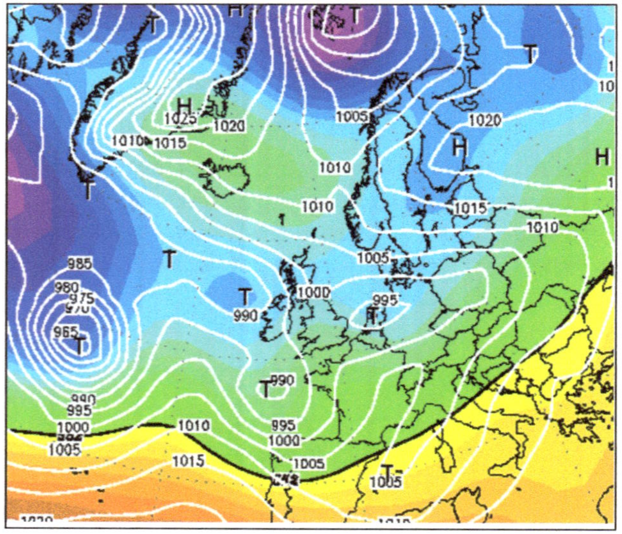

RAIN POTENTIAL

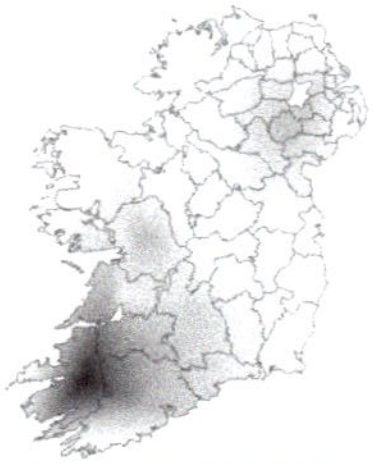

FROST/SNOW

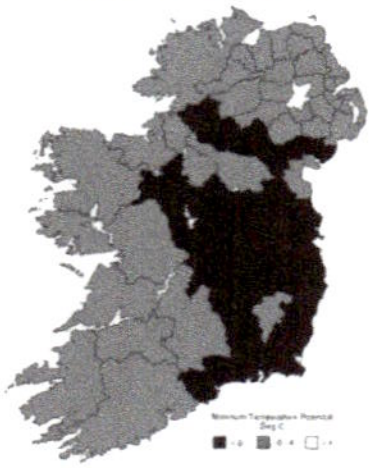

FEBRUARY 20 THURSDAY

A strong and unsettled southeasterly flow prevails over Ireland as a depression to the west of the country brings widespread cloud and bands of scattered wintry showers and drizzle patches in most districts. Snow likely in the north, while outbreaks of heavy precipitation possible in the south.

Belfast: Overcast, windy spells, wintry drizzly showers, snow likely
Dublin: Unsettled windy spells, overcast, cold, occasional showers
Cork: Overcast, rain, windy spells, chance of overnight patches of fog
Galway: Unsettled winds, overcast, cold, chance of frost, passing drizzly showers

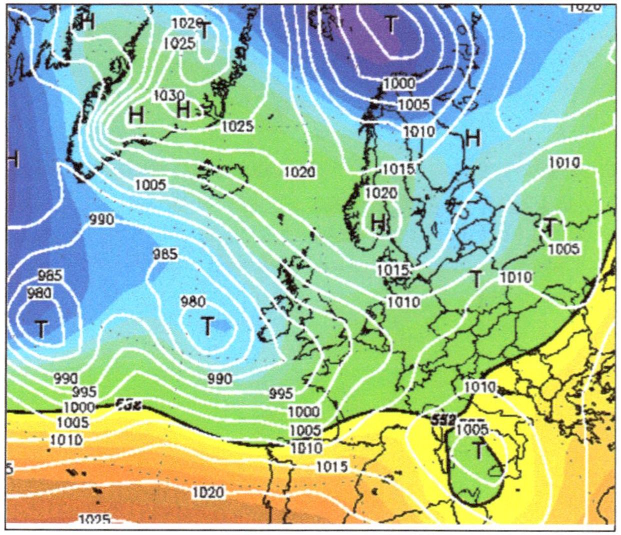

RAIN POTENTIAL

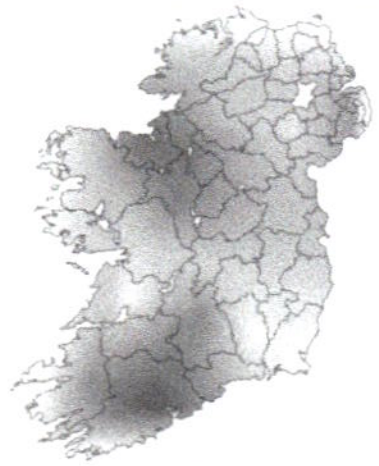

FROST/SNOW

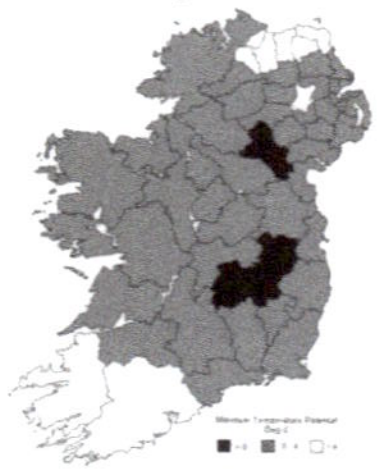

FEBRUARY 21 FRIDAY

An intense depression centred to the northwest of Ireland brings strong south to southwest winds over the country, fresh and gusty at times. Unsettled and stormy conditions prevail throughout with a mix of some sunny spells and bands of rain crossing the country, with heavier rain at times in the southwest.

Belfast: Cloudy, wintry showers, blustery spells, chance of gale winds at times
Dublin: Changeable, mix of sunny spells and wintry showers, shifting breezy spells
Cork: Unsettled and blustery winds at times, mix of sunny spells and showers
Galway: Stormy air, cloudy, milder, scattered showers, windy with gusts

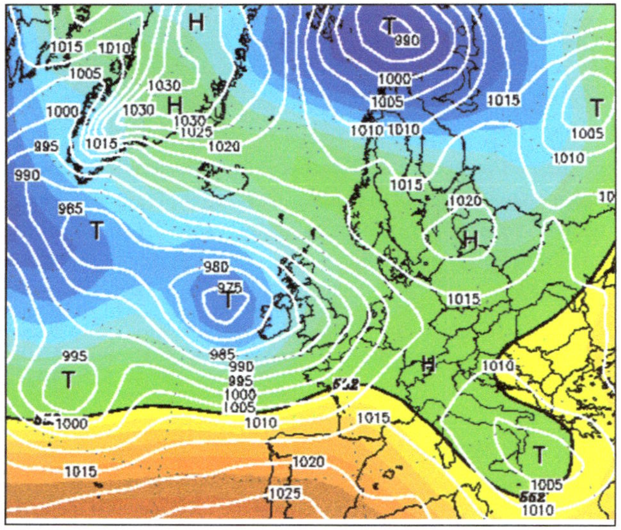

RAIN POTENTIAL

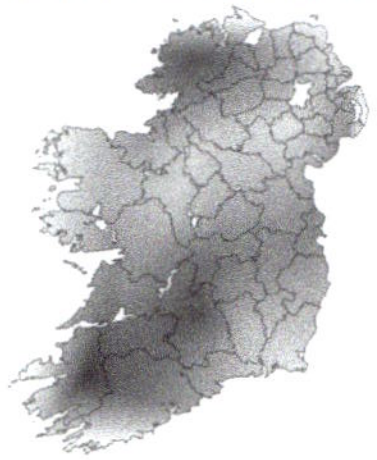

FROST/SNOW

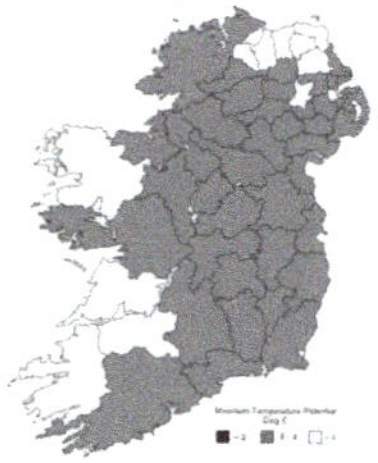

FEBRUARY 22 SATURDAY

A low pressure system crosses the country with light to moderate mostly westerly breezes prevailing, with occasional gusts at times around the coastal fringes. A mixed day of sunny spells and showers in most districts, as well as the chance of overnight patches of fog, mostly in the east. Frosts likely overnight.

Belfast: Cloudy, isolated showers, moderate breezes, cold
Dublin: Scattered cloudy spells, light showers and drizzle patches, chance of fog
Cork: Changeable, mix of sunny spells and showers, cold, fluctuating breezes
Galway: Cloudy, isolated showers of light rain, milder, fluctuating winds

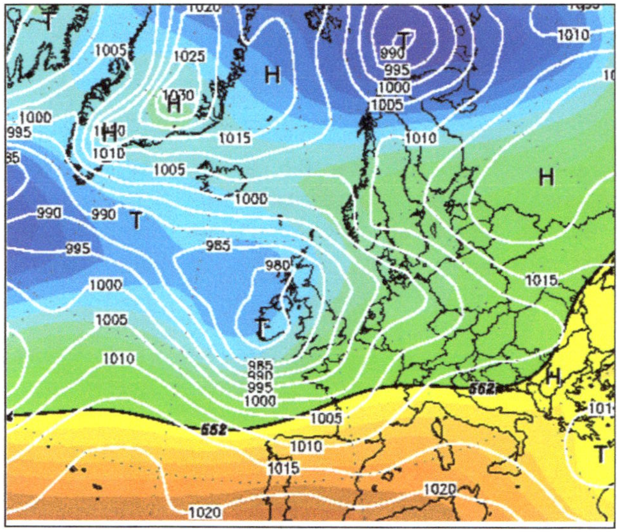

RAIN POTENTIAL

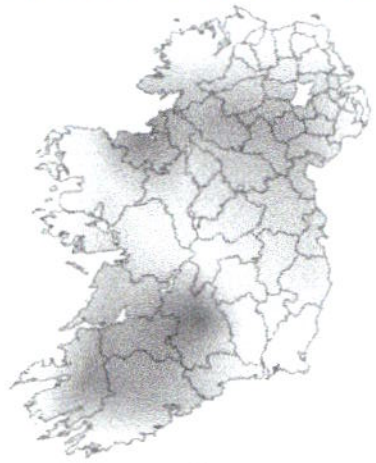

FROST/SNOW

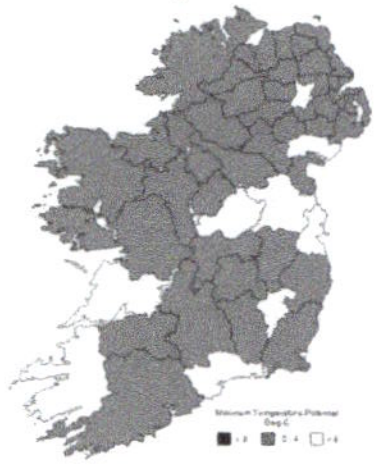

FEBRUARY 23 SUNDAY

Dull and gloomy everywhere as an intense depression moves away to the northwest of Ireland followed by a low pressure trough moving over the country in a moderate southwest direction. Patches of light showers and drizzle almost everywhere, as a band of rain crosses the country from the northwest at first and spreading down later into the southeast.

Belfast: Overcast, showers of light rain, shifting breezy spells, milder day cold night
Dublin: Overcast, breezy at times, dry with odd threat of rain
Cork: Cloudy, scattered showers, colder overnight, unsettled breezy spells
Galway: Overcast, brief showers and drizzle patches, frosts, fluctuating breezes

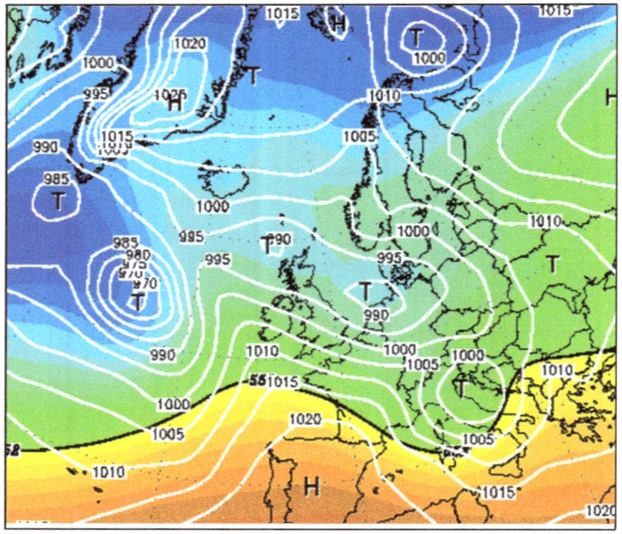

RAIN POTENTIAL

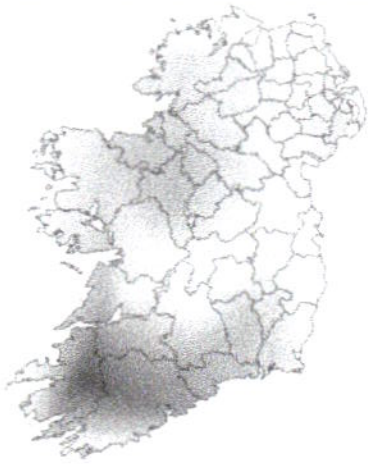

FROST/SNOW

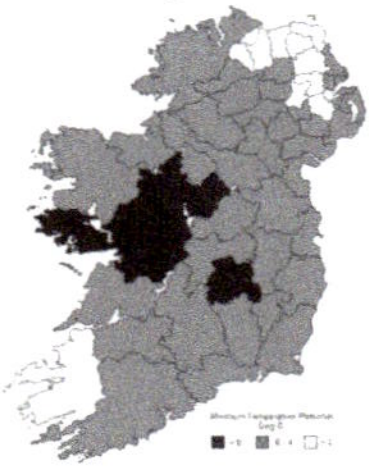

FEBRUARY 24 MONDAY

An intense depression to the northwest of Ireland directs a light and variable breeze over the country, strengthening later in the day with a southwest change. Mostly dry with occasional sunny spells in the east and south, while some showers likely elsewhere. Very cold with frosts overnight, and chance of fog patches in central districts.

Belfast: Light breezes, cloudy, mostly dry with passing threats of rain, colder
Dublin: Occasional sunny spells, dry, milder day colder night, mostly calm air
Cork: Light shifting breezes, odd sunny spells, mostly dry, chance of misty patches
Galway: Cloudy, milder day overnight frosts, scattered showers and persistent drizzle

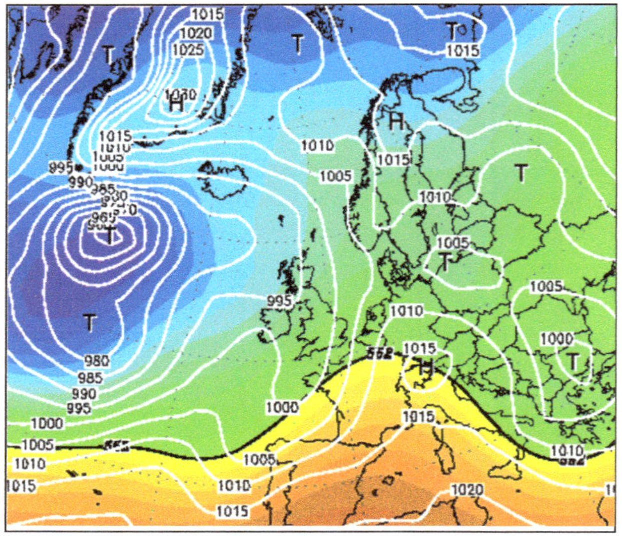

RAIN POTENTIAL

FROST/SNOW

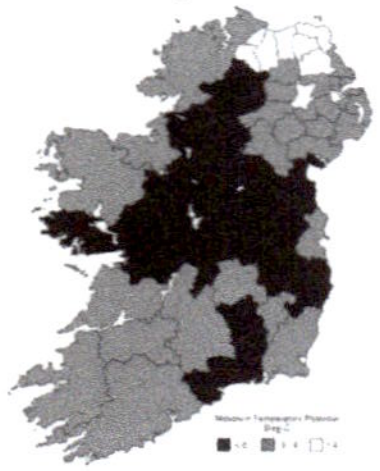

FEBRUARY **25** TUESDAY

The intense depression to the west of Ireland moves nearer to the country with strong and blustery south to southwest winds developing, up to gale force at times around the coasts. Widespread cloud and bands of squally showers throughout, with heavy falls in many western and southern parts of the country.

Belfast: Overcast, warmer day cold night, squally showers, strong gusty winds
Dublin: Unsettled and stormy air, overcast, milder temperatures, isolated showers
Cork: Stormy air with gale winds possible at times, dull and gloomy, rain
Galway: Blustery winds, overcast, squally showers, stormy air with gales possible

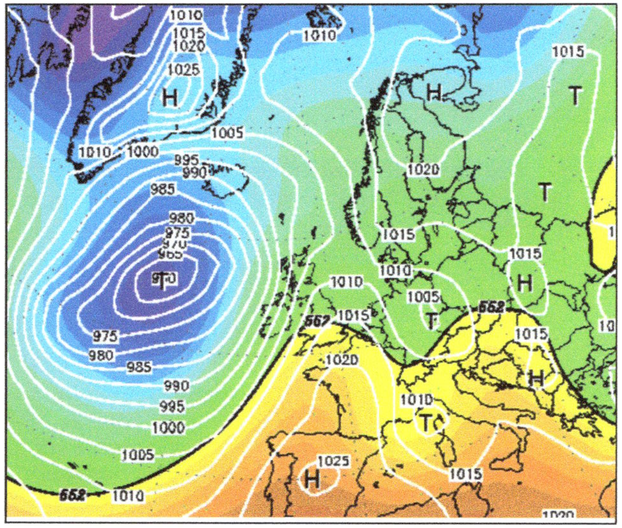

RAIN POTENTIAL

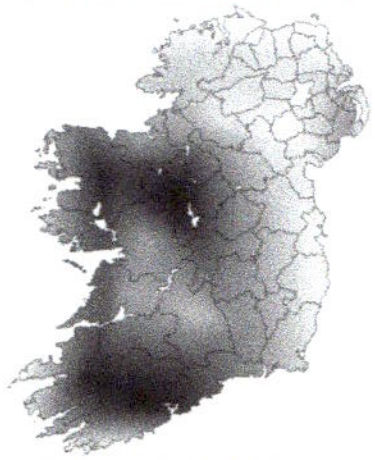

FROST/SNOW

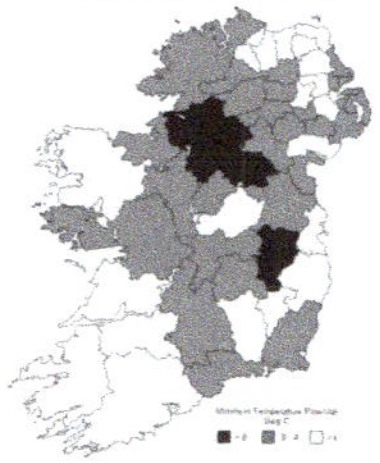

FEBRUARY **26** WEDNESDAY

The depression that brought stormy conditions across Ireland yesterday, contracts to the far north of the country, but with strong south to southwest winds continuing throughout, up to gale force again in the north. In contrast, lighter breezes to the south may see some patches of overnight fog here and there. Cloudy to overcast throughout, with lingering showers in the northeast and west, while clearing to mostly dry and frosty in the east.

Belfast: Unsettled and blustery, overcast, mostly dry with odd threat of rain, milder
Dublin: Overcast and gloomy air, cool, dry, strong winds easing later in the day
Cork: Overcast, isolated showers, winds easing, chance of fog patches
Galway: Mostly cloudy, dry, strong gusty winds at times, chance of overnight fog

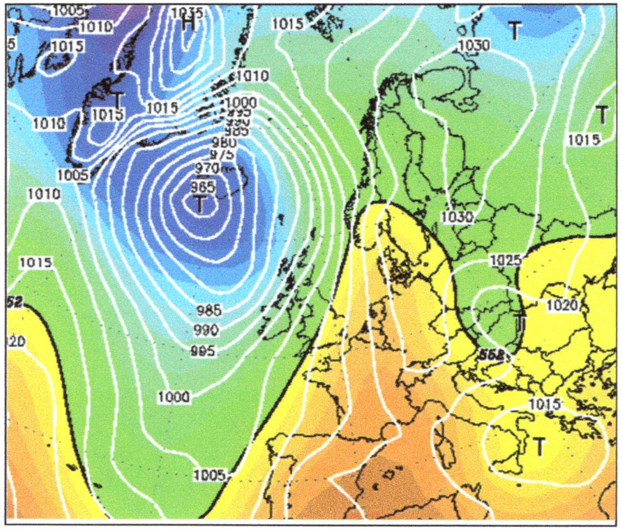

RAIN POTENTIAL

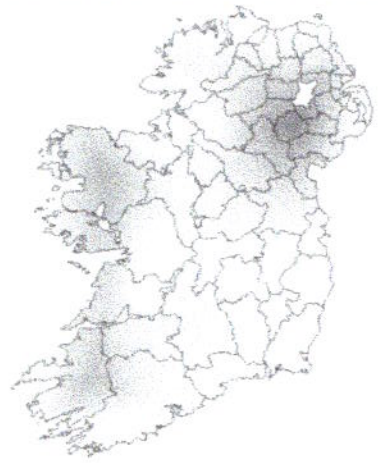

FROST/SNOW

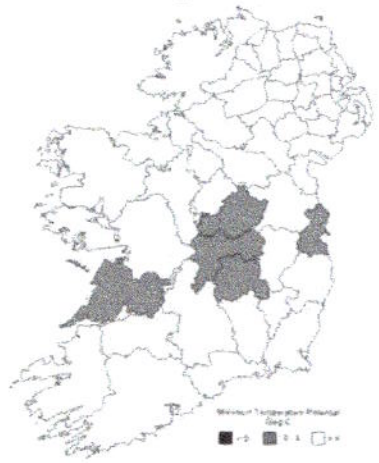

FEBRUARY **27** THURSDAY

A low pressure trough crosses the country brining mostly light and very cold north to northwest breezes throughout. Mostly dry with occasional sunny spells everywhere, as well as the likelihood of widespread patches of fog and frost. Chance of snow for a time in the north of the country.

Belfast: Mostly sunny and dry, mild day cold night, chance of overnight snow and fog
Dublin: Partly cloudy, dry, cold, gentle shifting breezes, chance of fog
Cork: Partly cloudy, mostly dry, cold overnight, light breezes, possible fog patches
Galway: Mostly sunny and dry, light shifting breezes, chance of freezing fog patches

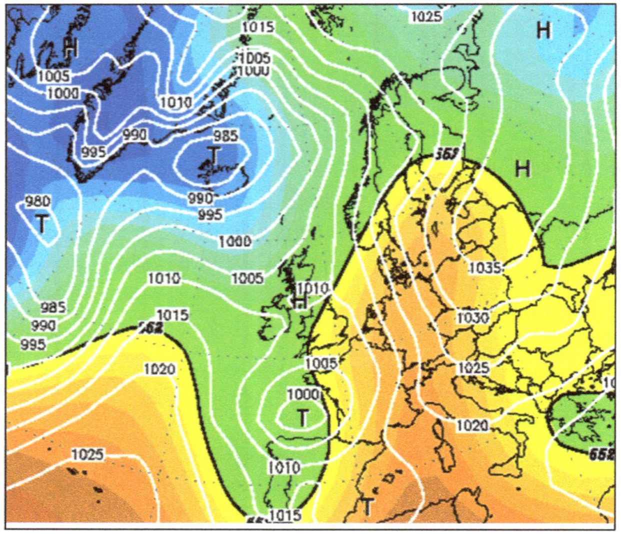

RAIN POTENTIAL

FROST/SNOW

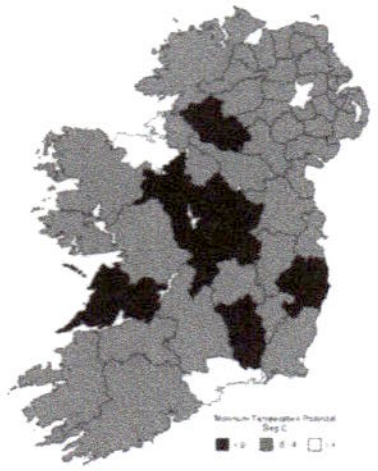

FEBRUARY **28** FRIDAY

A low pressure trough between a complex of depressions continues to bring a light, mostly southerly flow over the country. Cold everywhere, with frost and fog patches overnight, and some brief sunny spells during the day. Fog may be slow to clear in many parts of the country, before rain develops later in the day in the southwest and spreads slowly eastwards up through the country as the day progresses.

Belfast: Mostly calm and cloudy air, fog, frost, light wintry showers, chance of snow
Dublin: Dry, passing sunny spells, cold with frosts, gentle shifting breezes
Cork: Cloudy, isolated showers, colder, gentle shifting breezes
Galway: Cloudy, cold, frosts, patches of rain, moderate to light breezes, fog possible

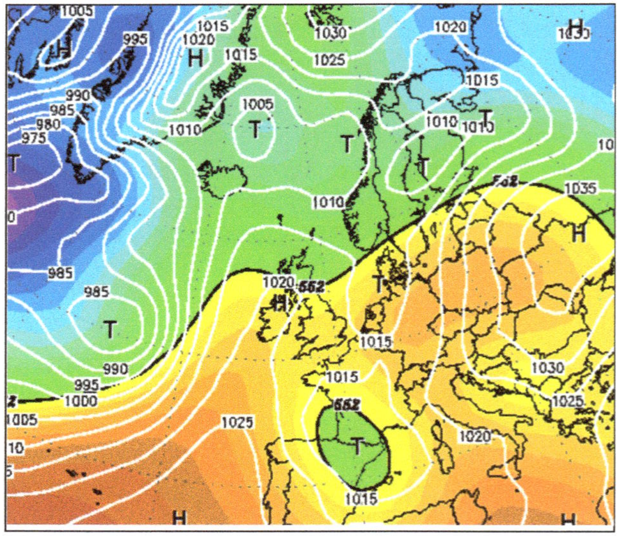

RAIN POTENTIAL

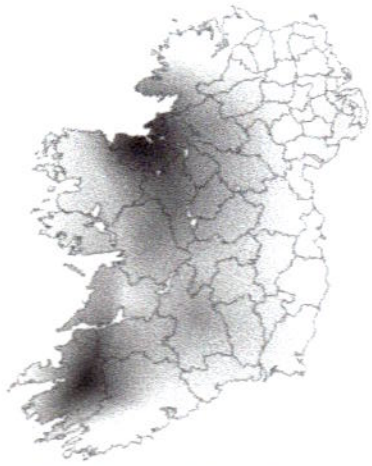

FROST/SNOW

MARCH

PHASES OF THE MOON

1st	Perigee #7
2nd	Crossing Equator
6th	First Quarter
7th	Northern Declination
14th	Full Moon
15th	Crossing Equator
17th	Apogee
22nd	Third Quarter
22nd	Southern Declination
29th	New Moon
29th	Crossing Equator
30th	Perigee #4

1	2	3	4	5	6	7	8	9	10	11	12	13	14	15	16	17	18	19	20	21	22	23	24	25	26	27	28	29	30	31
Fri	Sat	Sun	Mon	Tue	Wed	Thur	Fri	Sat	Sun	Mon	Tue	Wed	Thur	Fri	Sat	Sun	Mon	Tue	Wed	Thur	Fri	Sat	Sun	Mon	Tue	Wed	Thur	Fri	Sat	Sun
P7	XhN				1Q	^							F	XhS		A					3Q V							N XhN	P4	

MONTHLY SUMMARY

1st-3rd An area of low pressure crossing the country brings widespread showers.
2nd Wintry showers of hail likely.
3rd Chance of widespread fog. It may be warmer than normal with few frosts, especially in the south due to prevailing south-west winds.
6th-7th Fog patches in the north of the country likely.
8th-9th Fog likely in the south.
9th Unusually mild temperatures throughout, with many districts seeing maximas in the mid teens.
9th-10th An intense depression brings a very windy spell and a stormy air throughout, with possible gale gusts, and thunderstorms with hail mostly around the coastal fringes in the south and southwest.

Over the next four weeks low pressure systems will dominate the first half of this upcoming outlook period, before high pressure prevails for most of the second half. Overall, rainfall and sunshine hours will be fairly typical for this time of year, while, despite regular frosts, temperatures will be milder than normal. Most days will see some rain but the most rainfall can be expected over the first half of this outlook period, particularly towards the end of the first week and again, midway through the third week, while the least rainfall can be expected in the last ten days. Overall, most districts can expect around 25% of their average rainfall for this time of the year. It will be sunnier that normal throughout this outlook period, especially in the western half of the country, with the sunniest period being towards the end of the fourth week, but also unusually sunnier for this time of year over a number of days during the third week of this look-ahead period. This sunnier than normal period will also reflect in above average maximum temperatures during this outlook period, despite frequent ground frosts especially in inland parts. It will be especially warmer than normal in the west of the country, while minimum temperatures will be below average in the midlands to the eastern districts. During this outlook period, rain can be expected on 14-18 days, thunderstorms on 6 days, hail on 5 days, snow on 3 days, frosts on 15-20 days, fogs on 3 days, and gale winds on up to 4 days.

11th-27th A series of depressions near the west of Ireland bring frequent passages of rain or showers across the country, with particularly heavy falls at times mostly in the west and south.
11th Sunny spell with many districts seeing 8 hours of sunshine or more.
12th-13th Chance of fog in the north.
12th-15th A series of wet fronts cross the country bringing widespread rain, with very heavy falls at first in the east, and then a couple of days later in the west and south. During this spell, some wintry showers of hail, sleet or snow may occur, particularly in central and western areas. Temperatures likely to be unusually mild during the day but cold and frosty overnight.
13th A depression over Ireland may bring wintry showers of snow and hail, especially in the west. Despite this, many districts may see 8 hours or more of sunshine during the day.
14th-15th Very strong winds prevail, possibly reaching gale force with gusts at times.
14th-19th Bands of thunderstorms may cross the country during this period, with frequent hail showers likely around the 19th.
19th Mostly sunny day with many districts seeing up to 8 hours or more of sunshine.
20th-27th A high pressure system to the south of Ireland maintains a ridge over the country.
21st Another intense front brings a very wet day with especially heavy rain at times in the west and south of the country.
22nd Chance of snow in the north of Ireland but also a very sunny day in many districts.
27th-30th An intense depression to the north of Ireland brings a series of cold fronts and stormy conditions throughout the country with winds to storm force at times, hail and snow, and possible thunderstorms. The cold artic airstream may see snow accumulate in parts of the midlands, western, and higher reaches.
31st Widespread and some severe ground frosts throughout the country but milder daytime temperatures.

IRELAND RAINFALL ESTIMATES

March 2025

d=mainly dry, n= nonrecordable, l=light shrs, s=significant shrs, r=rain, h=hvy falls

		ULSTER PROVINCE					CONNAUGHT PROVINCE						LEINSTER PROVINCE				MUNSTER PROVINCE						
		Alderg	Hillsbo	Armag	Loughl	Malin H	Sligo	Ardtari	Belmul	Drums	Galway	Roscoi	Markre	Edende	Kilkenn	Dublin	Shannc	Newpo	Tralee	Killarne	Cork	All	Moon
1st	Sat	l	l	n	s	l	s	s	s	s	l	s	l	l	s	s	s	l	h	s	h	137	P7
2nd	Sun	l	s	l	h	d	l	h	l	d	l	d	d	s	r	s	l	l	l	l	l	86	XhN
3rd	Mon	s	s	l	l	s	r	l	s	s	s	s	s	s	s	l	l	s	h	r	r	132	
4th	Tues	l	l	l	l	d	s	l	l	l	l	l	n	d	d	l	l	l	r	h	n	50	
5th	Wed	l	s	l	s	l	r	s	l	l	s	l	s	s	s	l	s	s	r	h	s	99	
6th	Thur	l	l	l	s	l	s	s	l	h	l	s	l	l	s	s	s	l	s	s	l	88	1Q
7th	Fri	l	s	l	l	l	s	l	l	l	d	l	l	l	d	l	l	l	l	l	d	32	^
8th	Sat	s	r	d	s	l	s	s	s	r	h	r	s	r	l	s	l	r	s	s	l	141	
9th	Sun	s	l	s	l	h	l	l	s	s	s	s	s	l	l	s	l	r	s	s	l	125	
10th	Mon	l	l	l	s	s	l	s	l	s	s	s	s	l	l	s	l	s	s	s	l	79	
11th	Tues	l	d	l	d	l	l	d	l	n	d	l	l	l	l	d	l	l	d	l	d	20	
12th	Wed	l	l	d	l	l	s	l	s	s	l	l	l	s	h	r	l	l	s	l	h	94	
13th	Thur	l	l	d	l	l	r	l	l	l	s	l	s	l	s	l	l	s	l	r	l	75	
14th	Fri	l	l	l	s	l	r	s	r	s	s	s	s	l	s	s	s	s	r	l	r	118	F
15th	Sat	r	r	r	r	s	h	h	h	h	h	r	r	r	r	l	r	r	h	h	h	345	XhS
16th	Sun	s	l	s	s	s	s	s	s	l	s	l	n	n	s	l	l	l	s	s	l	68	
17th	Mon	l	s	d	r	l	r	r	r	s	l	r	r	l	l	l	l	s	s	l	l	105	A
18th	Tues	s	l	s	s	l	l	s	s	l	s	s	d	s	s	l	l	s	r	s	l	84	
19th	Wed	l	l	l	l	l	l	l	l	n	l	d	s	n	d	d	l	l	l	n	l	22	
20th	Thur	l	d	l	l	d	l	l	l	l	l	l	l	d	d	d	d	s	d	l	d	27	
21st	Fri	s	r	d	l	s	s	l	s	r	r	r	s	r	s	s	r	s	r	h	r	171	
22nd	Sat	s	d	r	l	l	l	l	l	s	s	l	l	d	d	l	l	s	l	l	d	54	3Q V
23rd	Sun	d	d	d	d	l	n	d	l	d	d	l	d	l	d	l	l	n	l	d	d	16	
24th	Mon	d	d	d	l	d	l	l	d	d	l	l	n	d	d	d	d	r	d	n	d	23	
25th	Tues	d	n	d	d	s	l	d	r	l	d	d	l	n	d	d	l	n	l	l	l	28	
26th	Wed	l	l	l	l	l	l	l	l	l	l	l	l	d	d	d	d	l	d	d	d	17	
27th	Thur	l	l	l	s	s	s	s	s	s	r	l	s	l	l	l	s	h	l	l	d	95	
28th	Fri	s	r	l	s	s	s	s	r	r	s	r	s	s	s	l	s	r	s	r	s	143	
29th	Sat	s	l	r	l	l	l	l	l	s	l	l	l	l	d	l	l	s	s	l	s	65	N XhN
30th	Sun	l	d	l	d	d	n	d	l	d	d	n	l	n	d	d	d	l	l	n	d	10	P4
31st	Mon	l	l	d	s	s	s	s	l	s	l	s	s	l	l	l	l	s	l	l	l	64	
	Estimate:	91	98	84	132	98	171	132	130	149	139	128	122	93	113	79	91	165	207	232	159	2612	
	Average:	67	77	66	96	88	109	111	99	88	95	109	109	77	63	53	79	141	116	84	98	1824	
	Trend:	w tr	w tr	w tr	w tr	av	w tr	w tr	w tr	w tr	w tr	w tr	w tr	w tr	w tr	w tr	w tr	w tr	w tr	w tr	w tr	w tr	

IRELAND SUNSHINE ESTIMATES
March 2025

F=fine (8-12 hours of sunshine), pc= partly cloudy (4-7 hours), c=cloudy (1-3 hours), o=overcast (0 hours)

		ULSTER PROVINCE					CONNAUGHT PROVINCE						LEINSTER PROVINCE				MUNSTER PROVINCE						
		Alderg	Hillsbo	Armag	Loughl	Malin H	Sligo	Ardtari	Belmul	Drums	Galway	Roscoi	Markre	Edende	Kilkenı	Dublin	Shannc	Newpo	Tralee	Killarne	Cork	All	Moon
1st	Sat	o	o	o	o	o	o	o	o	o	o	o	o	o	o	o	o	o	o	o	o	0	P7
2nd	Sun	pc	pc	pc	pc	pc	c	pc	pc	c	pc	pc	pc	c	c	pc	c	c	o	o	c	76	XhN
3rd	Mon	c	c	c	c	c	o	c	o	c	o	o	c	c	c	pc	o	c	o	o	o	26	
4th	Tues	c	c	c	c	c	c	c	c	c	c	c	c	c	c	c	c	c	c	c	c	42	
5th	Wed	o	o	o	o	o	o	o	c	o	o	c	o	o	o	o	o	o	o	o	o	6	
6th	Thur	o	o	o	o	o	o	o	o	o	o	o	o	o	o	o	o	o	c	c	c	13	1Q
7th	Fri	c	c	c	c	c	c	c	pc	c	pc	pc	c	c	c	c	c	c	pc	pc	pc	76	^
8th	Sat	c	c	c	c	c	o	c	o	c	o	o	c	c	c	pc	o	c	o	o	c	30	
9th	Sun	c	c	c	c	c	c	c	c	c	c	c	c	c	c	c	pc	c	c	c	c	40	
10th	Mon	c	c	c	c	c	o	c	c	c	c	c	c	c	c	c	c	c	o	o	c	30	
11th	Tues	pc	pc	pc	pc	pc	pc	pc	pc	F	pc	pc	pc	F	F	F	pc	F	F	F	F	159	
12th	Wed	c	c	c	c	c	c	c	pc	o	c	pc	c	o	o	o	o	o	c	c	o	28	
13th	Thur	F	F	F	F	F	pc	F	pc	F	F	pc	F	F	F	F	F	F	pc	pc	F	165	
14th	Fri	c	c	c	c	c	c	c	c	c	c	c	c	c	c	o	pc	c	pc	pc	c	43	F
15th	Sat	o	o	o	o	o	o	o	o	o	o	o	o	o	o	o	o	o	o	o	o	5	XhS
16th	Sun	pc	pc	pc	pc	pc	pc	pc	pc	pc	pc	pc	pc	pc	pc	pc	c	c	c	c	c	105	
17th	Mon	c	c	c	c	c	o	c	o	c	o	o	c	c	c	pc	o	c	c	c	c	39	A
18th	Tues	c	c	c	c	c	o	c	c	c	o	c	c	c	c	c	o	o	o	o	o	26	
19th	Wed	pc	pc	pc	pc	pc	pc	pc	pc	F	pc	pc	pc	F	F	F	pc	pc	pc	pc	F	149	
20th	Thur	c	c	c	c	c	o	c	o	c	o	o	c	c	c	pc	o	c	o	o	o	25	
21st	Fri	o	o	o	o	o	o	o	o	o	o	o	o	o	o	o	o	o	o	o	o	0	
22nd	Sat	F	F	F	F	F	F	F	F	F	F	F	F	F	F	F	F	F	F	F	F	168	3Q V
23rd	Sun	pc	pc	pc	pc	pc	pc	pc	pc	pc	c	pc	pc	pc	c	pc	c	c	c	c	c	79	
24th	Mon	o	o	o	o	o	o	o	o	c	o	o	o	c	c	c	c	c	c	c	c	25	
25th	Tues	c	c	c	c	c	pc	c	pc	c	pc	pc	c	c	c	c	pc	c	pc	pc	c	79	
26th	Wed	o	o	o	o	o	o	o	o	o	o	o	o	o	o	o	o	o	o	o	o	1	
27th	Thur	o	o	o	o	o	o	o	o	c	o	o	o	c	c	c	o	c	o	o	pc	22	
28th	Fri	pc	pc	pc	pc	pc	c	pc	pc	pc	pc	pc	pc	pc	pc	pc	pc	pc	pc	pc	pc	94	
29th	Sat	pc	pc	pc	pc	pc	pc	pc	pc	pc	pc	pc	pc	pc	pc	pc	F	pc	pc	pc	pc	132	N XhN
30th	Sun	pc	pc	pc	pc	pc	F	pc	pc	pc	pc	pc	pc	pc	pc	pc	F	F	pc	pc	pc	149	P4
31st	Mon	o	o	o	o	o	o	o	o	o	o	o	o	o	o	1.0	o	o	o	o	o	6	
Estimate hours:		94	94	94	94	94	79	94	90	97	89	89	94	97	100	110	88	94	78	78	89	1839	
Average hours:		93	93	89	95	96	96	97	102	93	105	109	102	106	103	112	104	92	97	95	109	1987	
Trend:		av	av	av	av	av	less	av	less	av	less	less	av	av	av	av	less	av	less	less	less	less	

MARCH 1 SATURDAY

An intense depression to the northwest of Ireland directs a strong and blustery south to souwest flow over the country. Widespread cloud and scattered wintry rain activity prevails throughout with particularly heavy falls in southern parts, although mostly dry in the afternoon for a time. Milder than of late with frost patches not expected.

Belfast: Overcast, cold, light showers and drizzle patches, strong winds at times
Dublin: Overcast and blustery, milder, scattered showers
Cork: Strong southerly winds, overcast, heavy rain
Galway: Unsettled and blustery winds, overcast, mild day cold night, brief showers

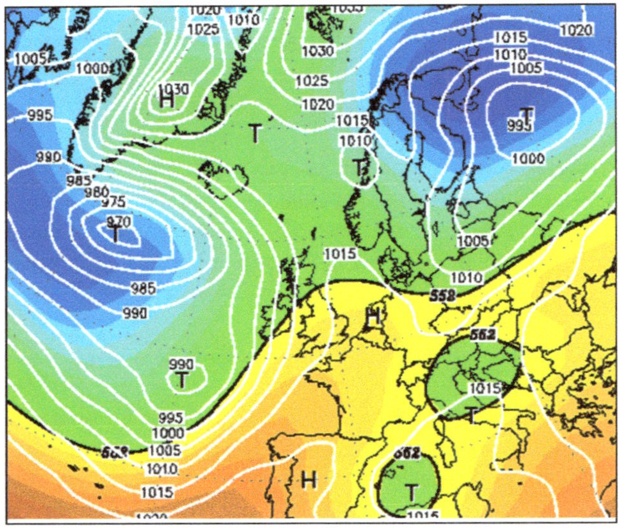

RAIN POTENTIAL

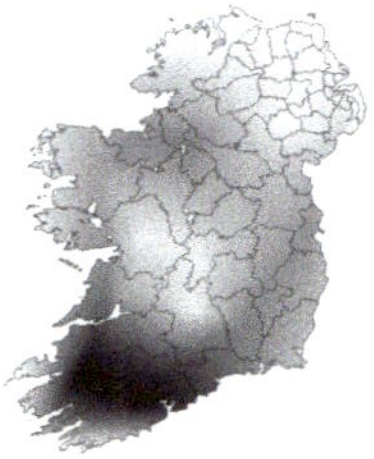

FROST/SNOW

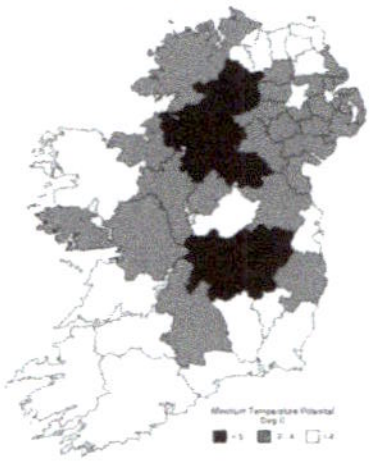

MARCH 2 SUNDAY

A depression centred to the northwest of Ireland directs fresh to strong southwesterly winds over the country. A mix of showery spells and sunny breakthroughs likely in most districts, with heavy falls at times in the northwest. Some patches of overnight fog and frost possibly, particularly in the east.

Belfast: Scattered cloudy spells, passing showers, strong blustery winds, milder
Dublin: Mix of sunshine and showers, windy spells, chance of overnight fog
Cork: Cloudy, winds easing to light, odd showers, chance of mist overnight
Galway: Partly cloudy, light showers, fluctuating winds

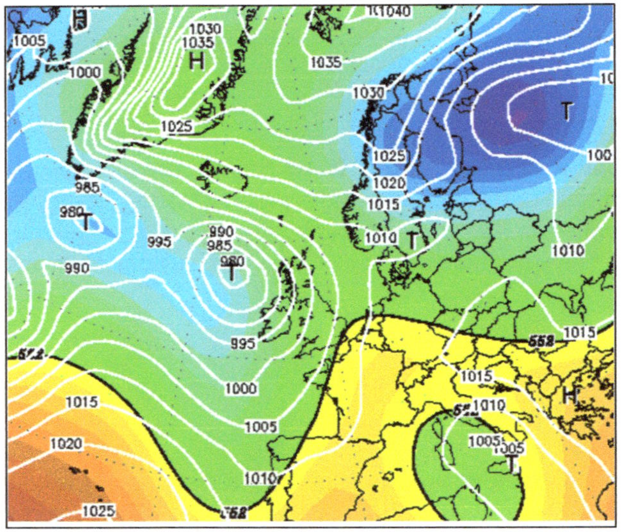

RAIN POTENTIAL

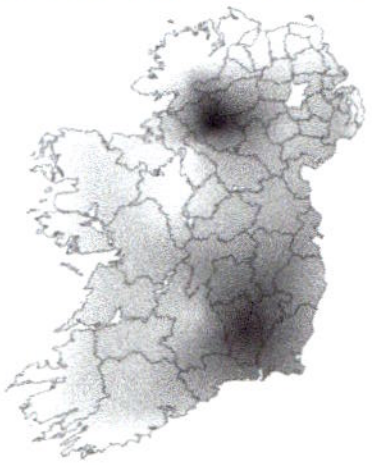

FROST/SNOW

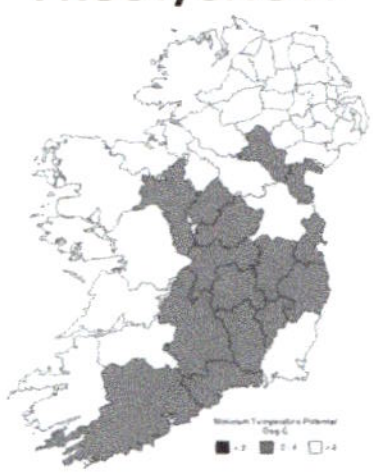

MARCH 3 MONDAY

The depression to the northwest of Ireland drifts further away to the west, allowing for winds to east to light throughout, of variable direction. Mostly cloudy and with outbreaks of showers everywhere, including heavy rain at times in the south as a wet front moves across the country from the west at first later moving into the east. Chance of fog and frost patches overnight again.

Belfast: Cloudy, milder, spells of rain, mostly calm, chance of mist, drizzle and fog
Dublin: Partly cloudy, light showers, mostly calm air, patches of shallow fog likely
Cork: Overcast, rain, light breezes freshening later, chance of overnight fog
Galway: Overcast, scattered showers, chance of frost overnight, mostly calm air

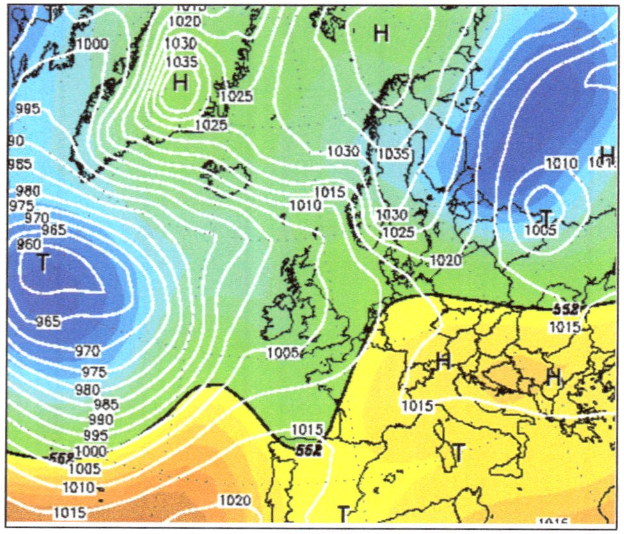

RAIN POTENTIAL

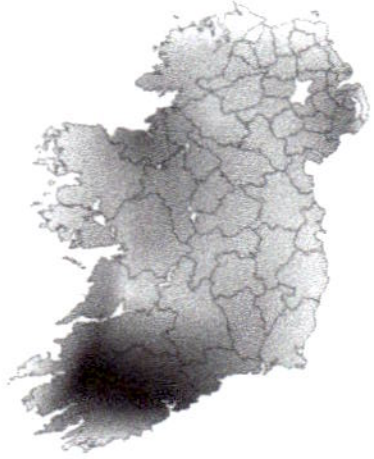

FROST/SNOW

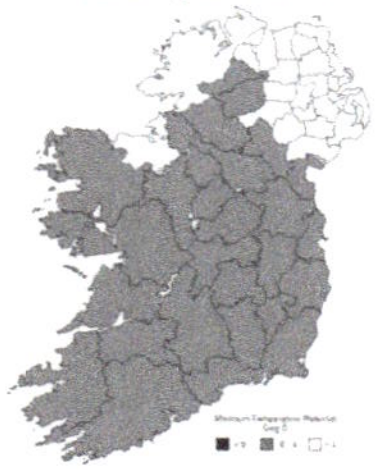

MARCH 4 TUESDAY

A large depression continues to linger to the northwest of Ireland returning fresh to moderate south to southwest winds throughout. Cloudy and with some showers almost everywhere, except for mostly dry around the southeast coastal fringes.

Belfast: Cloudy, freshening winds, isolated showers
Dublin: Cloudy, warmer, brief showers, freshening breezes
Cork: Cloudy, isolated showers, milder, strong southerly breezes at times
Galway: Cloudy, light showers, milder, unsettled winds

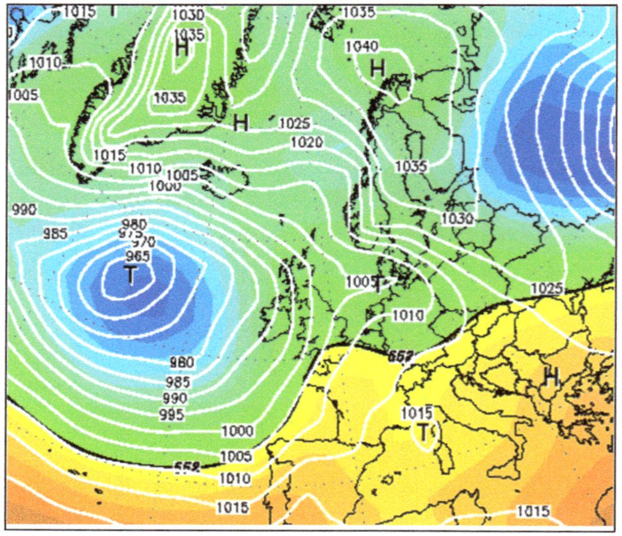

RAIN POTENTIAL

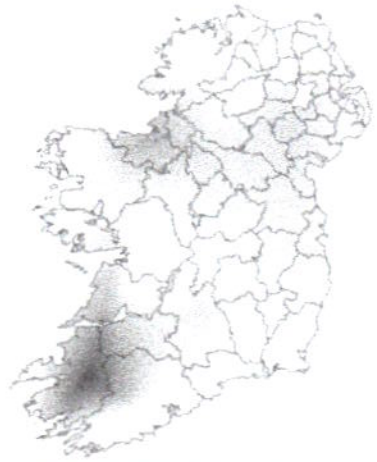

FROST/SNOW

MARCH 5 WEDNESDAY

Dull and gloomy throughout the country with scattered rain activity everywhere, as the centre of the depression to the west of Ireland slowly drifts closer. Strong winds in a south to southwest flow prevails everywhere, gusty at times in the north, but easing later in the day.

Belfast: Overcast, warmer, spells of light rain, fresh winds with chance of gale gusts
Dublin: Overcast, moderate winds, mild temperatures, brief showers
Cork: Overcast, scattered showers, blustery winds easing later, chance of gales
Galway: Overcast, squally showers, mild temperatures, chance of gales for a time

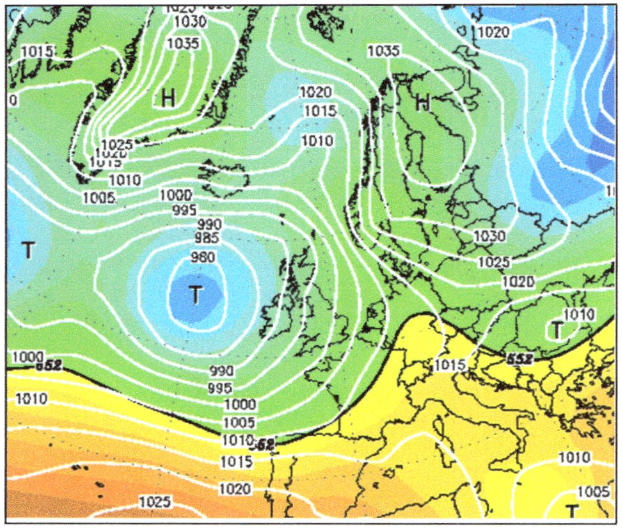

RAIN POTENTIAL

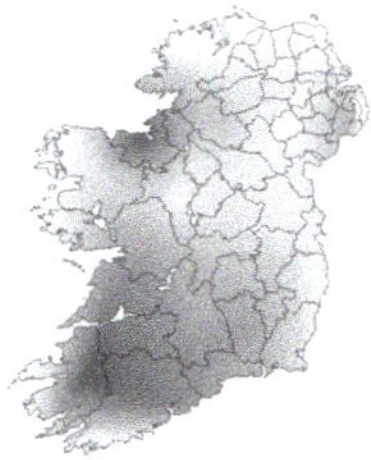

FROST/SNOW

MARCH 6 THURSDAY

A depression centres over the country bringing light to moderate shifting breezes, occasionally more northwesterly at times. Dull and gloomy everywhere with patches of fog overnight in many districts, before showers move into the north of the country at first, and then spread drizzle and showers into most districts as the day progresses before contracting back into the north again. Chance of some heavy falls for a time in the northwest.

Belfast: Overcast, isolated showers, mostly calm air, chance of fog patches
Dublin: Overcast, scattered showers and outbreaks of drizzle, light shifting breezes
Cork: Cloudy, brief showers and drizzle patches, cold night, moderate winds
Galway: Overcast, brief showers, moderate shifting breezes, chance of morning fog

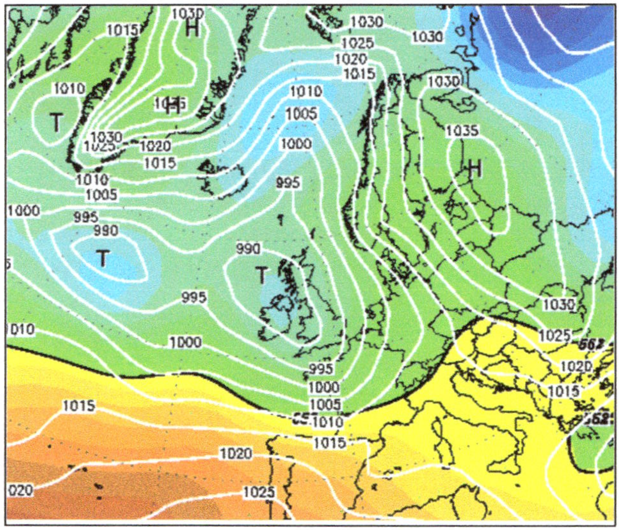

RAIN POTENTIAL

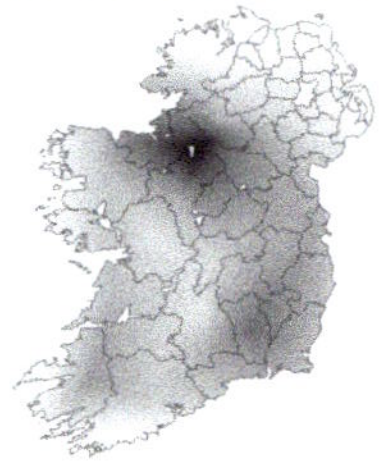

FROST/SNOW

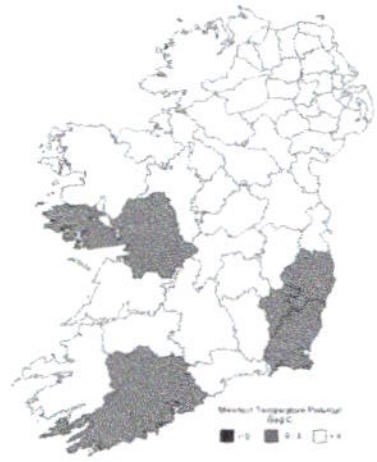

MARCH 7 FRIDAY

A low pressure system centred over Ireland brings light to moderate northwest breezes at first, before becoming stronger and more gusty for a time and then easing once more. Some sunny spells in most parts of the country at times, while occasional wintry showers prevail in most parts except around the southeast coastal fringes. Colder everywhere with the chance of some showers in the north turning to light snow.

Belfast: Cool, mix of wintry showers and sunny spells, light breezes, chance of fog
Dublin: Mix of sunny spells and light showers, gentle shifting breezes
Cork: Occasional sunny spells, mostly dry, moderate breezes
Galway: Mostly calm air, dry, occasional sunny spells

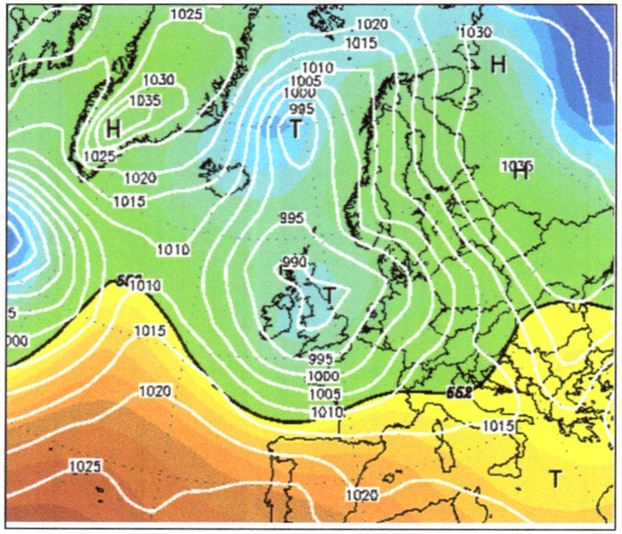

RAIN POTENTIAL

FROST/SNOW

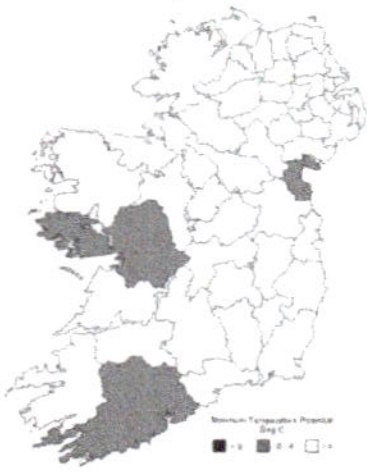

MARCH 8 SATURDAY

A weak high pressure ridge brings a brief dry spell at first, with light to moderate west to southwest breezes prevailing. Later a low pressure trough moves onto the country bringing increased cloud and scattered showers everywhere, starting in the south at first and steadily moving northeastward to reach the northern coasts later in the day and outbreaks of heavy rain likely for a time in the midlands. Chance of fog patches overnight in some southern parts.

Belfast: Cloudy, rain, mild day cold night, chance of overnight light snow
Dublin: Partly cloudy, occasional showers, cold overnight, light breezes
Cork: Cloudy, milder, brief showers, gentle breezes, chance of fog patches
Galway: Widespread cloud, heavy rain, light winds, cold overnight

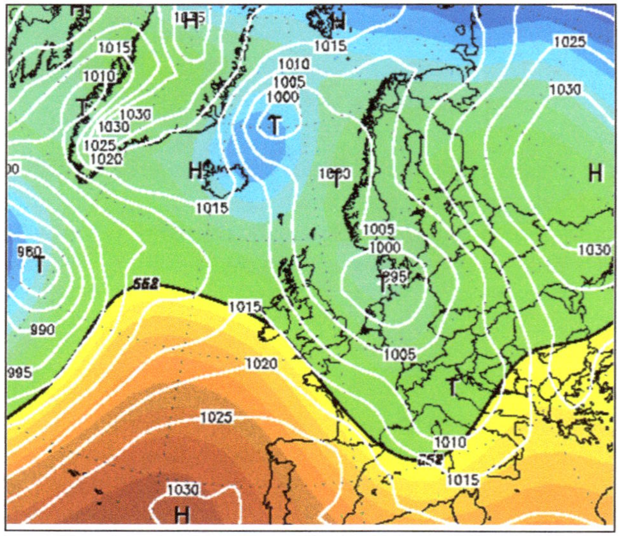

RAIN POTENTIAL

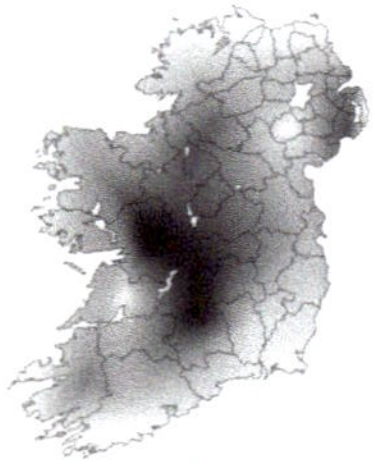

FROST/SNOW

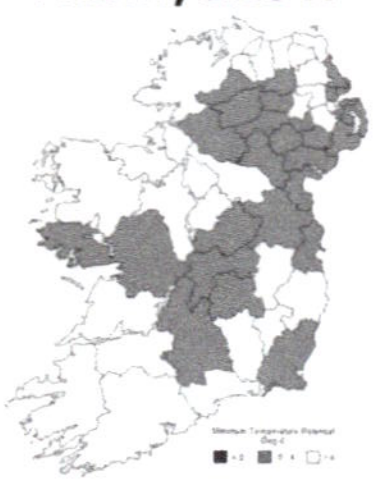

MARCH 9 SUNDAY

A large depression to the northwest of Ireland directs fresh to strong southwest winds over the country, blustery at times around the coasts. Cloudy and with showers everywhere, with chance of heavy rain at times in some northern parts, and possibly thunderstorms at times in southern parts. Chance of further fog patches overnight mostly in sheltered southern parts.

Belfast: Unsettled winds, cloudy, brief showers
Dublin: Cloudy, warmer, scattered showers, fresh winds strengthening later
Cork: Cloudy, brief showers, fresh breezes
Galway: Cloudy, scattered showers, milder, freshening winds

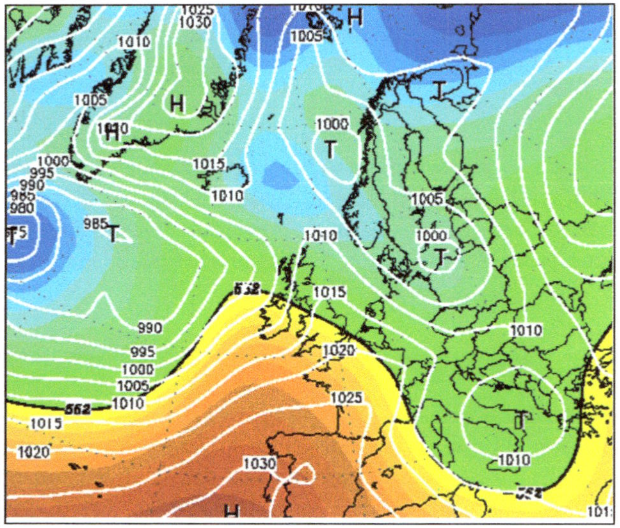

RAIN POTENTIAL

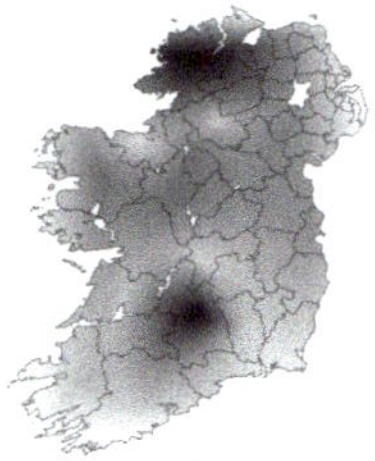

FROST/SNOW

MARCH 10 MONDAY

An intense depression builds to the north of Ireland bringing a stormy day throughout, with very strong and blustery southwest winds prevailing, easing with a northwest change later in the day. Mostly cloudy and with a band of rain crossing the country spreading showers almost everywhere. Chance of thunderstorms overnight, and winds possibly reaching to gale gust levels at times, especially around the coasts.

Belfast: Cloudy, warmer, isolated showery spells, strong winds with gale gusts likely
Dublin: Stormy air, cloudy, cooler, strong gusty winds, scattered showers
Cork: Cloudy, light rain, strong winds with forceful gusts, chance of gales
Galway: Cloudy, scattered showers, cold, stormy winds, gale gusts possible

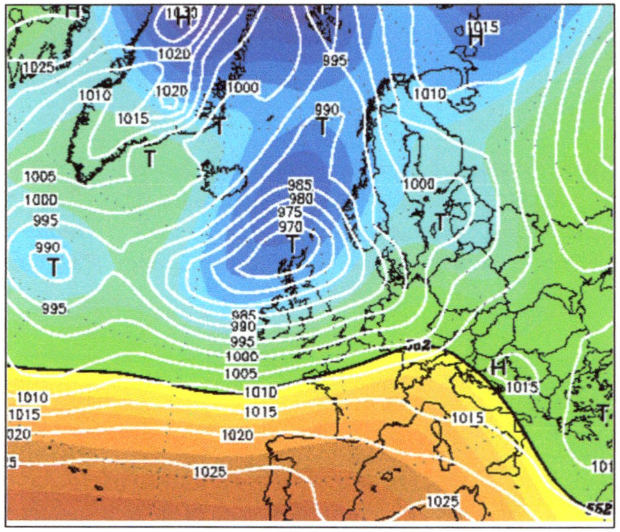

RAIN POTENTIAL

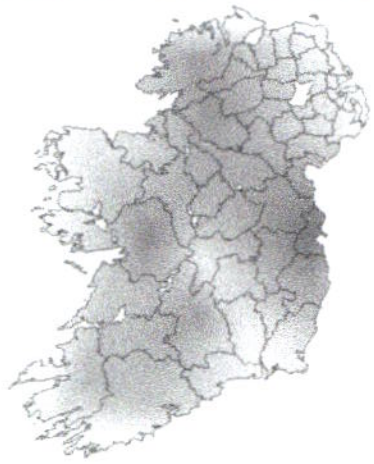

FROST/SNOW

MARCH **11** TUESDAY

Low pressure prevails over Ireland with a depression lingering to the northwest of the country while another remains almost stalled to the north. Strong and unsettled west to northwest winds throughout, easing later in the day. Some showers overnight in most districts, clearing to a mostly sunny and dry day in most districts. Chance of frosts returning after dark, while some showers of hail may occur briefly in central parts.

Belfast: Blustery and cooler, sunny, mostly dry, chance of brief overnight showers
Dublin: Very sunny, dry, cold overnight, strong blustery winds easing later in the day
Cork: Sunny, dry, mild day cold night, winds easing
Galway: Cloudy spells, cold overnight with frost, dry, strong winds easing later

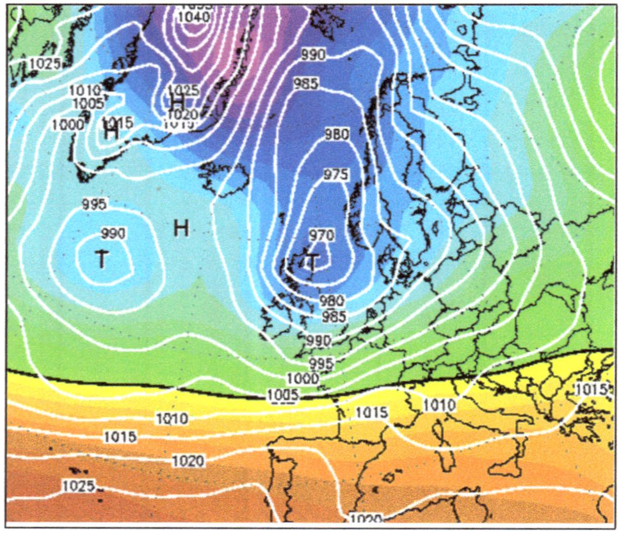

RAIN POTENTIAL

FROST/SNOW

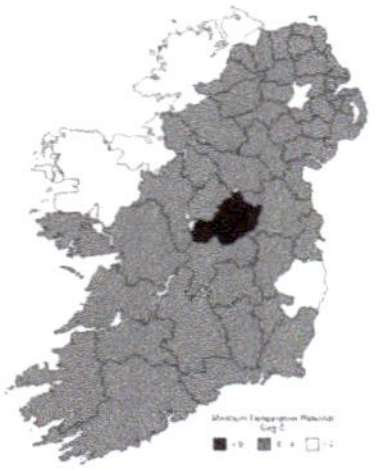

MARCH **12** WEDNESDAY

A trough between two low pressure centres lingers over Ireland bringing light south to southwest breezes throughout, with fog patches likely overnight mostly in the north, and frost in central districts. Bands of rain cross the country during the day, moving up from the south, with heavy rain in the east of the country while only isolated showers reach the far north.

Belfast: Cloudy, passing showers, cold, mostly calm air, possible fog
Dublin: Overcast, cold, rain, light breezes to mostly calm air
Cork: Overcast, colder, heavy rain, light breezes
Galway: Cloudy, light showers, overnight frosts, mostly calm air

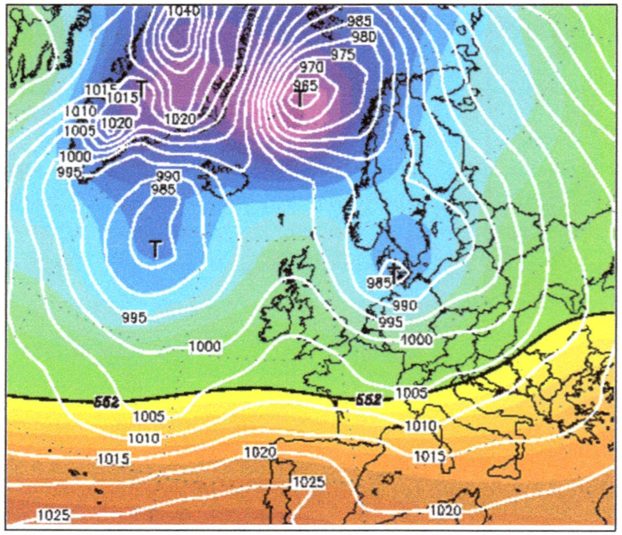

RAIN POTENTIAL

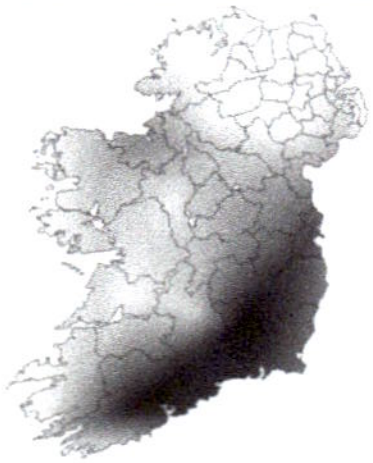

FROST/SNOW

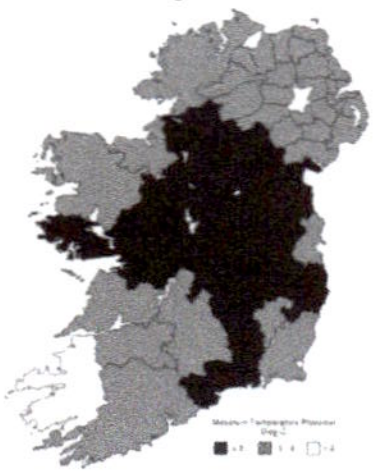

MARCH **13** THURSDAY

A weak depression forms over Ireland bringing west to southwest breezes throughout. A mixed day everywhere with sunny spells and the occasional wintry showers, some of hail or light snow, especially in the south of the country.

Belfast: Fluctuating winds, cooler, sunny day, some morning showers likely
Dublin: Fluctuating breezes, sunny, brief showers
Cork: Sunny, shifting breezes, brief wintry showers, chance of hail or light snow
Galway: Sunny day, overnight showers, fluctuating winds

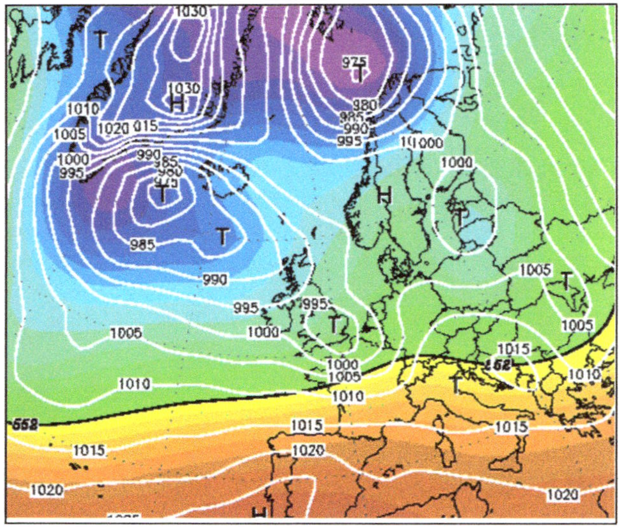

RAIN POTENTIAL

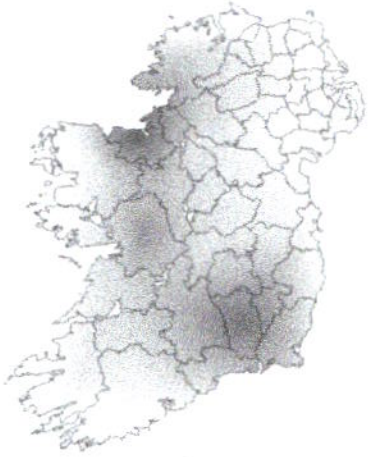

FROST/SNOW

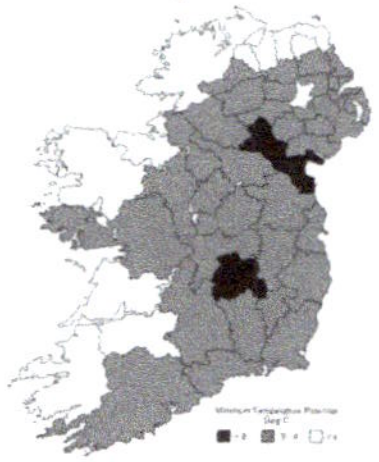

MARCH **14** FRIDAY

Low pressure prevails with a couple of depressions to the west of Ireland directing moderate southeasterly breezes at first, but strengthening later and becoming very strong and blustery especially around the northern coasts. Mostly cloudy and with rain everywhere, occasionally heavy and persistent for a time, moving up from the southwest at first and spreading throughout as the day progresses.

Belfast: Cloudy, light showers, fresh southerly breezes
Dublin: Overcast, occasional showers, warmer day cold night, moderate breezes
Cork: Mostly cloudy, rain, windy spells, chance of light snow
Galway: Unsettled and windy air, cloudy, rain

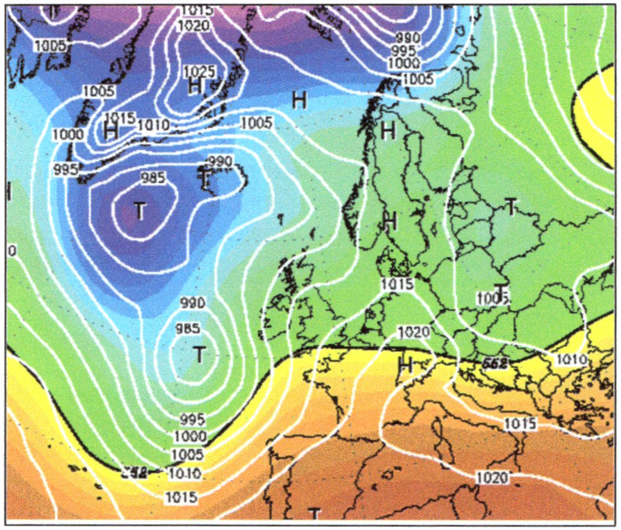

RAIN POTENTIAL

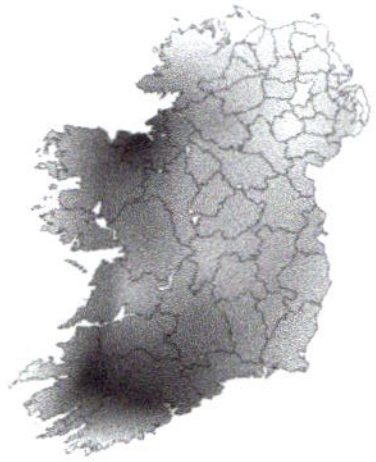

FROST/SNOW

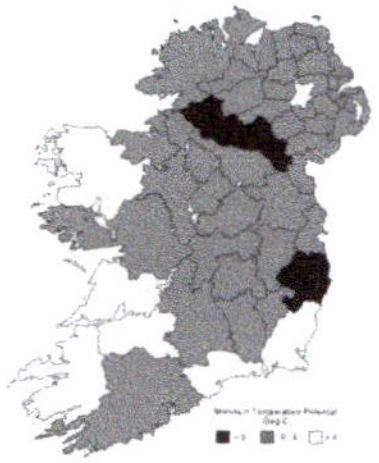

MARCH 15 SATURDAY

An intense depression centred to the immediate northwest of Ireland directs a stormy air over the country, with very strong and blustery south to southwest winds, up to gale force at times. Widespread cloud and rain spreading throughout with heavy falls in most areas, especially in the west and south. Chance of snow in inland parts, in the west and north.

Belfast: Stormy westerly change, strong gusty winds, overcast, rain
Dublin: Overcast, squally showers and drizzle patches, strong and blustery winds
Cork: Overcast, heavy rain, cold overnight, stormy air with strong gusty winds
Galway: Overcast, heavy rain, stormy winds, very cold overnight with chance of snow

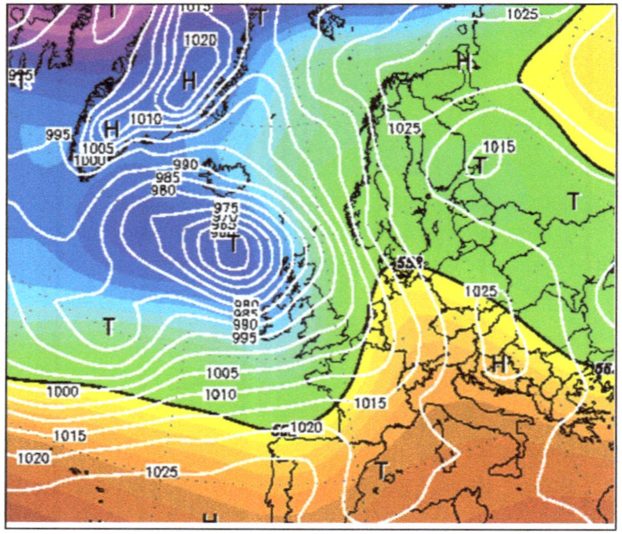

RAIN POTENTIAL

FROST/SNOW

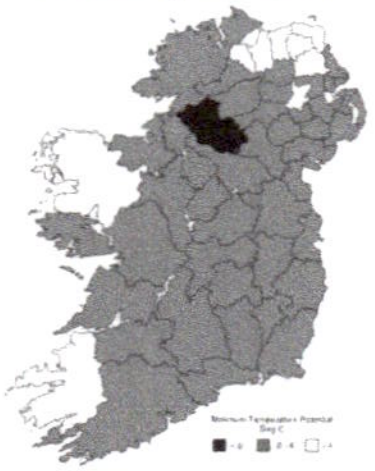

MARCH 16 SUNDAY

Stormy conditions continue as the depression to the northwest of Ireland contracts slightly further to the north of the country, with strong and blustery west to southwest winds prevailing throughout, up to gale force at times around the coasts. A changeable day with rain easing to showers everywhere, and increasing sunny spells developing. Chance of snow overnight in the west and possible frosts especially in inland parts.

Belfast: Strong stormy winds, squally showers, milder, good sunny breakthroughs
Dublin: Stormy winds, isolated showers, milder temperatures, good sunny spells
Cork: Stormy, cloudy, cold, squalls
Galway: Changeable, mix of sunny spells and showers, storm winds, chance of snow

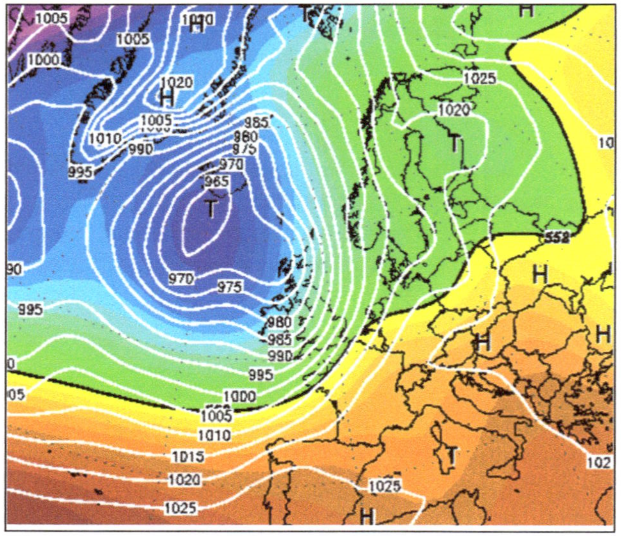

RAIN POTENTIAL

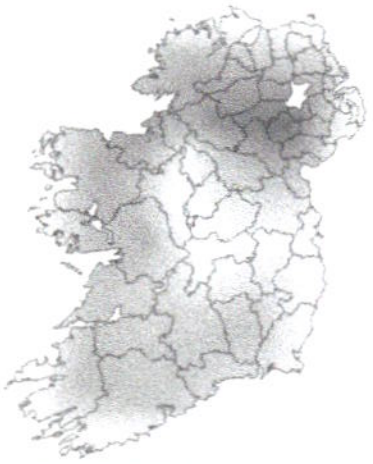

FROST/SNOW

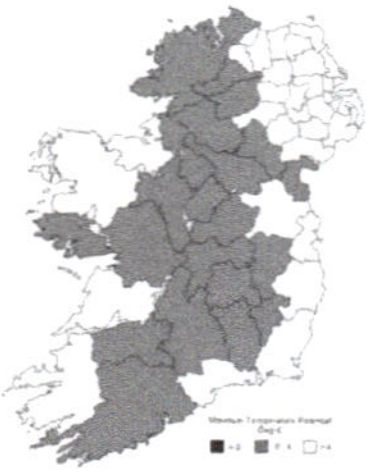

MARCH **17** MONDAY

Low pressure continues to prevail over Ireland from the depression to the north of the country, but with winds easing throughout in a lighter westerly flow. Mostly dry with scattered cloudy spells at first before a band of showers moves across the southwestern districts and up through the country as the day progresses. Some falls may be heavy for a time in the west to southern Ulster.

Belfast: Cloudy, scattered showers, winds easing, chance of odd gale gust
Dublin: Partly cloudy, cool, outbreaks of light rain, gentle shifting breezes
Cork: Cloudy and cold, winds easing to light, some brief showers
Galway: Overcast, light showers, cold, winds easing to fresh

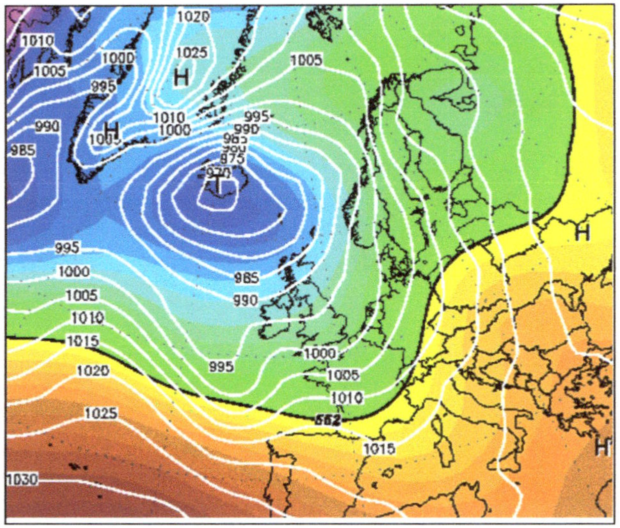

RAIN POTENTIAL

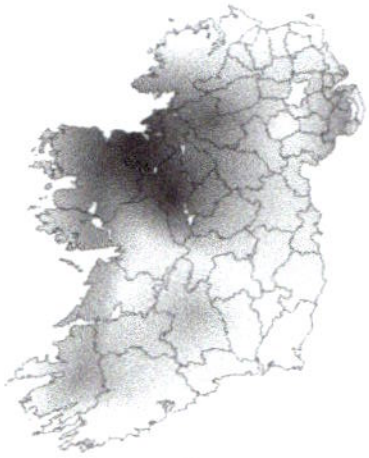

FROST/SNOW

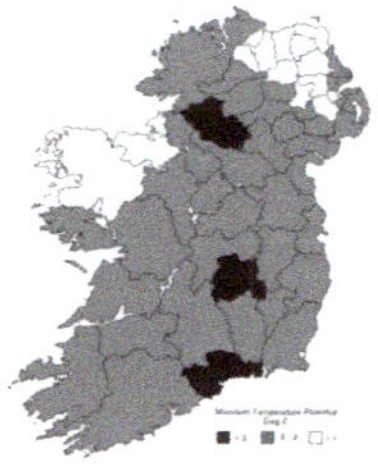

MARCH **18** TUESDAY

A low pressure trough continues to bring a light to moderate westerly flow over the country, while a high pressure ridge builds to the southwest. Mostly cloudy and with brief showers everywhere, as well as the chance of some light snow in the west. Longer sunny spells likely in the north and east. Chance of frost patches settling in overnight.

Belfast: Cloudy, brief showers, fluctuating breezes
Dublin: Cloudy, brief showers, cold overnight, gentle breezes
Cork: Overcast and dull, passing showers and drizzles, light shifting breezes
Galway: Fluctuating breezes, overcast, some showers, chance of frost or snow

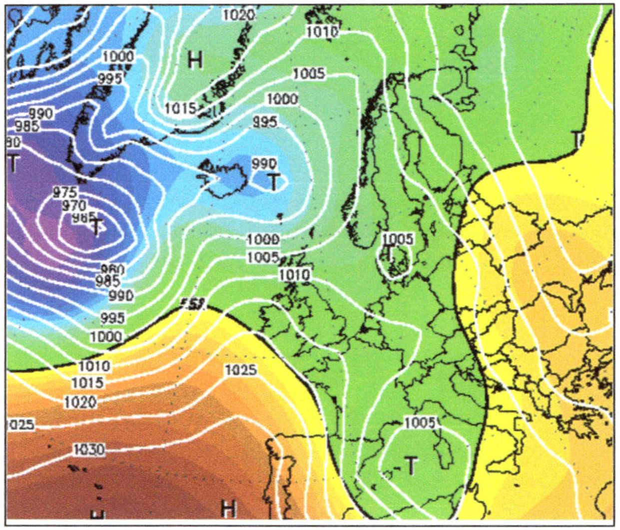

RAIN POTENTIAL

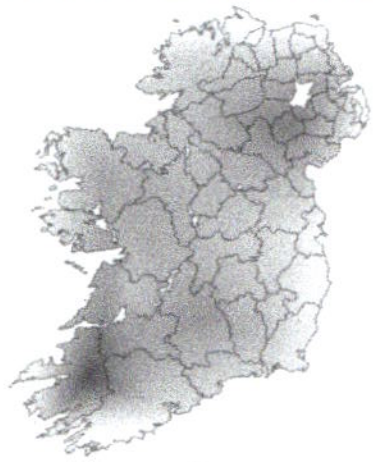

FROST/SNOW

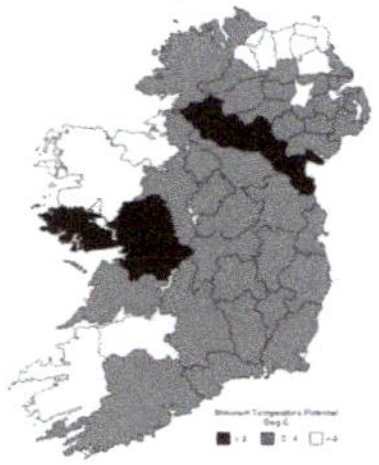

MARCH 19 WEDNESDAY

A large anticyclonic system centred to the south of the country extends a ridge over Ireland with moderate westerly breezes prevailing, stronger and with gusts to the north. Mostly dry and sunny in the east and south, while light showers linger overnight elsewhere, and return again in the evening. Once more there may be the odd outbreak of snow flurries overnight in the west.

Belfast: Mostly sunny, isolated showers, strong winds with occasional gusts likely
Dublin: Sunny, mostly dry, moderate breezes, chance of misty patches
Cork: Sunny, mild day cold night, isolated showers, moderate breezes
Galway: Cold, snow flurries possible, occasional sunny spells, windy at times

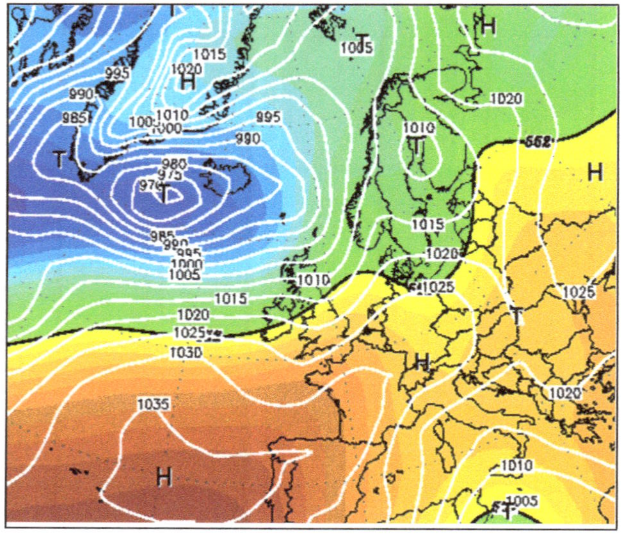

RAIN POTENTIAL

FROST/SNOW

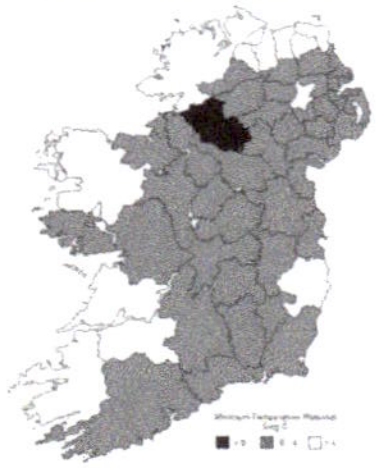

MARCH 20 THURSDAY

A large anticyclonic system passes to the south of Ireland bringing a fresh to strong southwest change over the country, stronger over the northern half of the country. Light drizzly showers likely mostly in inland parts, while southern districts and coastal parts remain mostly dry. Dull and gloomy in the south and west, with more frequent sunny breakthroughs in the north and east.

Belfast: Blustery, cloudy, mosty dry, threats of rain, chance of evening drizzle
Dublin: Partly cloudy, dry, warmer, moderate breezes
Cork: Dull and gloomy air, mostly dry, odd threat of rain, milder, moderate breezes
Galway: Overcast, light showers, milder, windy at times

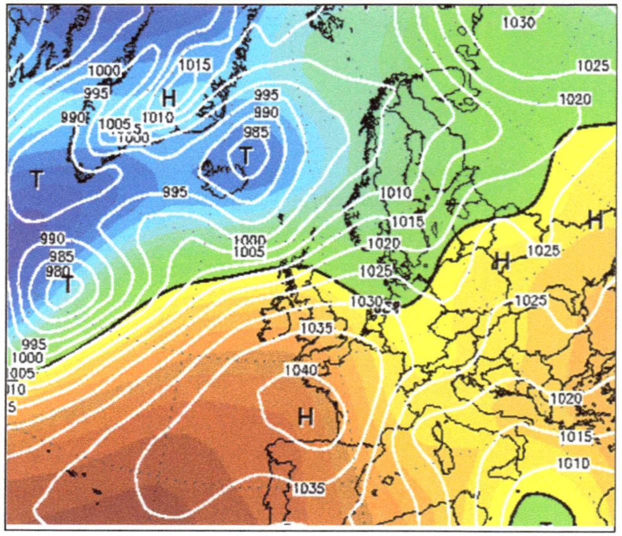

RAIN POTENTIAL

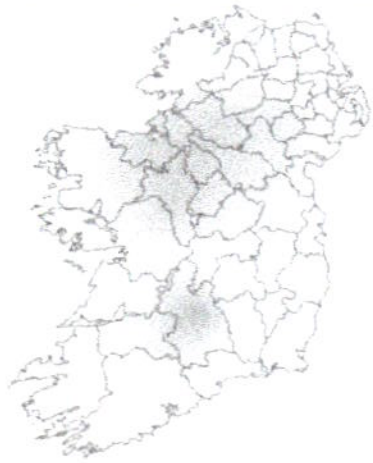

FROST/SNOW

MARCH **21** FRIDAY

High pressure to the south of Ireland drifts away to the east, while an intense depression to the northwest intensifies, with moderate to fresh southwest winds, widespread cloud, and a stormy air prevailing. Rain will be widespreads throughout with heavy falls in the west to inland parts. Milder temperatures.

Belfast: Wet, dull and gloomy, blustery winds, chance of overnight snow
Dublin: Overcast, cooler, scattered showers, windy at times
Cork: Stormy air, overcast, rain, fresh winds with occasional gusts
Galway: Stormy air, overcast, rain, cold, strong gusty winds

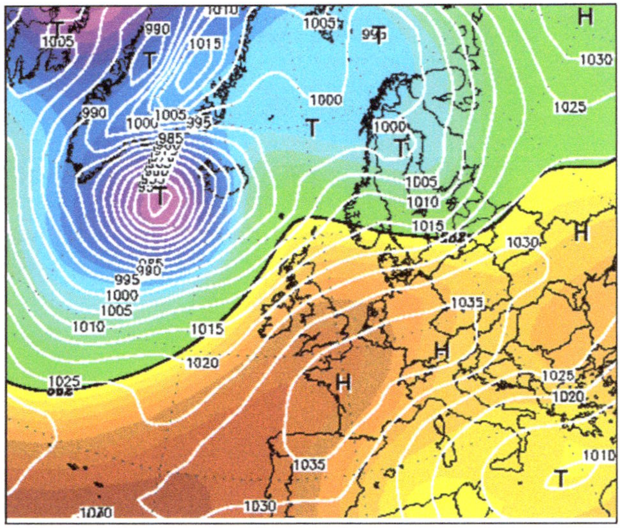

RAIN POTENTIAL

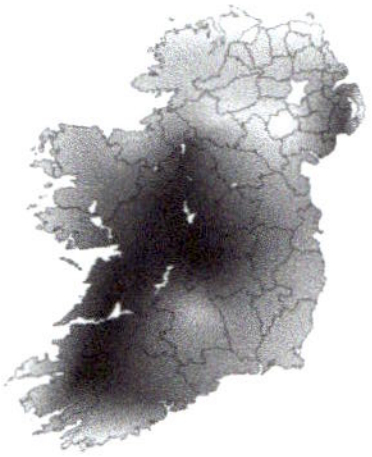

FROST/SNOW

MARCH **22** SATURDAY

A moderate to fresh west to southwest flow prevails over the country as a high pressure system ridges to the south of Ireland, and low pressure trough moves over the northern half of the country. Overnight heavy rain likely, particularly in the north, later clearing to a mostly dry and sunny day. Chance of overnight light snow in the north as well as the odd patch of fog.

Belfast: Sunny, dry, cooler, light breezes, chance of fog patches
Dublin: Sunny, mild, brief evening showers, odd breezy spell
Cork: Mostly sunny and dry, mild day cold night, moderate breezes
Galway: Sunny day, evening showers, mild day cold night, breezy at times

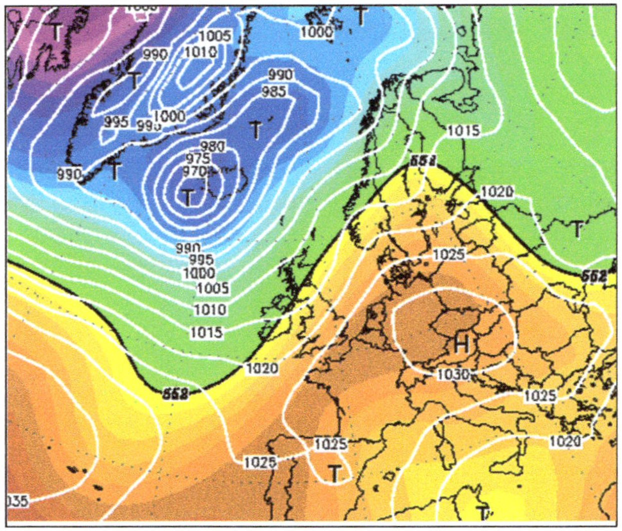

RAIN POTENTIAL

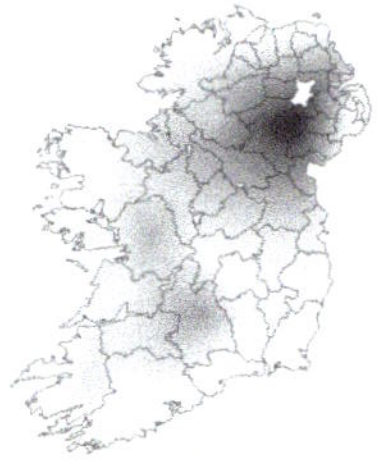

FROST/SNOW

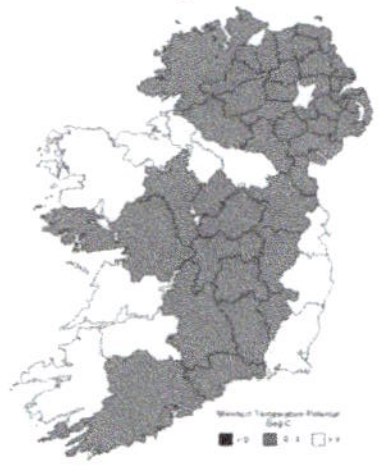

MARCH **23** SUNDAY

A high pressure ridge moves up over the country as a depression departs to the north. Some overnight patches of fog here and there, clearing to a mostly dry day with good sunny spells throughout except for the odd isolated shower here and there. Winds generally light to moderate and in a west to southwest flow.

Belfast: Partly cloudy, dry, fresh breezes at times
Dublin: Partly cloudy, light spotty showers, moderate breezes
Cork: Cloudy, mostly calm air, dry, odd threat of rain, chance of overnight fog
Galway: Moderate breezes, dry, mostly cloudy, milder

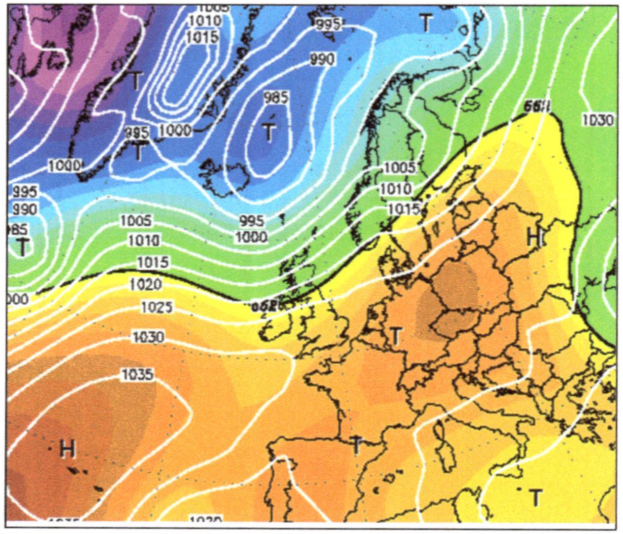

RAIN POTENTIAL

FROST/SNOW

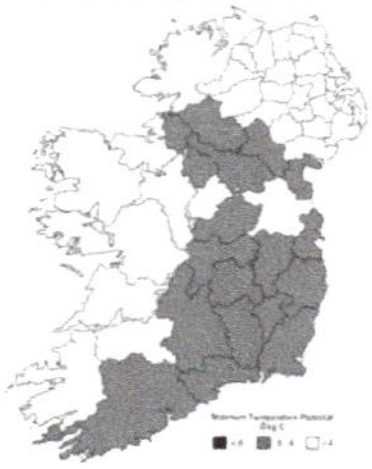

MARCH **24** MONDAY

A large anticyclonic system spreads over the country with gentle west to southwest winds prevailing, fresh for a time along the northern coasts. Widespread cloud and mostly dry throughout with sunnier spells over the southern half of the country. Chance of overnight misty patches in the north, and thicker fog in the south. Chance of brief showers in inland parts to the northwest for a brief time.

Belfast: Light breezes, overcast, dry, passing threats of rain, chance of misty patches
Dublin: Cloudy, dry, milder day cooler night, light breezes to mostly calm air
Cork: Cloudy, dry, gentle breezes, patches of fog likely
Galway: Light breezes to mostly calm air, overcast, milder, brief showers

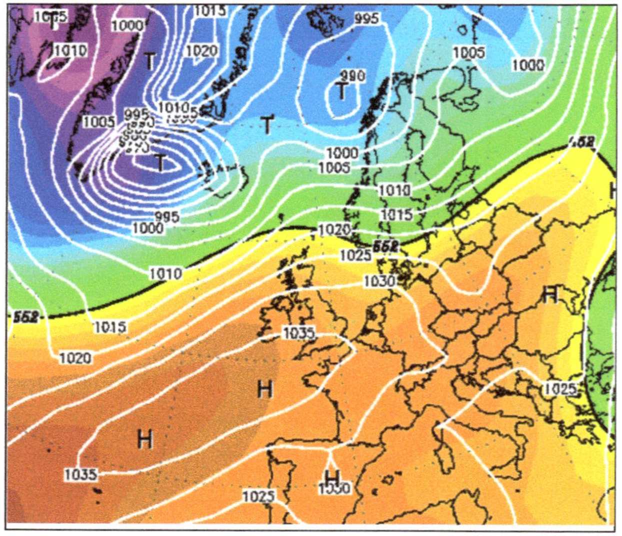

RAIN POTENTIAL

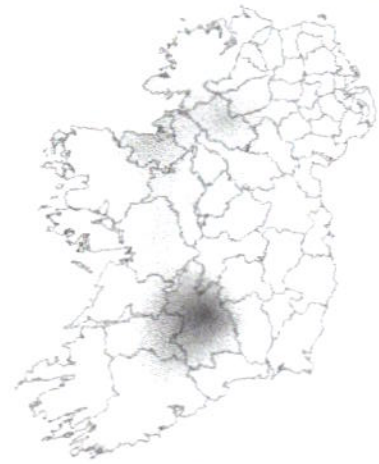

FROST/SNOW

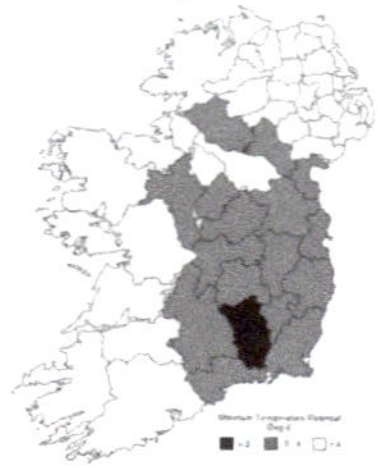

MARCH **25** TUESDAY

A large high moves over the country with moderate southwesterly breezes prevailing, stronger and gustier at times in the north. Mostly cloudy and dry in the eastern half of the country, while a mix of sunnier spells and the occasional shower prevails over the western half and in the north.

Belfast: Partly cloudy, isolated showers, moderate breezes
Dublin: Light breezes, cloudy, colder, mostly dry with chance of brief misty patches
Cork: Cloudy, brief showers, mild day cold night, some light evening rain likely
Galway: Partly cloudy and dry, moderate temperatures, unsettled windy spells

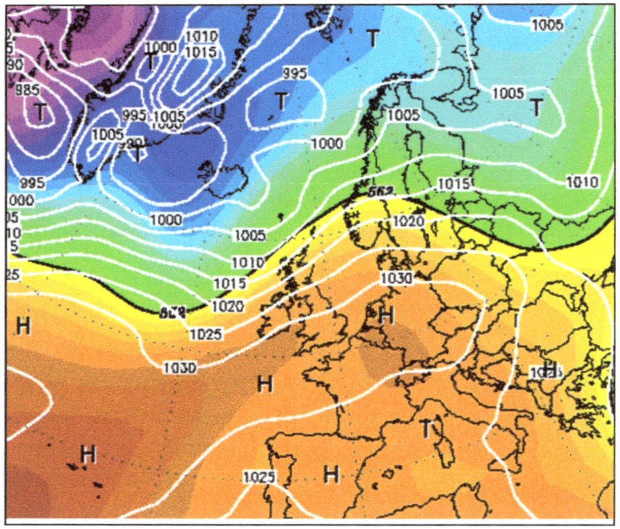

RAIN POTENTIAL

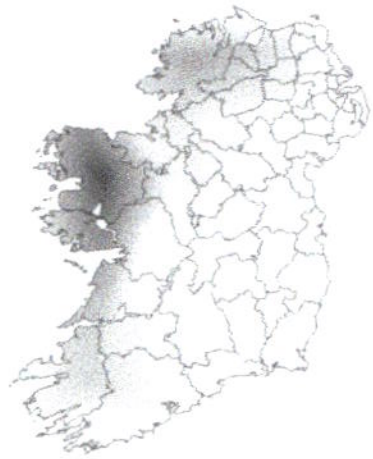

FROST/SNOW

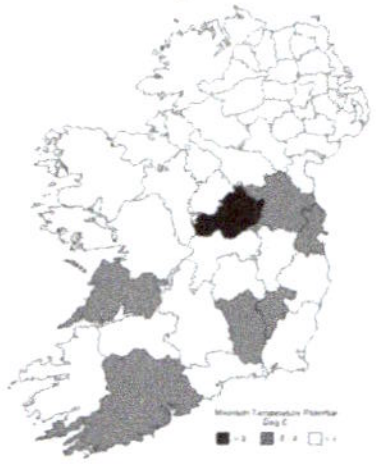

MARCH **26** WEDNESDAY

A large anticyclonic system to the west of Ireland directs a fresh to strong west to southwest flow over the country. Widespread cloud brings a dull and gloomy day throughout, with patches of frost and fog overnight and passing outbreaks of drizzle in many parts, although remaining mostly dry in the eastern and southern coastal districts.

Belfast: Light breezes, overcast, isolated showers, chance of frost and fog
Dublin: Mostly calm, dry, overcast and dull, cooler, chance of evening drizzle patches
Cork: Overcast, dry with passing threats of rain, gentle breezes
Galway: Overcast, colder with frosts, brief evening showers, moderate breezes

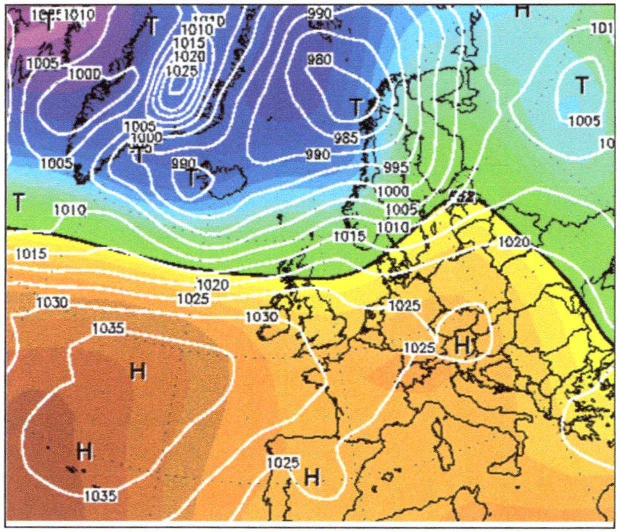

RAIN POTENTIAL

FROST/SNOW

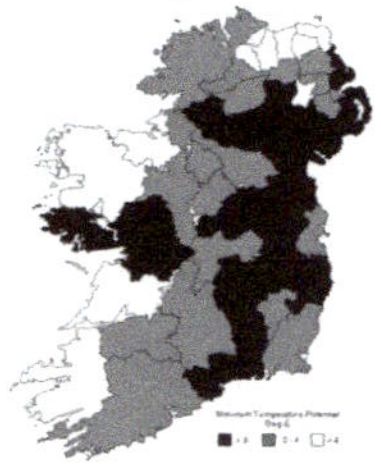

MARCH **27** THURSDAY

A large high to the southwest of Ireland strengthens while a depression to the north of the country moves down towards the northern coasts. Moderate to fresh westerly winds prevail at first, turning northwesterly and becoming stronger later in the day. A band of rain moves onto the western coasts at first and spreads across the country as the day progresses, with very little sunny spells likely anywhere. Rain may be heavy for a time in central parts.

Belfast: Moderate breezes, overcast, milder, light showers
Dublin: Cloudy, fresh breezes at times, milder, light drizzly showers
Cork: Partly cloudy, dry, milder, moderate breezes
Galway: Overcast, rain, windy spells

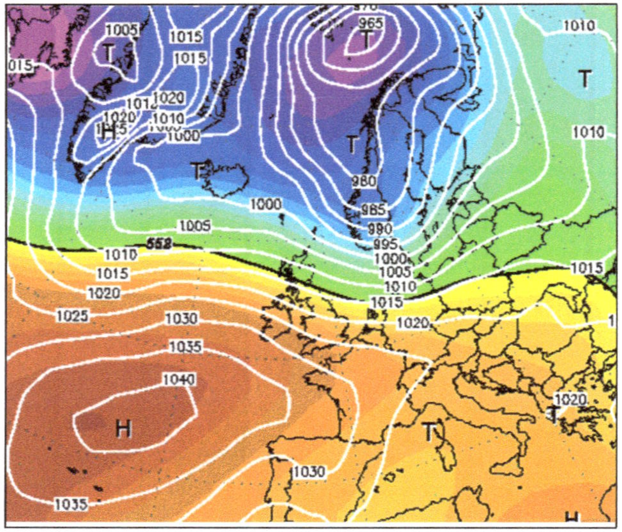

RAIN POTENTIAL

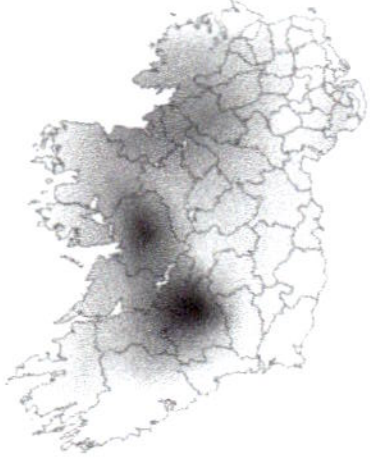

FROST/SNOW

MARCH **28** FRIDAY

An intense North Sea depression moves down onto Ireland bringing widespread stormy conditions throughout, with winds up to gale force in many areas, particularly in the north where winds may reach storm force for a time. A mixed day with bands of rain, possible thunderstorms and hail between and some sunny breakthroughs everywhere. Chance of overnight snow occuring.

Belfast: Stormy, rain, some sunny spells, chance of thunderstorms, hail, and snow
Dublin: Changeable, mix of showers and sunny spells, cold, overnight frost and snow
Cork: Mix of sunny spells and showers, gusty winds, chance of hail or snow
Galway: Stormy, gale winds, mix of sunny spells and showers, chance of hail or snow

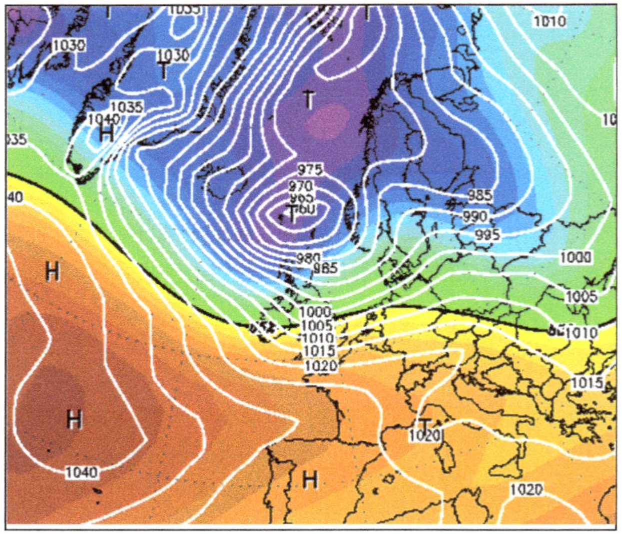

RAIN POTENTIAL

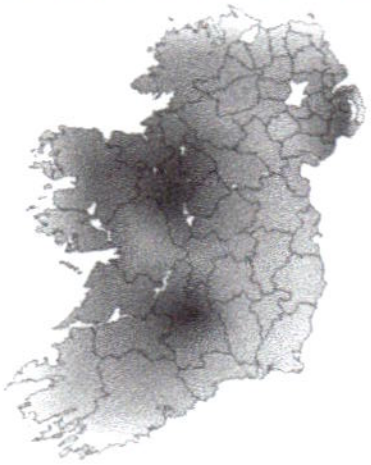

FROST/SNOW

MARCH **29** SATURDAY

An intense depression to the northeast of Ireland maintains stormy and cold conditions throughout Ireland, with fresh to strong north to northwest winds prevailing, up to gale force at times in northern parts. Wintry showers overnight, including the chance of snow and hail and scattered thunderstorms likely, clearing to a mostly dry and sunny day.

Belfast: Colder, partly cloudy, light wintry showers, stormy winds, chance of snow
Dublin: Mostly sunny, cold, mostly dry, chance of light snow overnight, strong winds
Cork: Colder, mostly sunny, some wintry showers, strong winds, chance of snow
Galway: Stormy winds, partly cloudy, brief wintry showers, chance of hail or snow

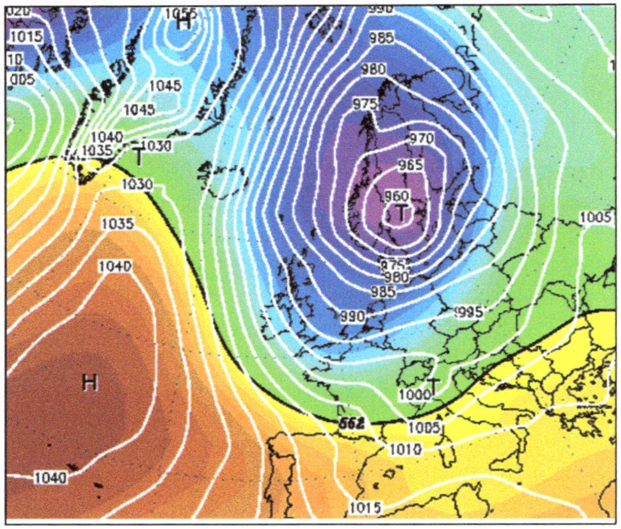

RAIN POTENTIAL

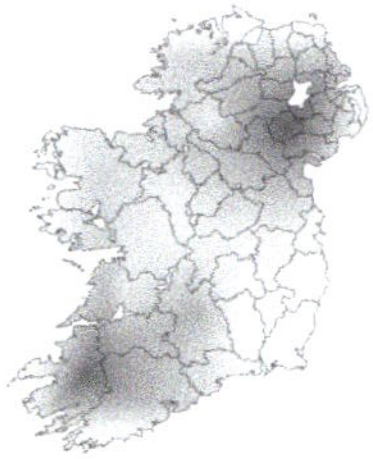

FROST/SNOW

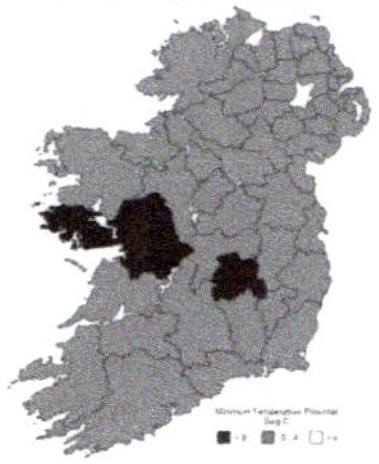

MARCH **30** SUNDAY

A large anticyclonic system to the west of Ireland ridges over the country while the depression that brought stormy conditions over the past couple of days contracts back into northeast of the country. Mostly sunny, cold, and dry day throughout with frosts overnight and the chance of late afternoon hail showers or snow flurries mostly in the west and north. Cold and blustery northwesterly winds prevail, possibly to gale force for a time.

Belfast: Sunny, cold, frosts, dry, fresh northerly winds, chance of snowflurries
Dublin: Mostly sunny, dry, frosts, strong and cold northerly winds
Cork: Mostly sunny, cold northerly winds, dry
Galway: Mostly sunny, dry, cold northerly winds, possible gales and overnight snow

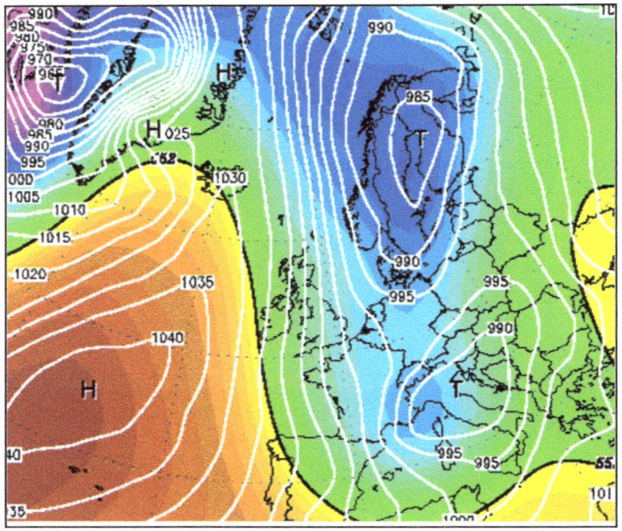

RAIN POTENTIAL

FROST/SNOW

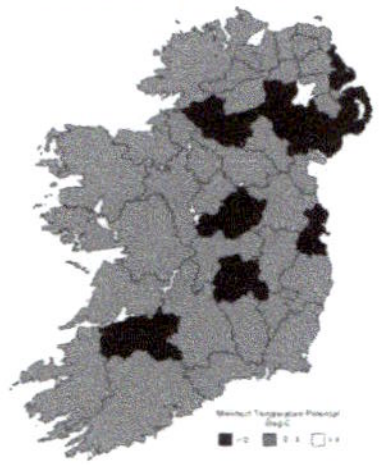

MARCH 31 MONDAY

Despite a high pressure ridge prevailing over the country directing light and variable breezes, the day is likely to be very dull and gloomy throughout. Cold with overnight frost patches mostly in the west and north. A band of rain will cross the country during the day bringing wintry showers to most districts, including possible showers of hail or light snow here and there.

Belfast: Overcast, winds easing, brief showers, frost patches, chance of mist and fog
Dublin: Cloudy, cold with sharp frosts, overnight drizzle patches, lighter breezes
Cork: Overcast, isolated showers and passing drizzle patches, moderate breezes
Galway: Overcast, some showers, chance of frosts, hail or snow, winds easing to light

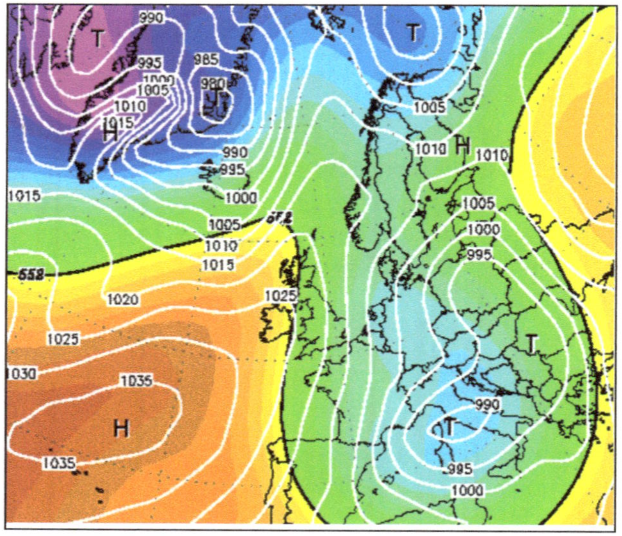

RAIN POTENTIAL

FROST/SNOW

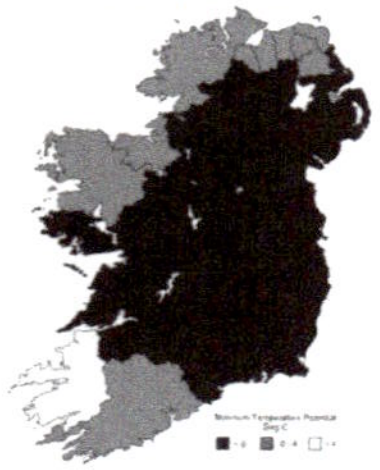

APRIL

PHASES OF THE MOON

3rd	Northern Declination
5th	First Quarter
11th	Crossing Equator
13th	Full Moon
13th	Apogee
18th	Southern Declination
21st	Third Quarter
26th	Crossing Equator
27th	New Moon
27th	Perigee #3

1	2	3	4	5	6	7	8	9	10	11	12	13	14	15	16	17	18	19	20	21	22	23	24	25	26	27	28	29	30
Mon	Tue	Wed	Thur	Fri	Sat	Sun	Mon	Tue	Wed	Thur	Fri	Sat	Sun	Mon	Tue	Wed	Thur	Fri	Sat	Sun	Mon	Tue	Wed	Thur	Fri	Sat	Sun	Mon	Tue
		^		1Q						XhS		F A					V			3Q					XhN	N P3			

MONTHLY SUMMARY

1st-10th A large high pressure system to the south of the country brings a rise in air pressure over the country during this period as it moves from the southwest to the northeast of Ireland. This will bring a more settled spell of mostly dry and sunny weather, except for an brief wet spell around the 7th. Light variable to northerly winds will prevail at first but later backing east to north-easterly.
3rd Unusually sunny day.
5th-7th Widespread fog likely which may be particularly dense in some places and slow to clear.
7th-8th A brief wet spell with wintry showers, scattered thunderstorms and hail.
8th Very bright and sunny day in the west of the country.
10th Very sunny, especially in the western half of the country.

Over the next four weeks persistent high pressure systems near or over Ireland will bring a prolonged settled period with warm, dry, and mostly sunny weather and light breezes prevailing everywhere especially over the first three weeks. During this period rain activity will be very infrequent, leading a number of areas to consider drought contingencies due to significant soil moisture deficits and the possible outbreaks of fire here and there. The only rain of any significance during this outlook period can be expected during the fourth week. Accompanying the very dry conditions will be significantly higher than normal sunshine hours, possibly record breaking in many parts of the country, where up to 50% more sunshine than is typical for this time of year is expected. Maximum temperatures may average up to 2-3°C more than normal for this time of year almost everywhere, and possibly resulting in new records in some districts. However, despite the very high maximum temperatures, overnight minimas will still be cold enough to develop frosts by up to a third more often than is typical for this time of year, especially in inland parts. Overall, there is only likely to be about 2-7 days when some rain occurs, 5 days with thunderstorms, 11 days when fog will be seen, 10-17 days with frosts, especially in inland parts, but no days with hail, snow, or gales.

11th-30th A series of anticyclonic systems near to Ireland prevail during this period bringing a prolonged period of mostly dry, sunny, and settled weather except for some light coastal showers now and then. Most days during this spell will be much warmer than is typical for this time of year, especially in inland parts, but night time temperatures will be much cooler than normal, especially in the first week, resulting in many areas seeing ground frosts.

14th-15th Very cold overnight with chance of widespread frosts, especially cold in northern parts. Very sunny during the day. Chance of snow flurries overnight in the north.
15th-18th Widespread fogs at first, later clearing to very sunny weather.
23rd-25th Unusually warmer than normal spell.
29th-30th High pressure prevails. Widespread fogs.

IRELAND RAINFALL ESTIMATES

April 2025

d=mainly dry, n= nonrecordable, l=light shrs, s=significant shrs, r=rain, h=hvy falls

		ULSTER PROVINCE					CONNAUGHT PROVINCE						LEINSTER PROVINCE				MUNSTER PROVINCE						
		Alderg	Hillsbo	Armag	Loughl	Malin I	Sligo	Ardtarı	Belmul	Drums	Galway	Roscoı	Markre	Edende	Kilkenı	Dublin	Shannı	Newpo	Tralee	Killarne	Cork	All	Moon
1st	Tues	l	d	l	d	l	d	d	d	d	d	n	d	d	d	d	d	d	n	d	d	5	
2nd	Wed	d	d	d	d	d	d	d	d	d	d	d	d	d	d	d	d	d	l	d	d	2	
3rd	Thur	d	d	d	d	d	d	d	d	d	d	d	d	d	d	d	d	d	d	d	d	0	^
4th	Fri	d	d	d	d	d	d	d	d	d	d	d	d	d	d	d	d	d	d	d	d	0	
5th	Sat	d	d	d	d	d	d	d	d	d	d	d	d	d	d	d	d	d	d	d	d	0	1Q
6th	Sun	d	d	d	d	d	l	d	l	d	d	d	l	d	s	d	d	d	d	d	d	9	
7th	Mon	d	d	d	s	l	s	s	s	l	n	l	l	l	n	d	d	l	l	l	d	33	
8th	Tues	l	l	n	l	d	l	l	l	l	l	n	l	l	l	l	l	d	l	d	d	20	
9th	Wed	l	d	s	d	d	d	d	d	d	d	d	d	d	d	d	d	d	d	d	d	7	
10th	Thur	d	d	d	d	d	d	d	d	d	d	d	d	d	d	d	d	d	d	d	d	0	
11th	Fri	d	d	d	d	d	d	d	d	d	d	d	d	d	d	d	d	d	d	d	d	0	XhS
12th	Sat	d	d	d	d	d	d	d	d	d	d	d	d	d	d	d	d	d	d	d	d	0	
13th	Sun	d	d	d	d	d	d	d	d	d	d	d	d	d	d	d	d	d	d	d	d	0	F A
14th	Mon	d	d	d	d	d	d	d	d	d	d	d	d	d	d	d	d	d	d	d	d	0	
15th	Tues	d	d	d	d	d	d	d	d	d	d	d	d	d	d	d	d	d	d	d	d	0	
16th	Wed	d	d	d	d	d	d	d	d	d	d	d	d	d	d	d	d	d	d	d	d	0	
17th	Thur	d	d	d	d	d	d	d	d	d	d	d	d	d	d	d	d	d	d	d	d	0	
18th	Fri	d	d	d	n	l	d	n	d	d	d	d	d	d	d	d	d	d	d	d	d	1	V
19th	Sat	d	n	d	n	l	d	n	d	d	d	d	d	d	d	d	d	d	d	d	d	2	
20th	Sun	n	d	l	d	l	n	d	d	d	d	d	d	d	d	d	d	d	d	d	d	2	
21st	Mon	d	d	d	d	d	d	d	d	d	d	d	d	d	d	d	d	d	d	d	d	0	3Q
22nd	Tues	d	d	d	d	d	d	d	d	d	d	d	d	d	d	d	d	d	d	d	d	0	
23rd	Wed	d	d	d	d	d	d	d	d	d	d	d	d	d	d	d	d	d	d	d	d	0	
24th	Thur	d	d	d	d	d	d	d	d	d	d	d	d	d	d	d	d	d	d	d	d	0	
25th	Fri	d	d	d	d	n	n	d	l	d	d	d	d	d	d	d	d	d	d	d	d	2	
26th	Sat	d	d	d	d	d	n	d	d	d	d	d	n	d	d	n	d	d	d	d	d	1	XhN
27th	Sun	d	d	d	l	l	d	l	l	d	d	d	l	d	d	d	d	d	d	d	d	4	N P3
28th	Mon	n	d	l	d	d	d	d	d	d	d	d	d	d	d	d	d	d	d	d	d	1	
29th	Tues	d	d	d	d	s	d	d	d	d	d	d	d	d	d	d	d	d	d	d	d	4	
30th	Wed	n	d	l	l	d	n	l	d	d	d	d	l	d	d	d	d	d	n	d	d	5	
	Estimate:	6	4	9	8	9	7	8	8	3	1	2	8	4	9	2	1	3	5	3	1	100	
	Average:	58	61	58	70	65	78	80	72	67	72	81	78	66	56	54	59	97	110	81	77	1439	
	Trend:	drr	drr	drr	drr	drr	drr	drr	drr	drr	drr	drr	drr	drr	drr	drr	drr	drr	drr	drr	drr	drr	

IRELAND SUNSHINE ESTIMATES

April 2025

F=fine (8-12 hours of sunshine), pc= partly cloudy (4-7 hours), c=cloudy (1-3 hours), o=overcast (0 hours)

		ULSTER PROVINCE					CONNAUGHT PROVINCE						LEINSTER PROVINCE				MUNSTER PROVINCE						
		Alderg	Hillsbo	Armag	Loughl	Malin H	Sligo	Ardtar	Belmul	Drums	Galway	Roscoi	Markre	Edende	Kilkenn	Dublin	Shann	Newpo	Tralee	Killarn	Cork	All	Moon
1st	Tues	c	c	c	c	c	pc	c	pc	c	pc	pc	c	c	c	c	c	c	pc	pc	c	72	
2nd	Wed	c	c	c	c	c	c	c	pc	c	pc	pc	c	c	c	c	pc	pc	c	c	c	69	
3rd	Thur	F	F	F	F	F	pc	F	o	pc	c	c	F	pc	pc	F	pc	pc	o	o	pc	121	^
4th	Fri	pc	pc	pc	pc	pc	pc	pc	F	pc	pc	F	pc	pc	pc	pc	pc	pc	F	F	pc	131	
5th	Sat	pc	pc	pc	pc	pc	c	pc	F	pc	pc	pc	pc	pc	pc	pc	pc	pc	F	F	pc	120	1Q
6th	Sun	pc	pc	pc	pc	pc	pc	pc	F	pc	F	F	pc	pc	pc	pc	F	F	F	F	F	154	
7th	Mon	pc	pc	pc	pc	pc	pc	pc	F	pc	F	F	pc	pc	pc	o	F	pc	pc	pc	F	115	
8th	Tues	pc	pc	pc	pc	pc	F	pc	F	pc	F	F	pc	pc	pc	c	F	F	F	F	F	155	
9th	Wed	c	c	c	c	c	pc	c	pc	c	pc	pc	c	c	o	o	c	c	pc	pc	c	68	
10th	Thur	F	F	F	F	F	F	F	F	F	F	F	F	F	F	F	F	F	F	F	F	190	
11th	Fri	F	F	F	F	F	F	F	F	F	F	F	F	F	F	F	F	F	F	F	F	221	XhS
12th	Sat	F	F	F	F	F	F	F	F	F	F	F	F	F	F	F	F	F	F	F	F	195	
13th	Sun	pc	pc	pc	pc	pc	pc	pc	F	pc	pc	pc	pc	pc	pc	pc	pc	pc	F	F	pc	139	F A
14th	Mon	F	F	F	F	F	F	F	F	F	F	F	F	F	F	F	F	F	F	F	F	229	
15th	Tues	F	F	F	F	F	F	F	F	F	F	F	F	F	F	F	F	F	F	F	F	229	
16th	Wed	F	F	F	F	F	F	F	F	F	F	F	F	F	F	F	F	F	F	F	F	210	
17th	Thur	F	F	F	F	F	F	F	F	F	F	F	F	F	F	F	F	F	F	F	F	206	
18th	Fri	pc	pc	pc	pc	pc	pc	pc	c	F	pc	pc	pc	F	F	F	pc	F	pc	pc	F	158	V
19th	Sat	c	c	c	c	c	o	c	o	pc	c	o	c	pc	pc	pc	c	c	c	c	pc	55	
20th	Sun	c	c	c	c	c	c	c	c	c	c	c	c	c	c	c	c	c	c	c	c	48	
21st	Mon	pc	pc	pc	pc	pc	c	pc	c	pc	pc	c	pc	pc	F	pc	pc	F	pc	pc	F	129	3Q
22nd	Tues	F	F	F	F	F	F	F	F	F	F	F	F	F	F	F	pc	F	F	F	F	193	
23rd	Wed	F	F	F	F	F	F	F	F	F	F	F	F	F	F	F	F	F	F	F	F	221	
24th	Thur	F	F	F	F	F	pc	F	F	pc	pc	pc	F	pc	pc	F	c	pc	pc	pc	o	146	
25th	Fri	c	c	c	c	c	o	c	c	c	c	c	c	c	pc	pc	pc	pc	c	c	pc	62	
26th	Sat	pc	pc	pc	pc	pc	pc	pc	F	F	F	F	pc	F	F	F	F	F	F	F	F	179	XhN
27th	Sun	pc	pc	pc	pc	pc	pc	pc	pc	pc	pc	pc	pc	pc	F	pc	F	F	F	F	F	138	N P3
28th	Mon	F	F	F	F	F	F	F	F	F	F	F	F	F	F	F	F	F	F	F	F	245	
29th	Tues	pc	pc	pc	pc	pc	pc	pc	c	F	pc	pc	F	F	F	F	F	F	F	F	F	168	
30th	Wed	pc	pc	pc	pc	pc	c	pc	c	pc	c	c	pc	pc	pc	pc	c	pc	c	c	pc	81	
Estimate hours:		214	214	214	214	214	202	214	218	225	223	221	214	225	235	225	227	231	238	238	246	4448	
Average hours:		151	151	133	152	153	138	140	158	137	149	155	145	151	146	156	148	140	150	150	156	2957	
Trend:		more	more	more	more	more	more	more	more	more	more	more	more	more	more	more	more	more	more	more	more	more	

APRIL **1** TUESDAY

The high pressure ridge contracts to the southwest of the country while low pressure prevails to the north. Winds generally very light with a cool northwest flow throughout. Mostly cloudy and dry everywhere, except for some drizzly showers in the northeast and patches of fog in the north, lighter misty patches in central and eastern parts. Cold with chance of patches of frost overnight in sheltered parts.

Belfast: Mostly cloudy, dry, cool, light shifting breezes, chance of fog pockets
Dublin: Cloudy, warmer, dry, mostly calm air, chance of misty patches
Cork: Cloudy, dry, warmer, light shifting breezes
Galway: Partly cloudy and dry, fluctuating breezes, cool

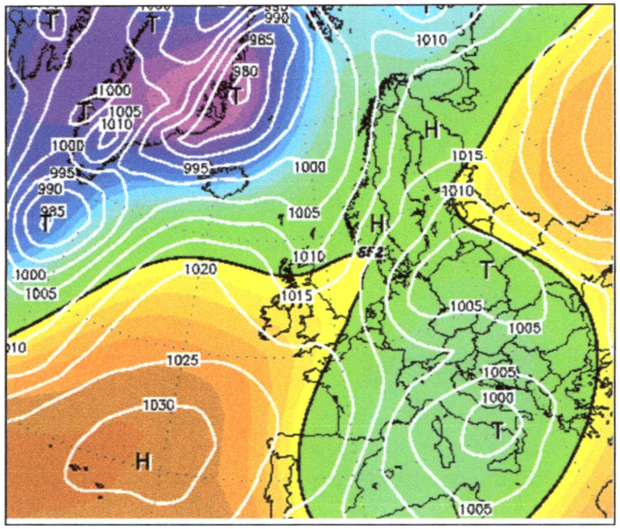

RAIN POTENTIAL

FROST/SNOW

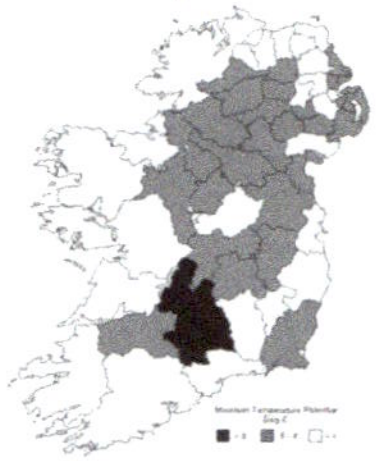

APRIL **2** WEDNESDAY

Mostly calm atmosphere and dry throughout the country with patches of frost and fog in many parts overnight. Occasional sunny spells more frequent in the north and west. Chance of brief drizzle patches in some southeastern parts.

Belfast: Partly cloudy, dry, warmer, calm and serene air, chance of overnight mist
Dublin: Mostly cloudy, dry, colder, calm and serene air, chance of overnight mist
Cork: Cloudy, dry, milder, mostly calm air, chance of overnight fog
Galway: Partly cloudy, dry, calm air, cold overnight with frost and fog likely

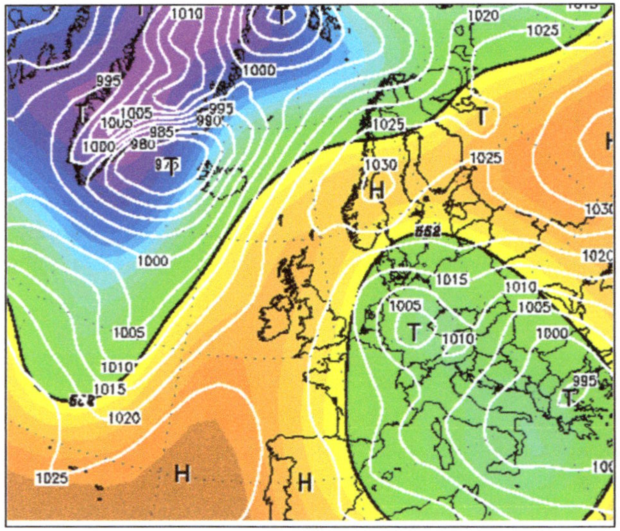

RAIN POTENTIAL

FROST/SNOW

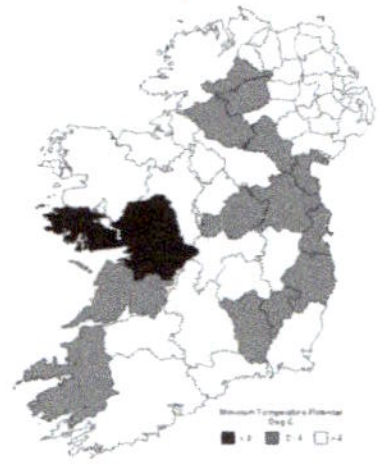

APRIL 3 THURSDAY

A high pressure ridge sits over the country in a gentle southerly to variable flow. Mostly dry throughout with good sunny spells in the north and east, and increased scattered cloud in the south and west. Patches of overnight fog expected in many sheltered areas as well as frosts, with severe frosts likely in eastern parts.

Belfast: Sunny, dry, mild day cold night, mostly calm air, light mist patches likely
Dublin: Sunny, dry, cold, chance of frost patches, gentle southerly flow
Cork: Partly cloudy and dry, cooler, gentle southerly breezes, shallow fog possible
Galway: Mostly cloudy, cool and dry, fresh breezes

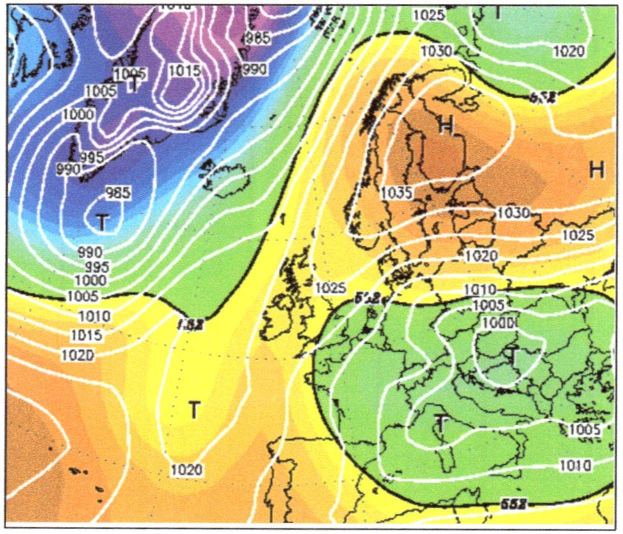

RAIN POTENTIAL

FROST/SNOW

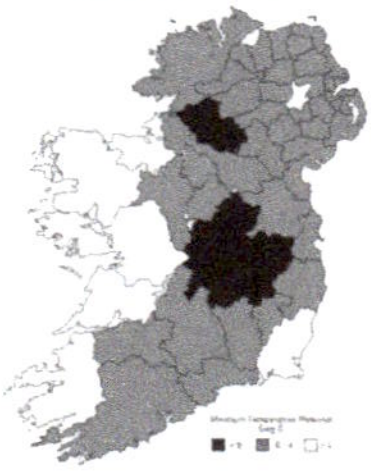

APRIL 4 FRIDAY

A trough between two high pressure systems, one to the southwest and another to the northeast moves over the country, with light to moderate variable breezes. Cold overnight with frost and fog patches in many areas, with severe frosts likely in central parts. Mostly dry day with good sunny spells everywhere.

Belfast: Partly cloudy, dry, mild day, overnight frost and mist patches, gentle breezes
Dublin: Partly cloudy, dry, cool, light shifting breezes
Cork: Partly cloudy, mostly dry, odd misty patch likely, cool
Galway: Sunny, cooler, dry, gentle breezes, chance of misty patches

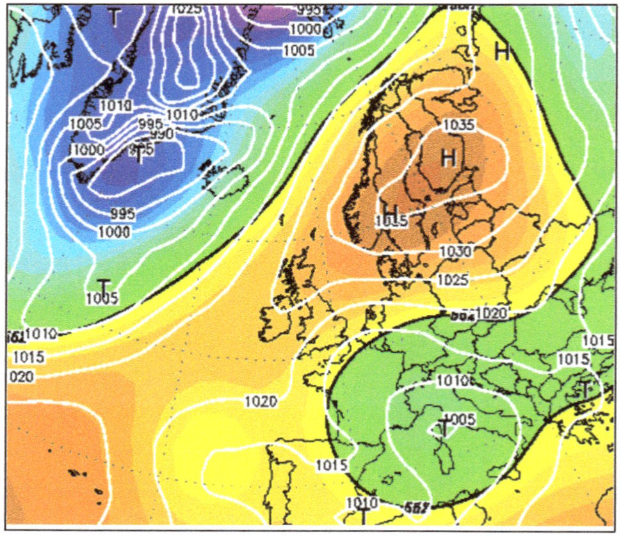

RAIN POTENTIAL

FROST/SNOW

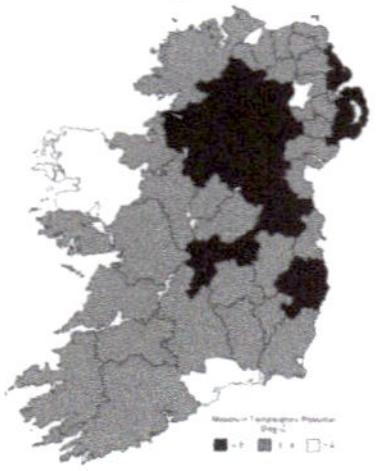

APRIL 5 SATURDAY

High pressure prevails over Ireland in a gentle east to southeast flow. Fairly widespread patches of overnight mist, fog, and frosts. Dry and mostly sunny day throughout, sunniest around the coasts, while cloudier in inland parts and to the west.

Belfast: Partly cloudy, dry, mild day cold night, serene air, chance of haze and fog
Dublin: Mostly sunny and dry, cold overnight with chance of frost and fog patches
Cork: Mostly sunny and dry, milder, gentle breezes, misty patches possible
Galway: Partly cloudy, dry, warmer day cold night, shifting breezes, chance of fog

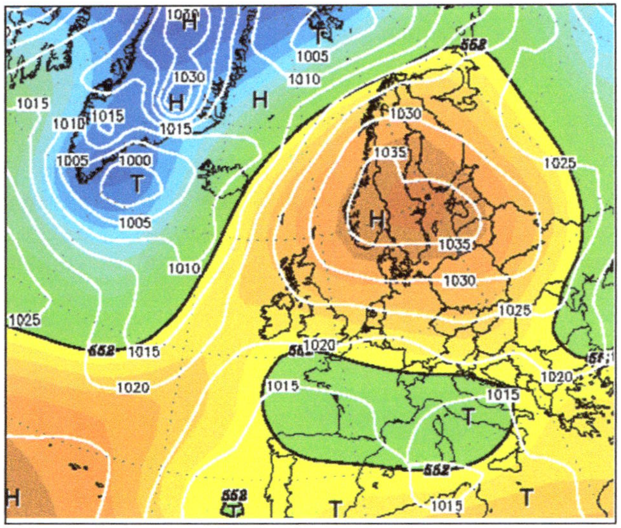

RAIN POTENTIAL

FROST/SNOW

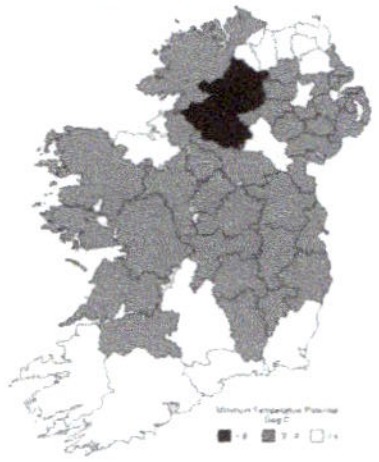

APRIL 6 SUNDAY

A very gentle flow prevails over Ireland in variable directions, as a trough between to highs drifts over the country. Cold and frosty overnight with widespread patches of fog, with some very dense patches of fog in inland parts that will be slow to clear. Mainly dry everywhere with good sunny spells, although cloudier in inland parts. Chance of light patches of drizzle developing towards evening in the southeast.

Belfast: Dry, cold overnight, frosts, light scattered cloud, calm air, fog and haze likely
Dublin: Partly cloudy, dry, warmer day cold night, serene air, chance of frost and fog
Cork: Sunny, dry, warmer day colder night, serene atmosphere, fog likely
Galway: Sunny, dry, light breezes, overnight patches of frost and fog likely

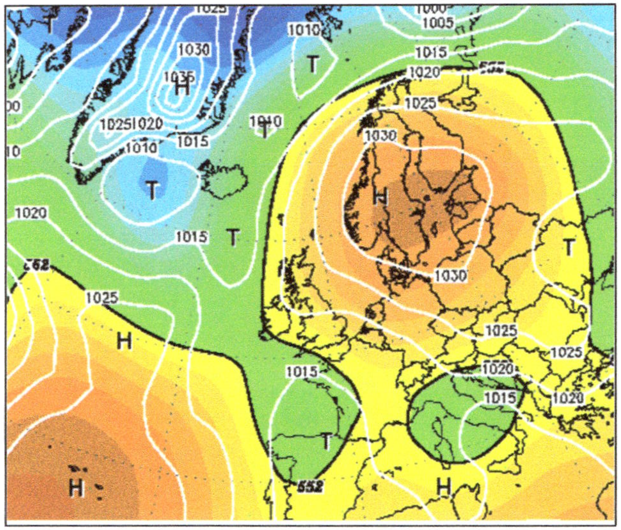

RAIN POTENTIAL

FROST/SNOW

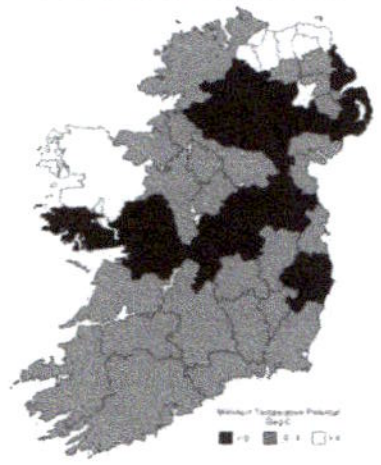

APRIL 7 MONDAY

A high pressure ridge moves onto the west of the country while low pressure departs to the east, with a light to moderate mostly northerly flow prevailing throughout. Widespread overnight fog and frost on either side of the day. Mostly sunny and dry in the south and west with increased cloud elsewhere and some showers briefly mostly in the western half of the country overnight. Showers in the north may be wintry and form flurries for a time.

Belfast: Slightly warmer, dry, occasional sunny spells, calm air, fog likely
Dublin: Overcast, dry, increasing threats of rain, colder, mostly calm air, possible fog
Cork: Sunny and cold, dry, gentle breezes
Galway: Sunny, mostly dry, chance of brief showers, cool, fluctuating breezes

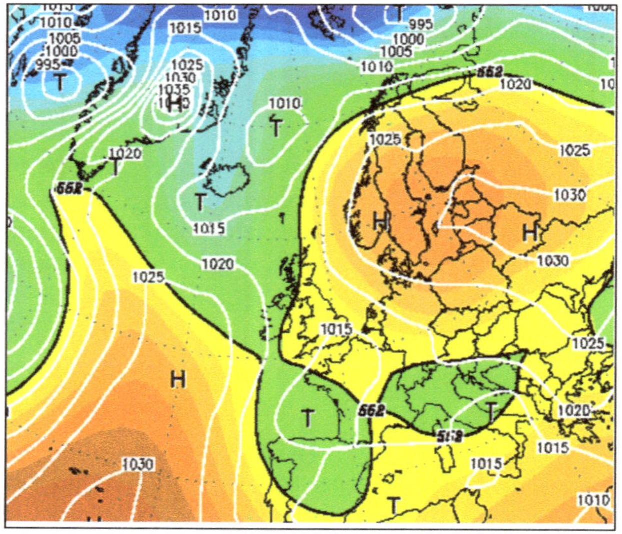

RAIN POTENTIAL

FROST/SNOW

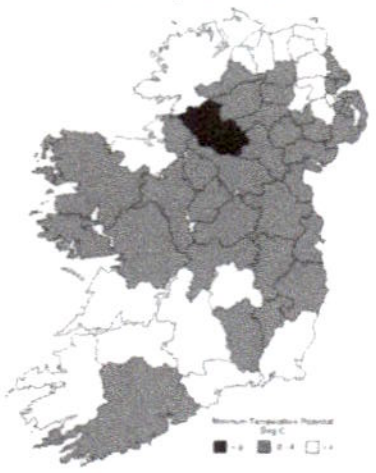

APRIL 8 TUESDAY

A weak low pressure system to the east of Ireland directs a light to moderate north to northeast flow over the country, while a high pressure ridge sits to the west. Overnight showers prevail over the northern half of the country with some wintry and possibly of hail or snow. Mostly dry and sunny during the day before evening showers return mostly in the east to the midlands, but remaining mostly dry along the eastern coasts.

Belfast: Mostly sunny, colder, brief showers, chance of light snow overnight
Dublin: Mostly cloudy, overnight frost, brief showers, chance of hail or light snow
Cork: Mostly dry and sunny, colder, gentle breezes
Galway: Sunny, brief showers, chance of frost or light overnight snow

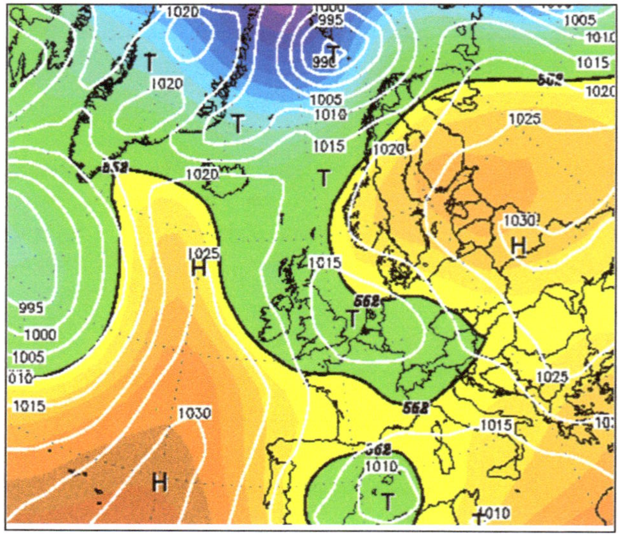

RAIN POTENTIAL

FROST/SNOW

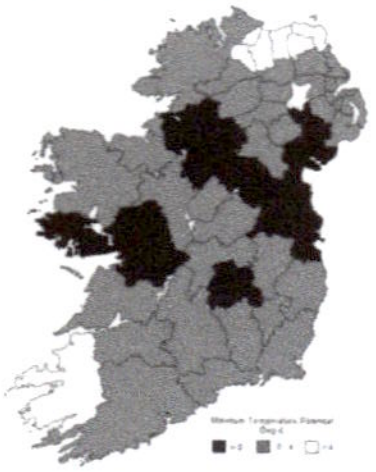

APRIL **9** WEDNESDAY

The high pressure ridge moves to the northwest of Ireland bringing light, mostly north to northeast breezes throughout. Mostly dry and cloudy everywhere, but with some sunnier spells in the west, and isolated showers for a time in the northeast. Some patches of fog likely overnight in the central to eastern districts.

Belfast: Dry, cold, serene air, scattered cloudy spells
Dublin: Widespread cloud, mostly dry, milder, odd threat of rain, mist patches likely
Cork: Cloudy and dry, cold, gentle shifting breezes
Galway: Partly cloudy, dry, slightly warmer, gentle breezes

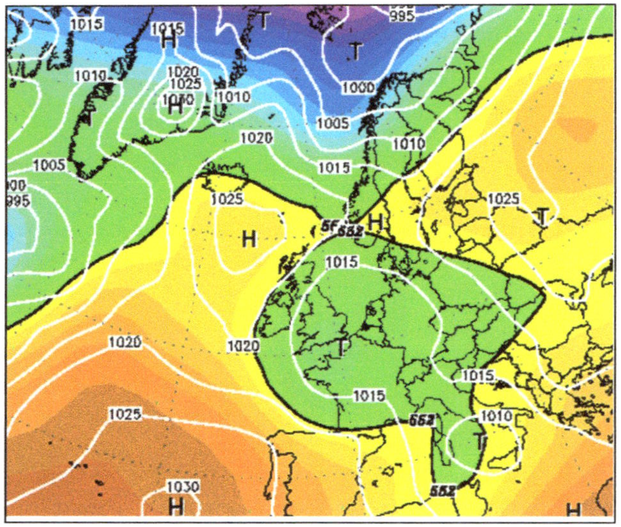

RAIN POTENTIAL

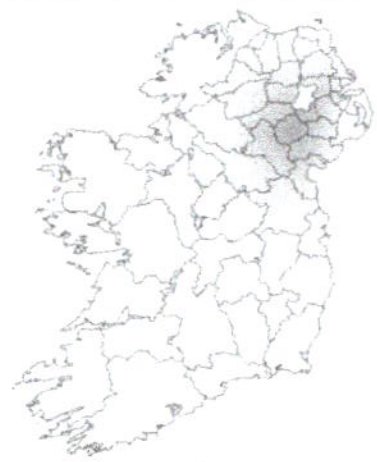

FROST/SNOW

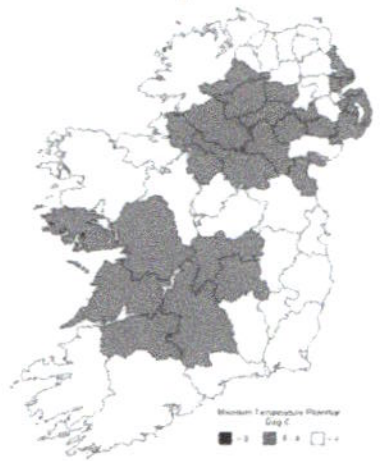

APRIL **10** THURSDAY

An anticyclonic system remains almost stalled to the immediate north of Ireland directing a gentle northesterly flow throughout. Overnight patches of misty in many parts, clear to a mostly sunny and dry day throughout.

Belfast: Sunny, dry, warmer day cold night, gentle breezes, chance of misty patches
Dublin: Gentle northerly air, sunny, dry, chance of misty patches
Cork: Warmer, sunny, dry, misty patches likely, gentle breezes
Galway: Dry, warmer and sunny day colder night, fresh breezes at times

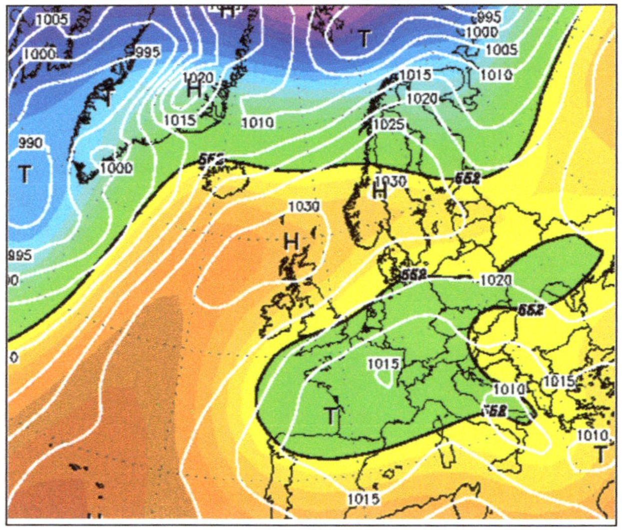

RAIN POTENTIAL

FROST/SNOW

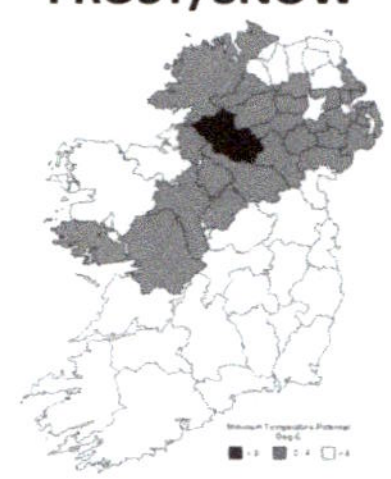

APRIL **11** FRIDAY

Anticyclonic. A large high sits over Ireland bringing a light northeasterly flow throughout. Very sunny and dry everywhere with some patches of mist and frost overnight.

Belfast: Bright and sunny, dry, warm day cold night, mostly calm atmosphere
Dublin: Very sunny, dry, cool, gentle northeasterly flow
Cork: Very sunny and dry, mild day cool night, moderate breezes
Galway: Sunny, dry, mild day cold night, fresh breezes

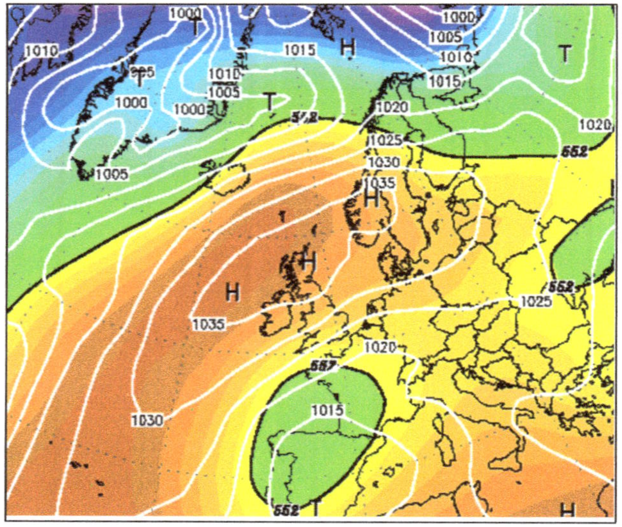

RAIN POTENTIAL

FROST/SNOW

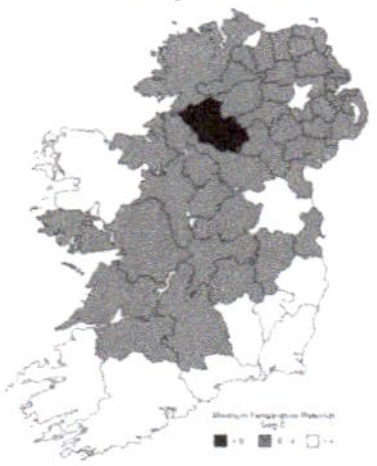

APRIL **12** SATURDAY

A large anticyclonic system remains almost stationary over Ireland with light to moderate east to northeast breezes prevailing, fresher at times in the south of the country. Sunny and dry almost everywhere, with patches of frost overnight in many areas.

Belfast: Sunny, dry, mild day cold night, gentle breezes to mostly calm air
Dublin: Sunny, warmer, dry, moderate breezes
Cork: Sunny, warmer and dry, freshening easterlies at times
Galway: Sunny, dry, warmer day cold night, fresh breezes with occasional gusts

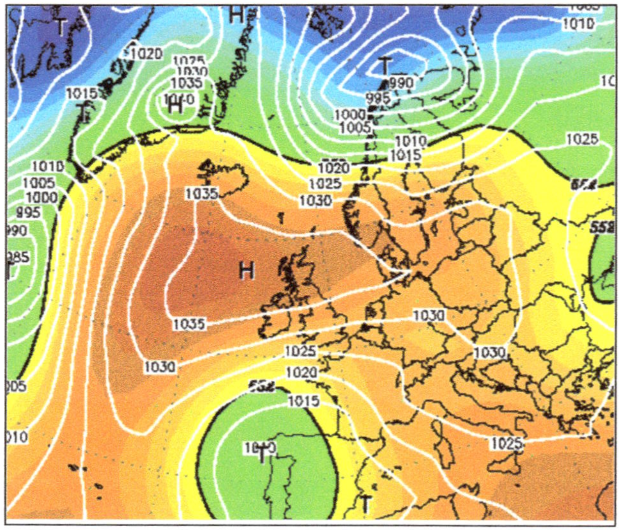

RAIN POTENTIAL

FROST/SNOW

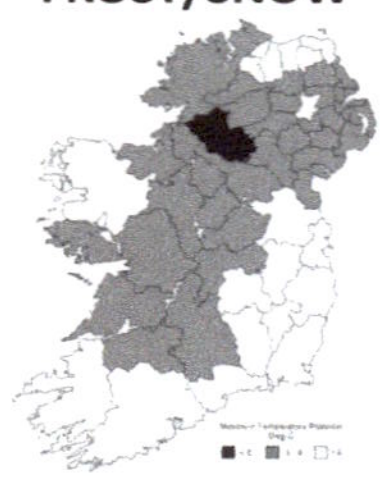

APRIL 13 SUNDAY

High pressure continues to prevail over Ireland with light to moderate east to northeast breezes prevailing, fresher at times in the southeast especially around the coasts. Mostly sunny and dry throughout. Chance of frost patches overnight mainly in inland parts.

Belfast: Mostly sunny and dry, warmer day cold night, light breezes
Dublin: Mostly sunny, cooler and dry, moderate breezes
Cork: Partly cloudy, colder, dry, fresh breezes
Galway: Unsettled and breezy air, mostly sunny, dry, cooler

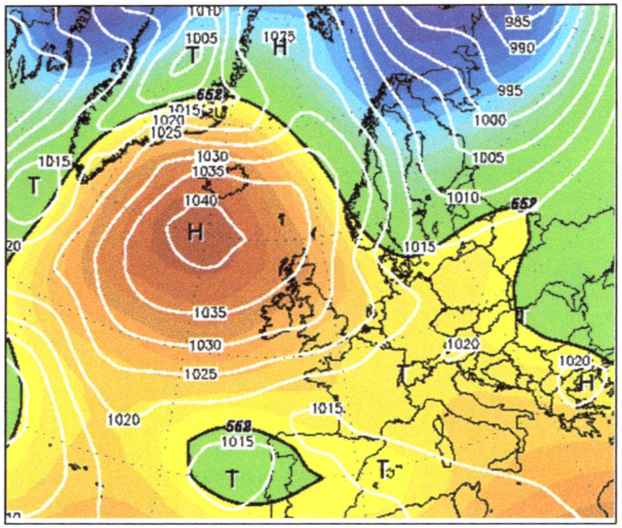

RAIN POTENTIAL

FROST/SNOW

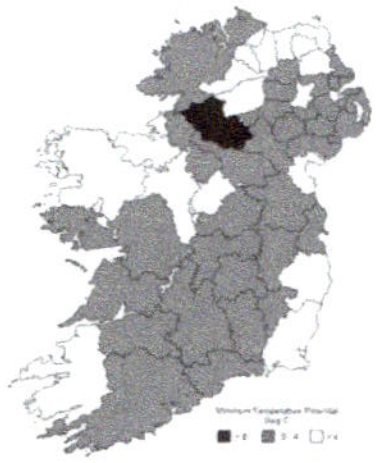

APRIL 14 MONDAY

A large stationary anticyclonic system continues to bring a dry and very sunny day throughout the country, with patches of mist in northern to central districts overnight and severe frosts in inland parts. Light to moderate winds prevail, of variable directions, northeasterly for a time, but turning northwesterly later in the day.

Belfast: Very sunny, dry, cooler, gentle breezes, chance of mist and overnight flurries
Dublin: Very sunny, dry, warmer day cold overnight, chance of frost, light breezes
Cork: Warmer day colder night, sunny, dry, light breezes
Galway: Sunny, dry, mild day, overnight frosts, mostly calm air, chance of fog

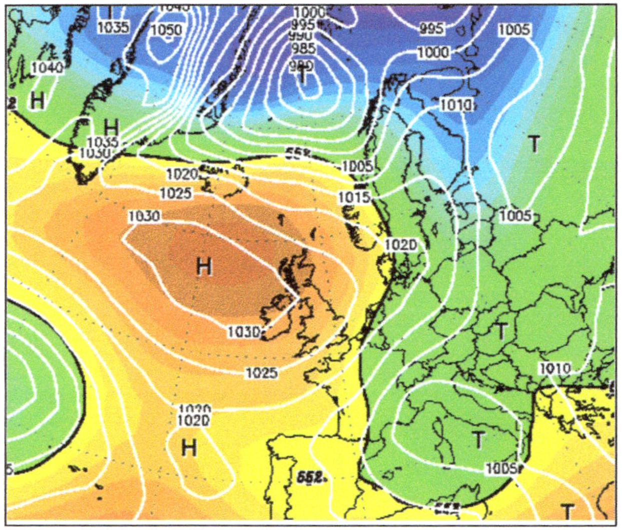

RAIN POTENTIAL

FROST/SNOW

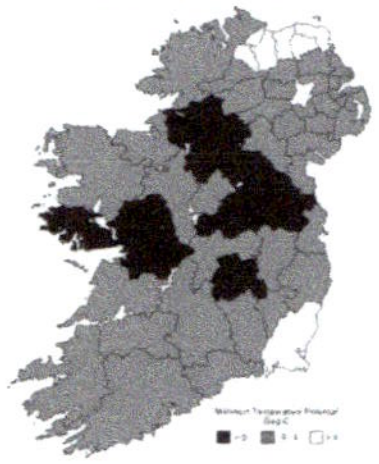

APRIL 15 TUESDAY

A weakening anticyclonic system remains stalled over Ireland directing a light north to northwest flow over the country. Very sunny and dry conditions continue everywhere, with patches of fog overnight mostly in the north to central districts, and widespread frosts in inland parts.

Belfast: Very sunny, dry, warmer day cold night, serene air, overnight fog possible
Dublin: Very sunny, dry, warm day cold night, gentle shifting breezes
Cork: Sunny, warmer, dry, pleasant and serene atmosphere
Galway: Sunny, dry, patches of frost and fog, calm air

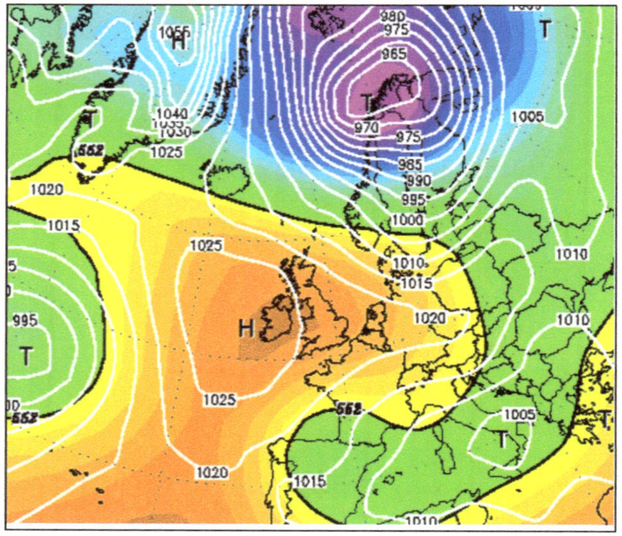

RAIN POTENTIAL

FROST/SNOW

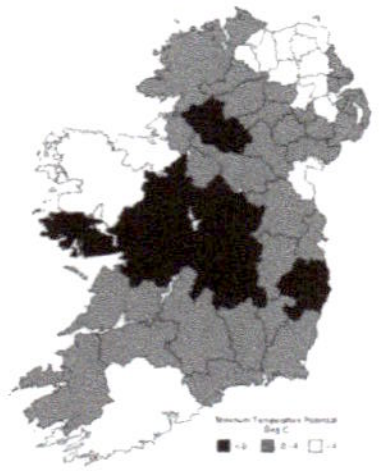

APRIL 16 WEDNESDAY

A high pressure system remains stationary over Ireland bringing light winds throughout. Pleasant, sunny, and dry everywhere, with widespread fog overnight, particularly dense in some places, and scattered patches of frost.

Belfast: Sunny, dry, warm day cold night, serene atmosphere, fog likely
Dublin: Sunny, dry, cooler, calm and pleasant air, chance of patches of fog
Cork: Sunny, dry, calm and pleasant air
Galway: Mostly calm, sunny, dry, milder, patches of fog

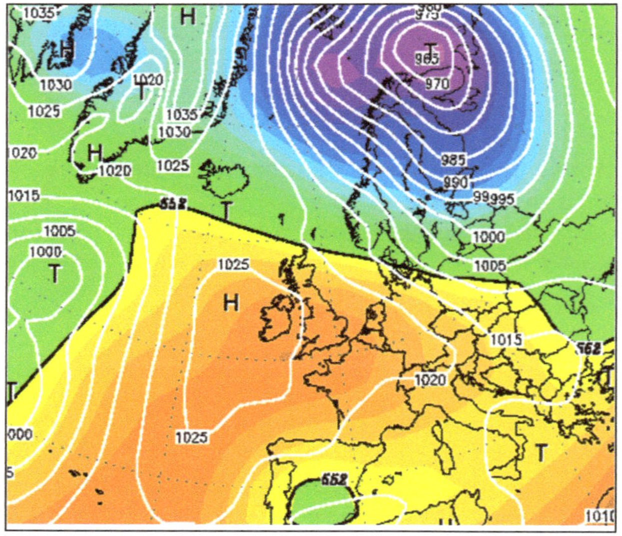

RAIN POTENTIAL

FROST/SNOW

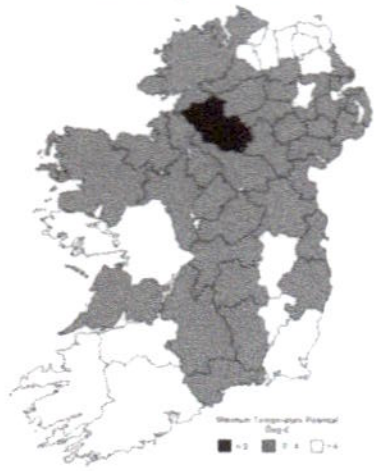

APRIL 17 THURSDAY

Anticyclonic conditions prevail throughout the country as the anticyclonic system remains centred over Ireland. A mostly calm and serene air prevails throughout. Patches of frost and overnight fog in many areas, dense in places, but quick to clear to a sunny, warmer, and dry day everywhere.

Belfast: Calm and pleasant air, dry, sunny, cold overnight with chance of fog patches
Dublin: Sunny and dry, mild day cold night, calm and serene air, fog patches likely
Cork: Calm atmosphere, sunny, dry, patches of shallow fog likely
Galway: Sunny, dry, mild day cool night, calm and serene air, fog pockets likely

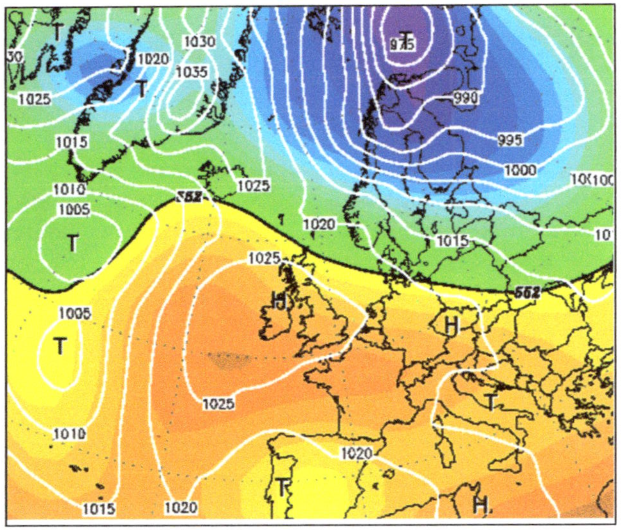

RAIN POTENTIAL

FROST/SNOW

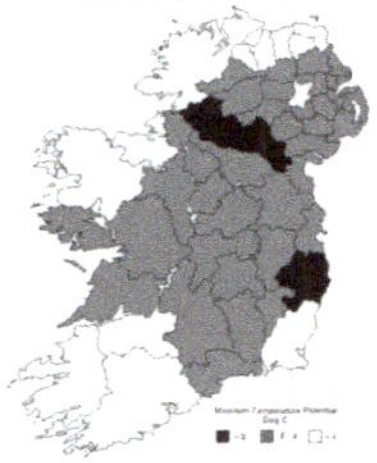

APRIL 18 FRIDAY

Anticyclonic. Mostly sunny and dry everywhere with gentle westerly flow and mostly serene atmosphere prevailing. Patches of frost and fog once more overnight, clearing quickly. Cloudier in the north, where some misty patches may briefly be drizzly for a spell.

Belfast: Mostly sunny and dry, slightly warmer, calm atmosphere
Dublin: Dry, warmer, sunny, calm atmosphere
Cork: Sunny, warmer, dry, calm atmosphere, fog likely
Galway: Partly cloudy, colder, dry, calm and serene air, chance of fog

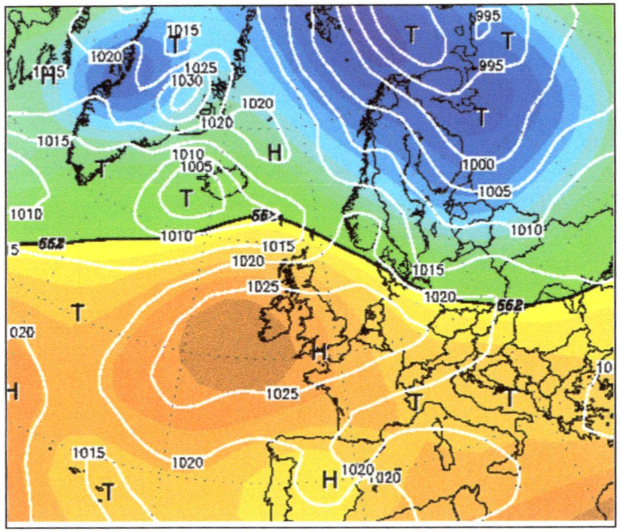

RAIN POTENTIAL

FROST/SNOW

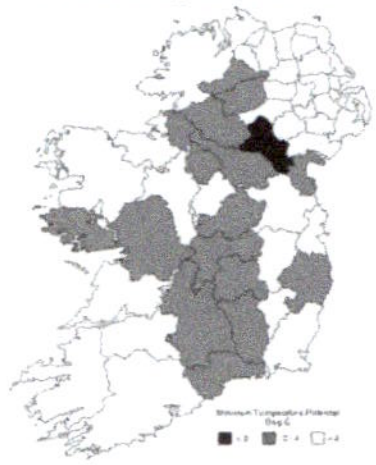

APRIL **19** SATURDAY

The large anticyclonic system that has dominated of recent slowly drifts to the west of the country in a light to moderate west to southwest flow. Increasing scattered cloud throughout, with patches of fog and frost overnight mostly in the north and west. Chance of some isolated showers briefly in the far north.

Belfast: Cloudy, isolated showers, gentle breezes, mild temperatures, chance of fog
Dublin: Mostly cloudy and dry, mild temperatures, mostly calm air
Cork: Mostly sunny, cooler and dry, shifting breezes
Galway: Cloudy, dry, moderate shifting breezes, chance of overnight patches of fog

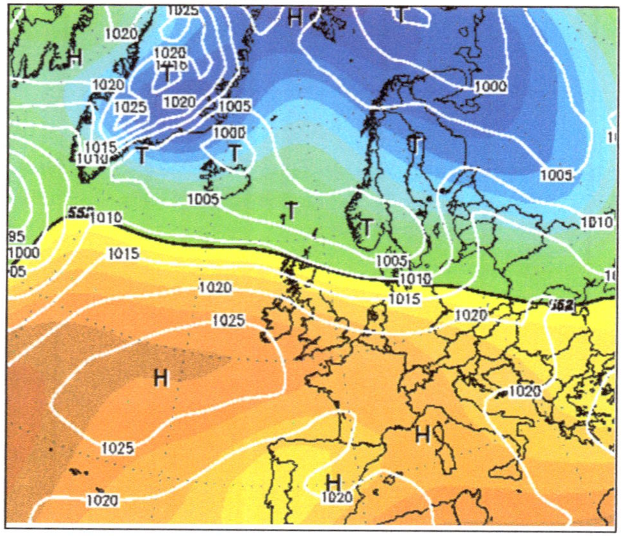

RAIN POTENTIAL

FROST/SNOW

APRIL **20** SUNDAY

The large high pressure system slowly drifts further southwest of Ireland with the ridge remaining over the southern half of the country, while low pressure troughs near the northern coasts. Light winds prevail in a mostly westerly flow, but fresher at times around the northern coasts. Milder and cloudier everywhere with the chance of fog patches overnight mostly in the north. Mostly dry except for some isolated showers likely in the north.

Belfast: Dry, mild temperatures, gentle southwesterly flow, fog patches likely
Dublin: Cloudy spells, dry, mild temperatures, gentle breezes to mostly calm air
Cork: Cloudy, warmer, dry, mostly calm air, chance of overnight fog
Galway: Cloudy and dry, milder temperatures, gentle shifting southwesterly flow

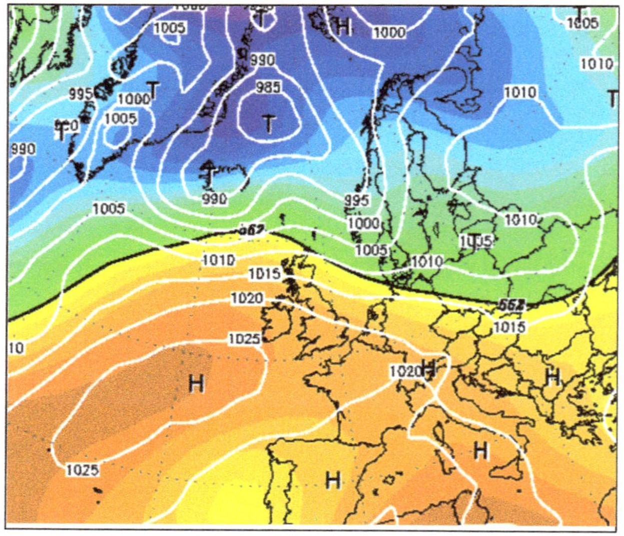

RAIN POTENTIAL

FROST/SNOW

APRIL **21** MONDAY

High pressure moves back up over Ireland returning light variable winds to a mostly calm and serene atmosphere throughout. Mostly sunny and dry everywhere with scattered patches of fog overnight. Chance of frost overnight in sheltered parts in inland parts to the east.

Belfast: Pleasant and serene air, cool, passing scattered cloudy spells, dry
Dublin: Mostly sunny and dry, mild temperatures, calm atmosphere
Cork: Calm and pleasant air, very sunny, dry, mild temperatures, fog patches likely
Galway: Partly cloudy, dry, warmer, calm atmosphere

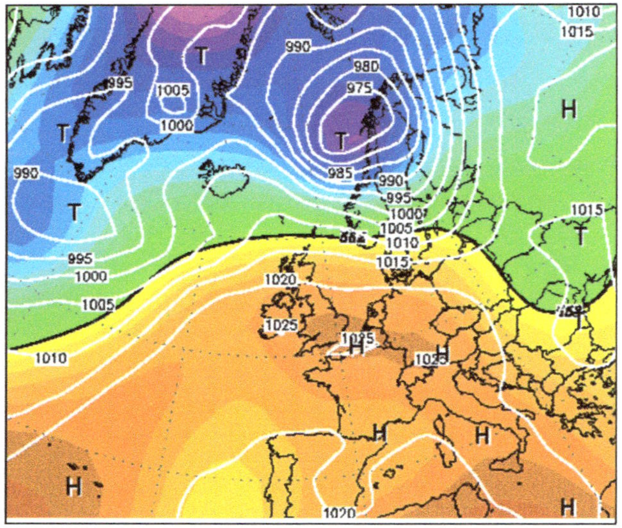

RAIN POTENTIAL

FROST/SNOW

APRIL **22** TUESDAY

High pressure maintains a pleasant air throughout the country in a gentle southwesterly flow. Sunny and dry everywhere. Patches of fog and mist overnight here and there, and chance of frost pockets in the east to inland districts. Winds may be fresher for a time along the western coastal fringes.

Belfast: Calm, sunny, warmer day cooler night, dry
Dublin: Pleasant and serene air, very sunny, dry, warm day cold night
Cork: Sunny, mild temperatures, dry, calm atmosphere, fog likely
Galway: Sunny, cooler, dry, mostly calm air with fresher breezes later in the day

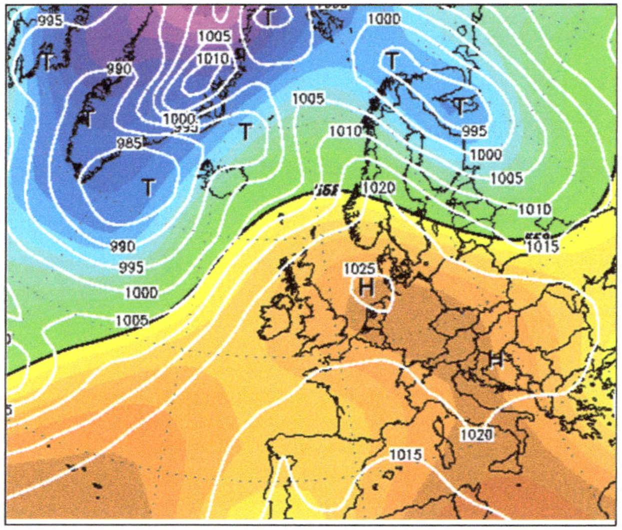

RAIN POTENTIAL

FROST/SNOW

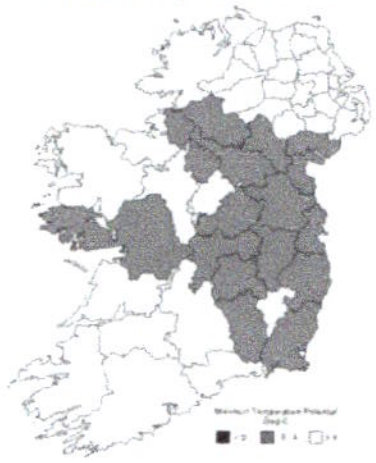

APRIL **23** WEDNESDAY

A high pressure ridge maintains a light south to southwest flow over the country, while low pressure systems linger in the North Atlantic Ocean. Cold and dry overnight with sharp frosts in some inland districts as well as scattered light fog in many districts. Dry, sunny and warmer than normal day everywhere.

Belfast: Warmer, sunny, dry, light southerly air, chance of misty patches and haze
Dublin: Sunny, dry, calm atmosphere, warmer, shallow fog patches likely
Cork: Sunny, dry, pleasant light air with patches of shallow fog likely
Galway: Very sunny, dry, warmer day cold overnight, freshening breezes

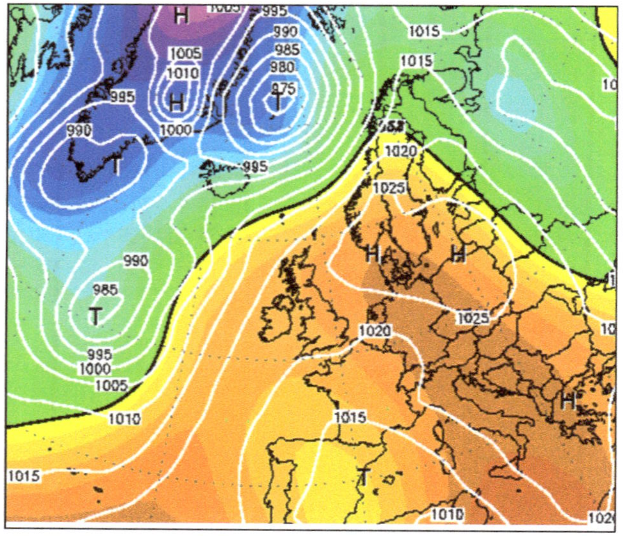

RAIN POTENTIAL

FROST/SNOW

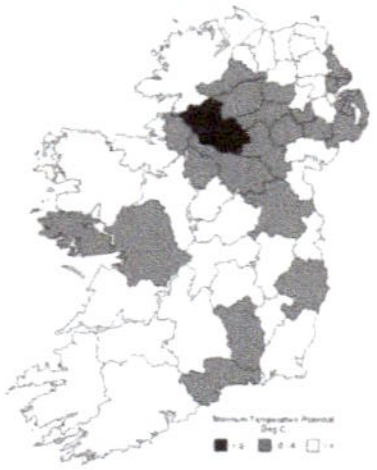

APRIL **24** THURSDAY

The high pressure ridge remains almost stationary over Ireland directing a light mostly southwest flow over the country with fresher breezes along the west and northern coasts. Warmer and dry everywhere. Patches of fog and mist in many districts overnight. Sunny in the north and east, while cloudier in the west and south.

Belfast: Sunny, warmer, dry, moderate breezes, chance of mist and haze patches
Dublin: Sunny, warmer, dry, gentle easterly flow, chance of mist in sheltered areas
Cork: Overcast, dry, odd threats of rain, milder, gentle breezes, chance of fog
Galway: Dry, warmer day cold night, cloudy spells, fresh breezes, occasional gusts

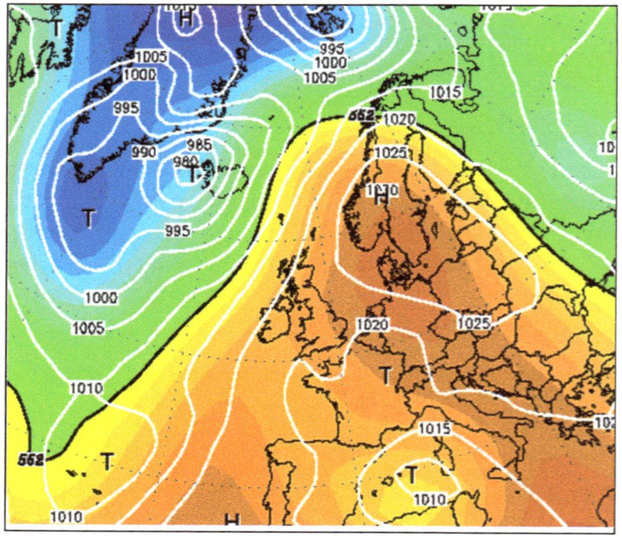

RAIN POTENTIAL

FROST/SNOW

APRIL **25** FRIDAY

Anticyclonic. Winds generallly light southwesterlies, fresher along the western coasts. Increased cloud throughout, and while most districts will be mainly dry, some isolated drizzly showers may move into the west of the country for a time, later drifting for a time into the southeast.

Belfast: Cloudy, dry, warm, gentle shifting breezes, chance of fog patches
Dublin: Partly cloudy, dry and mild, mostly calm air, mist patches possible
Cork: Calm and mostly dry air, passing scattered cloudy spells, fog likely, warmer
Galway: Calm air with odd breezy spell, mostly cloudy, colder day, warmer night, dry

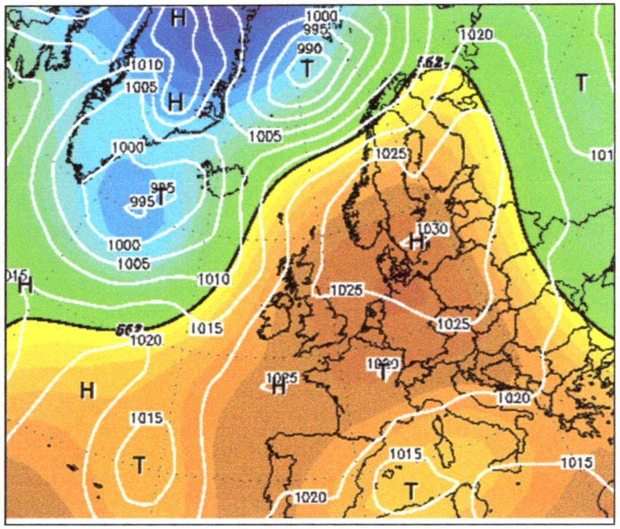

RAIN POTENTIAL

FROST/SNOW

APRIL **26** SATURDAY

High pressure stabilises once more over Ireland bringing another day with light northwesterly breezes throughout and further sunny and dry conditions in most districts. Patches of fog likely overnight particularly in the north and central districts. Chance of some frosts returning overnight in sheltered parts, and possible brief outbreaks of drizzle mostly in the east.

Belfast: Mostly sunny and dry, warm day cold night, calm atmosphere, fog likely
Dublin: Sunny and cooler, isolated showers, mostly calm air
Cork: Sunny, dry, pleasant temperatures, light breezes to mostly calm air
Galway: Sunny, dry, gentle breezes freshening later in the day

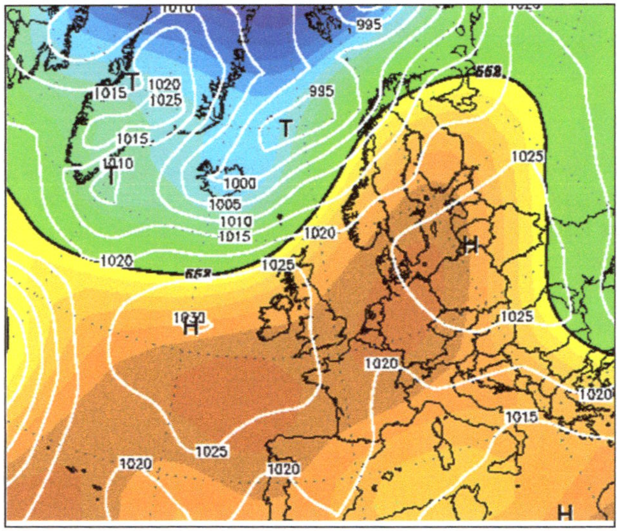

RAIN POTENTIAL

FROST/SNOW

APRIL **27** SUNDAY

An anticyclonic system strengthens as it lingers over Ireland bringing a mostly light northwesterly flow over the country, fresher at times along the northern coasts. Some frost patches likely overnight in inland parts, and possible light showers in the west and north, while remaining mostly dry elsewhere. Passing scattered cloud in all districts, but sunnier in the east and west.

Belfast: Partly cloudy, cooler, mostly dry, odd threat of rain, light shifting breezes
Dublin: Partly cloudy, dry, mild day cold overnight, moderate shifting breezes
Cork: Sunny, dry, light northerly air
Galway: Mostly sunny and dry, mild day cold night, fresh breezes at times

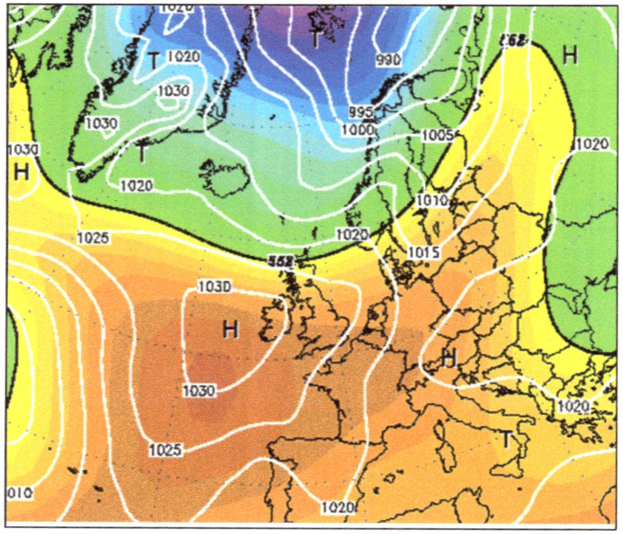

RAIN POTENTIAL

FROST/SNOW

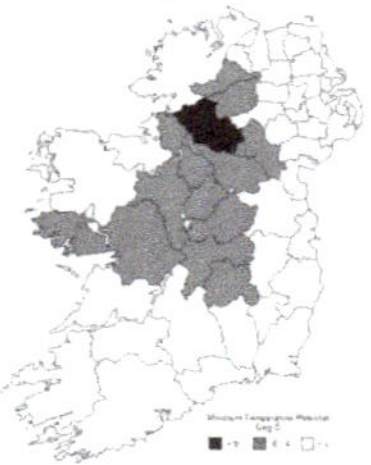

APRIL **28** MONDAY

Anticyclonic, with frost and fog patches overnight in many districts. Winds generally light and in variable directions. Very sunny and dry everywhere after fog lifts. Warmer than of recent, but cooler along the coasts and frosts returning after dark.

Belfast: Very bright and sunny, dry, cooler, calm atmosphere
Dublin: Very bright and sunny, dry, cold overnight, calm and serene atmosphere
Cork: Calm and pleasant air, very sunny, dry, mild temperatures
Galway: Sunny, dry, mild day, overnight frosts likely, calm air, chance of fog

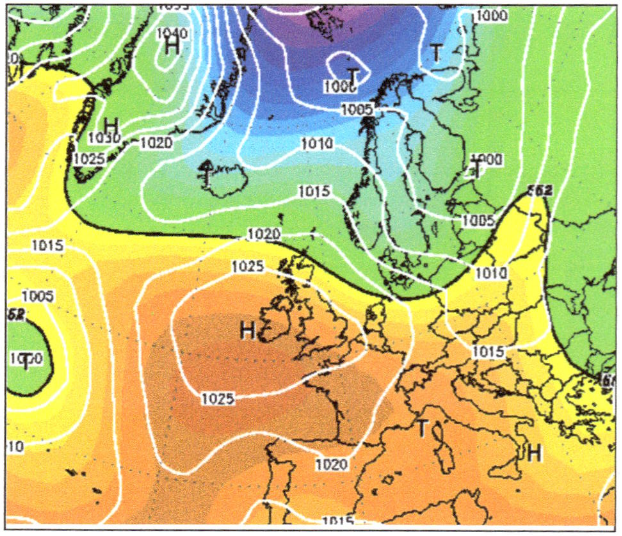

RAIN POTENTIAL

FROST/SNOW

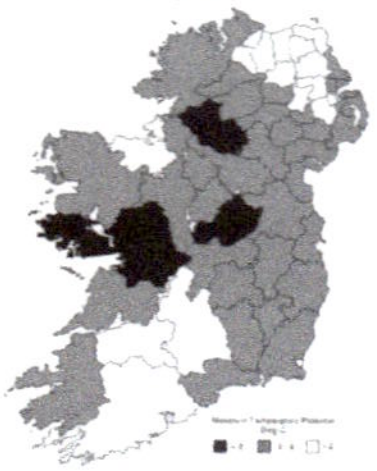

APRIL 29 TUESDAY

High pressure remains stalled over Ireland with gentle breezes prevailing in variable directions. Frost and fog patches once more overnight in many districts. Some light showers in the northwest of the country at first, clearing quickly, with a mostly dry and sunny day prevailing throughout.

Belfast: Mostly sunny, dry, light shifting breezes to mostly calm air
Dublin: Dry, sunny, mild temperatures, mostly calm air
Cork: Calm atmosphere, sunny, dry, patches of shallow fog likely
Galway: Mostly calm and pleasant air, partly cloudy, dry, possible patches of fog

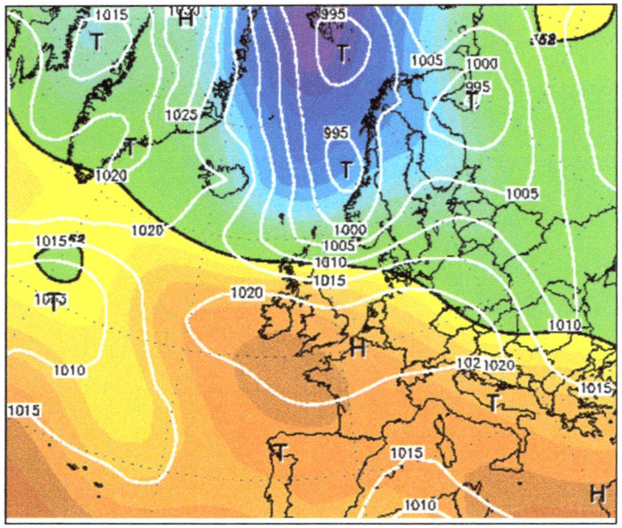

RAIN POTENTIAL

FROST/SNOW

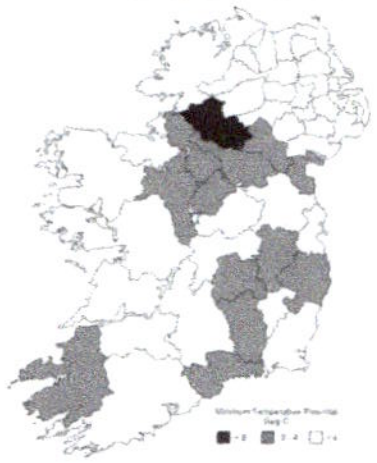

APRIL 30 WEDNESDAY

The large high that has maintained a steady influence over Ireland the past couple of weeks begins to weaken and drift away in a light to moderate south to southeast flow. Partly cloudy everywhere with the occasional threat of rain, and chance of some drizzly patches in the south and west for a time. Dry elsewhere, and sunniest in the east. Chance of fog and frost pockets overnight once more in the south, inland, and eastern parts.

Belfast: Partly cloudy, dry, cooler, gentle shifting breezes
Dublin: Partly cloudy, dry, mild day colder night, mostly calm air, fog patches likely
Cork: Partly cloudy, mostly dry, gentle breezes, fog likely
Galway: Cloudy, breezier, dry with increasing threats of rain, colder

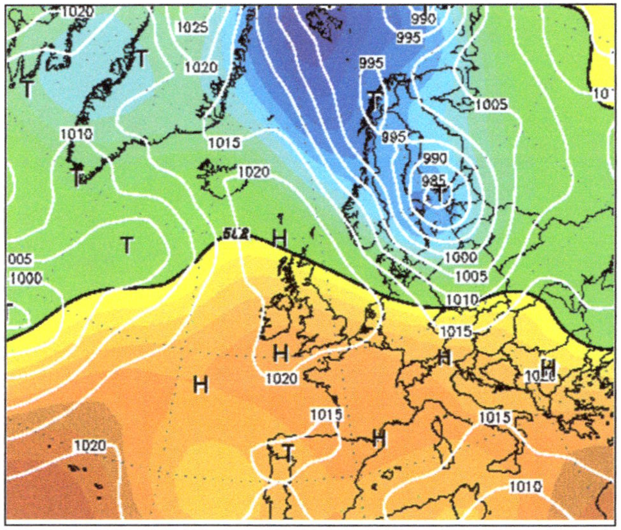

RAIN POTENTIAL

FROST/SNOW

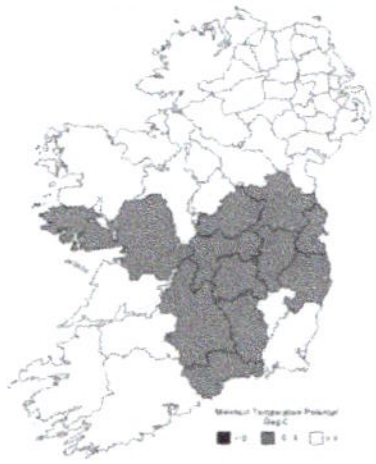

MAY

PHASES OF THE MOON

1st	Northern Declination
4th	First Quarter
9th	Crossing Equator
11th	Apogee
12th	Full Moon
15th	Southern Declination
20th	Third Quarter
23rd	Crossing Equator
26th	Perigee #5
27th	New Moon
28th	Northern Declination

1	2	3	4	5	6	7	8	9	10	11	12	13	14	15	16	17	18	19	20	21	22	23	24	25	26	27	28	29	30	31
Wed	Thur	Fri	Sat	Sun	Mon	Tue	Wed	Thur	Fri	Sat	Sun	Mon	Tue	Wed	Thur	Fri	Sat	Sun	Mon	Tue	Wed	Thur	Fri	Sat	Sun	Mon	Tue	Wed	Thur	Fri
^			1Q					XhS		A	F			V					3Q			XhN			P5	N	^			

MONTHLY SUMMARY

1st-5th High pressure is replaced by cloudy and changeable weather with frontal systems crossing the country bringing some rain or showers on most days. Some significant rain may occur at times, especially in the south and west. Chance of widespread fogs. Warmer than normal everywhere.

1st Very wet spell particularly in the northwest, with very heavy rain for a time likely in the Maam Valley region. Despite this, temperatures will be unusually milder.

2nd Scattered thunderstorms likely. Chance of funnel formation winds in some inland parts.

3rd Very strong winds in the northwest. Some heavy rain possible overnight mostly in the southwest.

4th-5th Spell of very heavy rain and scattered thunderstorms, especially in the west of the country.

5th-7th Cold overnight with frost patches here and there, especially in inland parts.

6th Very bright and sunny day.

6th-10th High pressure builds over Ireland, bringing a mostly settled period of dry and sunny conditions throughout with well above average temperatures in the midlands and western parts of the country, while cooler in the east.

8th Very warm day.

9th-10th A showery trough spreads a band of thunderstorms cross the south of the country, later moving into the west.

Over the next four weeks the highest pressure will prevail for the first week, and the lowest pressure over the last week and the weather won't be as settled as it was over the previous four weeks, despite a settled spell with the high pressure at the start of this outlook period. Overall, it will be drier than average for this time of year, warmer and sunnier than normal. Winds will be close to normal with the strongest winds accompanying a deep depression towards the middle of the third week. Many parts of the south and east of Ireland will be drier than normal, while the north will be closer to their norms, and the west slightly wetter than average. The first week is likely to be mostly dry, with every other remaining day during this outlook period likely to record some rain with the heaviest rain expected over the second week. It will be very sunny over the first half of this outlook period, with many areas being much sunnier than has been the norm over the past few years. The sunniest periods will be at the start of this outlook period, the end of the third week, and towards the end of the fourth week. Maximum temperature will be slightly above the average during this lookahead period, with parts of the southeast up to 2°C more than average but inland parts only warmer by around half a degree above their average. Overnight temperatures however, may still see some unusually late frosts in the first couple of weeks as well as the odd frost in the last week of this outlook period. It will be particularly warm at the start of this outlook period, while the lowest temperatures overnight will arrive in the last few days. During this outlook period, there will be around 10-15 days of rain in the east and south but up to 15-20 rain days elsewhere. Thunderstorms are likely on 3 days and hail on 5 days. Fogs can be expected on 9 days while in contrast gale winds are likely on up to 7 days. Frosts can be expected on between 9-13 days, which is almost double what is typical for this time of year.

11th-15th Anticyclonic and a light easterly breeze prevails. Very dry and settled conditions prevail. Warm and sunny particularly in the western districts to the midlands where up to 10 hours of sunshine may be recorded each day, slightly cooler in the east. Cooler temperatures overnight. Chance of cloud building later in the last couple of days bringing some showers into the northern districts.

13th-15th Scattered fogs likely.

16th-30th A series of Atlantic frontal systems cross the country bringing outbreaks of rain and showery spells each day, especially in the north and western parts of the country.

16th-20th Strong blustery winds prevail, possibly up to gale force at times.

17th-19th Very wet spell.

23rd Heavy rain likely.

24th Chance of thunderstorms.

26th-28th Chance of fog patches.

26th-30th A wet front embedded with thunderstorms cross the country bringing a particularly wet spell in the northwest of the country.

27th Very warm day throughout the country.

28th-29th A deep depression near the northwest of Ireland directs very strong southwest winds across the country, with gale gusts throughout. Chance of thunderstorms and hail showers, and outbursts of heavy rain.

29th-4th June A ridge of high pressure moves onto the country in a light west to southwest flow, and returning very sunny and mostly dry conditions everywhere, with up to 10 hours or more of sunshine many day. Chance of some briefly along the western and northern coasts.

31st Cloud begins to rebuild over the country.

IRELAND RAINFALL ESTIMATES
May 2025

d=mainly dry, n= nonrecordable, l=light shrs, s=significant shrs, r=rain, h=hvy falls

		ULSTER PROVINCE					CONNAUGHT PROVINCE						LEINSTER PROVINCE				MUNSTER PROVINCE						
		Alderg	Hillsbo	Armag	Loughl	Malin H	Sligo	Ardtar	Belmul	Drums	Galway	Roscoi	Markre	Edende	Kilkenn	Dublin	Shann	Newpo	Tralee	Killarne	Cork	All	Moon
1st	Thur	n	l	d	s	s	s	s	h	s	s	l	l	n	l	d	d	h	s	l	d	73	^
2nd	Fri	l	l	l	l	l	l	l	d	l	s	s	l	l	l	l	n	s	s	r	l	64	
3rd	Sat	l	l	l	l	l	l	l	s	s	r	s	l	l	s	l	l	l	h	r	h	98	
4th	Sun	s	r	l	h	s	r	h	l	r	s	r	r	s	r	r	s	h	s	r	l	207	1Q
5th	Mon	s	d	r	l	r	l	l	h	l	d	l	s	l	d	l	l	l	r	d	l	83	
6th	Tues	d	d	d	d	d	d	d	d	d	d	d	d	d	d	d	d	d	d	d	d	0	
7th	Wed	d	d	d	d	d	d	d	d	d	d	d	d	d	d	d	d	d	d	d	d	0	
8th	Thur	d	d	d	d	d	d	d	d	d	d	d	d	d	d	d	d	d	d	d	d	0	
9th	Fri	d	d	d	d	d	d	d	d	d	l	d	d	l	s	d	d	d	n	d	l	9	XhS
10th	Sat	d	d	d	d	d	d	d	n	d	d	d	d	d	l	d	d	d	d	s	l	8	
11th	Sun	d	d	d	d	d	d	d	d	d	d	d	d	d	d	d	l	d	d	d	d	1	A
12th	Mon	d	d	d	d	d	d	d	d	d	d	d	d	d	d	d	d	d	d	d	d	0	F
13th	Tues	d	d	d	d	d	d	d	d	d	d	d	d	d	d	d	d	d	d	d	d	0	
14th	Wed	d	d	d	d	d	d	d	d	d	d	d	d	d	d	d	d	d	d	d	d	0	
15th	Thur	d	d	d	l	l	l	l	l	l	l	l	l	d	d	d	l	s	d	l	d	22	V
16th	Fri	d	d	d	l	s	l	l	l	l	s	l	n	l	l	l	l	l	s	l	s	42	
17th	Sat	l	s	l	s	r	r	s	s	s	r	l	r	l	l	l	s	r	l	l	l	108	
18th	Sun	s	s	s	l	s	l	l	l	l	l	l	s	d	d	l	l	l	s	l	l	57	
19th	Mon	s	s	l	s	s	r	s	s	s	r	r	s	s	l	l	s	r	s	l	s	135	
20th	Tues	s	s	s	l	l	l	l	s	s	l	s	l	n	d	l	l	l	l	s	d	53	3Q
21st	Wed	l	l	l	l	d	l	l	l	l	s	l	l	r	r	l	r	s	s	s	s	80	
22nd	Thur	s	s	l	l	d	l	l	l	s	s	l	l	l	l	d	d	l	d	d	l	39	
23rd	Fri	l	d	l	d	d	d	d	d	d	l	d	l	d	l	d	d	n	d	d	s	14	XhN
24th	Sat	l	l	d	l	l	l	l	l	s	s	s	s	s	s	l	s	s	s	s	l	74	
25th	Sun	l	l	s	s	d	s	s	n	s	l	l	r	l	d	l	l	l	l	l	l	52	
26th	Mon	l	s	l	h	h	h	h	r	r	s	r	r	s	l	s	s	h	s	s	l	201	P5
27th	Tues	l	l	s	l	s	l	l	l	l	s	s	l	l	l	d	l	l	l	s	d	47	N
28th	Wed	l	s	l	r	s	r	r	s	r	r	s	s	l	l	l	s	r	l	s	l	124	^
29th	Thur	l	d	l	l	r	l	l	s	l	l	l	l	l	d	l	l	s	l	l	n	48	
30th	Fri	l	d	l	l	r	s	l	l	l	d	l	l	d	d	d	d	l	l	l	s	42	
31st	Sat	d	d	d	n	n	d	n	l	n	d	n	d	n	d	d	d	s	d	d	d	8	
	Estimate:	55	57	53	94	121	109	94	114	103	104	94	97	54	46	36	56	155	92	85	65	1687	
	Average:	57	59	58	68	58	81	83	70	71	75	82	81	69	60	60	65	95	89	59	82	1423	
	Trend:	av	av	av	w tr	w tr	w tr	w tr	w tr	w tr	w tr	w tr	w tr	drr	drr	drr	av	w tr	av	w tr	drr	w tr	

IRELAND SUNSHINE ESTIMATES

May 2025

F=fine (8-12 hours of sunshine), pc= partly cloudy (4-7 hours), c=cloudy (1-3 hours), o=overcast (0 hours)

		ULSTER PROVINCE					CONNAUGHT PROVINCE						LEINSTER PROVINCE				MUNSTER PROVINCE						
		Alderg	Hillsbo	Armag	Loughl	Malin H	Sligo	Ardtari	Belmul	Drums	Galway	Roscoi	Markre	Edende	Kilkenn	Dublin	Shannc	Newpo	Tralee	Killarne	Cork	All	Moon
1st	Thur	o	o	o	o	o	o	o	o	o	o	o	o	o	o	o	c	c	o	o	c	9	^
2nd	Fri	o	o	o	o	o	o	o	o	c	o	o	o	c	c	c	o	o	o	o	o	13	
3rd	Sat	o	o	o	o	o	o	o	o	o	o	o	o	o	o	o	o	o	o	o	o	1	
4th	Sun	c	c	c	c	c	pc	c	F	c	pc	pc	c	c	c	c	o	o	pc	pc	o	68	1Q
5th	Mon	pc	pc	pc	pc	pc	c	pc	pc	pc	F	pc	pc	pc	F	pc	F	F	F	F	F	135	
6th	Tues	F	F	F	F	F	F	F	F	F	F	F	F	F	F	F	F	F	F	F	F	246	
7th	Wed	F	F	F	F	F	F	F	F	F	F	F	F	F	F	F	F	F	F	F	F	210	
8th	Thur	F	F	F	F	F	pc	F	pc	F	pc	pc	F	F	F	F	F	F	pc	pc	pc	152	
9th	Fri	pc	pc	pc	pc	pc	c	pc	c	pc	o	o	pc	pc	pc	F	o	c	c	c	o	84	XhS
10th	Sat	F	F	F	F	F	F	F	F	F	pc	pc	F	F	pc	F	c	pc	o	o	o	153	
11th	Sun	F	F	F	F	F	F	F	F	F	F	F	F	F	F	F	F	F	pc	pc	F	233	A
12th	Mon	F	F	F	F	F	F	F	F	F	F	F	F	F	F	F	F	F	F	F	F	275	F
13th	Tues	F	F	F	F	F	F	F	F	F	F	F	F	F	F	F	F	F	F	F	F	277	
14th	Wed	F	F	F	F	F	F	F	F	F	F	F	F	F	F	F	F	F	F	F	F	233	
15th	Thur	c	c	c	c	c	o	c	o	pc	o	o	c	pc	pc	pc	o	c	o	o	pc	58	V
16th	Fri	c	c	c	c	c	c	c	pc	c	c	pc	c	c	o	c	o	o	o	o	o	40	
17th	Sat	pc	pc	pc	pc	pc	F	pc	pc	pc	pc	pc	pc	pc	F	pc	pc	pc	pc	pc	F	154	
18th	Sun	pc	pc	pc	pc	pc	pc	pc	F	pc	pc	pc	pc	pc	c	c	c	c	o	o	c	89	
19th	Mon	c	c	c	c	c	c	c	pc	c	pc	pc	c	c	c	c	c	c	c	c	c	56	
20th	Tues	pc	pc	pc	pc	pc	pc	pc	F	pc	F	F	pc	pc	pc	pc	F	pc	pc	pc	pc	145	3Q
21st	Wed	c	c	c	c	c	c	c	c	c	c	c	c	c	c	c	c	c	pc	pc	c	66	
22nd	Thur	c	c	c	c	c	o	c	c	pc	pc	c	c	pc	pc	pc	pc	pc	F	F	F	94	
23rd	Fri	pc	pc	pc	pc	pc	pc	pc	F	c	pc	F	pc	c	o	c	c	c	c	c	o	80	XhN
24th	Sat	F	F	F	F	F	pc	F	pc	F	pc	pc	F	F	F	F	pc	pc	pc	pc	F	143	
25th	Sun	o	o	o	o	o	o	o	pc	c	c	c	o	c	c	o	c	c	o	o	pc	25	
26th	Mon	o	o	o	o	o	o	o	o	c	o	o	o	c	c	o	o	c	o	o	pc	15	P5
27th	Tues	pc	pc	pc	pc	pc	c	pc	o	pc	c	o	pc	pc	F	pc	c	pc	o	o	F	95	N
28th	Wed	pc	pc	pc	pc	pc	pc	pc	pc	F	F	pc	pc	F	F	pc	F	F	pc	pc	F	163	^
29th	Thur	F	F	F	F	F	F	F	F	F	F	F	F	F	F	F	F	F	F	F	F	220	
30th	Fri	F	F	F	F	F	F	F	F	F	F	F	F	F	F	F	F	F	F	F	F	225	
31st	Sat	F	F	F	F	F	F	F	F	F	F	F	F	F	F	F	F	F	F	F	F	268	
Estimate hours:		203	203	203	203	203	194	203	215	206	203	209	203	206	209	212	192	201	176	176	206	4024	
Average hours:		189	189	165	191	192	189	188	190	153	180	183	172	176	170	189	179	168	180	180	187	3611	
Trend:		more	more	more	more	more	av	more	more	more	more	more	more	more	more	more	more	more	av	av	more	more	

MAY **1** THURSDAY

A depression to the west of Ireland advances onto the country bringing a southwest change throughout, with winds light to moderate, and fresher at times along the western coasts. Mostly cloudy everywhere, with outbreaks of rain developing in the northwest and first and later reaching the southeast, while the eastern coasts remain mostly dry. Chance of fog overnight mostly in southern parts.

Belfast: Breezy spells, overcast, warmer, drizzly showers, chance of mist or haze
Dublin: Overcast, warmer, dry with occasional threats of rain, moderate breezes
Cork: Cloudy, dry, mild temperatures, gentle breezes, chance of overnight fog
Galway: Overcast and dull, occasional showers, moderate winds, occasional gusts

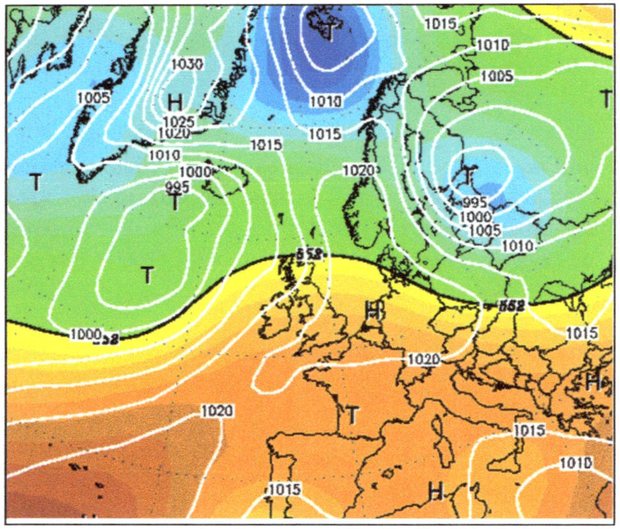

RAIN POTENTIAL

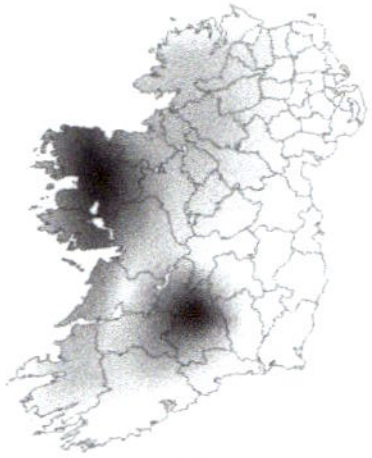

FROST/SNOW

MAY **2** FRIDAY

A fresh to strong southwest flow prevails over Ireland as a depression builds to the north of the country. Widespread cloud and patchy drizzle outbreaks in most districts develop, with showers more frequent in inland to western parts. Some brief sunny spells possible in the afternoon, mostly in the east. Pockets of fog possible overnight.

Belfast: Dull and gloomy, windy at times, light showers and passing drizzle outbreaks
Dublin: Cloudy, warmer, brief showers, windy spells
Cork: Gloomy air, passing showers, freshening winds, chance of overnight fog
Galway: Overcast, cooler, scattered showers, fresh winds with occasional gusts

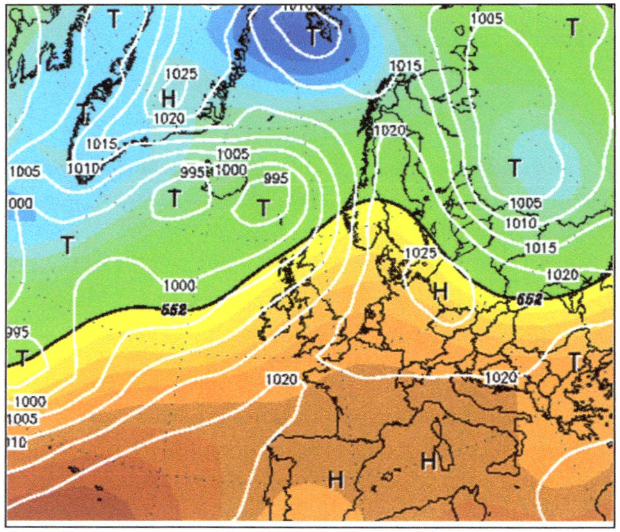

RAIN POTENTIAL

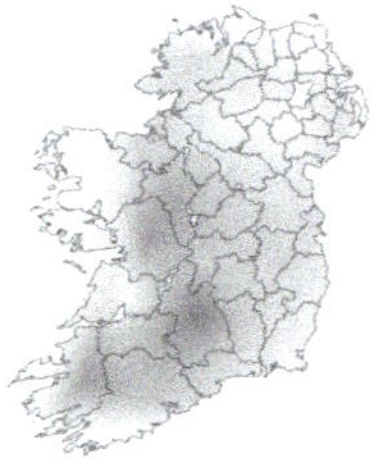

FROST/SNOW

MAY 3 SATURDAY

A depression consolidates to the west of Ireland bringing fresh to strong south to southwest winds throughout, gusty and up to gale force at times likely around the western coasts. A stormy air prevails everywhere, with widespread cloud and outbreaks of rain almost everywhere. Rain likely to be very heavy for a time in the south of the country.

Belfast: Warmer, overcast, light showers, fresh to strong breezes at times
Dublin: Overcast, isolated showers, cooler, fresh winds, gusts likely
Cork: Stormy air, overcast, heavy rain, blustery at times
Galway: Stormy air, overcast, rain, strong blustery winds

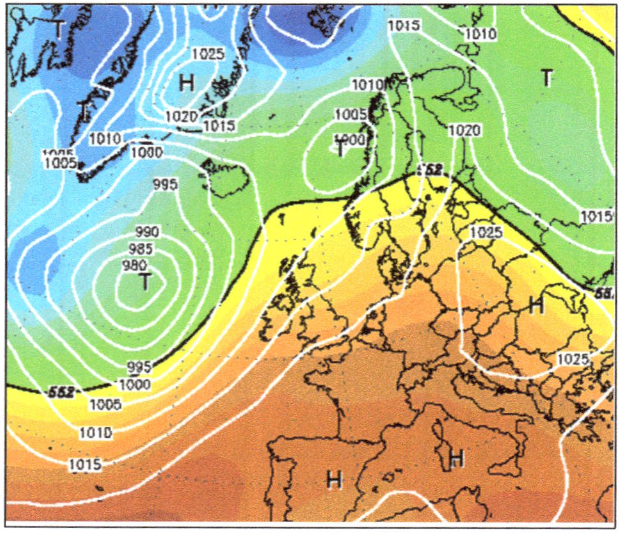

RAIN POTENTIAL

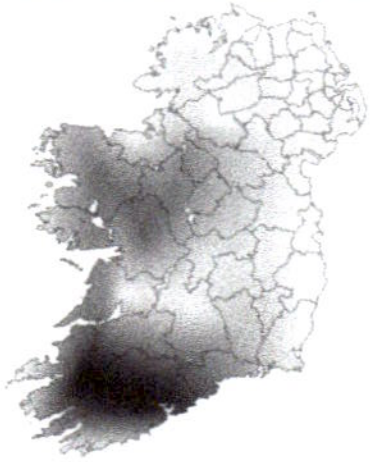

FROST/SNOW

MAY 4 SUNDAY

A depression to the northwest of Ireland intensifies bringing a stormy air across the country for a time, with southwest winds prevailing, easing later with a west to northwest change. Cloudy to mostly overcast everywhere with widespread outbreaks of rain and drizzle. Rain likely to be heavy in the northwest to the midlands for a time. Chance of fog overnight in some southern parts.

Belfast: Cloudy, rain, breezy with occasional gusts
Dublin: Cloudy, warmer, rain, strong blustery winds at times
Cork: Overcast, cool, patches of light rain and drizzle, fresh winds at times
Galway: Unsettled winds, mix of showers and sunny spells, slightly warmer

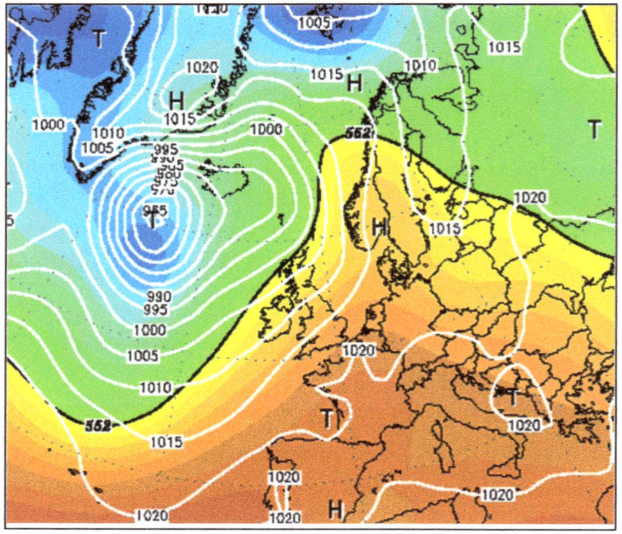

RAIN POTENTIAL

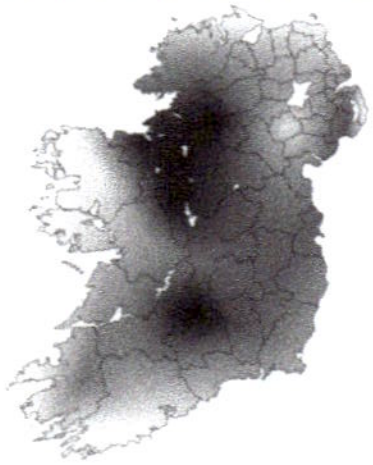

FROST/SNOW

MAY 5 MONDAY

A weak low pressure centre crosses the country, with light variable winds prevailing, Partly cloudy in the north and east, while mostly sunny in the south and west. Some rain at first overnight, with heavy falls at times along the western coastal fringes, and lighter showers elsewhere, clearing later to mostly dry throughout. Chance of fog developing after dark.

Belfast: Partly cloudy, dry, slightly cooler, light shifting breezes, chance of fog
Dublin: Partly cloudy, brief morning showers, cooler, unsettled winds at times
Cork: Sunny, cool, fluctuating winds, brief morning showers, chance of evening fog
Galway: Dry, sunny, cooler, shifting breezes

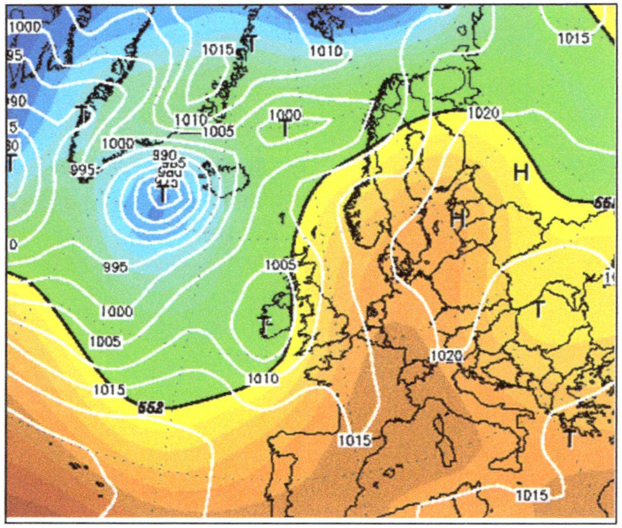

RAIN POTENTIAL

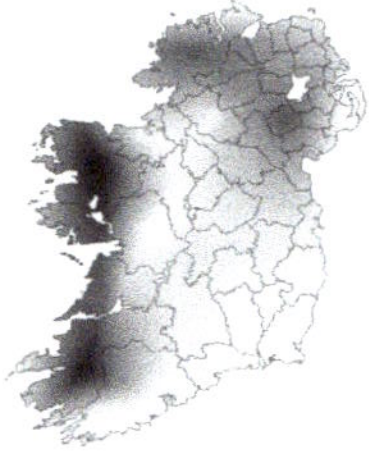

FROST/SNOW

MAY 6 TUESDAY

A weak high pressure ridge moves up over the country bringing light winds of variable direction. Very cold and dry overnight with frost and fog in some places, some patches of frost likely to be severe. Very sunny and dry throughout the day with warmer temperatures, and with evening cloudy patches moving in towards evening mainly in the southeast.

Belfast: Very sunny, dry, cold overnight, light shifting breezes, chance of morning fog
Dublin: Sunny, cool, dry, light shifting breezes, chance of fog patches
Cork: Sunny and dry, mild day cool night, gentle breezes, possible patches of fog
Galway: Very sunny, dry, mild day cold night, possible frost, shifting breezes

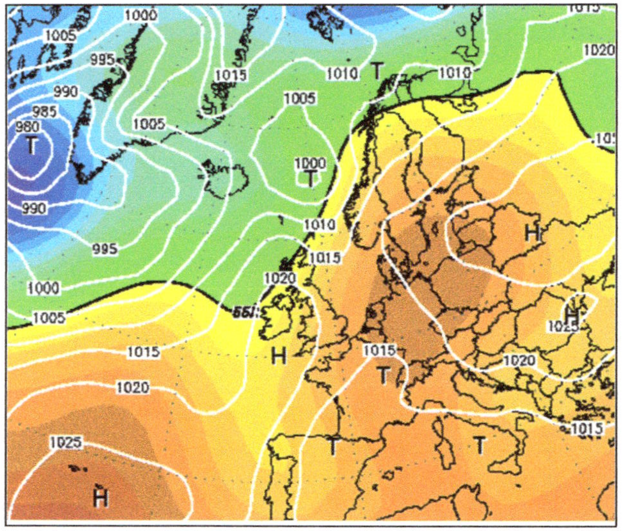

RAIN POTENTIAL

FROST/SNOW

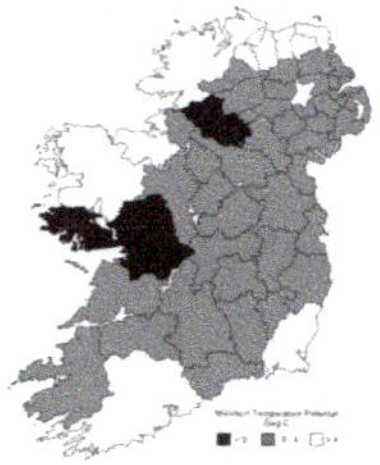

MAY **7** WEDNESDAY

Anticyclonic. Light variable breezes to a mostly calm and serene atmosphere prevail throughout. Sunny and dry everywhere. Some patches of frost overnight. Warmer almost everywhere during the day.

Belfast: Sunny, warmer, dry, calm and pleasant air
Dublin: Sunny and dry, gentle breezes
Cork: Sunny and dry, slightly warmer, mostly calm air
Galway: Light breezes to mostly calm air, very sunny, warmer, dry

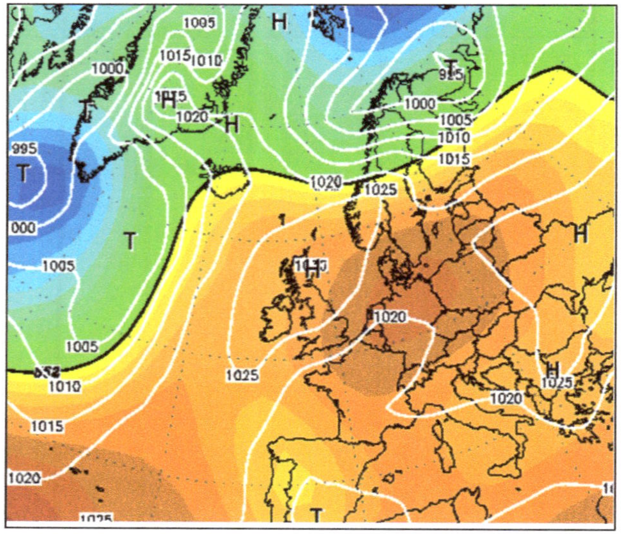

RAIN POTENTIAL

FROST/SNOW

MAY **8** THURSDAY

High pressure prevails over Ireland with light east to southeast breezes prevailing. Sunny and dry everywhere, with milder temperatures and no frosts expected. Chance of some patches of fog overnight, mostly in the west.

Belfast: Sunny, dry, gentle breezes to mostly calm air
Dublin: Sunny, dry, mild temperatures, gentle southeasterly breezes
Cork: Partly cloudy, dry, gentle breezes
Galway: Mostly sunny, dry, warmer day cool night, light breezes, chance of mist

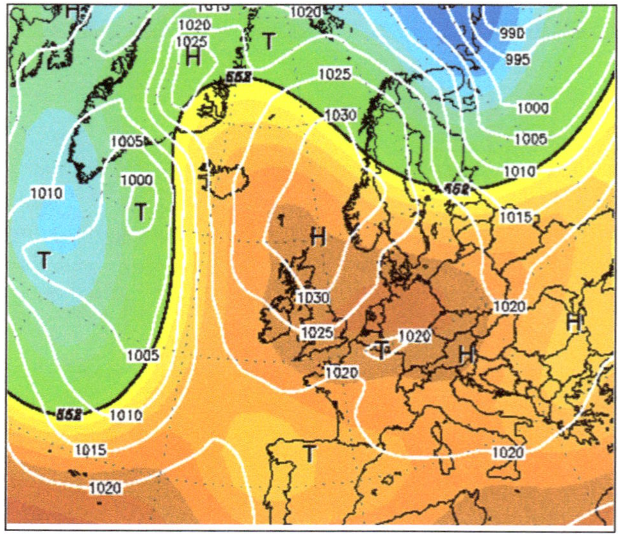

RAIN POTENTIAL

FROST/SNOW

MAY **9** FRIDAY

A high pressure ridge is maintained over the country while a weak depression lingers to the west of Ireland. Partly cloudy and dry in the north, with increased cloud to the south, becoming overcast and dull in the west and south. Some showers likely at first in the southeast and south, possibly accompanied by thunderstorm activity. Isolated showers likely to move later into the west for a brief time.

Belfast: Partly cloudy, slightly warmer, dry, moderate breezes
Dublin: Mild temperatures, dry, sunny, moderate breezes
Cork: Overcast and dull, light drizzly showers, cooler, fresh onshore breezes likely
Galway: Overcast, cooler, isolated showers, fresh breezes

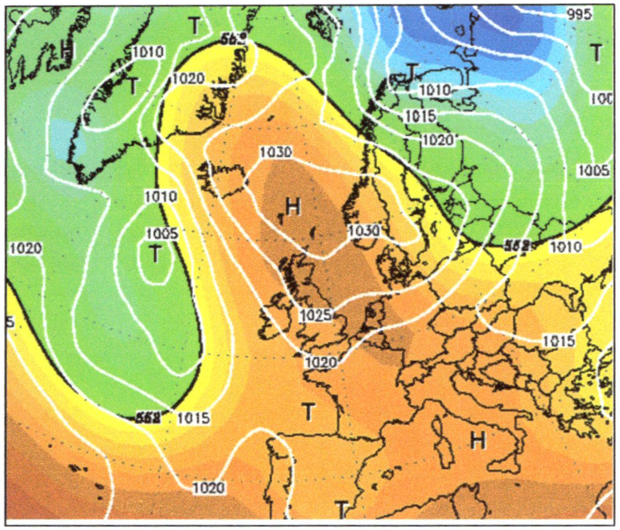

RAIN POTENTIAL

FROST/SNOW

MAY **10** SATURDAY

The high pressure ridge contracts to the north of Ireland while a low pressure trough moves over the country with winds generally light to moderate easterly in flow, with a mostly calm atmosphere later developing. Sunny and dry in the north and east, with increasing cloud in the west and south. A showery trough may briefly bring some isolated showers into the south and east, possibly accompanied by thunderstorms.

Belfast: Sunny, dry, moderate breezes
Dublin: Very sunny and dry, mild temperatures, moderate breezes
Cork: Overcast, cool, mostly dry with odd drizzle patch, chance of fog patches
Galway: Partly cloudy, dry, moderate breezes, chance of misty patches overnight

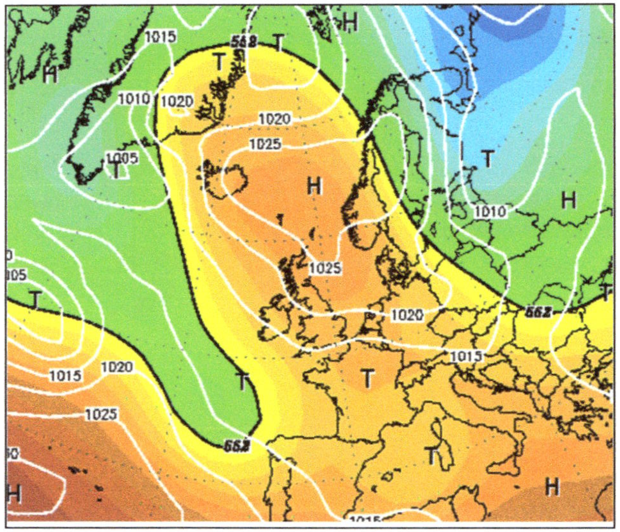

RAIN POTENTIAL

FROST/SNOW

MAY **11** SUNDAY

A low pressure trough lingers over Ireland maintaining a light to moderate east to southeasterly flow. Very bright and sunny throughout and mostly dry, except for the chance of some spotty drizzly spells possible overnight in southern parts

Belfast: Very bright and sunny, warm day cool night, dry, mostly calm air
Dublin: Very bright and sunny, dry, mild temperatures, gentle shifting breezes
Cork: Sunny, warmer, dry, light breezes to mostly calm air, chance of fog overnight
Galway: Very sunny, dry, warm day cool night, moderate breezes

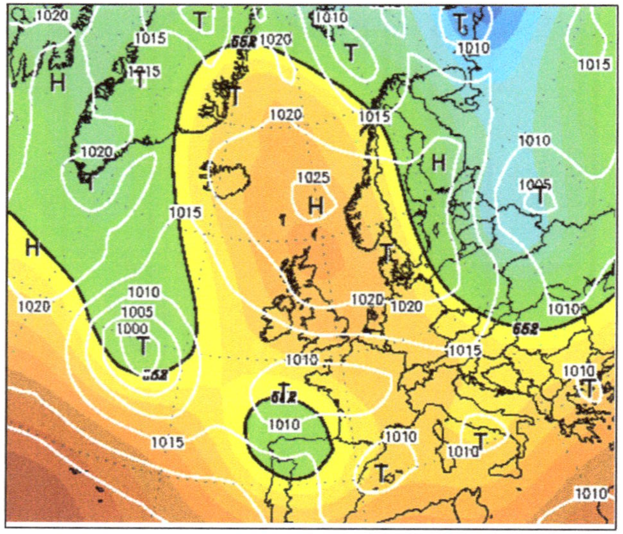

RAIN POTENTIAL

FROST/SNOW

MAY **12** MONDAY

A low pressure trough lingers over the southern half of the country while a weak high pressure ridges over the northern half. Sunny and dry throughout the country, with light east to northeast breezes prevailing, a little fresher at times around the coastal fringes. Chance of misty patches overnight in central parts.

Belfast: Pleasant and warm, dry, very sunny, gentle breezes
Dublin: Pleasant and dry, very sunny, mild day cool night, moderate breezes
Cork: Very sunny, dry, warmer, gentle breezes
Galway: Very sunny and dry, warm day cool night, moderate breezes

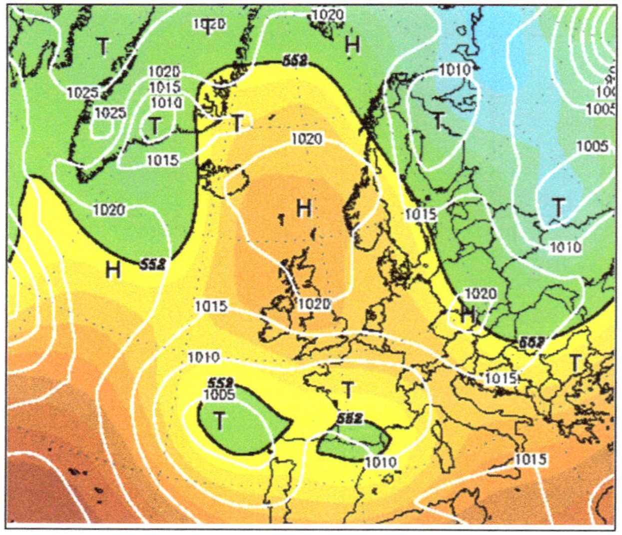

RAIN POTENTIAL

FROST/SNOW

MAY **13** TUESDAY

Anticyclonic conditions prevail over Ireland bringing light breezes to a mostly calm and pleasant air throughout. Warm, dry, and very sunny everywhere. Chance of frost patches overnight in central parts, and scattered fog in the west and north.

Belfast: Hot day cool night, very sunny, dry, calm and pleasant air
Dublin: Very bright and sunny, dry, mild day cold night, gentle breezes
Cork: Warm day cool night, very bright and sunny, dry, calm and pleasant air
Galway: Warm and pleasant air, very sunny, dry, mostly calm, chance of shallow fog overnight

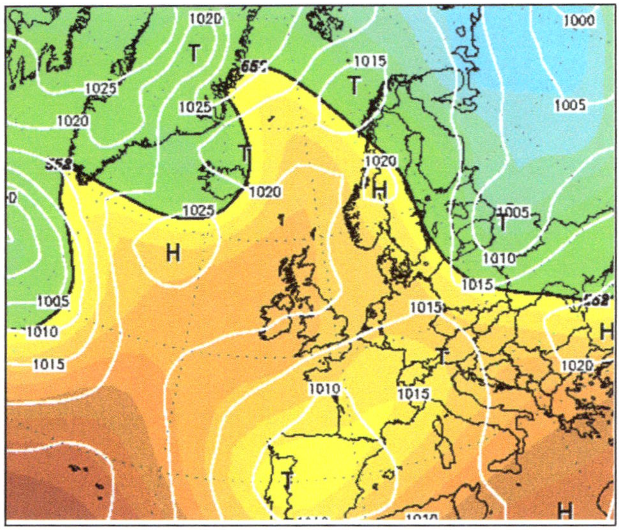

RAIN POTENTIAL

FROST/SNOW

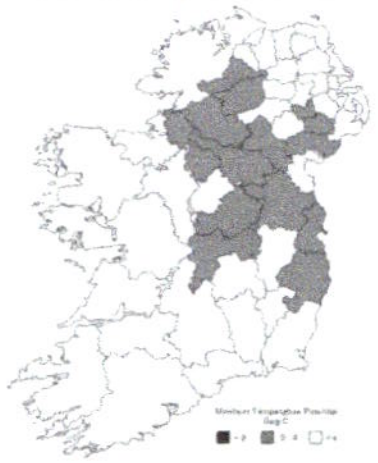

MAY **14** WEDNESDAY

Dry, sunny, and pleasant conditions continue to prevail as an anticyclonic system lingers over Ireland, directing light variable breezes throughout. Chance of fog patches overnight in the north and west, and frost patches in inland parts. Warmer than normal in the afternoon.

Belfast: Sunny, dry, gentle breezes, very hot day cool night, chance of overnight fog
Dublin: Sunny and dry, cooler, calm and pleasant air
Cork: Very sunny, bright and dry, warm day cool night, calm and serene air
Galway: Sunny, warm, dry, mostly calm air, chance of shallow fog in places

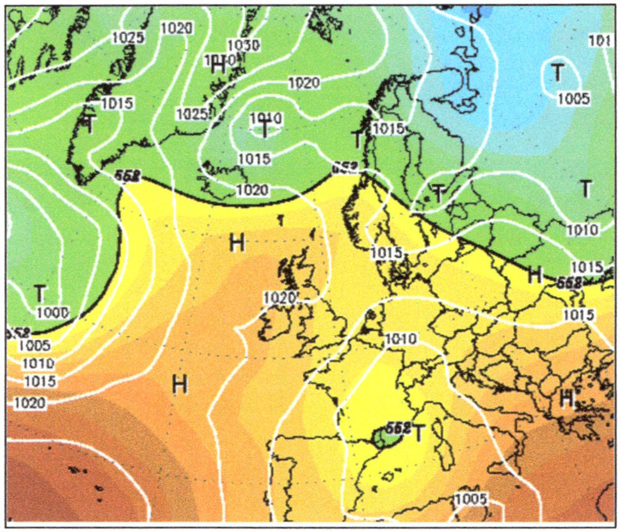

RAIN POTENTIAL

FROST/SNOW

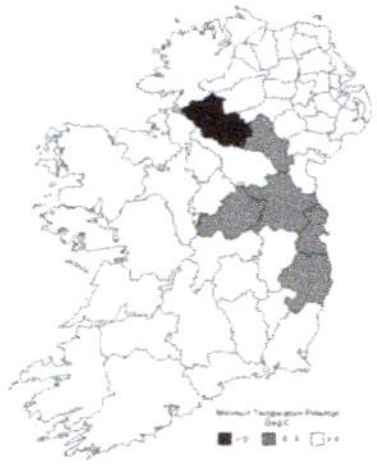

MAY **15** THURSDAY

A complex of high and low pressure systems surround Ireland with a very light and variable flow over the country, later freshening with a south to southwest change. Generally cloudy to overcast throughout, except for some sunnier spells in the south. Patches of overnight fog in many districts, as well as possible frost paches in some inland parts. Dry at first before showers move into the western half of the country later in the day.

Belfast: Cloudy, dry, cooler, calm atmosphere, fog patches possible
Dublin: Warmer day cold night, scattered cloudy spells, dry, calm air, chance of fog
Cork: Partly cloudy, dry, cooler, mostly calm air, chance of fog
Galway: Cooler, overcast and dull, light showers, calm air, chance of misty patches

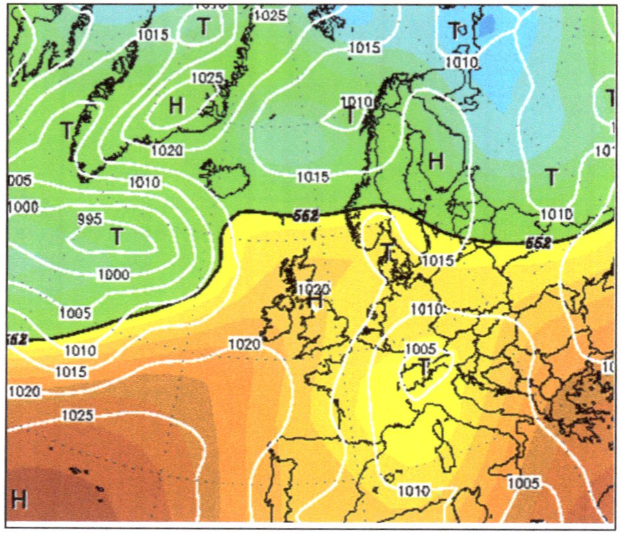

RAIN POTENTIAL

FROST/SNOW

MAY **16** FRIDAY

Unsettled and blustery west to southwest winds spread across Ireland bringing a cloudy and stormy air throughout, with winds to gale force at times. The depressions to the north of the country bring scattered rain activity throughout, crossing the northern coasts at first before spreading down through the country as the day progresses.

Belfast: Freshening winds with occasoinal gusts, cloudy, dry, odd threat of rain
Dublin: Cloudy, cooler, brief showers, unsettled winds, chance of gale gusts at times
Cork: Overcast, colder, scattered showers, strong winds with chance of gale gusts
Galway: Cloudy, cool, occasional showers, unsettled winds, chance of gales at times

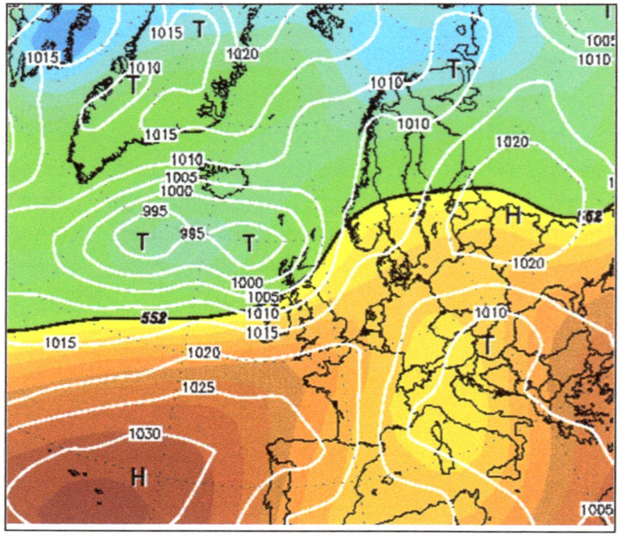

RAIN POTENTIAL

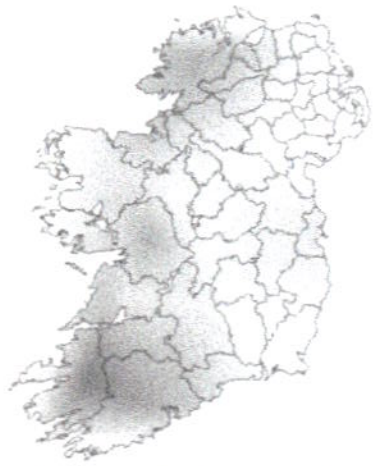

FROST/SNOW

MAY 17 SATURDAY

A couple of depressions linger to the north of Ireland while high pressure ridges to south. An unsettled and changeable day throughout, with a mix of sunny spells and widespread showers, some heavy at times particularly in the north. Winds generally fresh to strong in southerly flow, with chance of reaching to gales and gusts at times.

Belfast: Mostly sunny, cooler, blustery at times, occasional showers
Dublin: Partly cloudy, strong westerlies, brief showers
Cork: Unsettled and blustery at times, sunny, some brief morning showers likely
Galway: Wet and windy, cool, occasional sunny spells

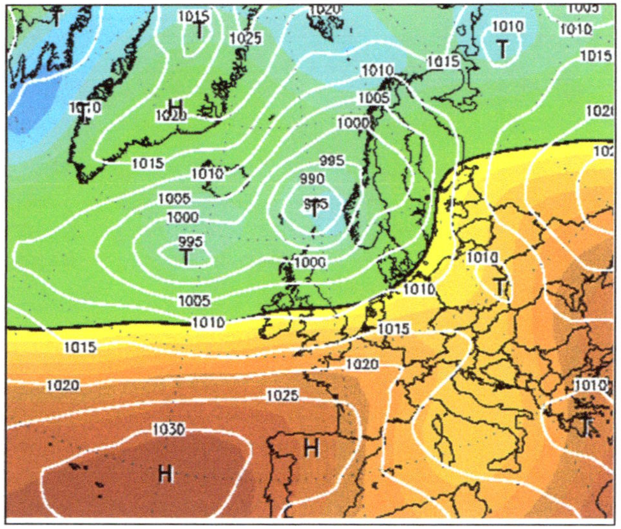

RAIN POTENTIAL

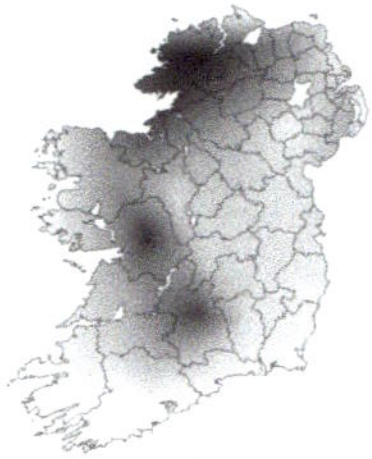

FROST/SNOW

MAY 18 SUNDAY

Two low pressure centres merge into one large depression centred o the northeast of Ireland bringing fresh to strong and occasionally gusty westerly winds. Another unsettled day with a mix of scattered showers, particularly in the west overnight, and a few sunny breakthroughs, as well as the chance of some heavy showers for a time in inland areas.

Belfast: Scattered cloudy spells, passing showers, blustery with chance of gale gusts
Dublin: Strong and blustery winds, mostly cloudy, light showers
Cork: Strong winds with occasional gusts, cloudy, light showers, cooler
Galway: Partly cloudy, occasional showers, blustery winds, chance of gales at times

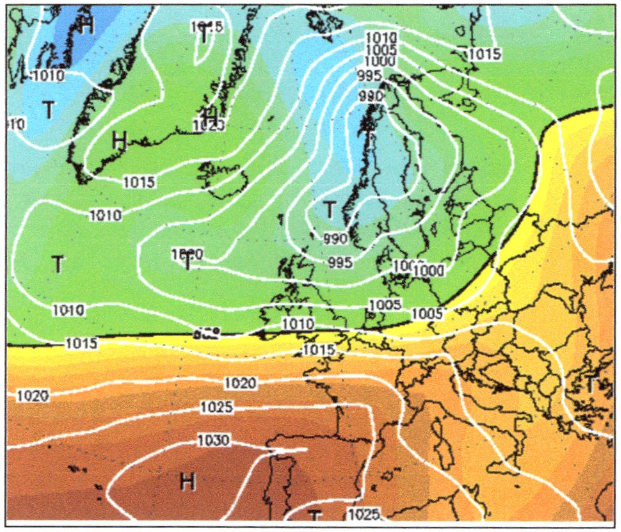

RAIN POTENTIAL

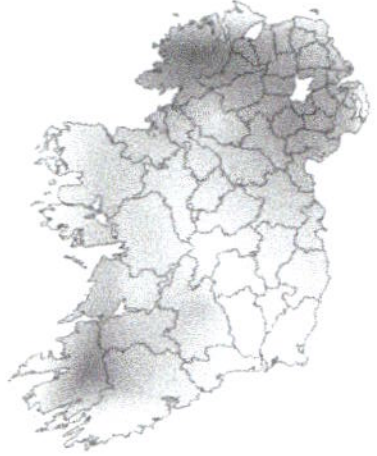

FROST/SNOW

MAY **19** MONDAY

A depression to the west of Ireland extends a wet front over the country with winds easing to a lighter southwesterly flow, strengthening again later in the day. Cloudy everywhere with rain crossing the southwest at first and with showers moving up through the country to the northern coasts as the day progresses. Some falls likely to be heavy for a time in western parts to the midlands.

Belfast: Cloudy, scattered showers, breezy spells
Dublin: Cloudy, winds easing, occasional showers
Cork: Cloudy, cooler, scattered showers and drizzle patches
Galway: Partly cloudy, rain, cooler, breezy at times

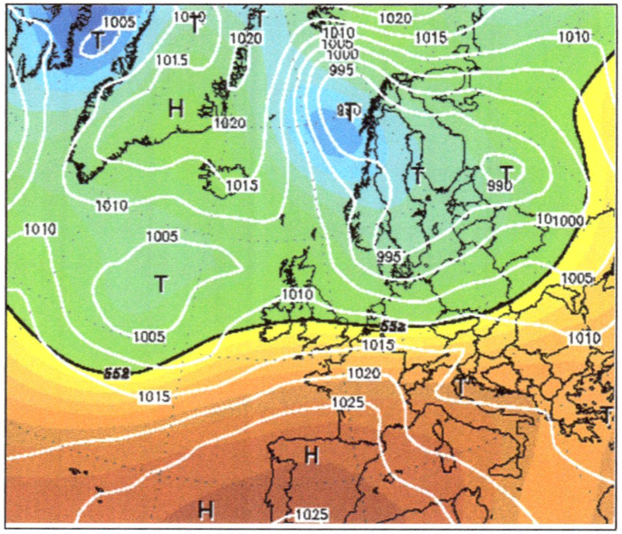

RAIN POTENTIAL

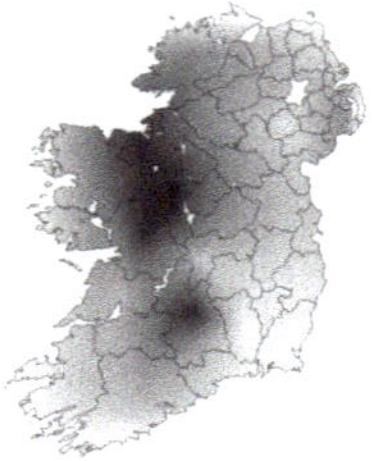

FROST/SNOW

MAY **20** TUESDAY

A weak depression sits over Ireland bringing an unsettled day with moderate to fresh winds westerly throughout. Some showers in the west and north, mostly overnight, clearing to a mostly dry day everywhere, except for lingering showers mostly in the west. Mostly sunny and dry in the east and south, cloudier elsewhere.

Belfast: Partly cloudy, occasional showers, chance of gales and gusts
Dublin: Mostly sunny, dry, chance of isolated showers, strong gusty winds
Cork: Partly cloudy, dry, strong westerly winds at times
Galway: Unsettled and blustery, possible gales, sunny, brief evening showers likely

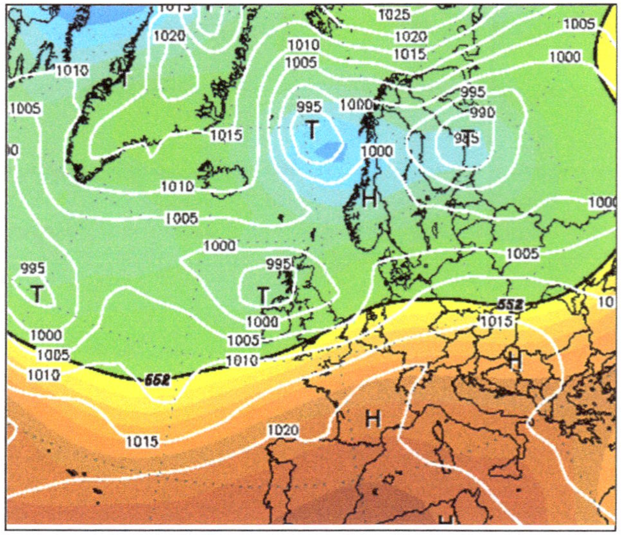

RAIN POTENTIAL

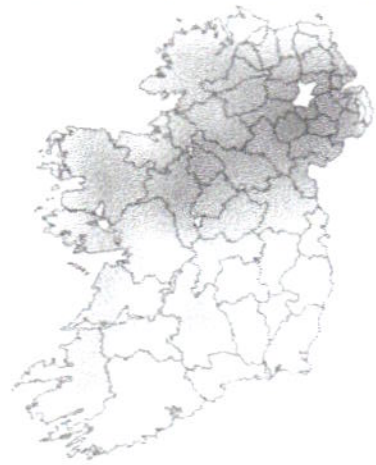

FROST/SNOW

MAY 21 WEDNESDAY

A depression continues to track away from Ireland with a low pressure trough remaining between two low pressure centres with generally light variable winds prevailing. A changeable day with a mix of showers and sunny spells in most districts, with rain moving into the southwest of the country at first and spreading northwards as the day progresses. Some rain may be heavy for a time in the southeast.

Belfast: Cloudy, moderate breezes, showers of light rain and drizzle patches
Dublin: Mix of sunny spells and showers, cooler, fluctuating winds
Cork: Changeable, mix of showers and sunny spells, unsettled winds
Galway: Mostly cloudy, occasional showers, colder, moderate winds

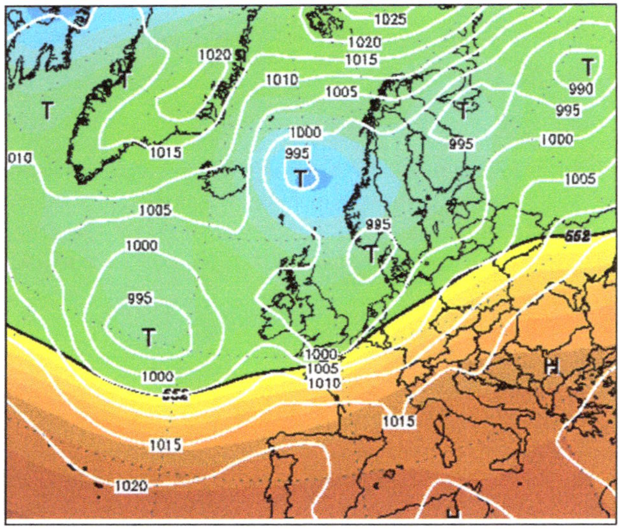

RAIN POTENTIAL

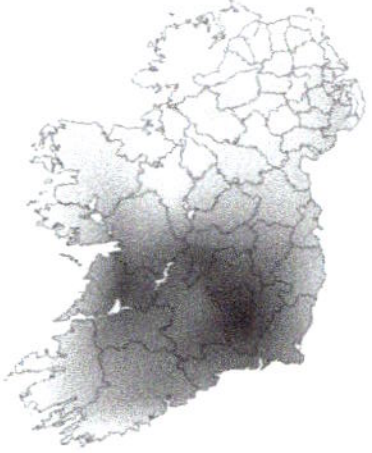

FROST/SNOW

MAY 22 THURSDAY

A weak depression sits over Ireland generating light variable winds throughout, but becoming unsettled and gusty for a time later in the day. Some patches of mist or fog overnight, clearing to a changeable day with a mix of sunny spells and the occasional spotty showery spells mostly in the north and west, with the east and south mostly dry. Colder overnight with the chance of some pockets of frost returning, mostly in inland parts.

Belfast: Cloudy, slightly cooler, scattered showers, shifting breezes, chance of fog
Dublin: Partly cloudy, dry, slightly warmer, gentle breezes
Cork: Sunny, mostly dry with brief evening showers likely, light shifting breezes
Galway: Changeable, mix of showers and sunny spells, light shifting breezes

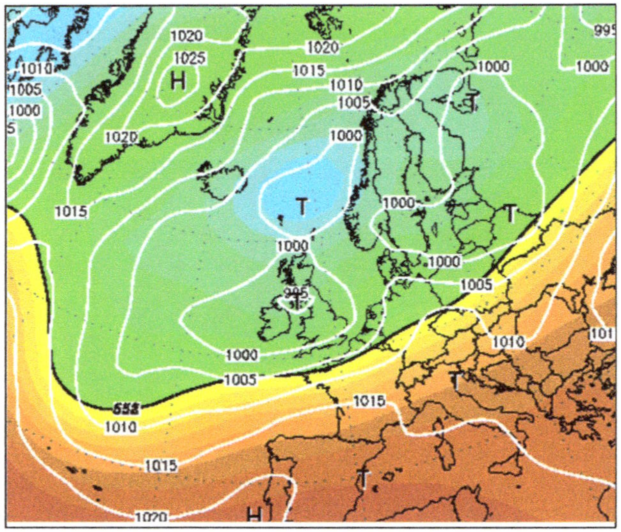

RAIN POTENTIAL

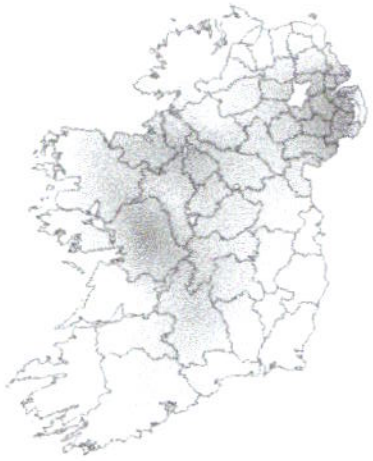

FROST/SNOW

MAY **23** FRIDAY

Low pressure continues to linger over Ireland bringing a light northerly air throughout, with winds freshening for a time especially around the coasts. Mostly dry at first with overnight with patches of fog in many districts. Good sunny spells in the north and west, while cloudier in the east and south. Some showers likely to cross the southern coasts later in the day. Chance of frosts overnight again in central parts.

Belfast: Partly cloudy, dry, mild temperatures, gentle breezes, chance of fog
Dublin: Cloudy, dry, colder, mostly calm air, chance of fog patches
Cork: Overcast, rain, colder, gentle breezes
Galway: Partly cloudy, light showers, mostly calm air, chance of patches of fog

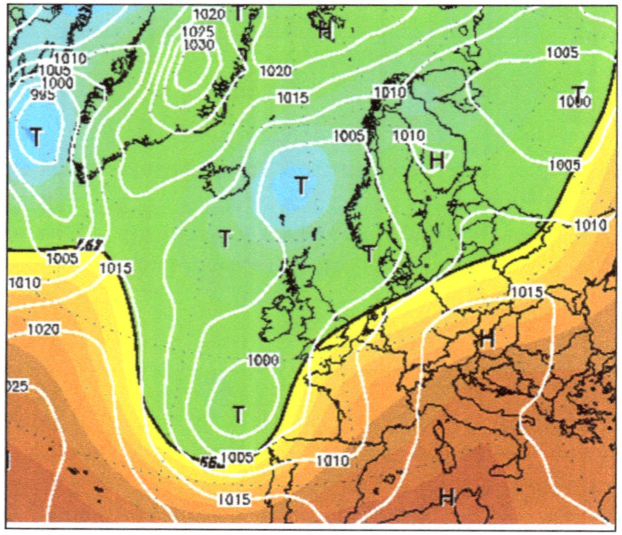

RAIN POTENTIAL

FROST/SNOW

MAY **24** SATURDAY

Low pressure continues to bring a changeable day, with a mix of long sunny spells and passing showers in many parts of the country, sunniest in the east half of the country. Light northeasterly winds prevail throughout for much of the day, but become fresher and more gusty later with a southwest change. Chance of further frosts overnight again in central parts of the country.

Belfast: Sunny day, evening showers, fluctuating winds with chance of gusts
Dublin: Sunny, mostly dry, chance of evening showers, fluctuating breezy spells
Cork: Changeable, sunny, windy, isolated evening showers likely, cold overnight
Galway: Changeable, rain, passing sunny spells, cold overnight, gusty winds at times

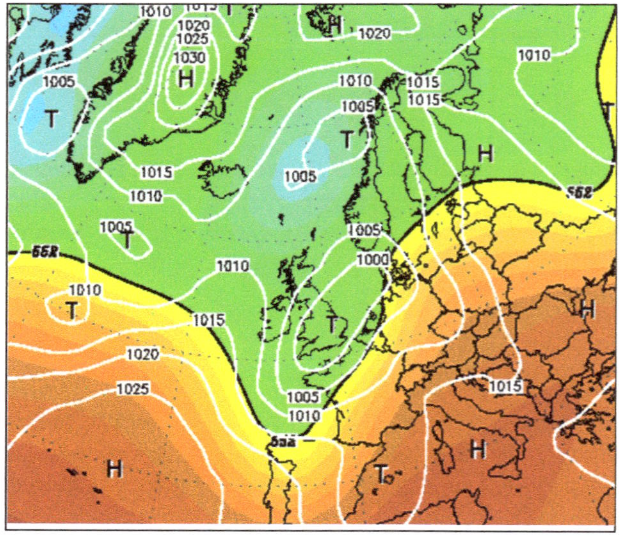

RAIN POTENTIAL

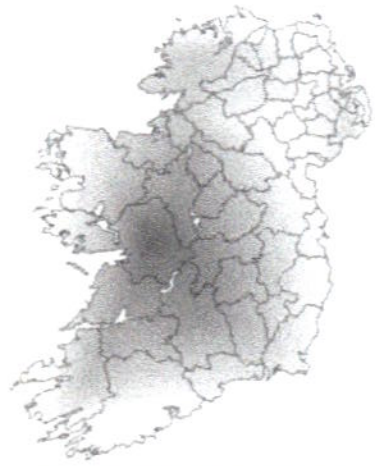

FROST/SNOW

MAY 25 SUNDAY

A weak depression lingers over Ireland bringing light variable winds throughout, unsettled at times along the coasts with a northwest change. Mostly cloudy everywhere, with patches of fog overnight and outbreaks of rain or drizzle developing in many areas, becoming more isolated as the day progresses before returning in the evening.

Belfast: Dull and gloomy air, changeable winds, brief showers
Dublin: Overcast, light drizzle at times, shifting windy spells
Cork: Partly cloudy, milder, brief showers, unsettled breezes
Galway: Cloudy, isolated showers, changeable winds

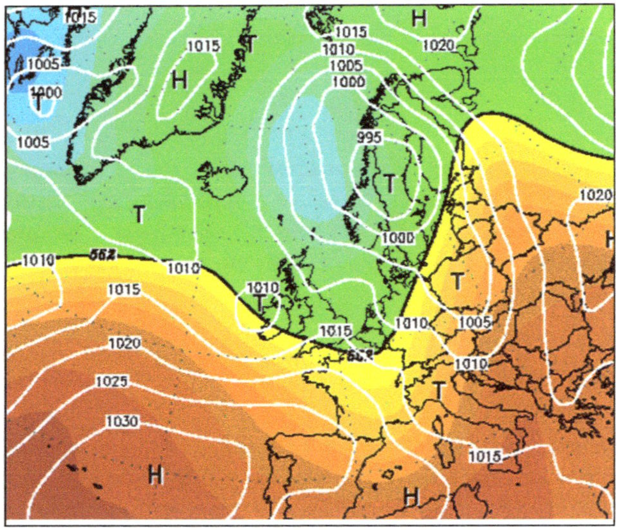

RAIN POTENTIAL

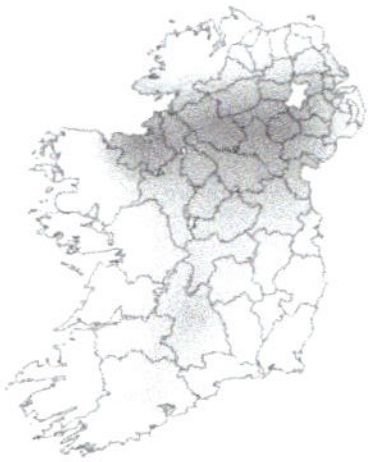

FROST/SNOW

MAY 26 MONDAY

A low pressure trough is maintained over Ireland while the high pressure ridge sits to the south of the country. Light southwest winds prevail at first, but later freshen with a westerly change, becoming gusty for a time. Mostly overcast throughout, with patches of rain in most districts, with rain particularly heavy in the northwest of the country as well as in some central parts, but remaining mostly dry along the southern coastal fringes.

Belfast: Overcast, slightly warmer, scattered showers, chance of overnight fog
Dublin: Overcast, scattered showers and drizzle patches, gentle breezes
Cork: Partly cloudy, warmer, isolated showers, fresh breezes at times
Galway: Dull and gloomy, scattered showers, blustery at times, chance of gales

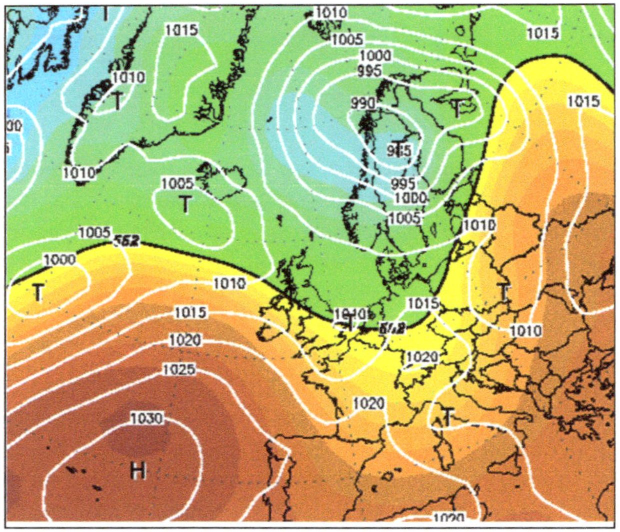

RAIN POTENTIAL

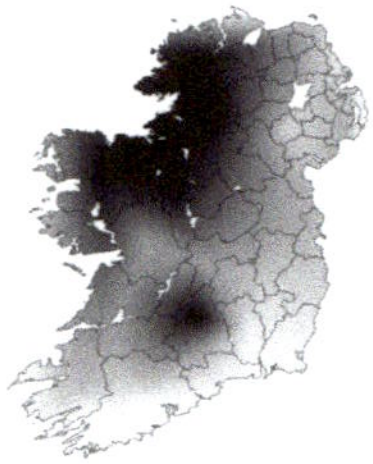

FROST/SNOW

MAY **27** TUESDAY

Low pressure prevails over Ireland as the high pressure ridge drifts nearer to the south of the country. Winds generally moderate in a westerly flow, changing to northwesterly later in the day. Mostly dry and sunny in the south and east of the country, with cloud increasing and occasional showers and drizzle patches elsewhere.

Belfast: Partly cloudy, light breezes strengthening later in the day, brief showers
Dublin: Fresh breezes, mostly dry and sunny, hotter
Cork: Unsettled and windy, very sunny, hot, mostly dry
Galway: Cloudy, scattered showers, strong winds, possible gales

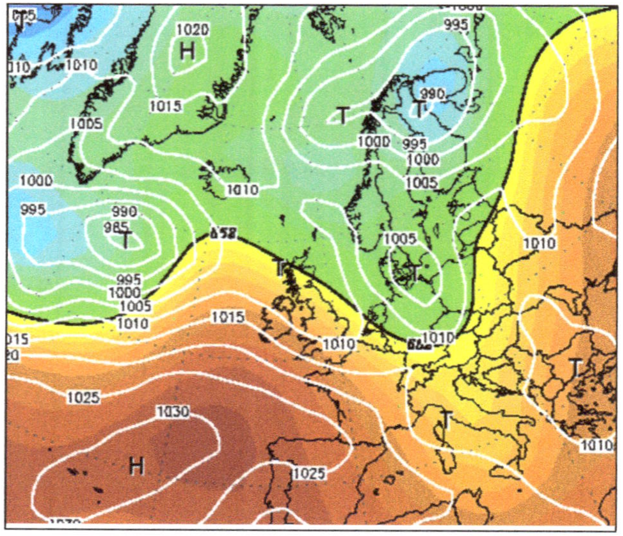

RAIN POTENTIAL

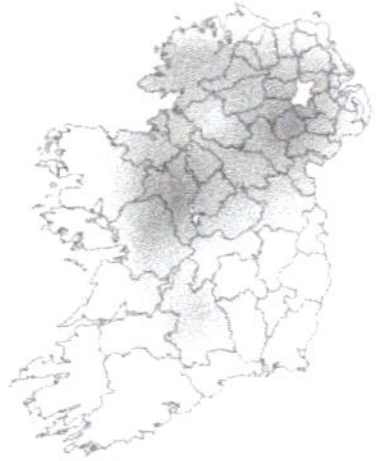

FROST/SNOW

MAY **28** WEDNESDAY

A depression to the northwest of Ireland intensifies bringing very strong and gusty west to southwest winds across the country and a stormy air throughout. Outbreaks of rain moves into the west at first and spreads eastward to reach most parts of the country as the day progresses, although good sunny spells are still likely in many parts. Chance of thunderstorms, heavy downpours, and gale gusts in the west.

Belfast: Very warm day, mild night, mostly sunny, overnight showers, building winds
Dublin: Partly cloudy, brief outbreaks of rain, unsettled and strengthening winds
Cork: Sunny, colder, light rainy spells, gusty winds at times
Galway: Stormy, sunny day, overnight rain, gusty winds, chance of thunderstorms

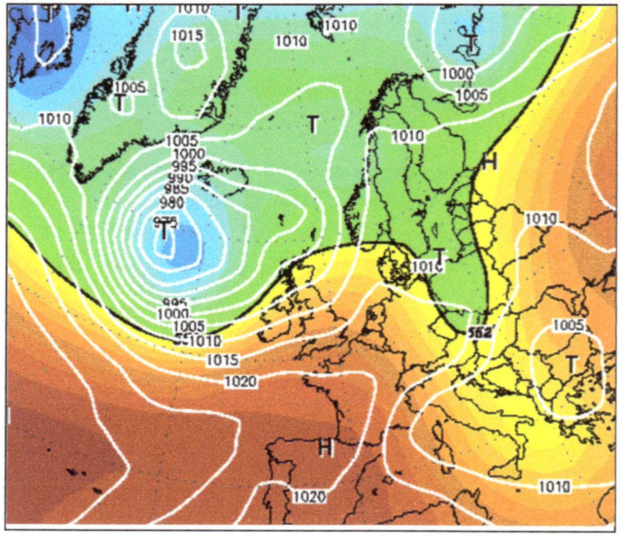

RAIN POTENTIAL

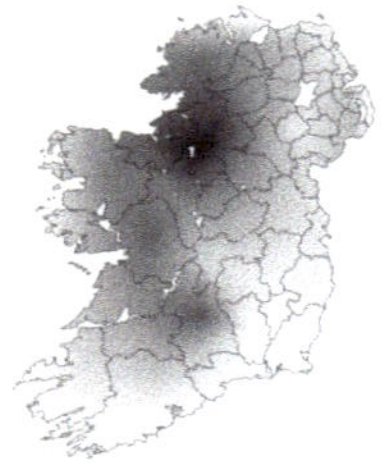

FROST/SNOW

MAY **29** THURSDAY

An intense depression to the north of Ireland continues to bring unsettled and blustery conditions throughout, with strong west to southwest winds prevailing, sometimes to gale force in places. A mixed and unsettled day with some sunny spells and passing squally showers, some possibly accompanied by thunderstorm activity, particularly in the west and north. Colder with possible patches of frost or hail showers in the east.

Belfast: Stormy winds, sunny, cooler, dry, gales likely
Dublin: Very sunny, isolated overnight showers, strong winds, gales likely
Cork: Stormy winds, sunny, isolated overnight showers
Galway: Sunny, cool, squally showers, gale winds likely, chance of thunderstorms

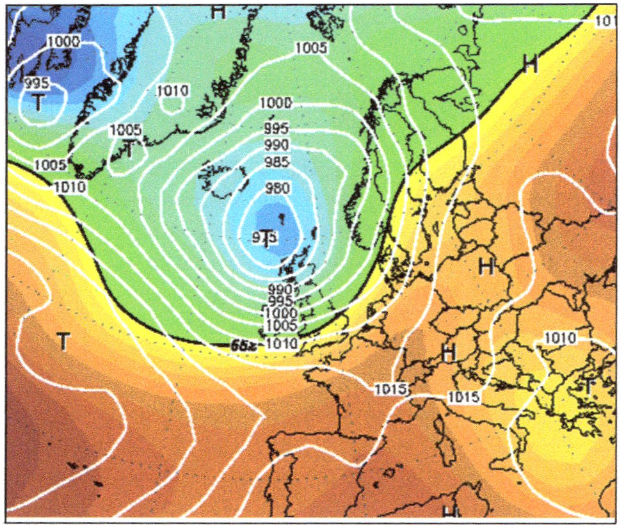

RAIN POTENTIAL

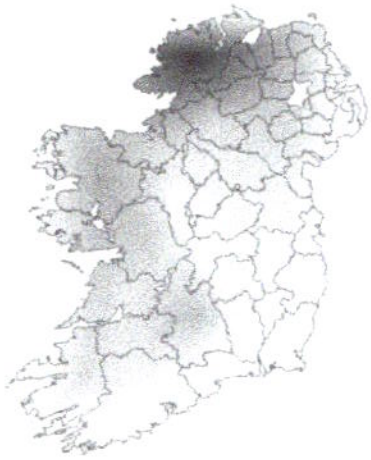

FROST/SNOW

MAY **30** FRIDAY

The depression to the north of Ireland decays as it slips further north and off the country. Winds ease to lighter westerlies throughout. Mostly dry and sunny in central parts of the country from the west coast to the east coast, while a mix of sunny spells and lingering showers prevail in the north and south. Chance of late frosts in the east overnight.

Belfast: Sunny, dry, cool, winds easing but remaining mostly fresh
Dublin: Winds easing, sunny, dry, cold overnight
Cork: Sunny, colder, scattered showers overnight, winds easing
Galway: Very sunny and mostly dry, cold overnight, strong winds easing

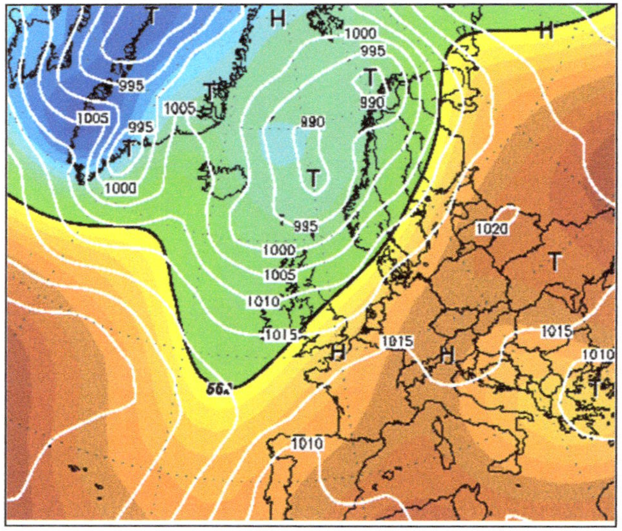

RAIN POTENTIAL

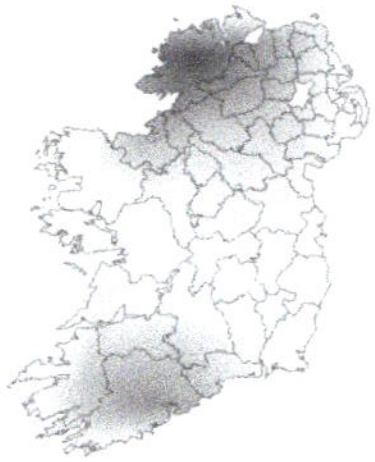

FROST/SNOW

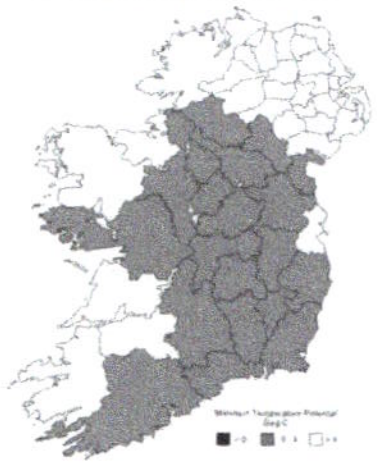

MAY **31** SATURDAY

A large Atlantic anticyclonic system extends a high pressure ridge up over Ireland bringing a light westerly flow at first, but fresher for a time with a northwest change. Very sunny and dry almost everywhere except for the chance of some isolated showers mostly in the far west and southern inland parts. Overnight patches of fog likely in sheltered parts and possible frost pockets once more in some inland parts.

Belfast: Very sunny, dry, mild day cool night, light breezes, chance of misty patches
Dublin: Very bright and sunny, mild day cold night, light breezes to mostly calm air
Cork: Very bright and sunny, dry, slightly warmer, moderate breezes
Galway: Sunny, dry, cold overnight, fluctuating winds, chance of overnight fog

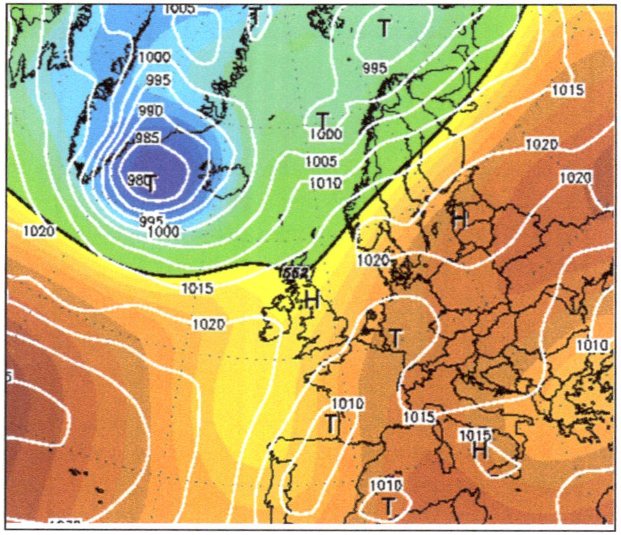

RAIN POTENTIAL

FROST/SNOW

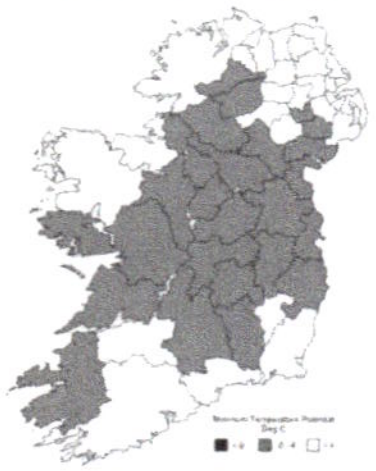

JUNE

PHASES OF THE MOON

1st	First Quarter
5th	Crossing Equator
7th	Apogee
11th	Full Moon
11th	Southern Declination
18th	Third Quarter
19th	Crossing Equator
23rd	Perigee #8
25th	New Moon
25th	Northern Declination

1	2	3	4	5	6	7	8	9	10	11	12	13	14	15	16	17	18	19	20	21	22	23	24	25	26	27	28	29	30
Sat	Sun	Mon	Tue	Wed	Thur	Fri	Sat	Sun	Mon	Tue	Wed	Thur	Fri	Sat	Sun	Mon	Tue	Wed	Thur	Fri	Sat	Sun	Mon	Tue	Wed	Thur	Fri	Sat	Sun
		1Q		XhS		A				F V							3Q	XhN				P8		N ^					

MONTHLY SUMMARY

1st-4th June A ridge of high pressure is maintained over the country bringing mostly dry conditions but increasing cloud and milder temperatures throughout the country. The high pressure ridge directs a light west to southwest flow, and returning very sunny and mostly dry conditions everywhere, with up to 10 hours or more of sunshine each day. Chance of some showers briefly along the western and northern coasts, and possible showers of snow grains in inland parts.

4th-6th Fog patches likely.

5th-10th Low pressure prevails throughout bringing mix weather with spells of sunshine and of rain or showers each day. Some showers likely to accompany thunderstorms and be of hail. Cooler throughout with chance of widespread frosts overnight.

6th-8th Very sunny with over 10 hours or more likely on each day. Chance of gale gusts at times mostly in the north and west.

7th Chance of scattered thunderstorms and hail showers in central to eastern parts, including the chance of snow flurries briefly.

8th Chance of scattered frosts, some severe in inland districts.

9th Cold overnight with chance of snow grains in central parts.

Over the next four weeks, there will be very little change to air pressure, with the highest pressure expected at the start of this outlook period, and the lowest at the end. However, much of the first week ahead will see low pressure prevailing before high pressure becomes the dominant weather influence over the next couple of weeks, then low pressure returns at the end. After a dry start at first, wetter than average conditions become established almost everywhere, with very high rainfall at times, particularly in the east and southeast, where up to twice as much rainfall than normal for this time of year can be expected. Some rain events over the next four weeks will be accompanied by thunderstorms and heavy rain possibly resulting in localised flooding. The prevalence of high pressure from the second week onwards may also see sunshine hours below the norm for this time of year with quite a variation across the country, and with the north and west possibly the sunniest. The sunniest period during the next four weeks may be over the first week to midway through the second week. Maximum temperatures may be up to one degree above the average for this time of year, but parts of the northwest possibly up to two degrees warmer than normal. The warmest period will be during the sunniest spell over the second week. Minimum temperatures in contrast, will be cooler than normal over the first and last weeks of this look-ahead period, including with the possibility of some frost patches overnight in the fourth week. During this outlook period, up to 10 days will see rain in the northwest, and 15-19 days elsewhere and thunderstorms can be expected on 10 days but no hail. Fogs will be observed on around 9 days, frosts on 3 days and snow grains possible on 2-3 nights, but no snow. Winds to gale force can be expected on at least 2 days.

11th-13th Bands of rain cross the country bringing heavy falls at times, especially in western parts at first, but later heaviest in the east. Warmer than average everywhere and despite the rain, fairly long sunny spells can be expected.

12th Very strong winds prevail in the west of the country in a south to southwest flow.

14th-21st A week of high pressure systems systems centred to the north of Ireland prevails. Widespread fogs almost daily throughout clearing to dry and sunny weather everywhere after the fogs clear. Temperatures will be above normal in the west and closer to norm in the east. Some areas may see up to 15 hours of sunshine for the first few days of this spell.

14th Chance of light frost patches in sheltered parts.

14th-16th Very bright and sunny everywhere, with many districts possibly seeing up to 15 hours of sunshine each day.

21st-25th A period when warmer air crosses to the south of Ireland while cooler artic airflows cross to the north. This combination will produce a series of active fronts crossing the country, with scattered thunderstorms, and very heavy falls which may see some localized flooding, especially in inland parts.

22nd-10th July A series of frontal systems associated with depressions cross the southern half of Ireland in an easterly trajectory, bringing widespread cloud and outbreaks of rain or showers daily, some very heavy at times and accompanying thunderstorm activity. Winds mainly easterly at first later backing northerly for a time before finally having a milder, southerly change. Temperatures likely to be mostly average to below average.

22nd-25th A band of thunderstorms cross the country.

24th-25th Very strong winds in the northwest.

28th-30th Very heavy rain with chance of localised flooding.

29th-30th Chance of thunderstorms mostly in the north and east.

IRELAND RAINFALL ESTIMATES

June 2025

d=mainly dry, n= nonrecordable, l=light shrs, s=significant shrs, r=rain, h=hvy falls

		ULSTER PROVINCE					CONNAUGHT PROVINCE						LEINSTER PROVINCE				MUNSTER PROVINCE						
		Alderg	Hillsbo	Armag	Loughl	Malin H	Sligo	Ardtarı	Belmul	Drums	Galway	Roscoı	Markre	Edende	Kilkenı	Dublin	Shannc	Newpo	Tralee	Killarne	Cork	All	Moon
1st	Sun	d	d	d	n	d	n	n	n	n	l	d	n	d	d	d	n	l	l	l	d	9	
2nd	Mon	d	d	d	l	n	l	l	l	l	l	l	l	d	d	d	d	l	l	d	d	12	
3rd	Tues	n	n	d	l	l	l	l	l	n	d	l	s	d	d	d	n	s	l	n	d	28	1Q
4th	Wed	l	l	d	l	l	l	l	l	n	n	s	d	d	d	d	d	l	d	l	d	21	
5th	Thur	l	l	d	l	l	l	l	s	s	s	r	d	s	s	s	s	s	s	s	s	86	XhS
6th	Fri	s	n	s	l	d	l	l	n	d	d	n	n	l	l	s	n	n	s	d	l	35	
7th	Sat	l	d	l	l	d	l	l	l	l	d	d	d	l	l	l	d	d	d	d	d	15	A
8th	Sun	d	d	d	d	n	d	d	d	d	d	d	d	d	l	d	d	d	n	r	d	15	
9th	Mon	n	l	d	l	d	l	l	d	s	s	s	s	l	l	l	r	l	r	s	r	80	
10th	Tues	l	l	d	s	d	l	s	d	d	d	d	l	d	d	d	l	l	l	r	l	36	
11th	Wed	l	l	d	l	d	s	l	l	l	r	s	s	l	s	d	l	s	s	r	l	81	F V
12th	Thur	r	h	l	l	r	s	l	h	r	s	r	r	r	h	s	s	r	h	r	s	253	
13th	Fri	r	s	r	d	s	d	d	d	d	d	l	n	r	l	h	l	d	d	d	l	87	
14th	Sat	s	d	r	d	d	d	d	d	d	d	d	d	n	d	l	d	d	d	d	d	19	
15th	Sun	d	d	d	d	d	d	d	d	d	d	d	d	d	d	d	d	d	d	d	d	0	
16th	Mon	d	d	d	d	d	d	d	d	d	d	d	d	d	d	d	d	d	d	d	d	0	
17th	Tues	d	d	d	d	d	d	d	d	d	d	d	d	d	d	d	d	d	d	d	d	0	
18th	Wed	d	d	d	d	d	d	d	d	d	d	d	d	d	d	d	d	d	d	d	d	0	3Q
19th	Thur	d	d	d	d	d	d	d	d	d	d	d	d	d	d	d	d	d	d	d	d	0	XhN
20th	Fri	d	d	d	d	d	d	d	d	d	d	d	d	d	d	d	d	d	d	n	d	1	
21st	Sat	d	d	d	r	d	s	r	s	d	l	d	h	l	d	d	s	h	l	l	l	92	
22nd	Sun	d	d	d	l	d	l	l	d	s	d	l	l	d	l	l	l	d	s	d	l	31	
23rd	Mon	r	s	h	s	d	s	s	l	s	s	h	r	l	r	s	l	r	l	s	s	150	P8
24th	Tues	r	r	r	s	d	l	s	l	l	h	l	s	l	s	s	s	s	l	h	h	155	
25th	Wed	r	s	h	l	d	l	l	l	s	d	s	l	l	r	r	l	l	l	l	r	115	N ^
26th	Thur	l	d	s	l	s	d	l	d	l	d	l	d	l	d	n	l	d	d	d	d	26	
27th	Fri	n	d	l	d	d	d	d	d	d	l	d	n	d	d	d	d	d	d	l	l	4	
28th	Sat	n	l	d	n	d	l	n	l	l	s	l	l	l	s	n	l	l	s	h	r	60	
29th	Sun	s	r	l	l	n	s	l	r	r	h	h	l	r	h	r	h	r	h	r	h	256	
30th	Mon	s	l	s	s	h	r	s	s	r	s	h	s	l	r	l	h	s	h	s	h	198	
	Estimate:	95	82	107	65	46	66	65	65	76	114	120	78	62	105	91	94	108	146	129	152	1864	
	Average:	62	66	58	71	70	82	84	72	74	80	82	82	74	61	67	70	90	108	72	81	1505	
	Trend:	wtr	wtr	wtr	av	drr	drr	drr	av	av	wtr	wtr	av	drr	wtr	wtr	wtr	wtr	wtr	wtr	wtr	wtr	

IRELAND SUNSHINE ESTIMATES

June 2025

F=fine (8-12 hours of sunshine), pc= partly cloudy (4-7 hours), c=cloudy (1-3 hours), o=overcast (0 hours)

		ULSTER PROVINCE					CONNAUGHT PROVINCE						LEINSTER PROVINCE				MUNSTER PROVINCE						
		Alderg	Hillsbo	Armag	Loughl	Malin H	Sligo	Ardtarı	Belmul	Drums	Galway	Roscoı	Markre	Edende	Kilkenı	Dublin	Shannı	Newpo	Tralee	Killarne	Cork	All	Moon
1st	Sun	pc	pc	pc	pc	pc	F	pc	pc	pc	pc	pc	pc	pc	c	pc	pc	pc	c	c	c	104	
2nd	Mon	o	o	o	o	o	o	o	o	c	o	o	o	c	c	o	o	c	o	o	pc	19	
3rd	Tues	o	o	o	o	o	o	o	o	c	o	o	o	c	c	c	o	c	o	o	pc	21	1Q
4th	Wed	pc	pc	pc	pc	pc	pc	pc	F	F	F	F	pc	F	F	pc	F	F	pc	pc	F	169	
5th	Thur	pc	pc	pc	pc	pc	pc	pc	pc	pc	c	pc	pc	pc	pc	pc	o	c	o	o	c	94	XhS
6th	Fri	F	F	F	F	F	F	F	F	F	F	F	F	F	pc	pc	F	F	F	F	F	209	
7th	Sat	F	F	F	F	F	F	F	F	F	F	F	F	F	F	F	F	F	F	F	F	256	A
8th	Sun	F	F	F	F	F	F	F	F	F	F	F	F	F	F	F	F	F	F	F	F	266	
9th	Mon	c	c	c	c	c	c	c	pc	c	c	pc	c	c	c	c	o	c	o	o	c	63	
10th	Tues	F	F	F	F	F	pc	F	pc	F	pc	pc	F	F	F	F	pc	pc	F	F	pc	171	
11th	Wed	F	F	F	F	F	pc	F	pc	F	pc	pc	F	F	F	F	c	pc	c	c	c	146	F V
12th	Thur	o	o	o	o	o	o	o	c	o	o	c	o	o	o	o	o	o	o	o	o	4	
13th	Fri	pc	pc	pc	pc	pc	F	pc	F	pc	F	F	pc	pc	pc	o	pc	pc	F	F	F	133	
14th	Sat	F	F	F	F	F	F	F	F	pc	F	F	F	pc	pc	c	F	pc	F	F	F	197	
15th	Sun	F	F	F	F	F	F	F	F	F	F	F	F	F	F	F	F	F	F	F	F	210	
16th	Mon	F	F	F	F	F	F	F	F	F	F	F	F	F	F	F	F	F	F	F	F	254	
17th	Tues	F	F	F	F	F	F	F	F	F	F	F	F	F	F	F	F	F	F	F	F	279	
18th	Wed	F	F	F	F	F	F	F	F	F	F	F	F	F	F	F	F	F	F	F	F	284	3Q
19th	Thur	F	F	F	F	F	F	F	F	F	F	F	F	F	F	F	F	F	pc	pc	F	249	XhN
20th	Fri	F	F	F	F	F	F	F	F	F	F	F	F	F	F	F	F	F	pc	pc	F	242	
21st	Sat	pc	pc	pc	pc	pc	pc	pc	F	pc	pc	F	pc	pc	pc	pc	c	c	c	c	c	118	
22nd	Sun	pc	pc	pc	pc	pc	c	pc	c	pc	c	c	pc	pc	pc	pc	c	c	c	c	c	76	
23rd	Mon	o	o	o	o	o	o	o	c	c	c	c	o	c	c	o	o	c	pc	pc	c	26	P8
24th	Tues	o	o	o	o	o	o	o	o	o	o	o	o	o	o	o	o	o	o	o	c	4	
25th	Wed	o	o	o	o	o	o	o	o	o	o	o	o	o	o	o	o	o	c	c	o	4	N ^
26th	Thur	o	o	o	o	o	o	o	F	o	pc	F	o	o	o	o	c	o	c	c	o	39	
27th	Fri	o	o	o	o	o	o	o	pc	c	pc	pc	o	c	c	o	F	pc	c	c	pc	56	
28th	Sat	c	c	c	c	c	o	c	c	c	c	c	c	c	pc	c	pc	pc	o	o	pc	51	
29th	Sun	o	o	o	o	o	o	o	o	o	o	o	o	o	o	o	o	o	o	o	o	0	
30th	Mon	pc	pc	pc	pc	pc	pc	pc	pc	c	c	c	pc	c	c	c	c	c	c	c	o	70	
Estimate hours:		188	188	188	188	188	199	188	235	187	210	222	188	187	186	176	184	185	168	168	196	3814	
Average hours:		167	167	142	169	171	162	162	161	137	152	159	155	158	155	180	154	146	152	154	172	3173	
Trend:		more	more	more	more	more	av	more	more	more	more	more	more	more	more	av	more	more	more	more	more	more	

JUNE **1** SUNDAY

A high pressure ridges to the southeast of Ireland while low pressure prevails to the north, with light to moderate westerly winds prevailing throughout, fresher with occasional gusts at times around the northern coasts. Mostly dry with passing scattered cloud everywhere, except for the chance of some isolated showers some southern midland counties.

Belfast: Unsettled blustery winds at tmes, dry, scattered cloudy spells, cool
Dublin: Partly cloudy, dry and mild, pleasant breezes
Cork: Cloudy and breezy, dry, mild day cool night
Galway: Strong westerly winds, isolated showers, passing cloudy spells

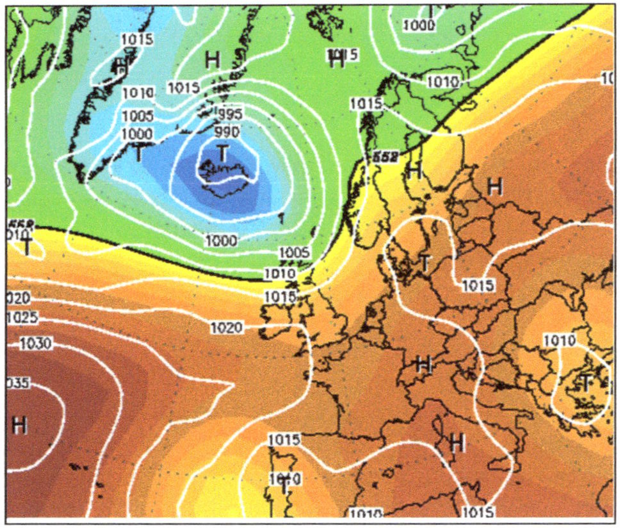

RAIN POTENTIAL

FROST/SNOW

JUNE **2** MONDAY

The high pressure ridge moves further onto the country in a light west to southwest flow throughout. Mostly cloudy everywhere, with patches of overnight fog particularly over the western half of Ireland. Mostly dry everywhere, with the odd threat of rain now and then, and possible isolated showers for a time in the west.

Belfast: Widespread cloud, dry with passing threats of rain, moderate breezes
Dublin: Overcast, dry with odd threat of rain, gentle breezes to mostly calm air
Cork: Partly cloudy, dry, warm, gentle westerly breezes
Galway: Overcast, windy spells, brief drizzly showers, cooler

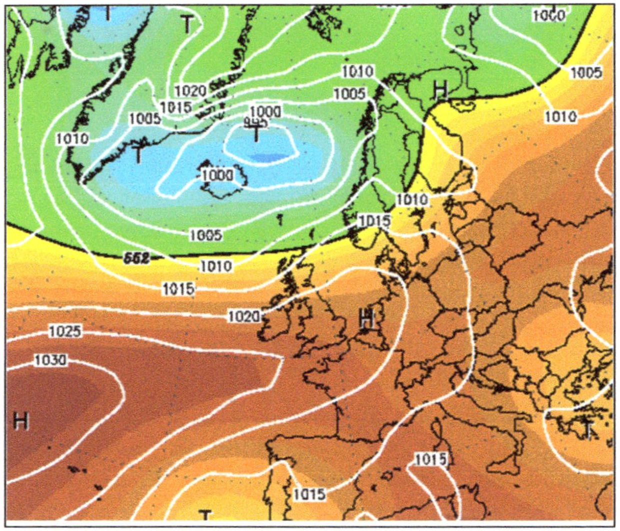

RAIN POTENTIAL

FROST/SNOW

JUNE **3** TUESDAY

The high pressure system to the west of Ireland ridges over the southern half of the country, while a weak depression flows over the northern half. Widespread cloud and patches of drizzle at first, clearing to a mostly dry but dull day. Chance of fog overnight in the north to the midlands areas, with light to moderate southwesterly air prevailing, turning northwesterly later in the day.

Belfast: Overcast, light breezes, mostly dry, chance of misty drizzle patches or fog
Dublin: Cloudy, dry, gentle breezes
Cork: Moderate breezes, scattered cloudy spells, dry and mild
Galway: Overcast, dry with passing threats of rain, moderate breezes

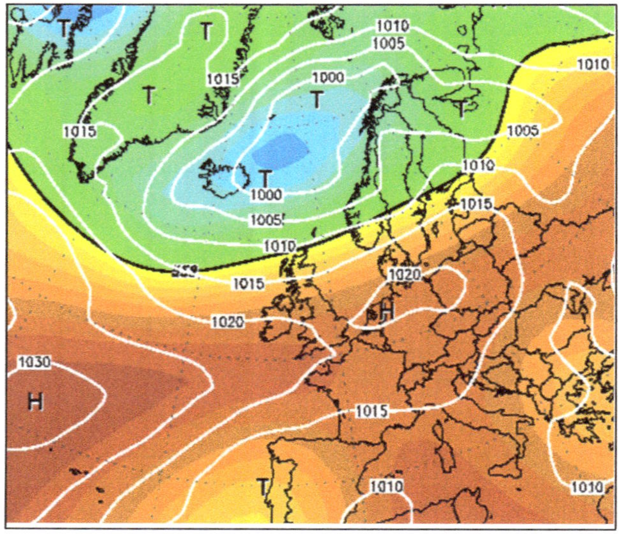

RAIN POTENTIAL

FROST/SNOW

JUNE **4** WEDNESDAY

A weak depression sits to the north of Ireland while high pressure sits to the west and east, with a low pressure trough directing a gentle to moderate northwest flow over the country, fresher at times around the northwestern coastal fringes. Mild temperatures throughout with a mix of sunny spells and the odd showery outbreak over the northern half of the country while remaining dry to the south.

Belfast: Mostly sunny, brief morning showers, gentle breezes
Dublin: Mostly sunny and dry, cooler, light breezes to mostly calm air
Cork: Sunny, dry, colder, light to moderate breezes
Galway: Sunny, isolated morning showers dry day, cold overnight, fluctuating winds

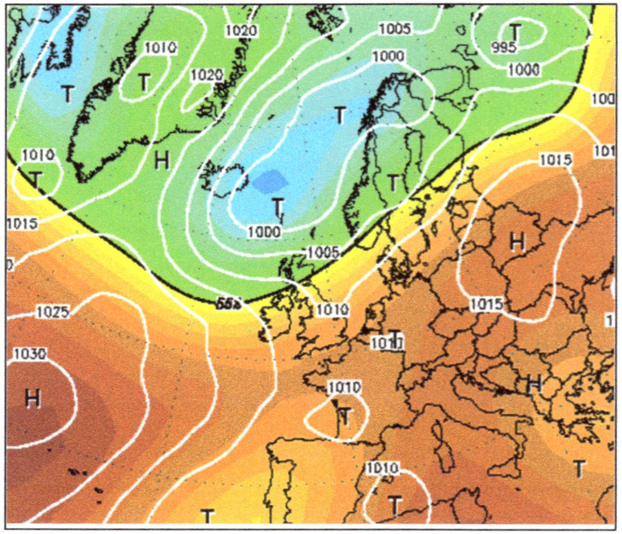

RAIN POTENTIAL

FROST/SNOW

JUNE **5** THURSDAY

A low pressure trough continues to flow over Ireland in a light to moderate westerly flow, backing northerly for a time and fresher around the coastal fringes. Cooler and mostly cloudy everywhere, with some showers now and then, including the chance of hail showers mostly in the east.

Belfast: Colder, partly cloudy, brief showers, moderate breezes
Dublin: Changeable, cooler, mix of showers and sunny spells, chance of hail showers
Cork: Cloudy, colder, scattered showers, unsettled breezes at times
Galway: Mostly cloudy, occasional showers, colder, chance of isolated frost patches

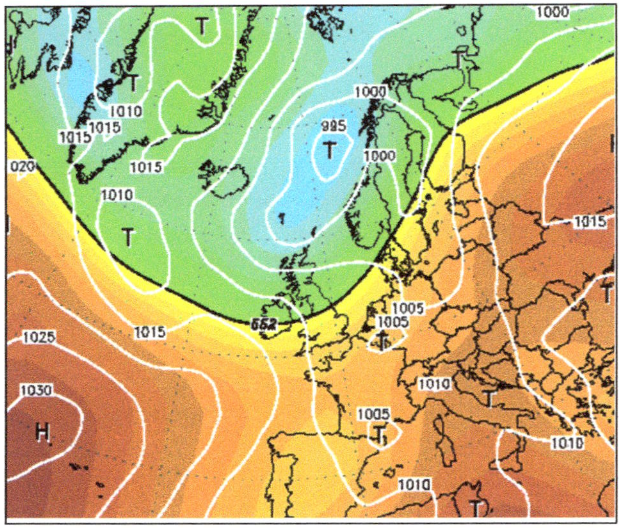

RAIN POTENTIAL

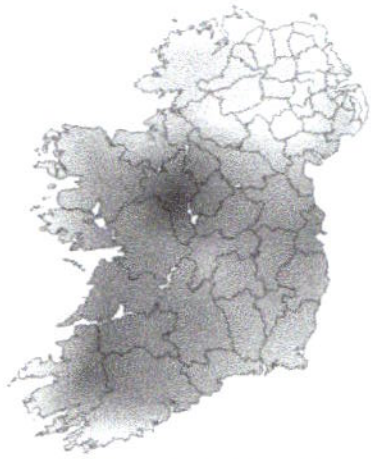

FROST/SNOW

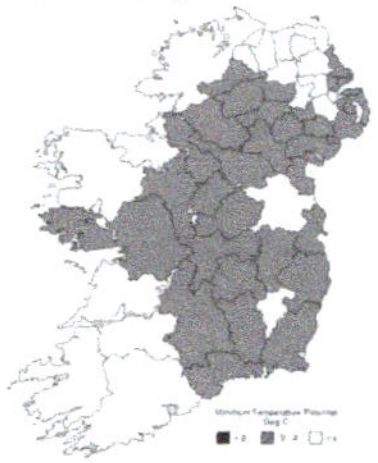

JUNE **6** FRIDAY

A weak depression crosses over the country, bringing a fresh to strong east to northeast flow throughout. An unsettled day with a mix of some sunny spells and passing showers and drizzle patches in most districts, but more isolated in the northwest. Chance of some thunderstorms and hail showers, particularly in the east and north.

Belfast: Sunny, cool, mostly dry, chance of morning drizzle patches, fluctuating winds
Dublin: Stormy, mix of showers and sunny spells, chance of thunderstorms and hail
Cork: Sunny, brief morning showers, cool, gusty winds
Galway: Very sunny, cool and dry, unsettled blustery winds

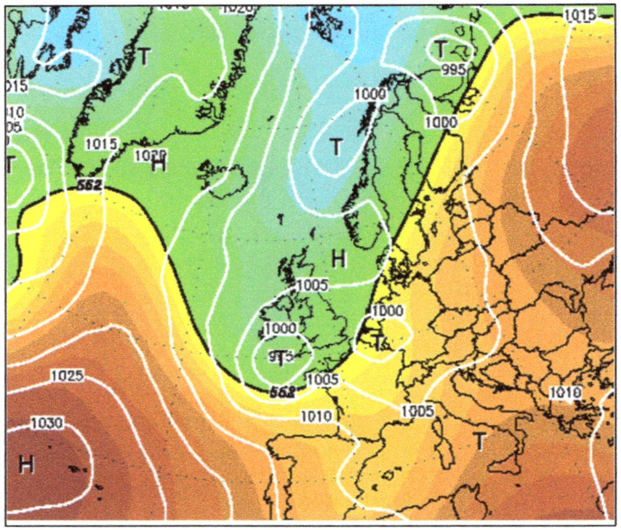

RAIN POTENTIAL

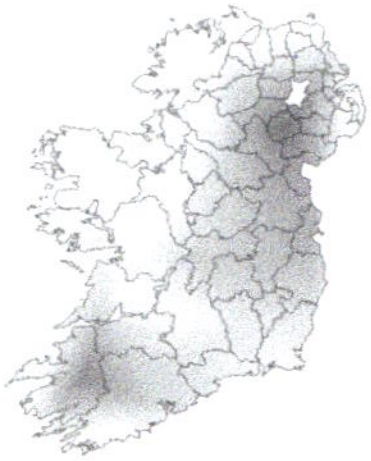

FROST/SNOW

JUNE **7** SATURDAY

An unsettled air associated with a low pressure trough continues to cross the country in a moderate north to northeasterly flow, with strong gusts at times particularly around the coasts. Overnight thunderstorms with hail and even the chance of some snow mostly in the east of the country possible, clearing everywhere to a very sunny and dry day throughout.

Belfast: Sunny, dry, mild day cold night, fresh northern winds
Dublin: Unsettled air, sunny, brief showers, chance of thunderstorms, hail, and snow
Cork: Very bright and sunny, dry, cold overnight, strong winds easing later
Galway: Very sunny, cool, dry, windy at times, chance of gale gusts at times

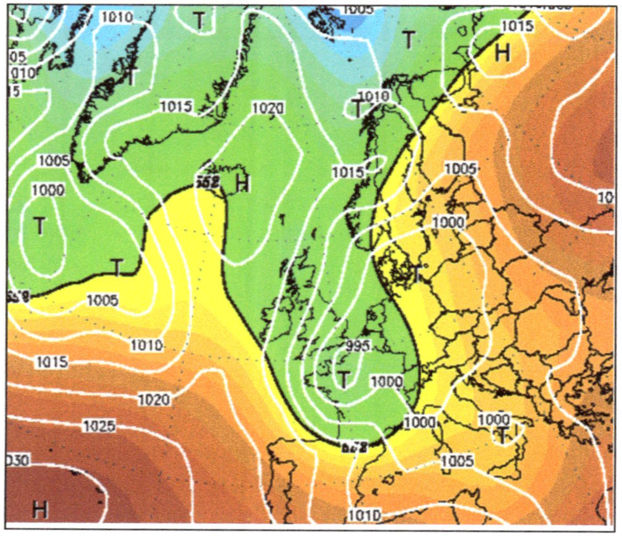

RAIN POTENTIAL

FROST/SNOW

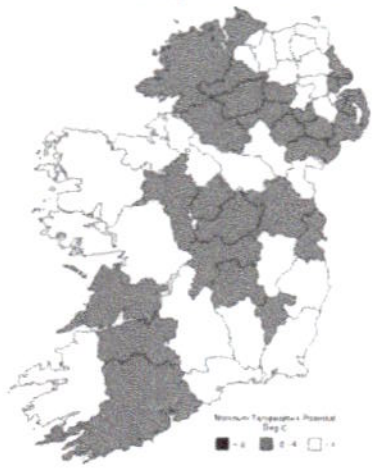

JUNE **8** SUNDAY

A high pressure ridge sits to the north of Ireland while depressions sit to the east and west of the country, with a low pressure trough dominating the country and directing a light to moderate southerly flow throughout, building to fresh and sometimes strong and accompanied by isolated showers later in the day around the southwest of the country. Very sunny and dry elsewhere with the chance of some patches of frost and fog overnight.

Belfast: Very sunny and dry, mild day cold night, chance of fog pockets in the region
Dublin: Very bright and sunny, mild day cold night, chance of frost patches, calm air
Cork: Sunny, cool, dry, mostly calm air
Galway: Very bright and sunny, dry, mild day cold night, possible overnight frost

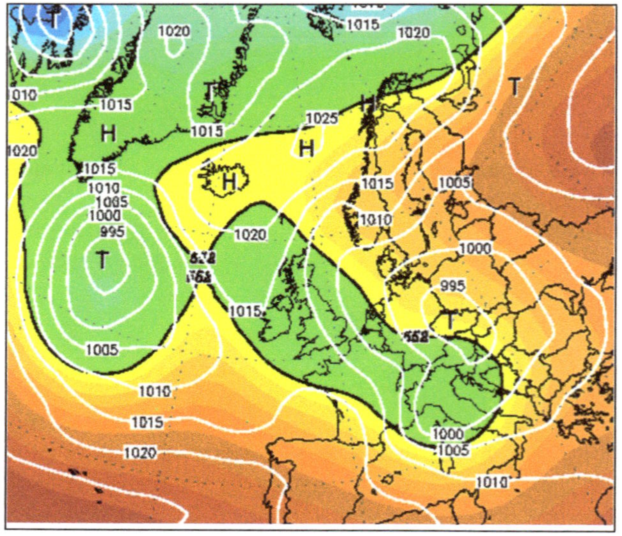

RAIN POTENTIAL

FROST/SNOW

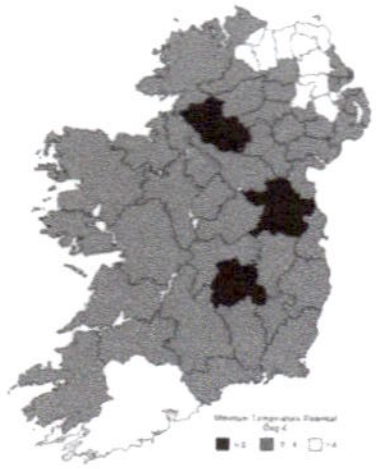

JUNE 9 MONDAY

A weak depression to the west of Ireland directs a light to moderate variable to southeasterly flow over the country. A wet front moves over the southern coastal fringes at first bringing heavy rain at times into many parts of Munster, with shower activity spreading northwards as the day progresses, and only bringing isolated showers into northern parts.

Belfast: Freshening windy spells, mostly cloudy, isolated showers, cold overnight
Dublin: Mostly cloudy, isolated showers, cold, chance of overnight frost patches
Cork: Unsettled winds, cloudy, rain
Galway: Mostly cloudy, cool, occasional showers, blustery at times

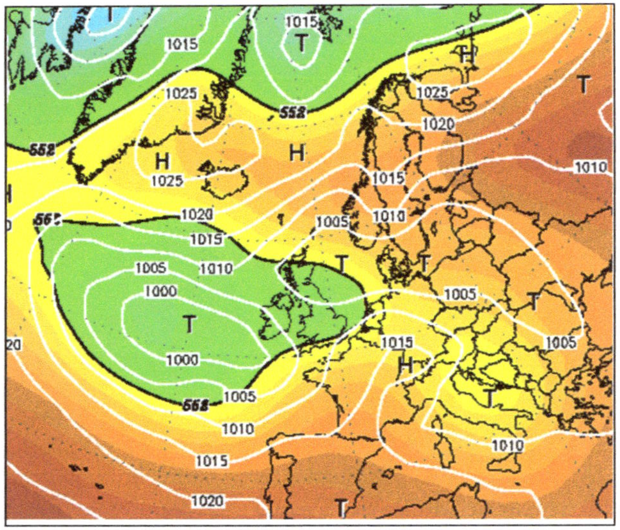

RAIN POTENTIAL

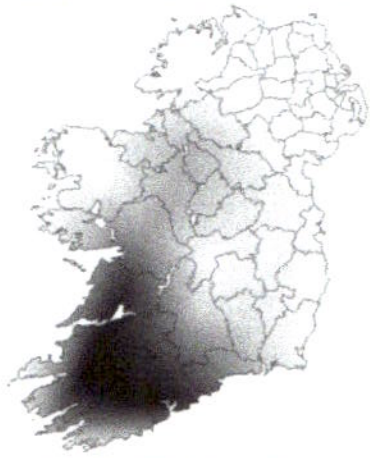

FROST/SNOW

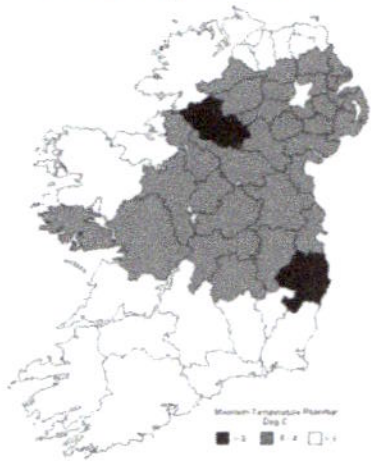

JUNE 10 TUESDAY

Low pressure over Ireland brings a light to moderate southerly flow over the country, more changeable and variable in northern parts. Some isolated showers here and there at first in the morning before clearing to a mostly sunny and dry day everywhere. Chance of thunderstorms in central parts.

Belfast: Sunny, warmer, mostly dry, fresh winds, chance of brief morning showers
Dublin: Fresh breezes, sunny, dry, warmer
Cork: Partly cloudy, light showers, windy spells with occasional gusts
Galway: Partly cloudy, warmer, dry, fresh breezes

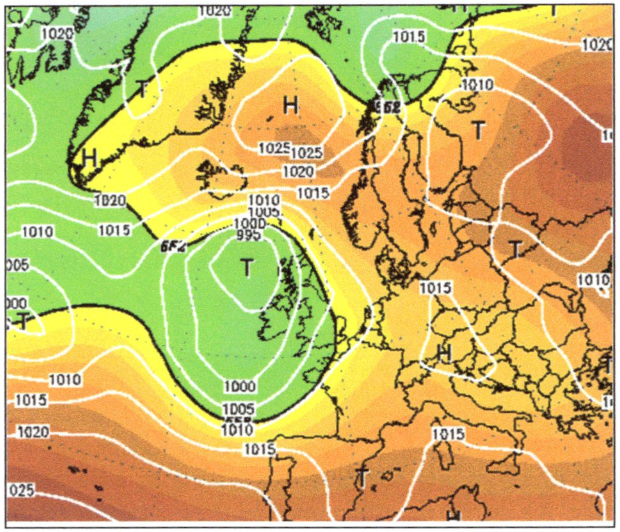

RAIN POTENTIAL

FROST/SNOW

JUNE 11 WEDNESDAY

A high pressure ridge sits to the northeast of Ireland while the depression moves away to the west, allowing for the low pressure trough to bring a light to variable south to southeasterly flow to prevail throughout with patches of fog overnight in many districts. Mostly sunny and dry in the north and east, with increased cloud and showers developing elsewhere.

Belfast: Sunny, light showers, warmer day cooler night, serene air, fog possible
Dublin: Sunny, warmer, dry, calm atmosphere, chance of shallow fog patches
Cork: Mostly cloudy, brief showers, mostly calm air, chance of light drizzle or fog
Galway: Partly cloudy, outbreaks of rain, mostly calm air, possible overnight fog

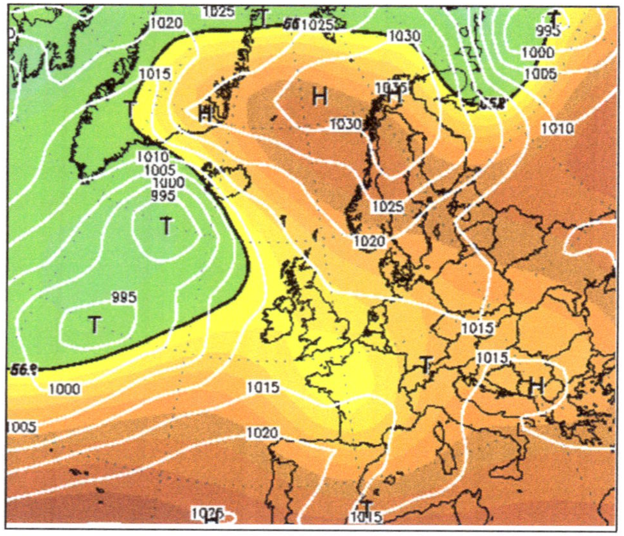

RAIN POTENTIAL

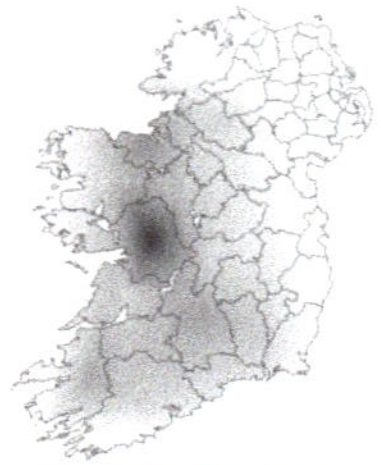

FROST/SNOW

JUNE 12 THURSDAY

A rain embedded trough associated with a large depression to the northwest of Ireland brings widespread cloud and rain throughout the country, with heavy falls in many districts, particularly in the southwest and far northeast. Moderate to strong southerly winds develop, with very strong gusts and the occasional spell of gale winds, mostly in the western half of the country.

Belfast: Stormy air, overcast, heavy rain, gale winds likely, warmer
Dublin: Overcast, occasional showers, freshening winds
Cork: Overcast, stormy winds, rain, cool
Galway: Overcast, scattered showers, strong and blustery winds

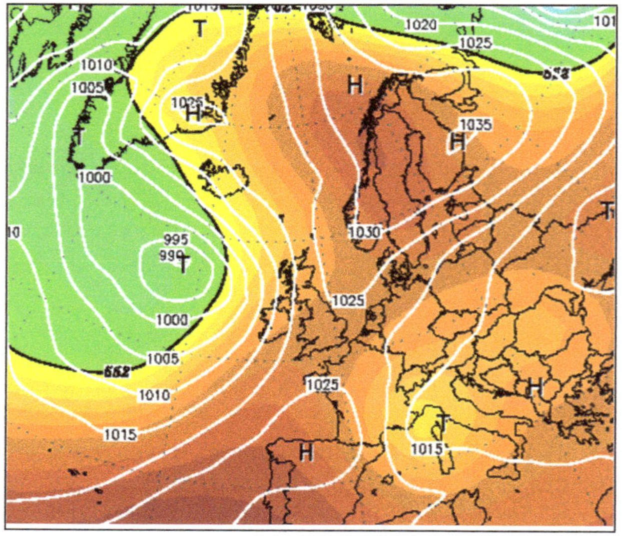

RAIN POTENTIAL

FROST/SNOW

JUNE **13** FRIDAY

High pressure prevails to the east and south of the country while the depression that brought widespread rain yesterday contracts to the northwest of Ireland. A mixed day with widespread rain persisting for much of the day in the east and north of the country, while clearing to mainly dry and mostly sunny in the west, and mostly sunny with isolated showers in the south. Light southerly breezes prevail.

Belfast: Changeable, partly cloudy, outbreaks of rain, winds easing to moderate
Dublin: Overcast, heavy rain, windy spells
Cork: Sunny, warmer, light showers, winds easing with occasional gusts
Galway: Sunny, dry, moderate breezes

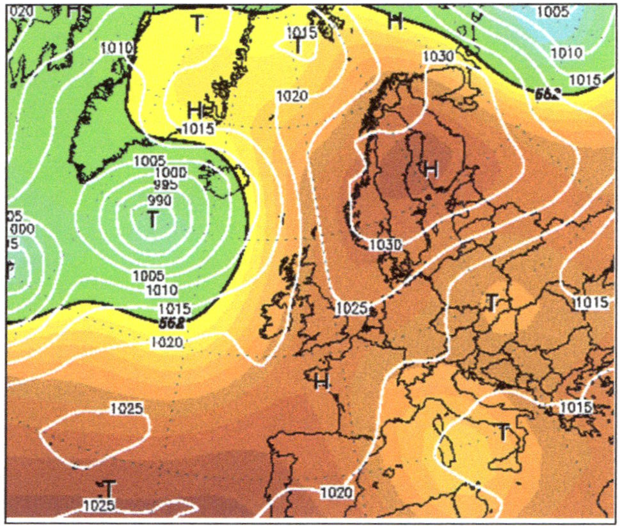

RAIN POTENTIAL

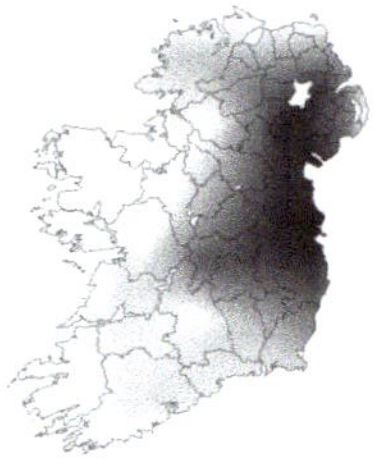

FROST/SNOW

JUNE **14** SATURDAY

A large anticyclonic system drifts over the country directing a light northerly flow throughout. Mostly sunny and dry everywhere except in the east to northeast, where cloudier conditions prevail and isolated showers persist. Chance of fog patches overnight in the west to midlands. Warmer.

Belfast: Sunny, cooler, mostly dry, light to moderate breezes
Dublin: Cloudy, isolated showers, cool northerly breezes
Cork: Sunny, dry, mild day cold night, light northerly breezes
Galway: Very sunny, dry, cooler, gentle shifting breezes, chance of fog overnight

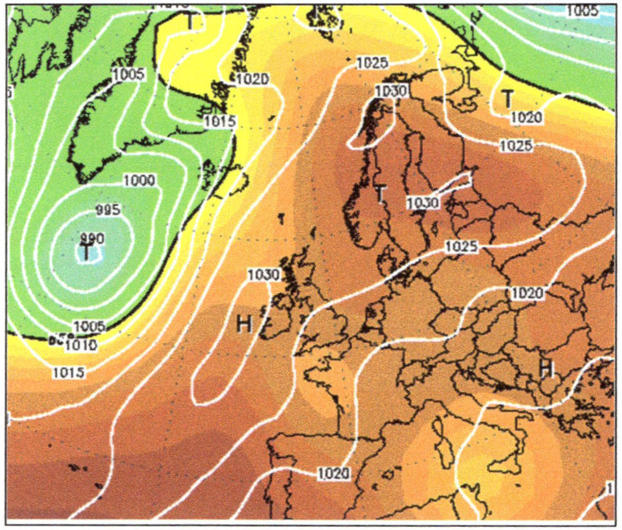

RAIN POTENTIAL

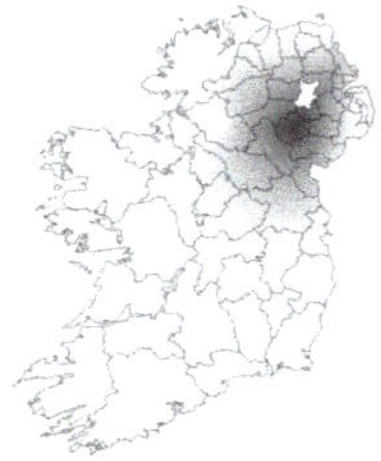

FROST/SNOW

JUNE **15** SUNDAY

Anticyclonic. Warm, dry, and sunny throughout the country with patches of overnight fog particularly in central, eastern and southern parts. A light easterly air prevails. Chance of some isolated showers briefly around the western coastal fringes in the evening.

Belfast: Sunny, warmer, dry, gentle breezes and pleasant air
Dublin: Sunny, dry, light breezes to mostly calm air, chance of patches of shallow fog
Cork: Very sunny and dry, milder, gentle breezes, shallow fog possible
Galway: Sunny, cooler, dry, mostly calm air

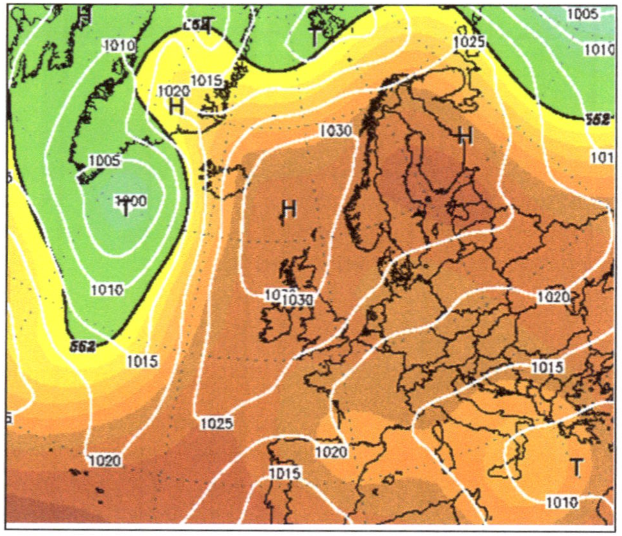

RAIN POTENTIAL

FROST/SNOW

JUNE **16** MONDAY

High pressure prevails over Ireland bringing a dry, warmer, mostly calm and serene day throughout the country, with plenty of sunshine and and patches of overnight fog in many districts.

Belfast: Very sunny, dry, mostly serene air, chance of misty patches overnight
Dublin: Mostly calm and dry, sunny, mild day cool night, fog patches likely
Cork: Sunny, dry, mild temperatures, mostly calm air, chance of fog
Galway: Very sunny, mild, dry, mostly calm and serene air, chance of misty patches

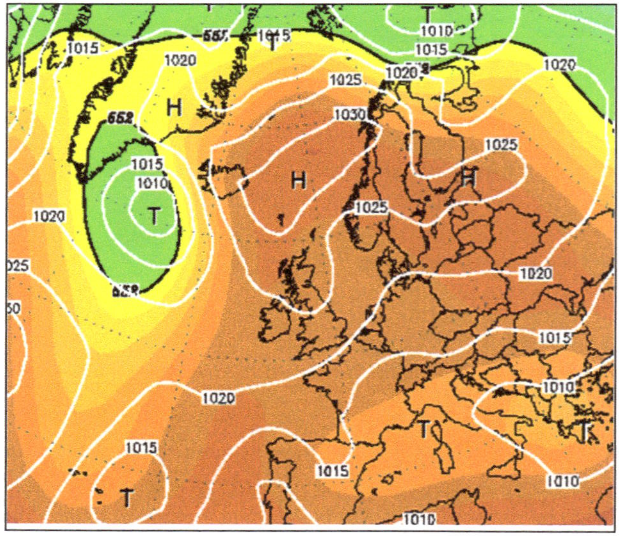

RAIN POTENTIAL

FROST/SNOW

JUNE **17** TUESDAY

A high pressure system continues to linger over Ireland maintaining the dry and pleasant conditions throughout. Very sunny and warm everywhere, with patches of overnight mist and fog in many districts.

Belfast: Very sunny, dry, warm, calm and pleasant air, chance of misty patches
Dublin: Very sunny, dry, mild day cool night, pleasant air, possible fog patches
Cork: Very sunny, dry, mild temperatures, calm and serene air, fog patches likely
Galway: Very bright and sunny, warmer, dry, calm air, chance of shallow fog patches

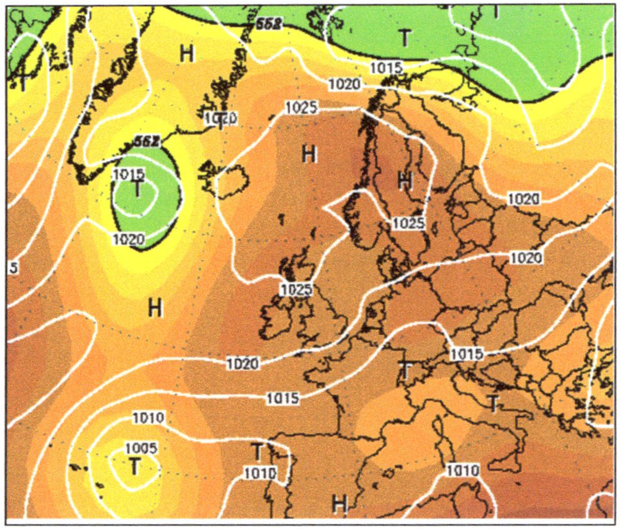

RAIN POTENTIAL

FROST/SNOW

JUNE **18** WEDNESDAY

Anticyclonic conditions continue to prevail throughout the country. A pleasant day with plenty of sunshine, warm temperatures and dry everywhere. Patches of overnight fog again in many districts.

Belfast: Very bright and sunny, warm day cool night, calm air, fog patches possible
Dublin: Very bright and sunny, warm day cool night, serene air, chance of fog
Cork: Mostly calm and pleasant air, very sunny, dry, chance of fog
Galway: Very bright and sunny, dry, mild temperatures, calm atmosphere

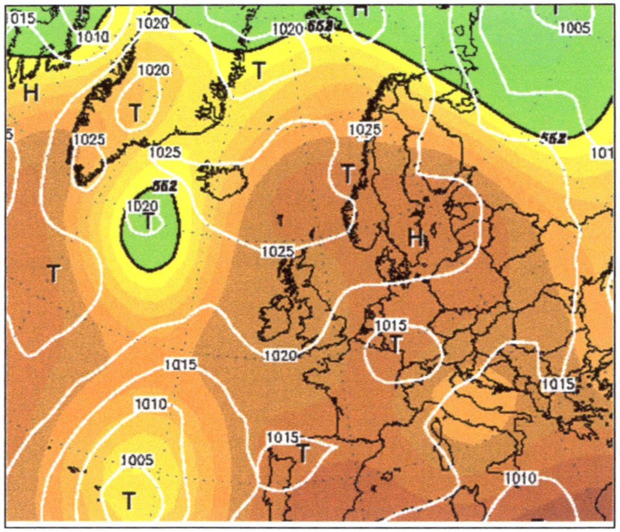

RAIN POTENTIAL

FROST/SNOW

JUNE 19 THURSDAY

The high pressure ridge remains fairly stable over Ireland maintaining a pleasant air throughout the country. Very sunny, warm, and dry everywhere with further patches of fog overnight.

Belfast: Warmer, dry, very sunny, calm atmosphere, chance of fog patches
Dublin: Warm day cool night, very sunny, dry, mostly calm air, fog patches possible
Cork: Pleasant and dry air, very sunny, chance of overnight fog
Galway: Sunny, dry, cooler, gentle shifting breezes

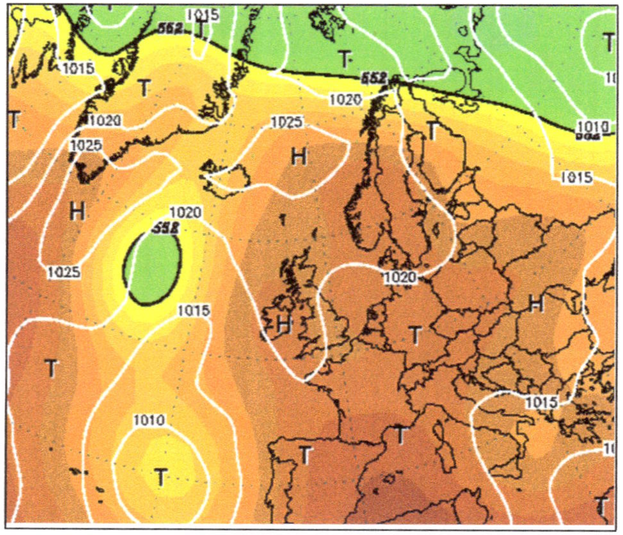

RAIN POTENTIAL

FROST/SNOW

JUNE 20 FRIDAY

The high pressure ridge begins to slowly drift away in a northern tract, while a weak depression to the southwest of Ireland nears the southeast of the country. Mostly sunny and dry after overnight fog clears, with very gentle southeasterly breezes prevailing. Cloud increases in the southwest later in the day.

Belfast: Very warm and sunny, dry, calm and pleasant air, chance of misty patches
Dublin: Sunny, dry, warm day mild night, serene atmosphere, fog patches likely
Cork: Sunny, dry, warmer, serene atmosphere, chance of misty patches overnight
Galway: Bright and sunny, warmer, dry, calm and pleasant air

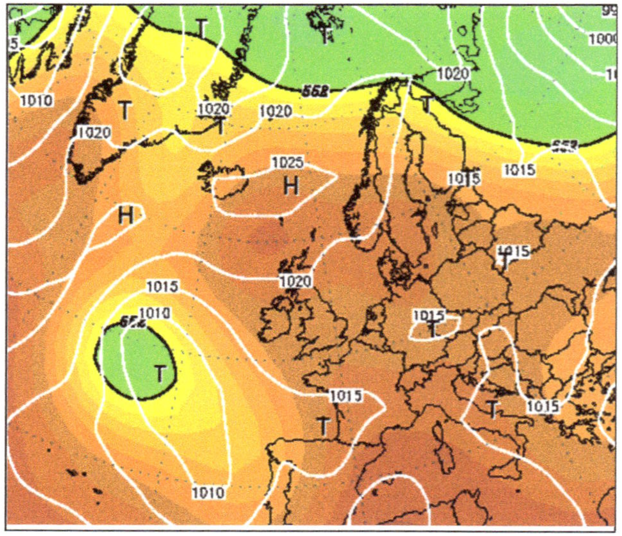

RAIN POTENTIAL

FROST/SNOW

JUNE **21** SATURDAY

A low pressure system to the southeast of Ireland directs a light southeasterly flow over the country, fresher at times around the coasts of Munster. A thunderstorm embedded trough associated with the depression to the southeast spreads increasing cloud throughout, and occasional showers and the odd thunderstorm in the south and southwest at first, before spreading in to central and northern parts, with much of the east and far northern coasts remaining mostly sunny and dry.

Belfast: Mostly sunny, very warm and dry, calm atmosphere, chance of shallow fog
Dublin: Partly cloudy, warm, humid, dry, mostly calm air, chance of shallow fog
Cork: Mostly calm and pleasant breezes, cooler, cloudy, brief misty showers
Galway: Partly cloudy, brief showers, chance of thunderstorms

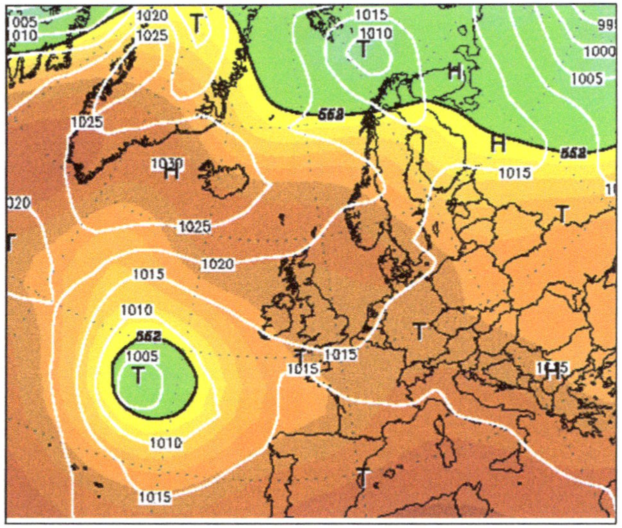

RAIN POTENTIAL

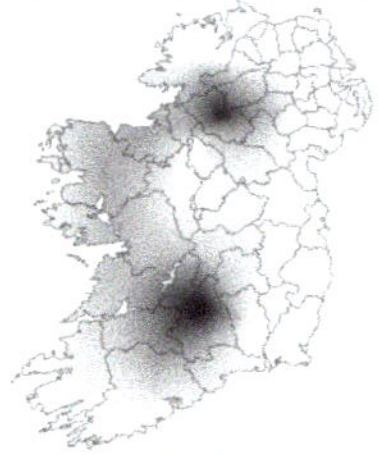

FROST/SNOW

JUNE **22** SUNDAY

A low pressure trough over Ireland brings a light south to southwest flow over the country, with fresher breezes accompanying the passage of occasional showers. Some overnight patches of mist and fog, particularly in the west and south. Generally mostly cloudy and humid, with some brief showers particularly in the north of the country, sometimes accompanied by thunderstorm activity.

Belfast: Hot, dry, cloudy spells, mostly calm and humid air, chance of thunderstorms
Dublin: Partly cloudy, warm and humid, light showers, calm and serene air
Cork: Cloudy, light showers, mild temperatures, serene air, chance of mist
Galway: Cloudy, calm air, warm, mostly dry, chance of fog patches

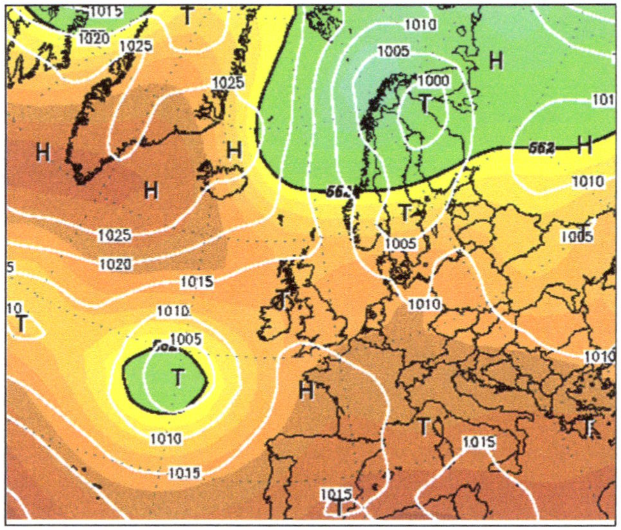

RAIN POTENTIAL

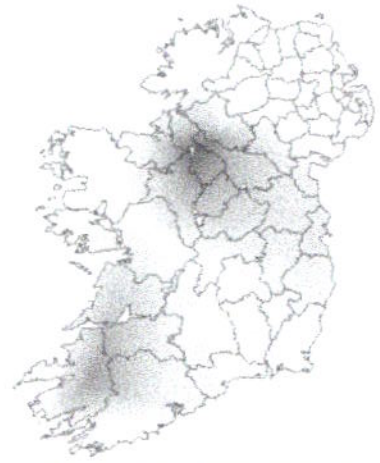

FROST/SNOW

JUNE 23 MONDAY

A thunderstorm embedded trough associated with a depression to the southeast of Ireland crosses the country bringing a mix of sunny spells and outbreaks of rain and some thunderstorms during the day in most districts. A very light and variable flow prevails with the chance of some overnight patches of mist and fog, particularly in the east and north. Slightly cooler.

Belfast: Warm, humid, overcast, occasional showers, calm and static air
Dublin: Overcast, cooler, scattered showers, mostly calm air, chance of mist
Cork: Mostly cloudy, occasional showers, light breezes
Galway: Cloudy, cooler, scattered showers, mostly calm atmosphere

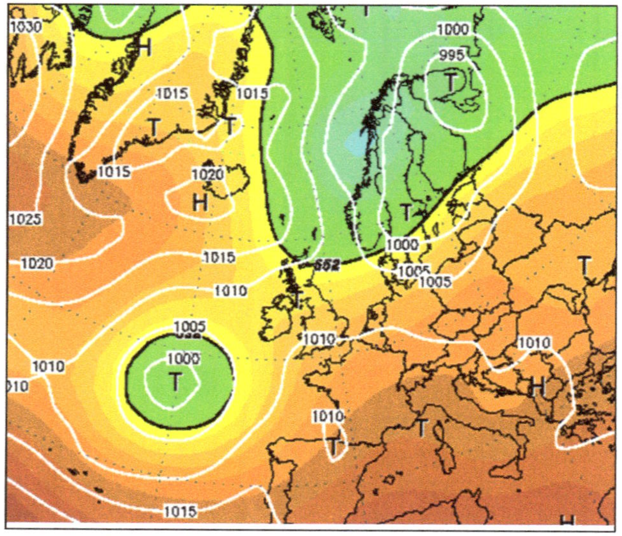

RAIN POTENTIAL

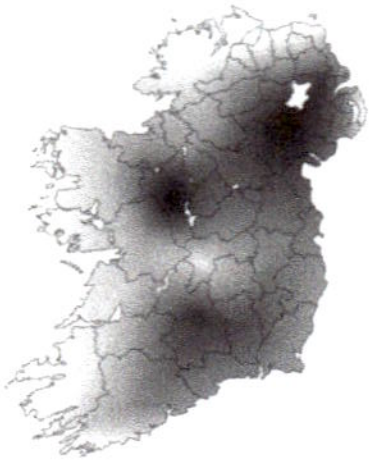

FROST/SNOW

JUNE 24 TUESDAY

The depression to the south of Ireland nears the southern coasts with light winds at first, but freshening later, with some gusty spells around the coastal fringes. Widespread cloud throughout with outbreaks of rain and scattered thunderstorms in most districts. Rain likely to be heavy for a time in some parts of the southwest and northwest.

Belfast: Overcast, cooler, rain, moderate breezes
Dublin: Overcast, colder, scattered showers, moderate breezes
Cork: Cloudy, heavy rain, breezy spells
Galway: Overcast, heavy rain, fresh winds with occasional gusts

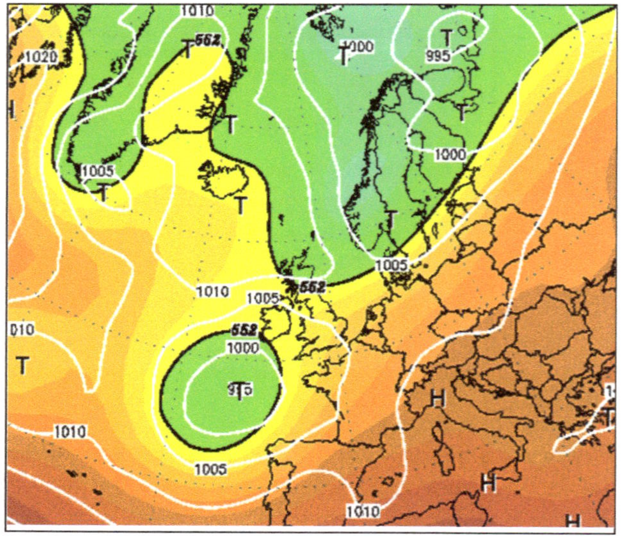

RAIN POTENTIAL

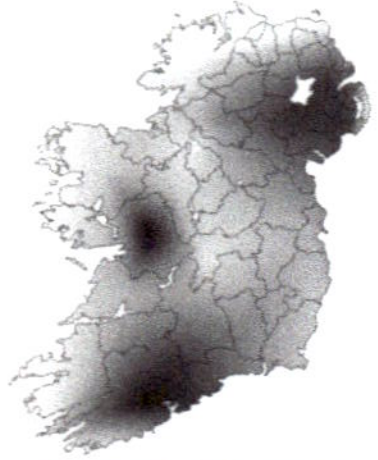

FROST/SNOW

JUNE 25 WEDNESDAY

A weak depression moves onto the country bringing a stormy air throughout, with widespread cloud and rain in most districts but lighter around the northwest, as well as the chance of some thunderstorms here and there. Moderate east to northeasterly winds prevail over the country, occasionally stronger and more blustery in the north.

Belfast: Overcast and colder, scattered showers, fluctuating winds
Dublin: Overcast, cold, rain, stormy air, fluctuating breezes
Cork: Overcast, colder, heavy rain, unsettled winds
Galway: Overcast, cool, dry with passing threats of rain, blustery at times

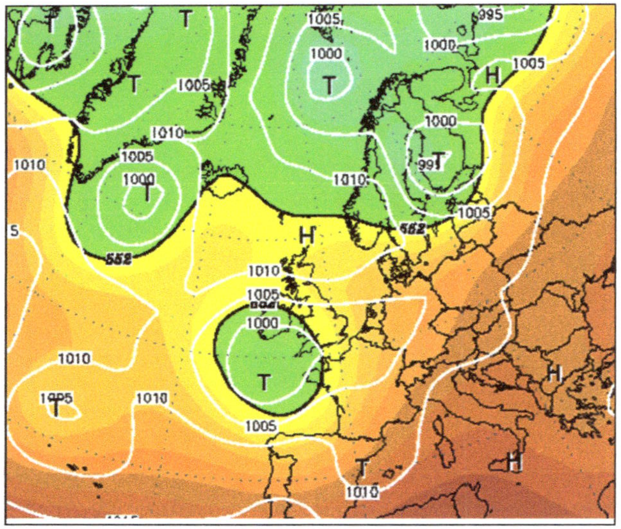

RAIN POTENTIAL

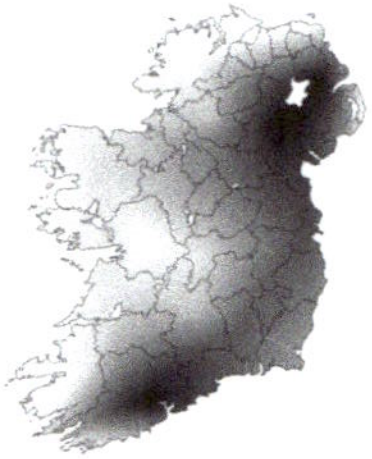

FROST/SNOW

JUNE 26 THURSDAY

A weak depression continues to linger over Ireland directing light to moderate north to northeast breezes. Widespread cloud prevails with brief outbreaks of rain or drizzle likely in most districts, but clearing quickly to mostly dry in the southern half of the country, and increasing sunny spells in the west.

Belfast: Cold, mostly dry, widespread cloud, threats of rain, shifting breezy spells
Dublin: Overcast, mostly dry, chance of isolated showers, cold, unsettled breezes
Cork: Overcast, cool, mostly dry with passing threats of rain, fluctuating winds
Galway: Partly cloudy, dry, cool, unsettled windy spells

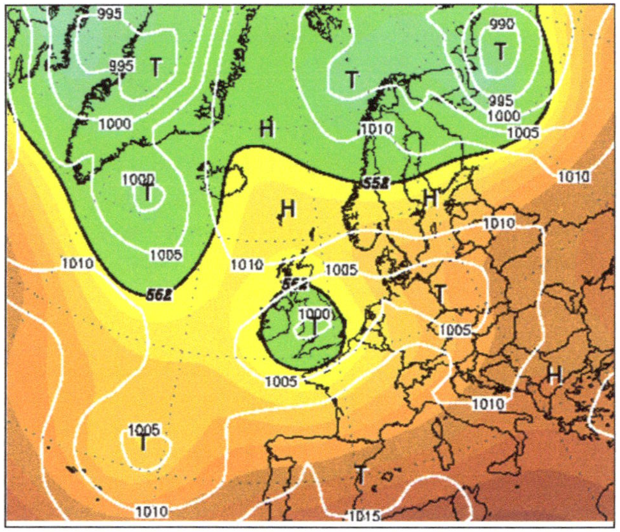

RAIN POTENTIAL

FROST/SNOW

JUNE **27** FRIDAY

Low pressure continues to prevail over Ireland directing a light variable flow throughout. Mostly dry everywhere except for some lingering isolated showers in the west, and patches of fog overnight in many places. Some good sunny spells in the south and west, while cloudier elsewhere.

Belfast: Overcast, dry, slightly warmer, calm air, chance of overnight fog patches
Dublin: Dull and gloomy, mostly calm air, dry with odd threat of rain
Cork: Mostly sunny, isolated showers, warmer day cooler night, mostly calm air
Galway: Light showers overnight, sunny day, cool, gentle breezes, chance of fog

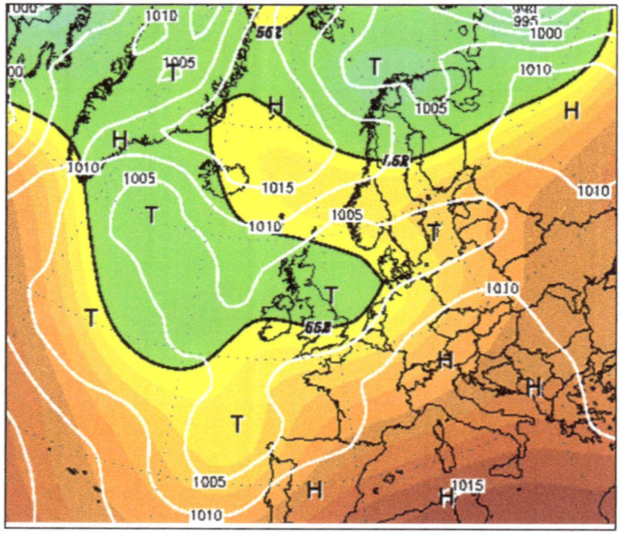

RAIN POTENTIAL

FROST/SNOW

JUNE **28** SATURDAY

Low pressure maintains a light east to northeast flow over the country, backing southeasterly later in the day. Mostly cloudy everywhere with a band of showers moving into the southern half of the country at first, and slowly crosses up through the country, spreading showers into the northern half of the country. Rain may be heavy for a time in the south of Munster.

Belfast: Cloudy, warm day cool night, isolated showers, light breezes, chance of fog
Dublin: Mostly cloudy, isolated showers, light shifting breezes, chance of shallow fog
Cork: Changeable, partly cloudy, overnight rain, unsettled windy spells
Galway: Cloudy, occasional showers, moderate shifting breezes

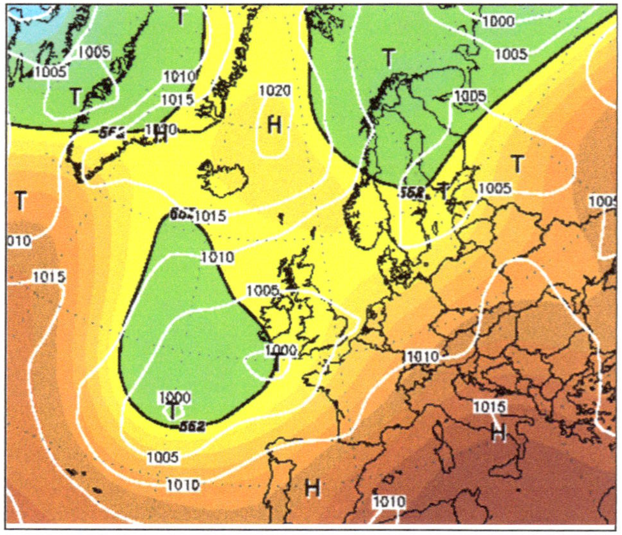

RAIN POTENTIAL

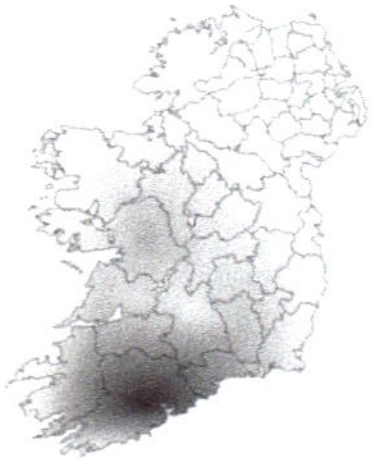

FROST/SNOW

JUNE **29** SUNDAY

A depression to the southeast of Ireland intensifies and brings a very unsettled and stormy day throughout the country. Dull and gloomy everywhere with widespread rain and scattered thunderstorms in most regions, although rain activity will be lighter along the northern coastal fringes. Winds will strengthen to strong at times, particularly in the south and west.

Belfast: Overcast, rain, warmer, strengthening winds
Dublin: Stormy, unsettled winds, overcast, heavy rain, chance of thunderstorms
Cork: Stormy air, overcast, heavy rain, strong winds at times, cooler
Galway: Stormy air, overcast, heavy rain, chance of gale gusts

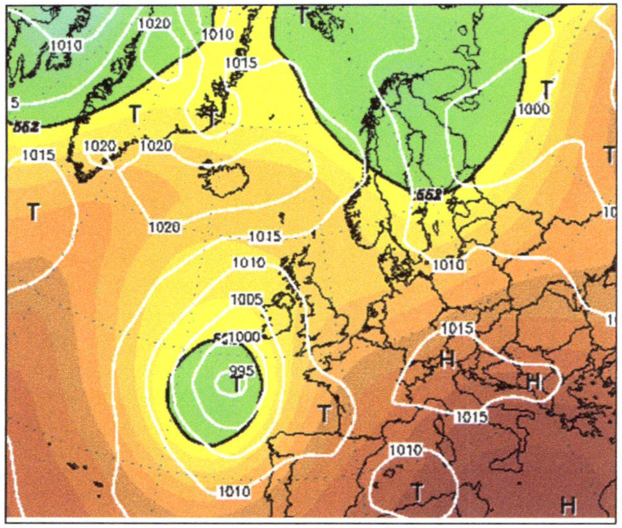

RAIN POTENTIAL

FROST/SNOW

JUNE **30** MONDAY

A stormy air prevails over Ireland as a depression lingers over the country. Widespread cloud, outbreaks of thunderstorms and rain likely in most districts, heavy at times particularly in the south and west.. Winds likely to be very unsettled in a fresh to strong south to southeast flow with gale gusts at times possible particularly around the coasts.

Belfast: Partly cloudy, light showers, strong and gusty winds with possible gale gusts
Dublin: Stormy air, strong gusty winds, light squally showers, frequent cloudy spells
Cork: Overcast, heavy rain, strong winds with gusts
Galway: Cloudy, rain, warmer, blustery winds, chance of thunderstorms and gales

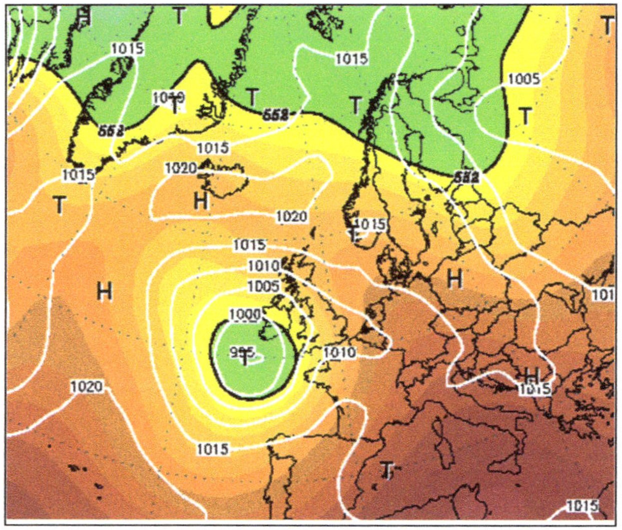

RAIN POTENTIAL

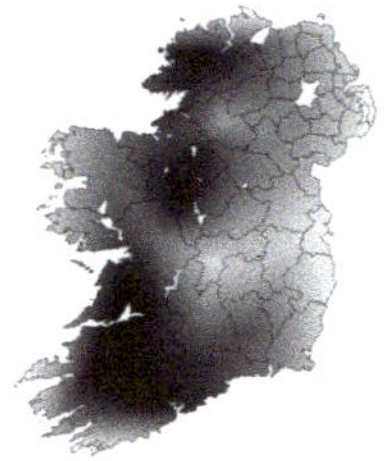

FROST/SNOW

JULY

PHASES OF THE MOON

2nd	First Quarter
2nd	Crossing Equator
5th	Apogee
9th	Southern Declination
10th	Full Moon
17th	Crossing Equator
18th	Third Quarter
20th	Perigee #11
22nd	Northern Declination
24th	New Moon
30th	Crossing Equator

1	2	3	4	5	6	7	8	9	10	11	12	13	14	15	16	17	18	19	20	21	22	23	24	25	26	27	28	29	30	31
Mon	Tue	Wed	Thur	Fri	Sat	Sun	Mon	Tue	Wed	Thur	Fri	Sat	Sun	Mon	Tue	Wed	Thur	Fri	Sat	Sun	Mon	Tue	Wed	Thur	Fri	Sat	Sun	Mon	Tue	Wed
	1Q XhS			A				V	F							XhN	3Q		P11		^		N						XhS	

MONTHLY SUMMARY

1st-10th Continuing from the end of last month, a series of frontal systems associated with depressions cross the southern half of Ireland in an easterly trajectory, bringing widespread cloud and outbreaks of rain or showers daily, some very heavy at times and accompanying thunderstorm activity. Winds mainly easterly at first later backing northerly for a time before finally having a milder, southerly change. Temperatures likely to be mostly average to below average.

1st-2nd Very heavy rain associated with thunderstorm activity in the east of the country may cause some localized flooding. Some showers may be of hail mostly in inland parts.

3rd Chance of snow grains being noted overnight in central parts.

4th A depression centred over the east of Ireland brings a showery day, some heavy at times, with chance of scattered thunderstorms into the southwest. Unstable north-westerly winds may produce funnel formation clouds.

7th-9th Chance of light frost overnight in the northwest to midlands and possible snow grains.

10th Chance of thunderstorms.

Over the next four weeks the first week begins with a depression, before high pressure dominates the following couple of weeks, with the fourth week dropping back into low pressure for a few days then ending the last couple of days on high pressure. Overall, this outlook period will be wetter than normal everywhere, especially in the eastern half of the country, where some districts may receive up to three times more rainfall than is typical for this time of year. Much of the rain activity during the next four weeks will be accompanied by thunderstorm activity and only a few days at the very end of this outlook period will be dry. Despite the frequent outbursts of rain however, sunshine hours will be above normal almost everywhere by the end of this four week look-ahead period, with the coastal counties in the west and north particularly sunnier than normal. Maximum temperatures will be below average almost everywhere except along the Atlantic coasts, where temperatures will be more typical for this time of year, while minimum temperatures in contrast, will be average to above average throughout most of this outlook period. During this outlook period, rain can be expected on between 15 and 27 days, with the most in the south and east, and the least rain days in the north and west. Thunderstorms can be expected on 6 days and hail showers on 4 days. Fogs are likely to be widespread on 4 days, while winds to gale force are likely on 2 days as well as up to 4 days when funnel clouds or water spouts may be observed. No frosts or snow are expected during this outlook period.

11th-29th Depressions centred near to Ireland bring a spell of very unsettled weather throughout the country. Some rain or showers can be expected daily.

11th-23rd Westerly winds are the predominant wind flow.

11th Thunderstorms likely and may accompany bursts of heavy rain.

15th-16th Very strong and gusty winds accompany a depression north of Ireland. Winds likely to reach gale force at times, particularly over the northern half of the country.

16th-18th Long sunny spells likely between the showers.

17th-18th Clear skies likely overnight bringing unusually low minimum temperatures. Chance of isolated patches of light frost in the east.

18th-19th Particularly wet spell with widespread heavy rain and possible thunderstorms and hail in many districts. Unsettled winds during this period may see the formation of funnel clouds.

21st-31st Prolonged wet spell when some rain can be expected every day in many parts of the country.

21st-22nd Chance of widespread fog overnight and unusually high maximum temperatures.

23rd Very heavy rain likely in most parts of the country.

24th-28th Long sunny spells can be expected between the showers.

25th Chance of widespread fog and a very sunny day with high maximum temperatures.

26th-30th Bands of thunderstorms and showers of hail cross the country. Very unsettled spell during which funnel clouds and water spouts may form.

29th Chance that heavy rain that accompanies thunderstorm and hail activity around this time may lead to flooding in some places, particularly in eastern parts.

30th Widespread fog likely.

31st Chance of funnel cloud formations being observed in the west of the country.

IRELAND RAINFALL ESTIMATES

July 2025

d=mainly dry, n= nonrecordable, l=light shrs, s=significant shrs, r=rain, h=hvy falls

		ULSTER PROVINCE					CONNAUGHT PROVINCE						LEINSTER PROVINCE				MUNSTER PROVINCE						
		Alderg	Hillsbo	Armag	Loughl	Malin H	Sligo	Ardtarı	Belmul	Drums	Galway	Roscoı	Markre	Edende	Kilkenı	Dublin	Shannı	Newpo	Tralee	Killarne	Cork	All	Moon
1st	Tues	l	l	l	l	d	l	l	l	d	l	l	l	l	l	s	s	l	s	l	l	40	
2nd	Wed	l	l	l	n	d	l	n	d	l	d	n	l	s	r	h	d	l	d	n	l	50	1Q XhS
3rd	Thur	r	r	r	s	s	s	s	l	l	d	s	r	l	l	s	l	l	l	l	d	88	
4th	Fri	s	l	s	l	n	d	l	l	l	s	s	l	l	s	l	r	l	s	h	r	93	
5th	Sat	l	d	s	d	d	d	d	l	d	d	d	d	d	d	d	d	d	d	d	l	13	A
6th	Sun	d	n	d	l	l	l	l	l	l	l	l	l	l	d	d	d	l	d	l	d	21	
7th	Mon	l	l	n	l	s	l	l	l	d	l	d	s	n	d	l	l	n	d	d	d	26	
8th	Tues	s	r	l	l	h	s	l	l	s	r	s	s	s	s	s	s	r	s	s	l	134	
9th	Wed	s	s	s	l	s	l	l	l	l	s	l	l	l	s	l	s	s	l	r	l	81	V
10th	Thur	s	s	l	s	s	s	s	s	r	s	l	l	s	s	s	s	s	h	h	s	140	F
11th	Fri	s	s	l	s	d	s	s	n	l	r	l	l	r	l	r	s	r	s	l	l	106	
12th	Sat	s	l	s	s	s	s	s	l	l	l	s	r	r	l	s	s	s	l	l	l	102	
13th	Sun	s	l	r	s	l	r	s	s	r	s	r	r	s	l	s	l	h	l	s	l	129	
14th	Mon	s	r	s	l	s	s	l	l	l	l	l	l	l	l	l	r	s	d	s	n	69	
15th	Tues	s	r	s	r	s	r	r	s	s	s	r	h	s	l	l	l	r	s	l	s	156	
16th	Wed	l	l	s	l	h	l	l	l	r	l	r	n	s	d	l	n	l	d	d	d	73	
17th	Thur	l	d	s	d	l	l	d	d	l	r	l	d	n	s	s	l	l	l	h	l	61	XhN
18th	Fri	s	r	d	s	l	r	s	l	s	l	l	s	s	l	l	l	s	s	s	h	117	3Q
19th	Sat	r	r	s	h	r	h	h	h	h	s	h	h	h	s	s	s	h	d	s	l	277	
20th	Sun	l	d	s	d	d	l	d	d	d	d	d	d	d	d	d	d	d	d	d	d	11	P11
21st	Mon	l	s	d	s	s	r	s	s	s	s	s	s	r	l	l	s	l	l	l	l	95	
22nd	Tues	l	l	l	l	l	l	l	n	d	d	d	l	l	s	l	l	d	l	l	n	31	^
23rd	Wed	s	r	n	l	r	s	l	l	s	l	s	s	r	r	h	s	l	s	l	r	133	
24th	Thur	r	s	h	n	l	l	n	s	l	d	h	l	s	r	l	l	d	d	n	d	104	N
25th	Fri	l	l	l	h	d	h	h	l	r	s	r	h	r	s	l	l	h	l	l	l	167	
26th	Sat	l	l	s	s	l	r	s	r	s	l	s	r	l	l	s	r	s	s	l	d	112	
27th	Sun	s	r	l	s	d	s	s	s	s	l	s	l	l	r	s	d	r	d	l	r	93	
28th	Mon	s	s	l	s	d	s	s	s	s	r	r	l	l	d	l	l	l	d	l	l	82	
29th	Tues	r	l	h	l	d	s	l	d	h	h	s	r	d	s	d	l	d	r	s	h	184	
30th	Wed	s	r	d	s	d	l	s	d	d	l	r	l	r	r	h	s	r	l	l	d	105	XhS
31st	Thur	l	d	s	l	l	l	l	d	l	l	l	l	l	l	l	l	s	d	d	s	47	
Estimate:		154	155	154	138	114	197	138	96	170	165	192	174	159	143	158	107	193	92	127	117	2942	
Average:		71	77	63	86	81	93	95	79	73	87	86	93	71	59	56	66	101	103	75	79	1593	
Trend:		w tr	w tr	w tr	w tr	w tr	w tr	w tr	w tr	w tr	w tr	w tr	w tr	w tr	w tr	w tr	w tr	w tr	drr	w tr	w tr	w tr	

IRELAND SUNSHINE ESTIMATES

July 2025

F=fine (8-12 hours of sunshine), pc= partly cloudy (4-7 hours), c=cloudy (1-3 hours), o=overcast (0 hours)

		ULSTER PROVINCE					CONNAUGHT PROVINCE						LEINSTER PROVINCE				MUNSTER PROVINCE						
		Alderg	Hillsbo	Armag	Loughl	Malin H	Sligo	Ardtar	Belmul	Drums	Galway	Roscor	Markre	Edend	Kilkenn	Dublin	Shann	Newpo	Tralee	Killarne	Cork	All	Moon
1st	Tues	c	c	c	c	c	c	c	F	o	pc	F	c	o	o	o	c	o	pc	pc	o	49	
2nd	Wed	o	o	o	o	o	o	o	c	c	c	c	o	c	pc	o	c	c	pc	pc	F	46	1Q XhS
3rd	Thur	c	c	c	c	c	o	c	o	c	c	o	c	c	c	c	c	c	c	c	c	35	
4th	Fri	pc	pc	pc	pc	pc	pc	pc	F	pc	pc	F	pc	pc	pc	pc	pc	pc	F	F	F	123	
5th	Sat	c	c	c	c	c	c	c	c	c	c	c	c	c	c	c	c	c	pc	pc	c	43	A
6th	Sun	c	c	c	c	c	c	c	c	c	c	c	c	c	c	pc	c	c	pc	pc	c	66	
7th	Mon	F	F	F	F	F	F	F	F	F	F	F	F	F	pc	F	F	F	F	F	pc	185	
8th	Tues	o	o	o	o	o	o	o	c	c	c	c	o	c	c	o	o	o	o	o	c	21	
9th	Wed	c	c	c	c	c	c	c	pc	pc	pc	pc	c	pc	pc	pc	pc	pc	c	c	c	90	V
10th	Thur	c	c	c	c	c	c	c	pc	c	pc	pc	c	c	c	c	pc	pc	c	c	c	72	F
11th	Fri	pc	pc	pc	pc	pc	pc	pc	pc	pc	pc	pc	pc	pc	pc	c	c	pc	o	o	F	86	
12th	Sat	pc	pc	pc	pc	pc	pc	pc	pc	pc	c	pc	pc	pc	pc	pc	c	c	c	c	pc	99	
13th	Sun	c	c	c	c	c	o	c	c	pc	pc	c	c	pc	pc	pc	pc	pc	pc	pc	F	84	
14th	Mon	c	c	c	c	c	o	c	c	pc	c	c	c	pc	pc	pc	pc	pc	c	c	pc	66	
15th	Tues	o	o	o	o	o	o	o	o	c	o	o	o	c	c	c	o	c	o	o	c	17	
16th	Wed	pc	pc	pc	pc	pc	pc	pc	F	F	F	F	pc	F	F	pc	F	F	F	F	F	178	
17th	Thur	F	F	F	F	F	F	F	F	F	F	F	F	F	F	F	F	F	F	F	F	247	XhN
18th	Fri	pc	pc	pc	pc	pc	c	pc	F	pc	F	F	pc	pc	pc	pc	pc	pc	F	F	pc	130	3Q
19th	Sat	c	c	c	c	c	o	c	o	c	o	o	c	c	c	pc	o	c	pc	pc	o	44	
20th	Sun	c	c	c	c	c	pc	c	F	c	F	F	c	c	pc	o	F	pc	c	c	pc	93	P11
21st	Mon	o	o	o	o	o	o	o	o	o	o	o	o	o	o	o	c	o	o	o	o	7	
22nd	Tues	c	c	c	c	c	pc	c	pc	pc	c	pc	c	pc	pc	c	c	pc	pc	pc	F	82	^
23rd	Wed	o	o	o	o	o	o	o	o	o	o	o	o	o	o	o	o	o	c	c	o	6	
24th	Thur	pc	pc	pc	pc	pc	pc	pc	F	F	F	F	pc	F	F	pc	pc	pc	F	F	F	158	N
25th	Fri	pc	pc	pc	pc	pc	pc	pc	pc	pc	pc	pc	pc	pc	pc	pc	F	pc	pc	pc	c	126	
26th	Sat	pc	pc	pc	pc	pc	c	pc	o	pc	c	c	pc	pc	F	pc	c	pc	F	F	F	112	
27th	Sun	pc	pc	pc	pc	pc	pc	pc	c	pc	pc	c	pc	pc	pc	c	F	pc	F	F	pc	121	
28th	Mon	c	c	c	c	c	c	c	pc	pc	pc	pc	c	pc	pc	pc	c	pc	F	F	F	111	
29th	Tues	pc	pc	pc	pc	pc	c	pc	o	F	c	c	pc	F	F	F	c	pc	c	c	pc	112	
30th	Wed	c	c	c	c	c	c	c	F	c	pc	F	c	c	c	o	o	o	c	c	c	51	XhS
31st	Thur	c	c	c	c	c	c	c	F	c	pc	F	c	c	c	o	pc	c	pc	pc	c	63	
Estimate hours:		119	119	119	119	119	107	119	169	134	148	159	119	134	149	131	128	138	164	164	168	2722	
Average hours:		152	152	135	143	133	121	122	133	129	144	157	142	155	144	167	142	121	142	140	167	2841	
Trend:		less	less	less	less	less	less	av	more	av	av	av	less	less	av	less	less	more	more	more	av	less	

JULY **1** TUESDAY

A depression sits over Ireland bringing a stormy air throughout the country with fresh and gusty winds for a time in many districts. Overnight rain prevails in most areas, with a mix of brief sunny spells and showers throughout the day. Cloudier in the south and east, with possible thunderstorms at times.

Belfast: Stormy, cloudy, squally showers, strong gusty winds
Dublin: Stormy air, overcast, squally showers, coolish
Cork: Overcast, passing drizzly showers, fresh breezes at times
Galway: Stormy air, partly cloudy, light squally showers, gale winds likely at times

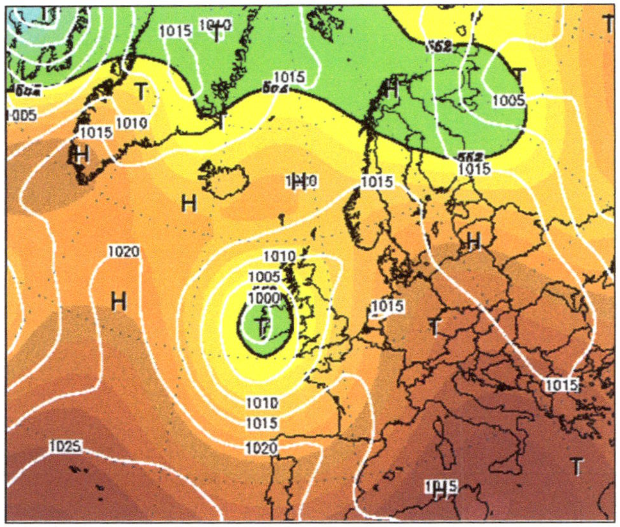

RAIN POTENTIAL

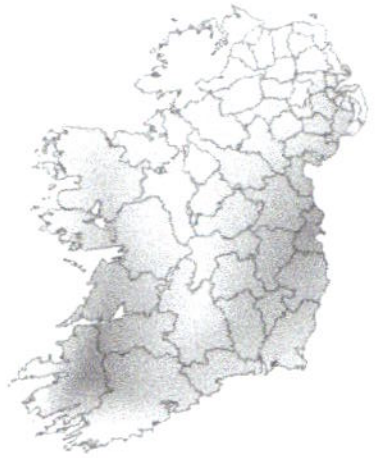

FROST/SNOW

JULY **2** WEDNESDAY

A weakening depression continues to linger over Ireland with light to moderate winds prevailing, generally in a northerly flow. Mostly dry in the south and west with passing sunny spells, while increased cloud and outbreaks of rain persist in the east, more isolated showers in the north.

Belfast: Winds easing, overcast, isolated showers, warm and humid
Dublin: Fluctuating breezes, dull and gloomy air with heavy rain likely
Cork: Sunny, warmer, brief showers, light breezes, chance of fog overnight
Galway: Cloudy, cooler, dry, strong winds easing

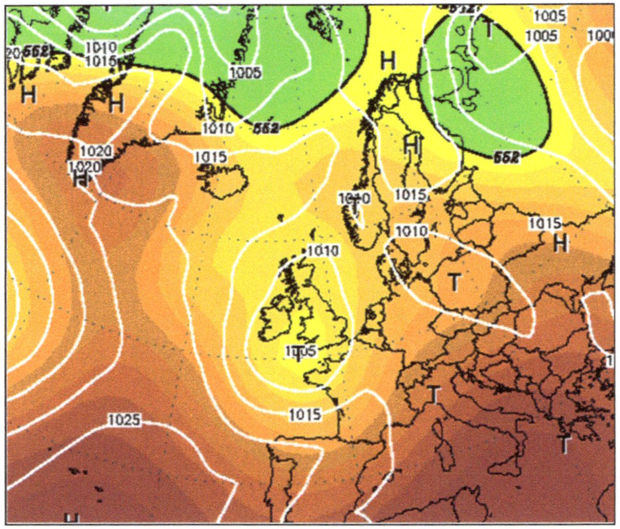

RAIN POTENTIAL

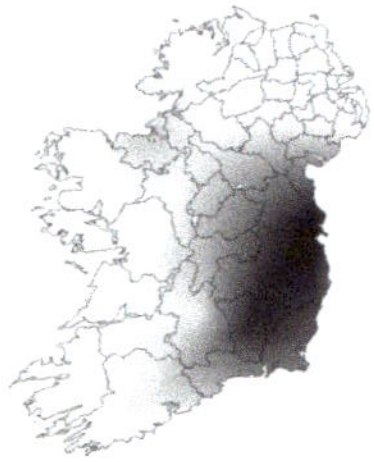

FROST/SNOW

JULY **3** THURSDAY

A gentle northwest flow prevails over Ireland directed from a low pressure trough crossing the country. Patches of overnight fog at first clear to a mixed day of some sunny spells and isolated showers in most parts, although mainly dry in the south of the country.

Belfast: Cloudy, rain, moderate breezes
Dublin: Partly cloudy, occasional showers, cooler, light breezes, chance of fog
Cork: Cloudy and cooler, dry, moderate breezes
Galway: Cloudy, cool, dry, light breezes

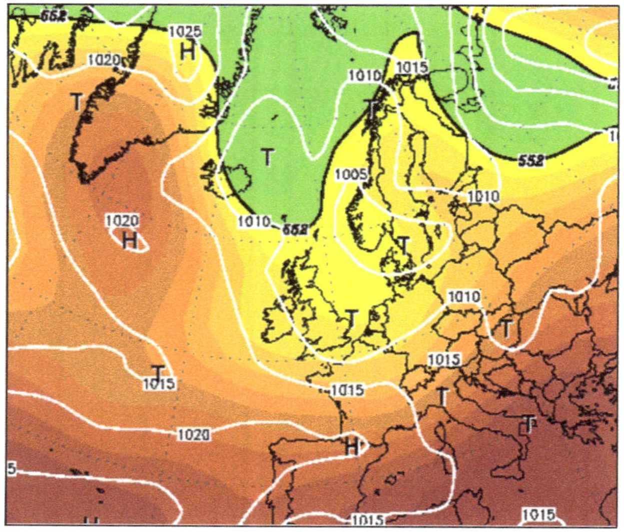

RAIN POTENTIAL

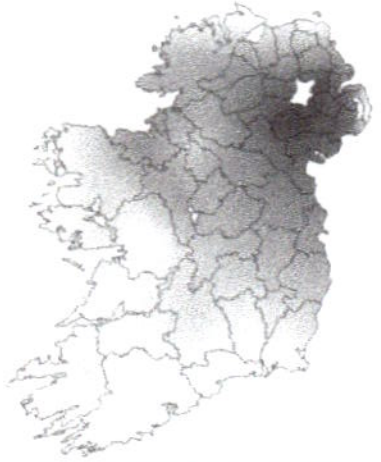

FROST/SNOW

JULY **4** FRIDAY

A weak low presure trough bring light to moderate north to northwest breezes over the country. Another changeable day with a mix of sunny spells and occasional showers almost everywyhere, although mostly dry in the far northwest of the contry. Some showers may be accompanied by thunderstorm activity and heavier downpours and possible hail showers for a time mostly in the east. Winds freshen later in the day.

Belfast: Partly cloudy, light showers, cooler, gentle breezes
Dublin: Partly cloudy, brief showers, cooler, mostly calm air
Cork: Changeable, mix of sunny spells and outbreaks of rain, fluctuating breezes
Galway: Mostly sunny, brief morning showers, fresh breezes at times

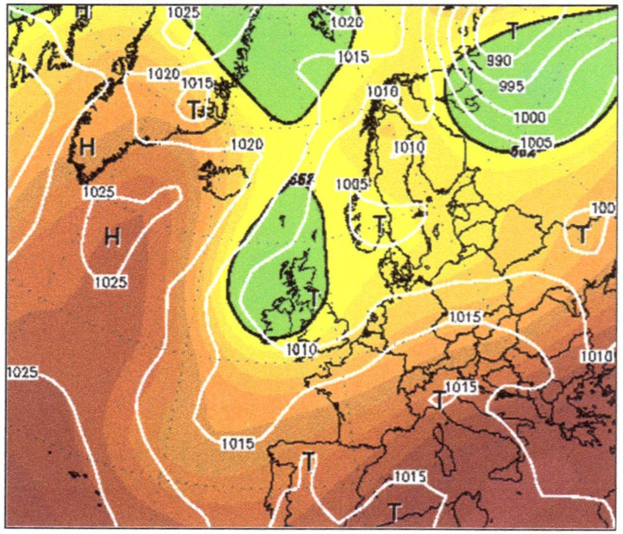

RAIN POTENTIAL

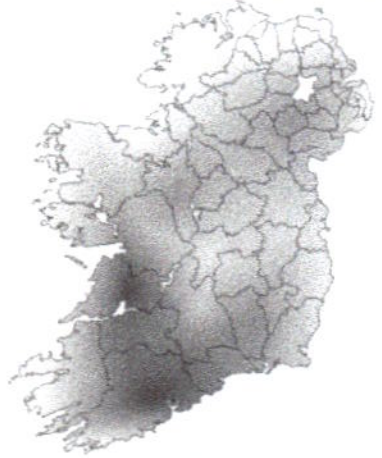

FROST/SNOW

JULY **5** SATURDAY

A weak depression builds over Ireland directing colder northerly flow, with winds becoming strong and gusty at times particularly around the coasts. Frequent scattered cloud everywhere with some isolated showers in the norteast and southeast, while remaining mostly dry elsewhere.

Belfast: Cloudy and dry, cool and blustery northerly winds at times
Dublin: Cloudy, mostly dry, odd threat of rain, strong breezes at times
Cork: Cloudy, shifting breezy spells, light showers
Galway: Cloudy, dry, strong gusty winds

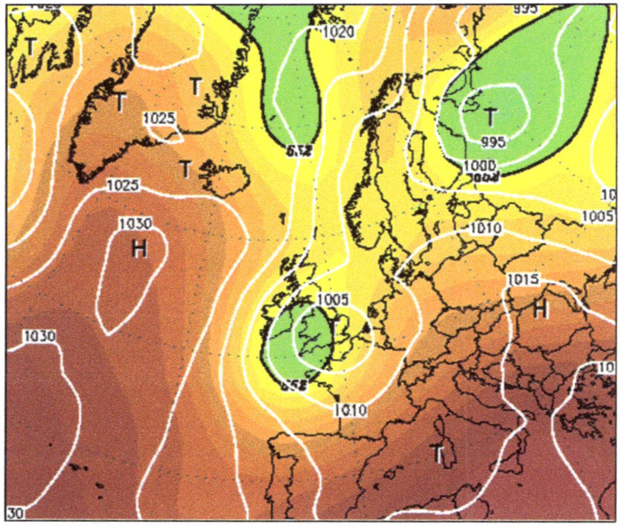

RAIN POTENTIAL

FROST/SNOW

JULY **6** SUNDAY

Low pressure continues to bring a colder than normal northerly flow over the country with blustery northwest breezes prevailing for a time, backing westerly later in the day as a high pressure system ridges near the western coasts. Frequent scattered cloud everywhere and mainly dry in the east and south, while isolated showers likely elsewhere.

Belfast: Mostly cloudy, dry, chance of morning drizzle patch, fresh northerly breezes
Dublin: Partly cloudy and mostly dry, strong blustery winds
Cork: Cloudy, mostly dry with threats of rain, cool, blustery winds, chance of gales
Galway: Cloudy, isolated showers, strong winds to near gales at times

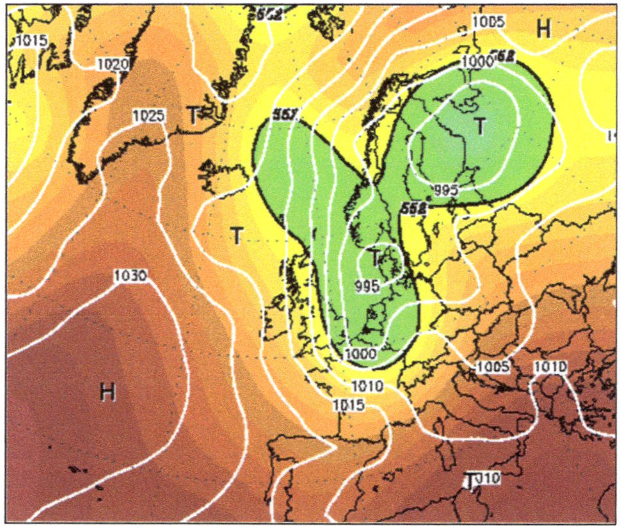

RAIN POTENTIAL

FROST/SNOW

JULY **7** MONDAY

The depression that brought some unsettled conditions of the previous few days departs to the northeast, while a high pressure ridge moves onto the country, directing a light to moderate northwest flow throughout. Mostly dry and sunny day everywhere, with some isolated overnight showers mostly over the northern half of the country.

Belfast: Sunny, fluctuating winds, brief showers overnight
Dublin: Sunny, mild day cool night, isolated showers, unsettled windy spells
Cork: Partly cloudy, mostly dry with odd threat of rain, unsettled windy spells
Galway: Sunny and dry day, windy, cloud evening with isolated showers likely

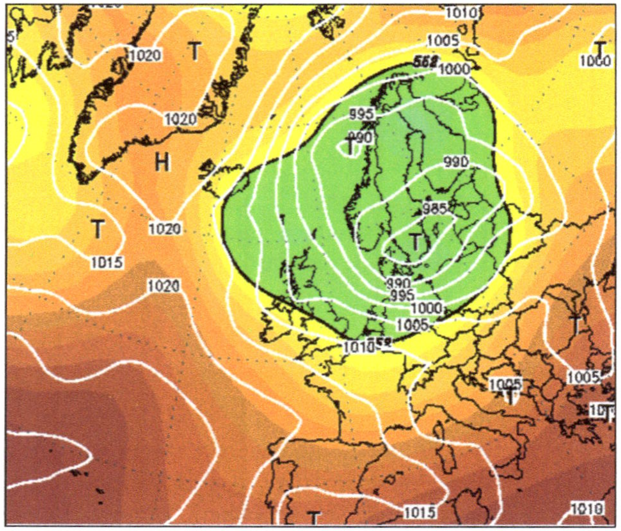

RAIN POTENTIAL

FROST/SNOW

JULY **8** TUESDAY

A low pressure trough replaces the high pressure ridge with a westerly change. Dull and gloomy day throughout with bands of showers crossing the country in northwesterly flow, light at first but becoming heavier for a time later in the day, particularly in the northwest and midlands.

Belfast: Overcast and dull, rain, breezy spells at times
Dublin: Overcast, occasional showers, fresh breezes
Cork: Cloudy, light showers, fresh winds at times
Galway: Cloudy, rain, fresh breezes at times

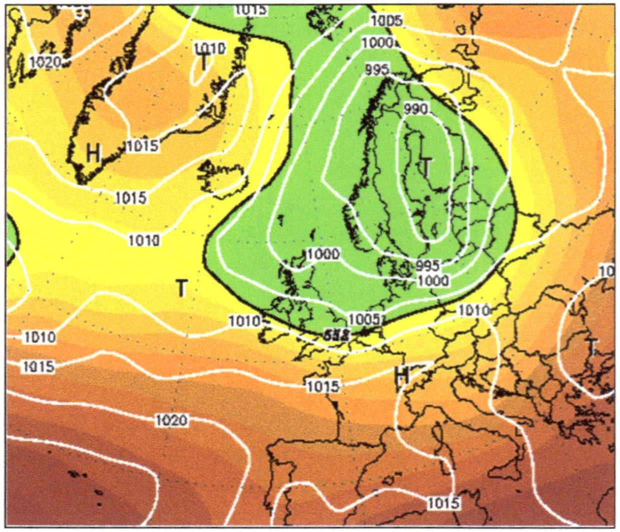

RAIN POTENTIAL

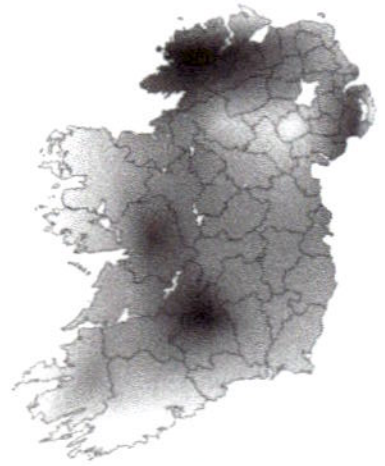

FROST/SNOW

JULY **9** WEDNESDAY

Low pressure continues to bring a cooler than normal spell across the country with a light northerly air prevailing, backing southeasterly later in the day. A changeable day throughout with a mix of occasional showers and passing sunny outbreaks, with showers at first in the north and central parts, and later another band of showers starting in the southwest and gradually spreads northwards with the southeasterly change.

Belfast: Changeable, light and cool northwest flow, mix of showers and sunny spells
Dublin: Unsettled winds, partly cloudy, isolated showers
Cork: Blustery at times, mostly cloudy, brief showers
Galway: Partly cloudy, scattered showers, fresh onshore westerly winds

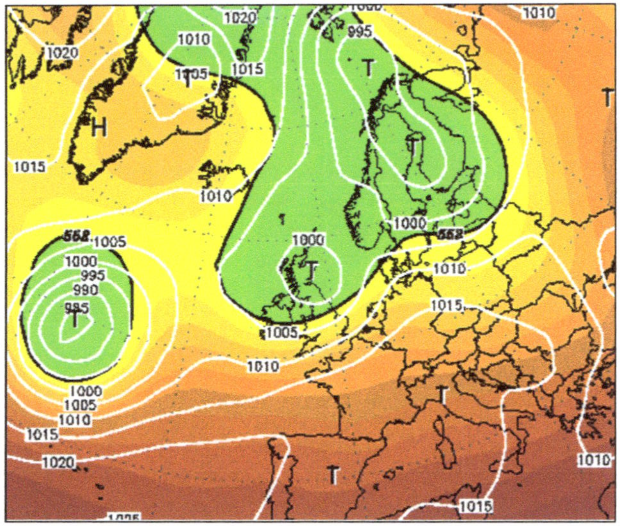

RAIN POTENTIAL

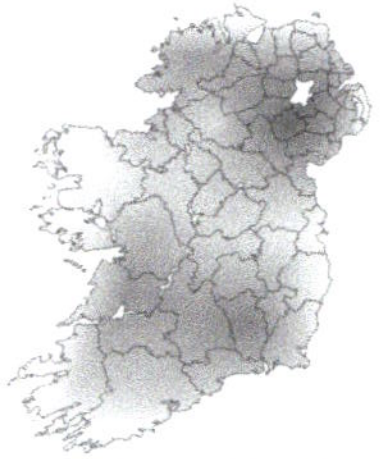

FROST/SNOW

JULY **10** THURSDAY

A depression builds to the west of Ireland while another weaker depression slips away to the northeast, with a low pressure trough between the two bringing a light to moderate flow at first, becoming stronger around the coasts later in the day. Unsettled conditions throughout with widespread cloud and scattered showers in most districts, some of which will be heavy at times, particularly in the southwest. Chance of thunderstorms in the west.

Belfast: Cloudy, rain, fresh winds strengthening later in the day
Dublin: Showery, cloudy, warmer, windy at times
Cork: Cloudy, outbreaks of rain, strong and blustery winds
Galway: Changeable, windy, occasional showers, chance of thunderstorms

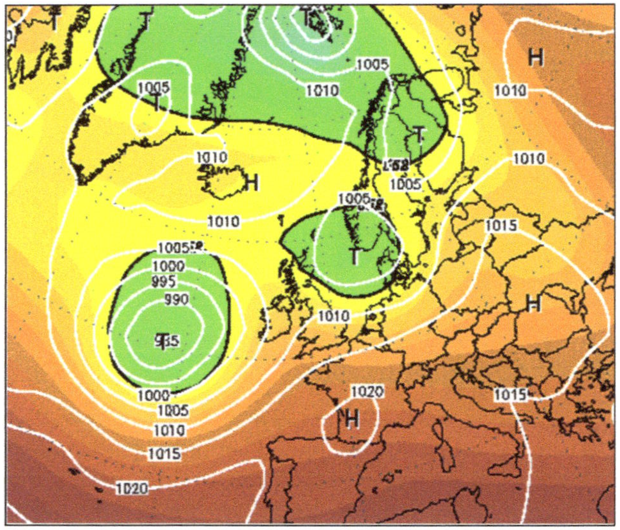

RAIN POTENTIAL

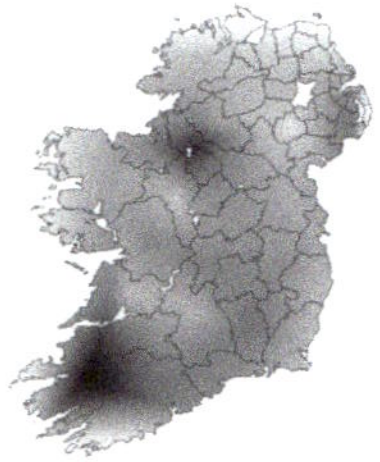

FROST/SNOW

JULY **11** FRIDAY

A depression moves onto the country bringing freshening southwest winds throughout. A changeable day with a mix of showery spells and sunny outbreaks. Showers may be heavy overnight, and occasional accompanied by thunderstorms, particularly in the east and south.

Belfast: Warmer, humid, some sunny spells, scattered showers, blustery at times
Dublin: Fresh to strong winds, cloudy, rain
Cork: Unsettled winds, brief overnight showers clearing to sunny day
Galway: Unsettled and blustery air, rain, some sunny spells, chance of thunderstorms

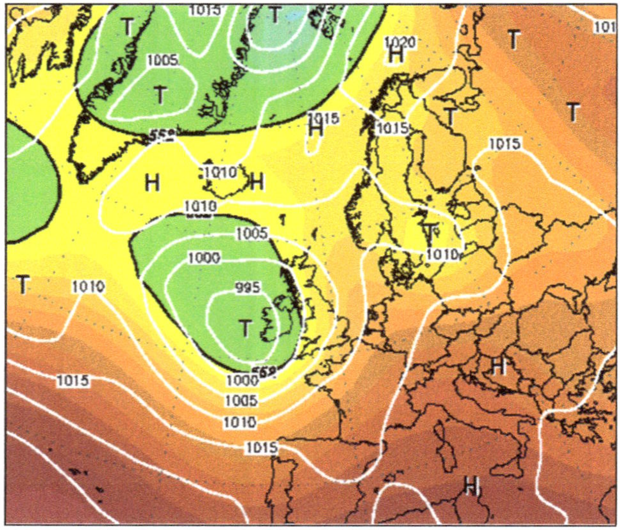

RAIN POTENTIAL

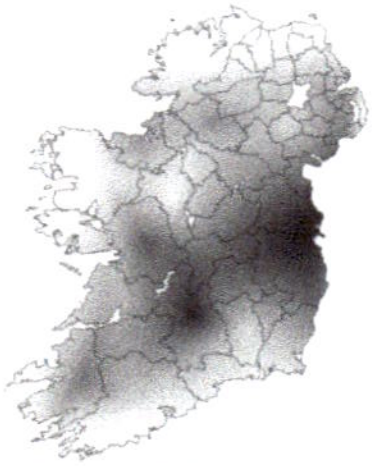

FROST/SNOW

JULY **12** SATURDAY

A depression continues to linger over Ireland with variable and moderate winds prevailing, strong at times around the northern coasts. Unsettled conditions everywhere with another day of a mixture of showers and sunny spells, as well as the likelihood of occasional thunderstorms, mostly in the east. Mostly dry around the southern coastal fringes and chance of brief patches of fog overnight in the far north.

Belfast: Partly cloudy, brief showers, fluctuating breezes, chance of shallow fog
Dublin: Changeable, sunny spells, showers and thunderstorms, unsettled breezes
Cork: Partly cloudy, cooler, isolated showers, fresh winds
Galway: Fresh northwest winds with gusts at times, mostly cloudy, light showers

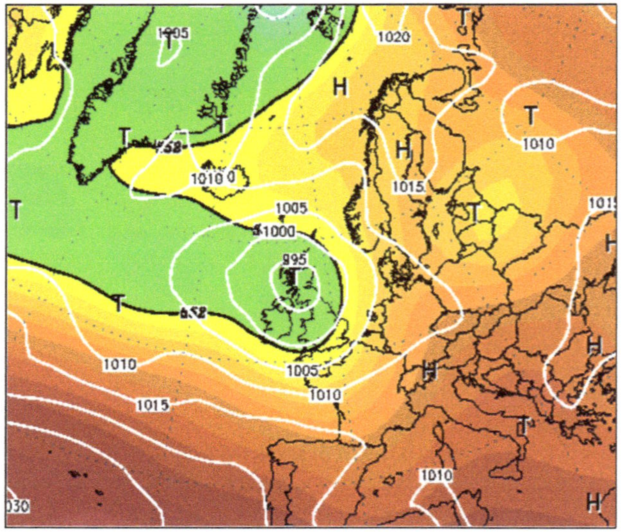

RAIN POTENTIAL

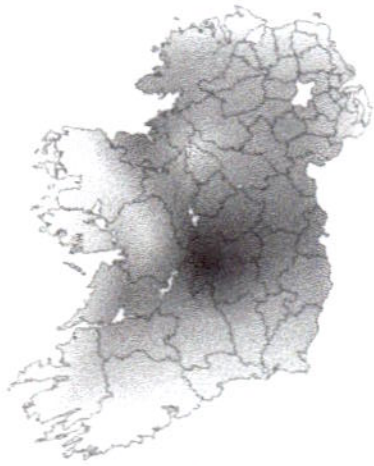

FROST/SNOW

JULY **13** SUNDAY

A low pressure system remains almost stalled over Ireland directing moderate to fresh west to northwest winds throughout, as well as a stormy air in all districts. A mixture of sunny spells and occasional showers and the odd thunderstorm prevails, especially in central parts, while remaining mostly dry along the southern coastal fringes. Cooler than normal temperatures for this time of year. Winds may get gusty for a time in some southern parts.

Belfast: Cloudy and cooler, light showers, windy spells
Dublin: Stormy air, partly cloudy, cooler, scattered showers
Cork: Sunny, cooler, strong winds, isolated showers
Galway: Stormy air, cooler, mix of showers and sunny spells

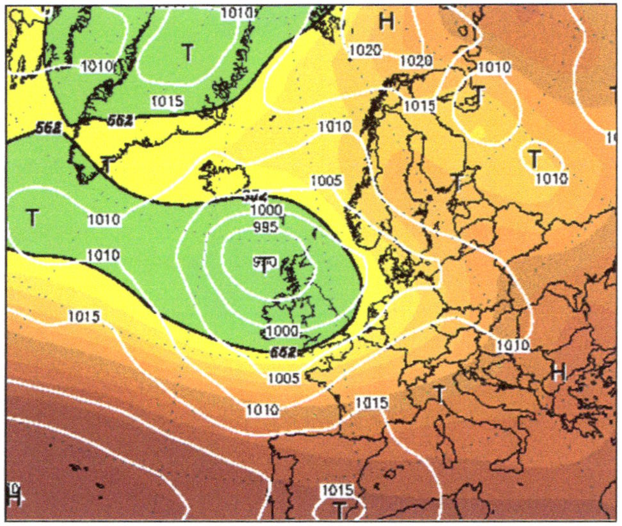

RAIN POTENTIAL

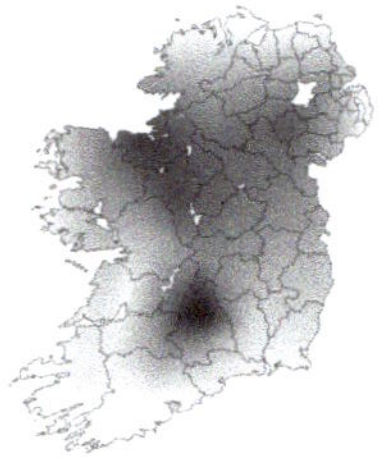

FROST/SNOW

JULY **14** MONDAY

A depression over Ireland begins to weaken bringing another mixed day of some sunny spells and the odd passing shower and drizzle patches. Winds generally light to moderate in a southwest flow, becoming stronger with a west to northwest change later in the day.

Belfast: Cloudy, cooler, rain, fresh breezes
Dublin: Partly cloudy, light showers, strong blustery winds, possble gales
Cork: Partly cloudy, isolated showers, strong and blustery winds
Galway: Cloudy, warmer, brief showers, strong winds, chance of gale gusts at times

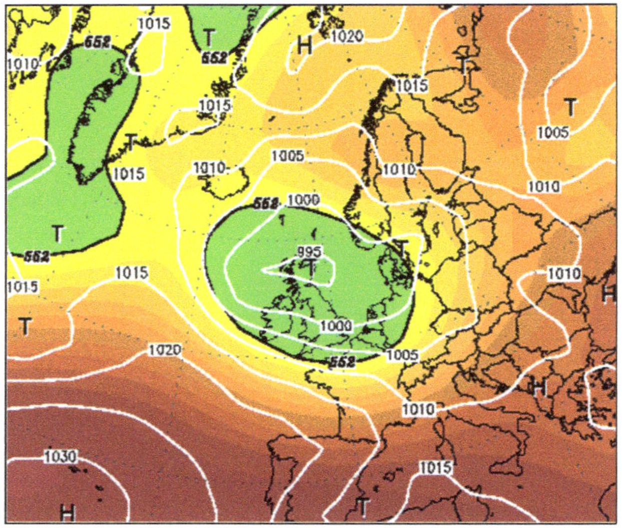

RAIN POTENTIAL

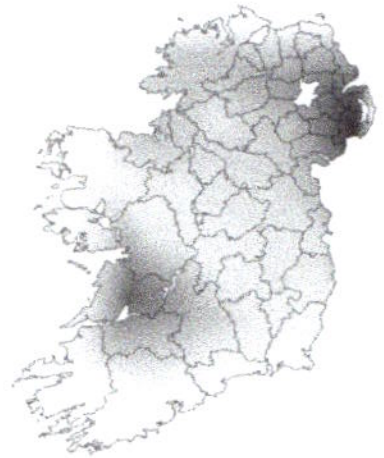

FROST/SNOW

JULY **15** TUESDAY

Low pressure continues to bring unsettled conditions over Ireland with widespread cloud prevailing and spotty showery spells now and then in most districts, as well as a band of heavy rain crossing the southwest of the country and moving up through the midlands and through the northwest by the end of the day. Light southwesterly breezes prevail at first, freshening to strong later with a southeasterly change.

Belfast: Overcast, rain, unsettled breezes at times
Dublin: Cloudy, light showers, unsettled and changeable winds
Cork: Cloudy, spells of rain, cool, fluctuating winds with occasional gusts
Galway: Overcas, scattered showers, windy spells

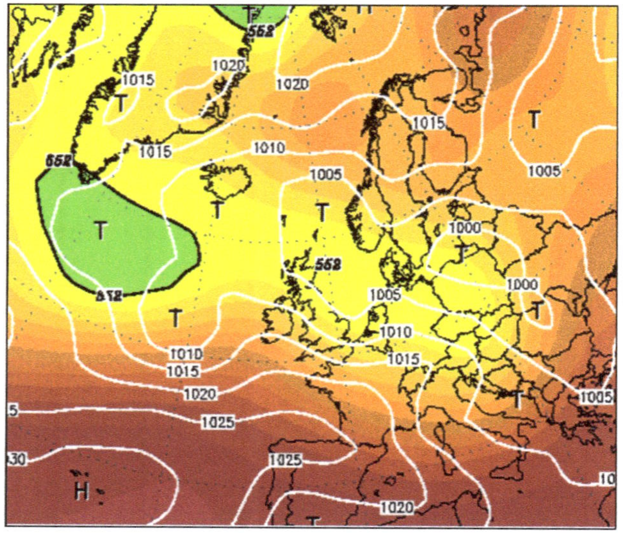

RAIN POTENTIAL

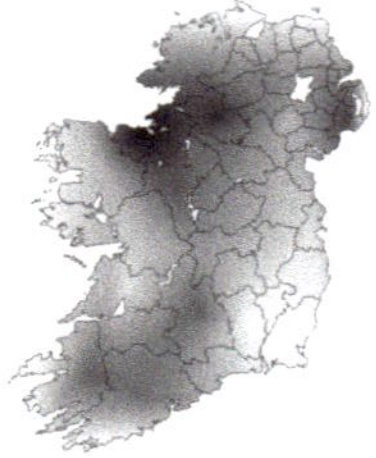

FROST/SNOW

JULY **16** WEDNESDAY

A weak depression continues to linger over Ireland bringing fresh to strong west to northwest winds and changeable conditions throughout. Despite spells of mostly sunny and dry conditions, particularly in the south, overnight outbreaks of showers and pockets of rain can be expected in most districts, mostly only in the northern half of the country in the morning, but becoming widespread in the evening and into the southern parts by tomorrow.

Belfast: Mostly sunny and dry, chance of brief showers, blustery, possible gales
Dublin: Mostly sunny, brief showers, strong and blustery winds, chance of gales
Cork: Strong and blustery winds, dry, very bright and sunny, warmer
Galway: Strong gusty winds to near gales at times, sunny day, brief evening showers

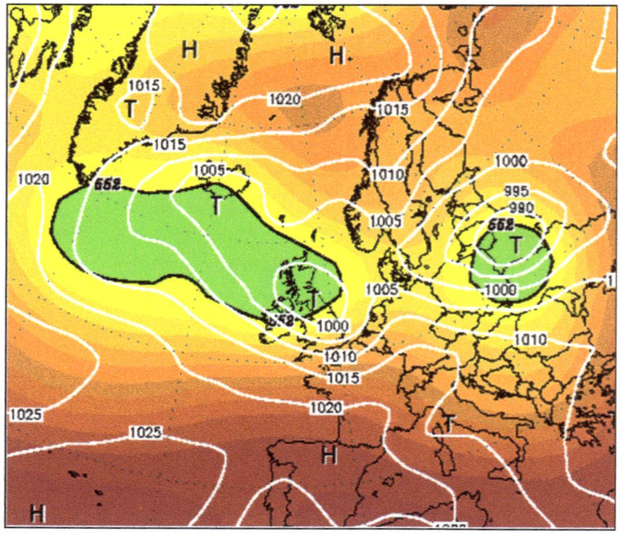

RAIN POTENTIAL

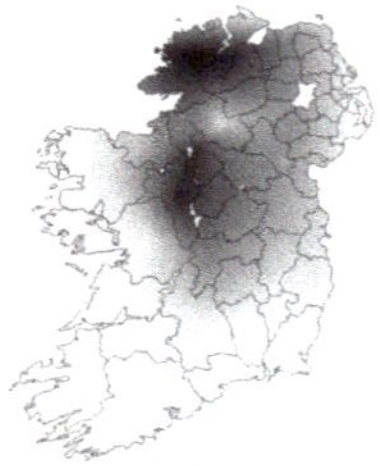

FROST/SNOW

JULY 17 THURSDAY

A low pressure trough directs a light to moderate northwest flow over the north of the country, while a high pressure system ridges to the south, with lighter breezes prevailing. Some brief morning showers in most districts clear early bringing a mostly sunny and dry day almost everywhere. Further isolated showers develop once more in the evening mostly in the east.

Belfast: Very sunny, dry, moderate breezes
Dublin: Sunny, evening showers, winds easing to light
Cork: Sunny, isolated evening showers, cooler, light breezes to mostly calm air
Galway: Winds easing to light, overnight rain, bright and sunny day, warmer

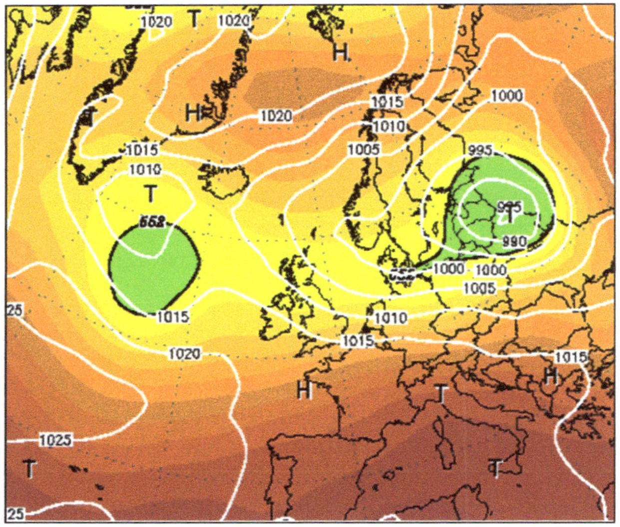

RAIN POTENTIAL

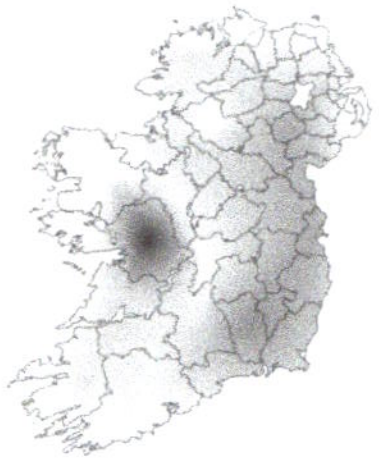

FROST/SNOW

JULY 18 FRIDAY

A high pressure ridge sits to the southwest of Ireland while a low pressure trough continues to prevail over the country. Generally light westerly winds throughout, but more squally at times around the showers. A band of thunderstorms cross the country bringing a mix of rain and some sunny spells. Thunderstorms may bring showers of hail as well as the chance of funnel formation winds, particularly in the west.

Belfast: Moderate breezes, mix of rain and some sunny spells
Dublin: Partly cloudy, isolated showers, mostly calm air
Cork: Partly cloudy, heavy rain, gentle breezes, chance of thunderstorms
Galway: Sunny, isolated showers, gentle breezes

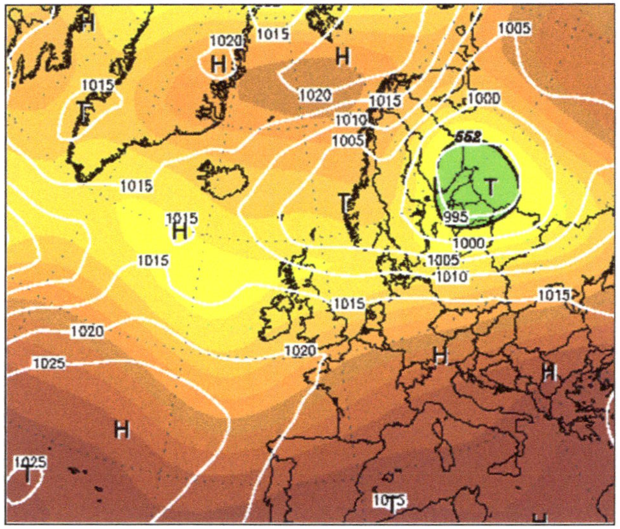

RAIN POTENTIAL

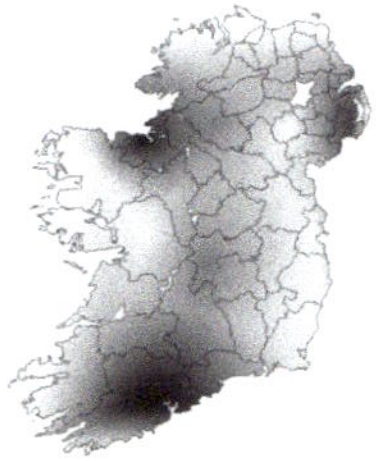

FROST/SNOW

JULY **19** SATURDAY

A changeable day as low pressure continues to prevail throughout, with moderate to fresher west to northwest winds. Mostly dull and overcast everywhere except for some brief sunny breakthroughs mostly in the east. Scattered showers in most districts, some of them heavy at times and possibly accompanied by thunderstorms, particularly through the western half of the country.

Belfast: Fresh breezes at times, cloudy, rain
Dublin: Changeable, mix of sunny spells and showers, moderate breezes
Cork: Overcast and dull, light showers, cooler, moderate breezes
Galway: Overcast, scattered showers, fresh breezes at times

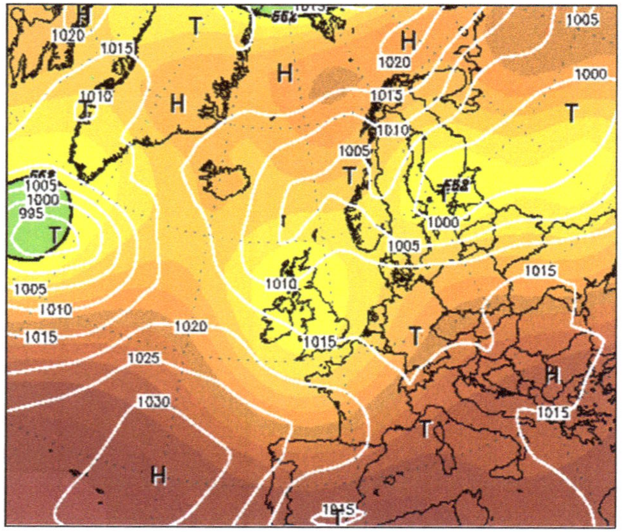

RAIN POTENTIAL

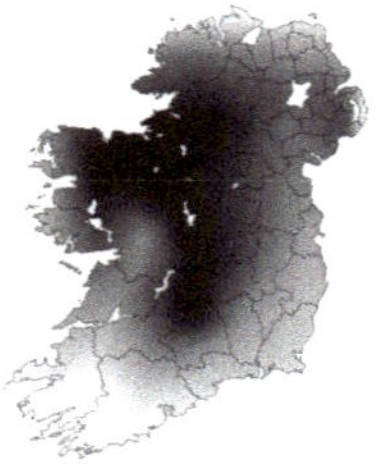

FROST/SNOW

JULY **20** SUNDAY

The high pressure ridge nears the southwest of Ireland while the low pressure trough departs to the northeast, with light to moderate west to northwest breezes prevailing. Dull and gloomy with widespread cloud and spotty showers at times in the north and east, while mostly dry and sunny elsewhere. Chance of patches of fog in central parts.

Belfast: Cloudy, cooler, light breezes, chance of fog patches overnight
Dublin: Dull and gloomy, mostly calm air, dry with odd threat of rain
Cork: Mostly sunny and dry, warmer, breezy spells
Galway: Sunny, warmer, dry, fresh breezes

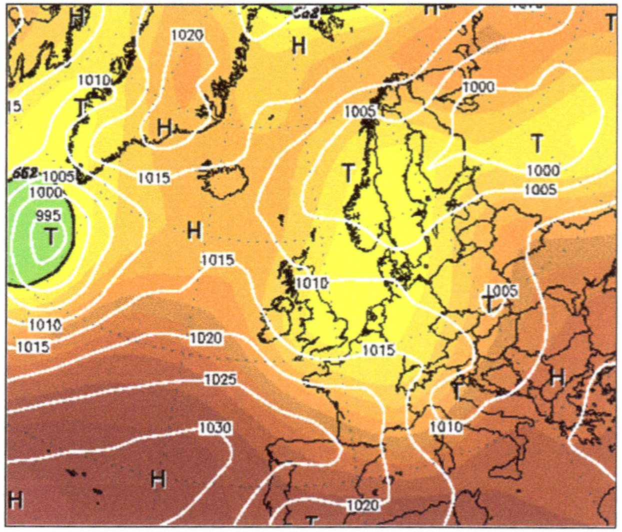

RAIN POTENTIAL

FROST/SNOW

JULY **21** MONDAY

A low pressure trough lingers over the country bringing widespread dull and gloomy skies throughout, light to moderate southwesterly breezes, some spotty showers and drizzle patches in most districts. Chance of patches of fog, particularly in the south and west.

Belfast: Overcast, occasional showers, moderate breezes
Dublin: Overcast, isolated showers, light breezes
Cork: Overcast, isolated showers, moderate breezes, chance of patches of fog
Galway: Overcast, occasional showers, light breezes, chance of fog patches

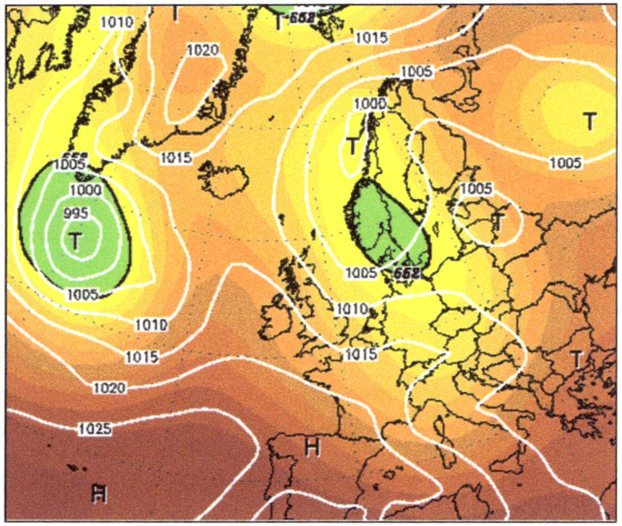

RAIN POTENTIAL

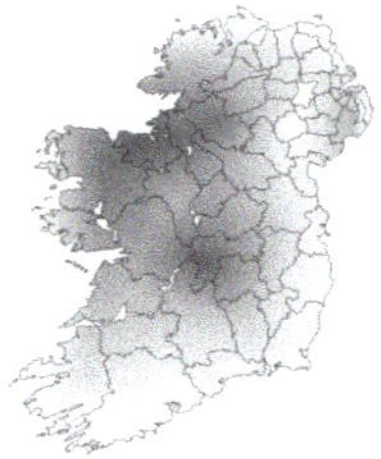

FROST/SNOW

JULY **22** TUESDAY

A depression to the northwest of Ireland maintains a low pressure trough over the country in a light westerly flow. Widespread cloud in most districts but some sunnier spells in the south. Mostly dry day with patches of morning fog here and there, before some light showers develop later in the day.

Belfast: Cloudy, slightly warmer, isolated showers, chance of drizzles and shallow fog
Dublin: Cloudy, warmer, light afternoon showers, rain threatening, gentle breezes
Cork: Warm, sunny, isolated evening showers, gentle breezes, chance of fog
Galway: Mostly cloudy and dry, warmer, fresh breezes at times

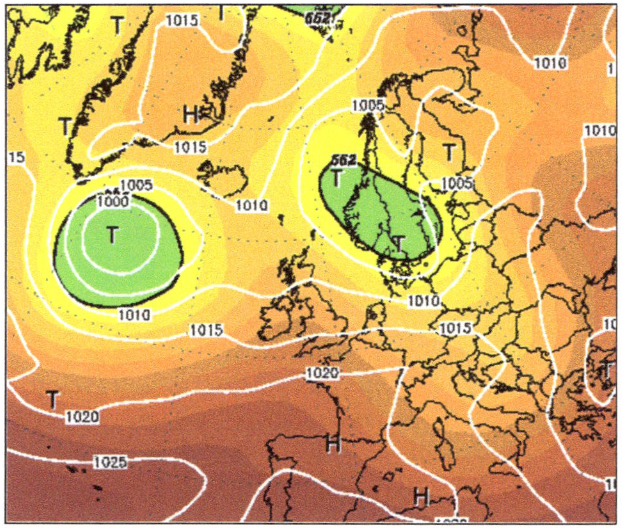

RAIN POTENTIAL

FROST/SNOW

JULY 23 WEDNESDAY

A light and variable flow prevails as a rain embedded low pressure trough sits over the country. Dull and gloomy everywhere with rain developing in the southwest at first and spreading eastward throughout the day and into northern parts later. Rain likely to be very heavy at times in the south and east.

Belfast: Dull and gloomy air, rain, light breezes, possible patches of fog overnight
Dublin: Overcast, heavy rain, cooler, light shifting breezes
Cork: Overcast, cooler, rain, gentle shifting breezes
Galway: Overcast, cooler, light showers, gentle fluctuating breezes

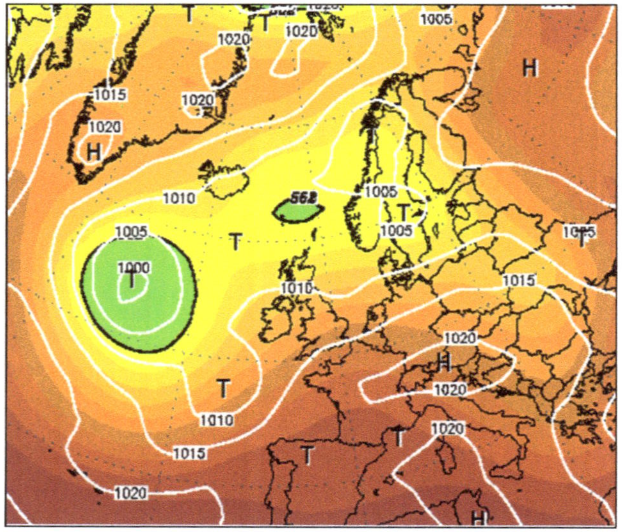

RAIN POTENTIAL

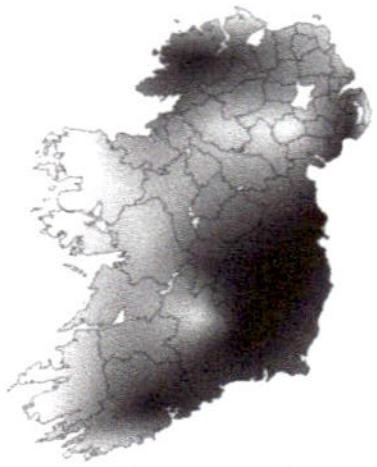

FROST/SNOW

JULY 24 THURSDAY

A weak depression moves onto Ireland bringing a changeable day with west to northwest breezes prevailing, becoming blustery at times around the showers. Long sunny spells can be expected in most districts, with the occasional passing showery spell. Showers will become more frequent in the afternoon, possibly accompanied by thunderstorms and sudden heavy falls, particularly in the northeastern and central districts as well as in the southeast of Leinster.

Belfast: Changeable, mix of sunny spells and showers, cooler, strong winds at times
Dublin: Fresh shifting breezes, mostly sunny, brief showers
Cork: Very sunny, mostly dry, threats of rain overnight, chance of fog, light breezes
Galway: Sunny and dry, warmer day cooler night, fluctuating winds, static air

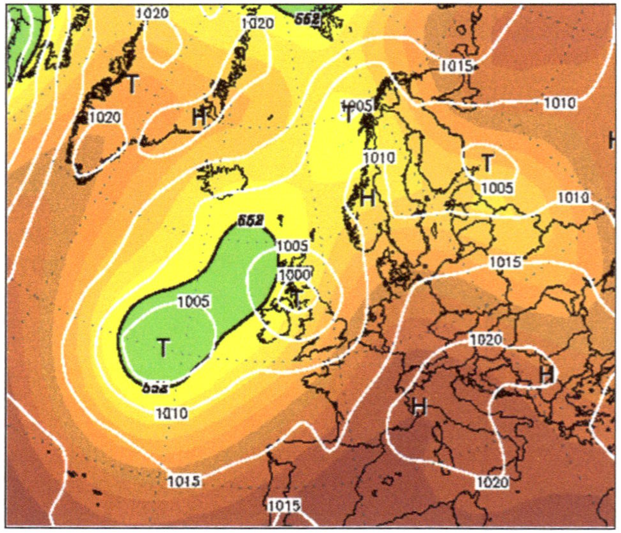

RAIN POTENTIAL

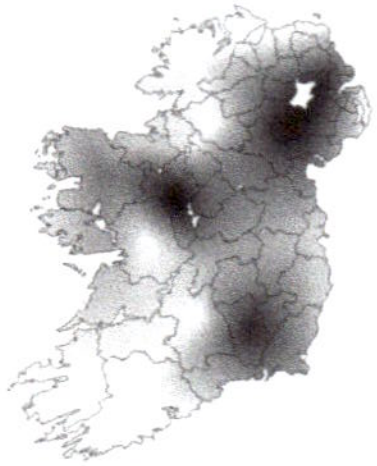

FROST/SNOW

JULY **25** FRIDAY

Low pressure continues to bring changeable conditions throughout the country, with a light southeasterly flow prevailing. Patches of fog at first in many districts clearing to some sunny spells, before showers return later in the day, occasionally accompanied by thunderstorm activity. Rain likely to be particularly heavy through the midlands to the northwest, with the chance of localised pockets of brief flooding.

Belfast: Partly cloudy, light showers, warmer day colder night, calm air, fog likely
Dublin: Partly cloudy, isolated showers, mostly calm air, misty patches possible
Cork: Cloudy, cooler, isolated showers, calm air, chance of fog
Galway: Warmer, mostly sunny, overnight shower, chance of thunderstorms

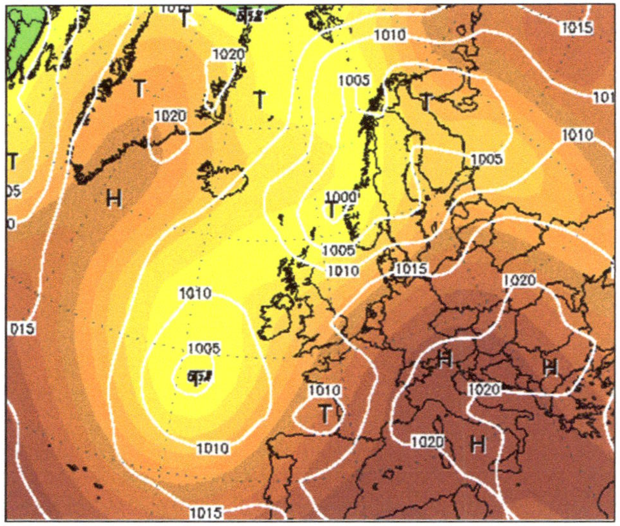

RAIN POTENTIAL

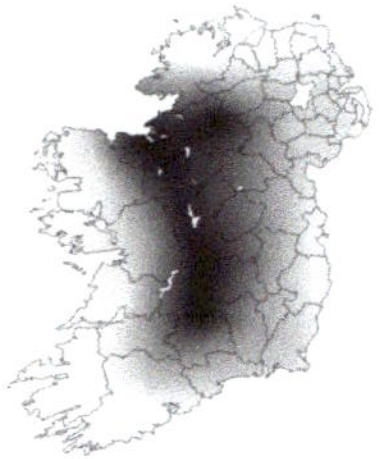

FROST/SNOW

JULY **26** SATURDAY

A weak depression centred to the southwest of Ireland maintains a light variable flow over the country with further patches of overnight fog in many districts. Mostly sunny at first, before another band of thunderstorms and showers cross the country starting in the southwest and passing up through the country into the northeast. Some falls likely to be heavy at times, particularly in the west. Chance of hail showers accompanying some thunderstorms.

Belfast: Warm, calm, sunny spells, isolated showers, chance of thunderstorms
Dublin: Mostly sunny, warmer, overnight showers, chance of thunderstorms
Cork: Sunny and warmer, dry, calm air, chance of overnight misty patches
Galway: Cloudy, light breezes, brief showers, chance of thunderstorms and hail

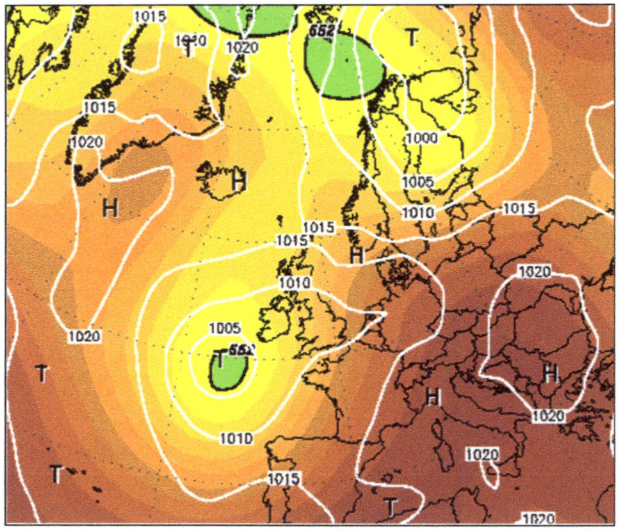

RAIN POTENTIAL

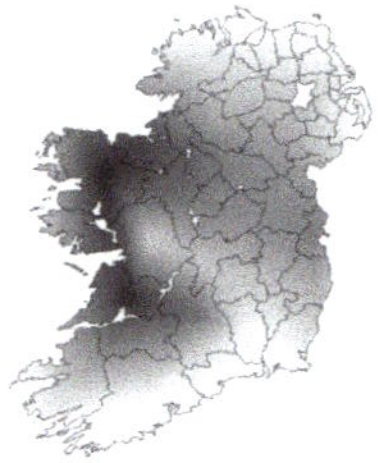

FROST/SNOW

JULY **27** SUNDAY

A weak depression continues to linger over Ireland bringing light shifting breezes throughout. Another mixed day of some sunny spells and passing showers, including further possible thunderstorms in many districts. Chance of patches of fog overnight mostly in the west and south.

Belfast: Warmer, partly cloudy, outbreaks of rain, chance of thunderstorms
Dublin: Mostly cloudy, occasional showers, cooler, thunderstorms possible
Cork: Overnight rain, cooler, some sunny spells, calm air, chance of thunderstorms
Galway: Partly cloudy, light showers, gentle breezes, cold overnight, chance of fog

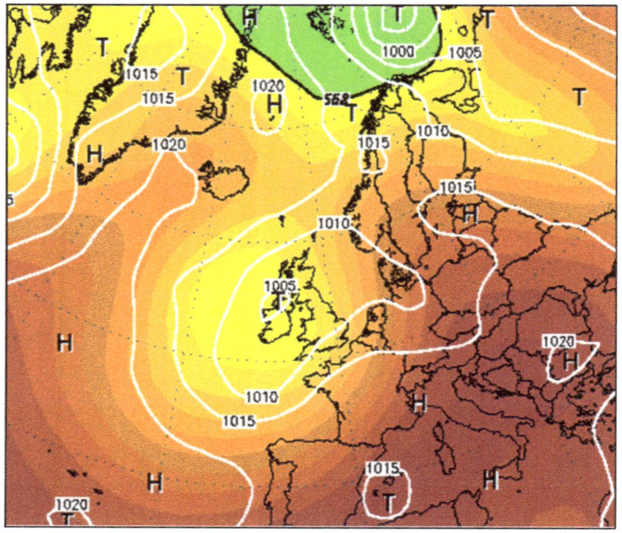

RAIN POTENTIAL

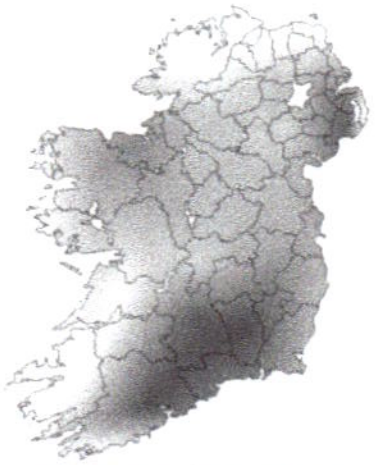

FROST/SNOW

JULY **28** MONDAY

The weak depression that has lingered over Ireland the past few days now drifts away to the northeast of the country, while a high pressure system ridges near the southern coasts. Mostly sunny and dry in the south, with increasing cloud and showers elsewhere. Winds likely to be gentle in a north to northwest flow.

Belfast: Mostly cloudy, rain, warm, static and humid air, gentle breezes
Dublin: Partly cloudy, light showers, gentle breezes
Cork: Gentle shifting breezes, brief overnight showers, sunny day
Galway: Partly cloudy, rain, moderate breezes

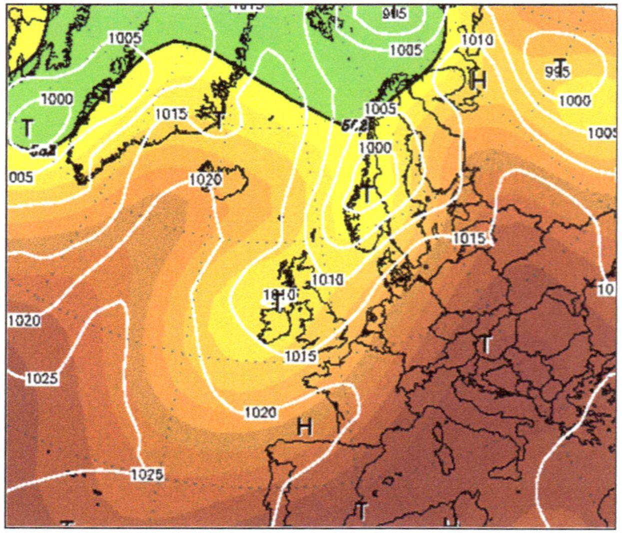

RAIN POTENTIAL

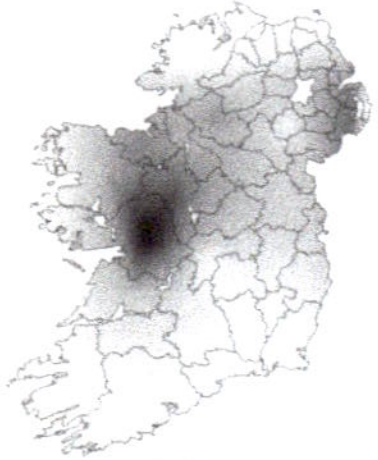

FROST/SNOW

JULY **29** TUESDAY

A high pressure system to the southwest of Ireland ridges over the country bringing a light variable flow throughout. Overnight fog patches likely, especially in the north and south, clearing to mostly sunny and dry, particularly in the east and southeast. Later in the day, thunderstorm embedded rain bands move up through the country, with very heavy downpours likely in the west and south, but some districts may remain dry, particularly in the northwest. Lighter rain activity elsewhere.

Belfast: Cooler, partly cloudy, light showers, chance of fog and thunderstorms
Dublin: Mostly calm air, very sunny and dry, cloud building later, chance of fog
Cork: Partly cloudy, heavy rain, gentle breezes, chance of shallow fog
Galway: Cloudy, cooler, heavy rain, unsettled breezes, chance of thunderstorms and flooding

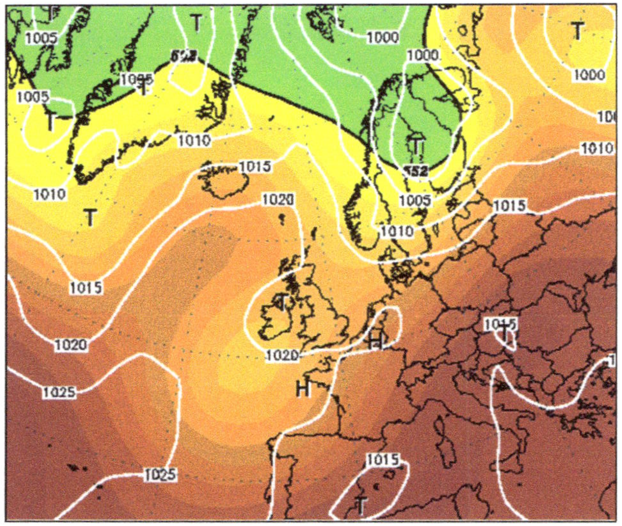

RAIN POTENTIAL

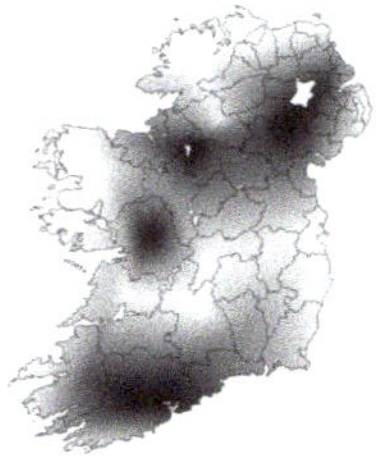

FROST/SNOW

JULY **30** WEDNESDAY

High pressure continues to direct a light and variable flow over the country. Widespread cloud throughout with patches of fog overnight in many districts. Rain likely to develop later in the morning, in the east at first, and spreading to inland parts throughout the day, while western fringes remain mostly dry with long sunny spells. Rain outbreaks likely to be heavy at times in central parts of the country and along the central eastern coasts.

Belfast: Cloudy, warm, humid, rain, thunderstorms possible, calm air, shallow fog
Dublin: Overcast, heavy rain, cooler, gentle breezes, chance of overnight fog
Cork: Cloudy, mostly dry and cooler, gentle breezes to mostly calm air, fog likely
Galway: Fluctuating breezes, mix of sunny spells and showers, possible fog patches

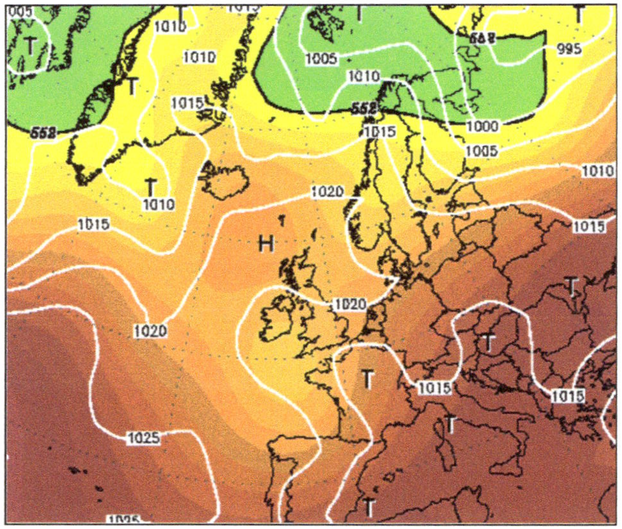

RAIN POTENTIAL

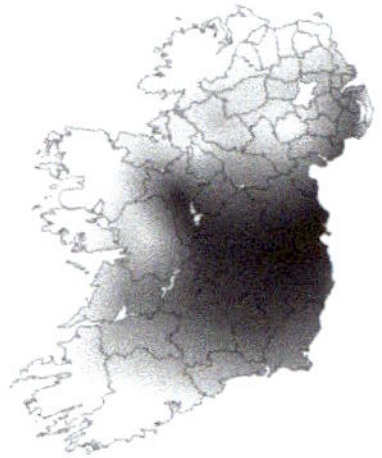

FROST/SNOW

JULY **31** THURSDAY

A complex of high and low pressure systems bring a changeable day over the country. High pressure ridges to the west of Ireland while low pressure sits to the east. Mostly dry and cloudy at first, sunnier in the south. Later, some light showers develop, occasionally accompanied by thunderstorms. Mostly dry and sunnier along the western coastal fringes.

Belfast: Cloudy, cooler, mostly dry, threats of rain, gentle breezes, misty patches
Dublin: Overcast, light showers, gentle breezes, fog patches possible
Cork: Cloudy, occasional showers, mostly calm air, shallow fog patches possible
Galway: Mostly sunny, brief showers, fluctuating breezes, chance of thunderstorms

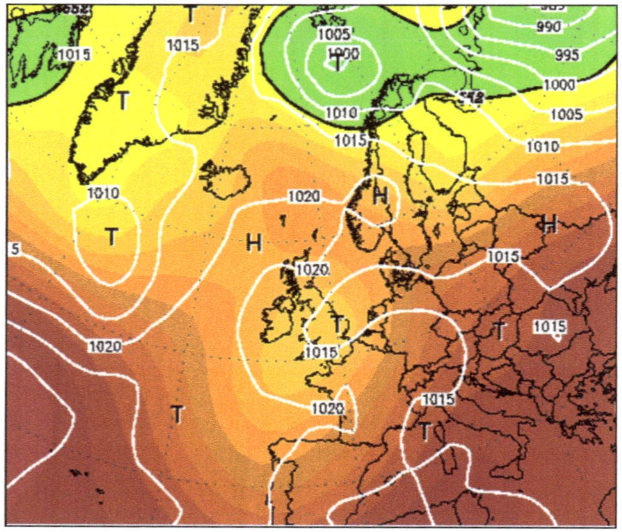

RAIN POTENTIAL

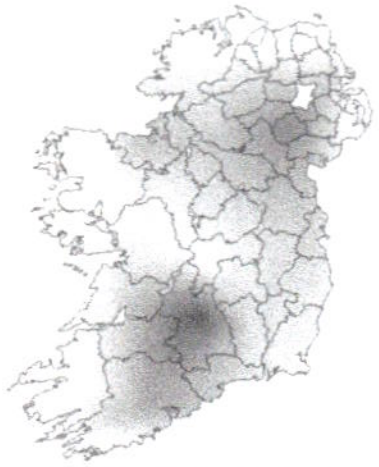

FROST/SNOW

AUGUST

PHASES OF THE MOON

1st	First Quarter
1st	Apogee
5th	Southern Declination
9th	Full Moon
13th	Crossing Equator
14th	Perigee #12
16th	Third Quarter
18th	Northern Declination
23rd	New Moon
26th	Crossing Equator
29th	Apogee
31st	First Quarter

1	2	3	4	5	6	7	8	9	10	11	12	13	14	15	16	17	18	19	20	21	22	23	24	25	26	27	28	29	30	31
Thur	Fri	Sat	Sun	Mon	Tue	Wed	Thur	Fri	Sat	Sun	Mon	Tue	Wed	Thur	Fri	Sat	Sun	Mon	Tue	Wed	Thur	Fri	Sat	Sun	Mon	Tue	Wed	Thur	Fri	Sat
1Q A				V				F				XhN	P12		3Q		^					N			XhS			A		1Q

MONTHLY SUMMARY

1st-8th A wet week when some rain can be expected every day in many parts of the country despite some long sunny spells.
4th-5th Scattered thunderstorms and hail likely, especially in the east. Strong and unsettled winds prevail with the chance of funnel formation clouds during this spell.
4th-6th Westerly winds are the predominant wind flow.
6th-10th High pressure builds over Ireland bringing dry, sunny, and settled weather with light winds and colder than normal overnight temperatures throughout and unusually high maximum temperatures during the day.

Over the next four weeks, fairly average air pressure will prevail for the first couple of weeks before rising to above average over the third and fourth weeks. Depressions and frontal systems will prevail through most of the first couple of weeks, before an anticyclonic influence becomes established and ridges to the west of Ireland bringing more settled conditions for the remainder of this outlook period. As such, most of the first three weeks will see some rain almost daily, especially in the eastern half of the country before the last week of this outlook period settles into mainly dry weather. Some heavy rain can be expected midway through this outlook period may result in flooding in some parts of the country, especially in the midlands. Overall, sunshine hours will be fairly typical for this time of year, with most days seeing some sunshine, despite the regular outbreaks of rain. Average maximum temperatures will be around half a degree warmer than is normal for this time of year overall, despite being cooler than normal in the middle of this outlook period due to the very unsettled weather at the time. In contrast, minimum temperatures will average around average to slightly above average over the next four weeks. During this outlook period, there is likely to be around 15-20 days with rain, with the west and northwest having more wet days than anywhere else, while the most rainfall in quantity will be recorded in the eastern to midlands districts. There will be thunderstorms on at least 6 days and hail on 1 day. Widespread fogs are likely on 4 days, and in contrast, gale winds likely on 7 days. There are no days with frost or snow expected.

11th-16th A series of weak fronts associated with a deep depression to the west of the country cross Ireland bringing outbreaks of showers and unsettled winds in most parts of the country.
12th-14th Warmer spell.
13th Strong southwest winds prevail bringing widespread cloud throughout, unusually high temperatures, and possible gale gusts in many districts.
13th-16th Very heavy rain likely in many parts of the country, except in the north.
14th Very heavy rain likely in central southern parts.
16th Thunderstorms with hail in the east to central parts. Near gales in the west.
17th-19th A ridge of high pressure moves up over the country bringing a brief spell of light northerly breezes, mostly dry and very sunny weather throughout the country with warm daytime temperatures and quite cool at night. Winds likely to back southwesterly later.
20th-30th Unsettled weather returns with a northwest change bringing outbreaks of rain or showers daily, some accompanying thunderstorm activity, and only brief sunny spells. Much cooler than usual.
21st-24th A depression brings thunderstorms mostly into the central districts with chance of hail or snow grains and chance of localised flooding in some parts of the midlands. Rain may be heavy at times in other parts of the country.
25th-31st Spell of unsettled and often strong winds in most districts.
26th Unusually cold overnight.
27th Heavy rain in many districts.
28th Widespread patches of fog overnight.
29th-30th Scattered thunderstorms cross the country sometimes bringing heavy rain in many parts of the country.
31st Chance of snow grains in central parts. More settled weather returns.

IRELAND RAINFALL ESTIMATES

August 2025

d=mainly dry, n= nonrecordable, l=light shrs, s=significant shrs, r=rain, h=hvy falls

		ULSTER PROVINCE					CONNAUGHT PROVINCE						LEINSTER PROVINCE				MUNSTER PROVINCE						
		Alderg	Hillsbo	Armag	Loughl	Malin H	Sligo	Ardtar	Belmul	Drums	Galway	Rosco	Markre	Edende	Kilkenn	Dublin	Shannc	Newpo	Tralee	Killarne	Cork	All	Moon
1st	Fri	l	l	s	d	d	l	d	s	s	l	l	d	r	d	s	l	d	l	s	s	50	1Q A
2nd	Sat	d	d	n	d	d	l	d	d	d	d	n	d	l	d	s	d	d	d	d	l	11	
3rd	Sun	l	s	d	l	l	s	l	s	s	s	s	s	l	s	l	s	s	r	s	s	96	
4th	Mon	s	r	l	r	h	r	r	s	r	r	s	r	l	n	l	s	h	s	s	l	166	
5th	Tues	l	l	l	s	s	r	s	s	l	l	s	s	l	l	l	l	l	s	l	s	73	V
6th	Wed	l	d	s	l	l	l	l	n	d	d	d	l	l	n	l	l	d	d	d	n	20	
7th	Thur	l	d	l	d	d	n	d	d	l	l	l	n	l	s	l	s	l	s	s	s	51	
8th	Fri	l	l	d	n	l	l	n	d	d	d	d	l	n	d	n	d	d	d	d	d	8	
9th	Sat	d	d	n	d	d	d	d	d	d	d	d	d	d	d	d	d	d	d	d	d	1	F
10th	Sun	d	d	d	l	n	s	l	d	d	l	l	d	d	d	d	d	s	d	d	d	22	
11th	Mon	l	l	l	l	s	n	l	s	l	l	l	l	d	l	l	d	d	d	l	l	29	
12th	Tues	l	d	l	d	d	d	d	l	d	l	l	d	d	l	d	s	d	l	l	n	20	
13th	Wed	l	l	d	l	l	l	l	r	s	s	s	l	l	l	d	s	s	s	s	l	75	XhN
14th	Thur	r	h	l	l	l	s	l	s	r	r	r	l	h	h	r	r	l	r	h	h	229	P12
15th	Fri	r	s	r	h	d	r	h	l	d	s	s	h	s	s	r	h	r	r	l	h	206	
16th	Sat	s	s	s	s	r	s	s	s	l	l	l	l	s	s	r	r	l	l	d	s	93	3Q
17th	Sun	s	d	s	d	l	d	d	d	d	d	d	l	d	d	n	d	d	l	d	d	16	
18th	Mon	d	d	d	d	d	l	d	d	d	s	d	l	d	d	d	d	s	d	d	d	11	^
19th	Tues	n	l	d	l	l	l	l	s	d	s	l	l	d	n	d	s	l	s	l	l	42	
20th	Wed	s	s	l	s	l	s	s	l	s	s	s	s	r	s	s	s	l	r	s	s	109	
21st	Thur	h	h	s	h	h	h	h	r	h	s	s	h	r	l	s	s	h	s	s	s	288	
22nd	Fri	r	l	h	l	l	l	l	s	l	l	l	l	s	l	l	l	l	l	l	l	80	
23rd	Sat	l	d	l	d	s	l	d	l	l	d	d	s	d	l	l	l	d	l	l	l	25	N
24th	Sun	l	l	l	s	l	s	s	l	r	h	s	s	r	h	s	r	s	r	s	s	151	
25th	Mon	s	l	s	l	l	s	l	l	l	l	l	s	l	d	s	s	l	l	s	l	57	
26th	Tues	l	d	l	l	s	l	l	l	l	l	l	l	l	l	l	l	l	l	l	d	25	XhS
27th	Wed	s	r	l	r	s	s	r	s	s	r	s	r	s	s	l	s	s	r	s	s	150	
28th	Thur	s	l	s	l	l	l	l	s	l	l	l	n	s	l	s	s	d	l	l	l	52	
29th	Fri	l	s	l	r	d	r	r	s	r	s	s	r	s	h	d	l	r	l	l	d	129	A
30th	Sat	r	d	h	d	h	n	d	l	d	n	l	l	d	d	h	s	n	l	d	l	76	
31st	Sun	d	d	d	d	d	d	d	d	d	d	d	n	d	d	d	d	d	d	d	d	1	1Q
	Estimate:	121	117	124	120	109	132	120	111	117	127	92	129	124	117	113	120	119	131	108	111	2362	
	Average:	84	87	76	101	95	102	104	102	88	108	109	102	86	72	73	82	133	106	77	97	1882	
	Trend:	wtr	wtr	wtr	wtr	wtr	wtr	wtr	av	wtr	wtr	drr	wtr	wtr	wtr	wtr	wtr	drr	wtr	wtr	wtr	wtr	

IRELAND SUNSHINE ESTIMATES

August 2025

F=fine (8-12 hours of sunshine), pc= partly cloudy (4-7 hours), c=cloudy (1-3 hours), o=overcast (0 hours)

		ULSTER PROVINCE					CONNAUGHT PROVINCE						LEINSTER PROVINCE				MUNSTER PROVINCE						
		Alderg	Hillsbo	Armag	Loughl	Malin H	Sligo	Ardtarı	Belmul	Drums	Galway	Roscoı	Markre	Edende	Kilkenı	Dublin	Shannc	Newpo	Tralee	Killarne	Cork	All	Moon
1st	Fri	c	c	c	c	c	c	c	c	c	c	c	c	c	c	c	c	c	c	c	c	33	1Q A
2nd	Sat	F	F	F	F	F	F	F	pc	pc	F	F	F	pc	pc	pc	F	F	F	F	pc	184	
3rd	Sun	c	c	c	c	c	c	c	c	pc	c	c	c	pc	pc	c	c	pc	pc	pc	F	73	
4th	Mon	pc	pc	pc	pc	pc	pc	pc	pc	pc	pc	pc	pc	pc	F	F	pc	pc	pc	pc	pc	135	
5th	Tues	pc	pc	pc	pc	pc	pc	pc	F	pc	F	F	pc	pc	pc	pc	F	pc	F	F	F	153	V
6th	Wed	pc	pc	pc	pc	pc	F	pc	F	pc	F	F	pc	pc	pc	pc	pc	pc	pc	pc	F	160	
7th	Thur	pc	pc	pc	pc	pc	pc	pc	pc	pc	c	c	pc	pc	pc	pc	o	c	o	o	c	85	
8th	Fri	F	F	F	F	F	F	F	F	F	F	F	F	F	F	F	F	F	F	F	F	248	
9th	Sat	F	F	F	F	F	F	F	F	F	F	F	F	F	F	F	F	F	F	F	F	244	F
10th	Sun	F	F	F	F	F	pc	F	o	F	pc	c	F	F	F	F	F	F	F	F	F	196	
11th	Mon	c	c	c	c	c	pc	c	pc	c	pc	pc	c	c	c	o	c	c	o	o	c	52	
12th	Tues	pc	pc	pc	pc	pc	pc	pc	F	pc	F	F	pc	pc	pc	pc	F	pc	pc	pc	pc	139	
13th	Wed	c	c	c	c	c	o	c	o	c	c	c	c	c	c	pc	pc	pc	o	o	o	41	XhN
14th	Thur	c	c	c	c	c	o	c	c	c	c	c	c	c	pc	pc	c	c	c	c	c	51	P12
15th	Fri	o	o	o	o	o	c	o	pc	c	pc	pc	o	c	c	o	pc	c	pc	pc	pc	60	
16th	Sat	pc	pc	pc	pc	pc	pc	pc	pc	pc	F	F	pc	pc	pc	pc	F	F	F	F	F	152	3Q
17th	Sun	F	F	F	F	F	F	F	F	F	F	F	F	F	F	pc	F	F	F	F	F	205	
18th	Mon	pc	pc	pc	pc	pc	c	pc	c	pc	c	c	pc	pc	F	pc	c	pc	c	c	F	96	^
19th	Tues	pc	pc	pc	pc	pc	c	pc	pc	c	c	pc	pc	c	c	pc	c	c	c	c	o	73	
20th	Wed	c	c	c	c	c	o	c	c	c	c	c	c	c	c	c	o	o	o	o	o	22	
21st	Thur	o	o	o	o	o	o	o	c	o	c	c	o	o	o	o	o	o	c	c	o	14	
22nd	Fri	F	F	F	F	F	pc	F	pc	F	F	F	F	F	F	F	F	F	pc	pc	pc	159	
23rd	Sat	pc	pc	pc	pc	pc	F	pc	pc	pc	pc	pc	pc	pc	pc	pc	pc	pc	c	c	c	113	N
24th	Sun	c	c	c	c	c	o	c	o	c	o	o	c	c	c	c	o	o	o	o	c	22	
25th	Mon	F	F	F	F	F	F	F	F	F	F	F	F	F	F	pc	F	F	pc	pc	F	166	
26th	Tues	F	F	F	F	F	pc	F	pc	F	pc	pc	F	F	F	F	pc	pc	pc	pc	pc	144	XhS
27th	Wed	o	o	o	o	o	o	o	o	o	o	o	o	o	o	c	o	o	o	o	o	9	
28th	Thur	o	o	o	o	o	o	o	pc	o	c	pc	o	o	c	o	c	c	c	c	c	29	
29th	Fri	c	c	c	c	c	c	c	c	pc	c	c	c	pc	pc	pc	c	c	c	c	c	61	A
30th	Sat	pc	pc	pc	pc	pc	F	pc	pc	pc	F	F	pc	pc	F	pc	F	F	F	F	F	160	
31st	Sun	pc	pc	pc	pc	pc	c	pc	c	pc	c	c	pc	pc	pc	pc	pc	pc	pc	pc	pc	95	1Q
Estimate hours:		163	163	163	163	163	162	163	171	169	175	173	163	169	175	164	179	177	165	165	186	3372	
Average hours:		146	146	133	140	133	127	127	144	122	140	152	136	148	140	158	138	124	140	140	159	2793	
Trend:		more	more	more	more	more	more	more	more	more	more	more	more	more	more	av	more	more	more	more	more	more	

AUGUST **1** FRIDAY

A depression to the west of Ireland with a secondary centre over Scotland brings a light flow over the country, variable in direction but predominately westerly. Cloudy everywhere with patches of overnight fog in many districts. Mostly dry in the far northwest of Ireland, with some showers elsewhere, as a showery front crosses the south coasts at first and slowly spreads northwards during the day. Some showers may be heavy and accompanied by thunderstorm activity.

Belfast: Cloudy, mild, isolated showers, fluctuating breezes, chance of fog
Dublin: Cloudy, light breezes, occasional showers, chance of patches of fog
Cork: Cloudy, cool, some showers, chance of fog, winds freshening later
Galway: Cloudy, light showers, warmer day colder night, fresh onshore breezes

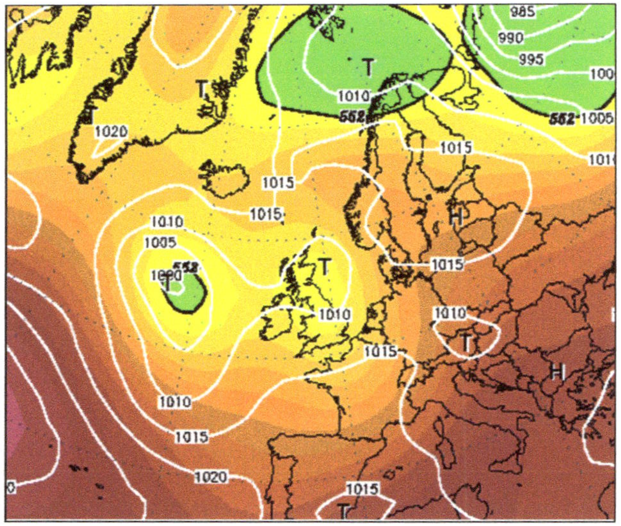

RAIN POTENTIAL

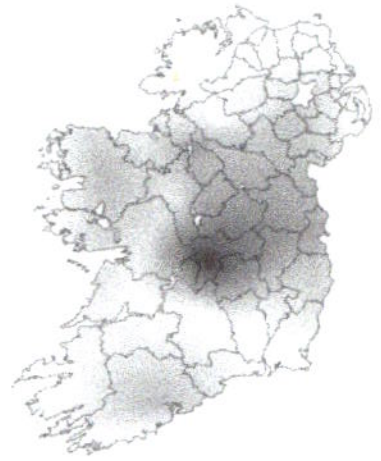

FROST/SNOW

AUGUST **2** SATURDAY

A low pressure system moves onto the country bringing unsettled conditions throughout. After patches of overnight fog clear, a changeable day can be expected with a mix of sunny spells and a few showers here and there, mostly in the east. Generally moderate northwest winds prevail, but fresher around the northern coasts.

Belfast: Sunny, warmer, dry, odd windy spell
Dublin: Increased sunny spells, some showers, warmer day cooler night, breezy
Cork: Partly cloudy, light showers, warmer, unsettled and changeable winds
Galway: Unsettled and changeable winds, sunny and dry

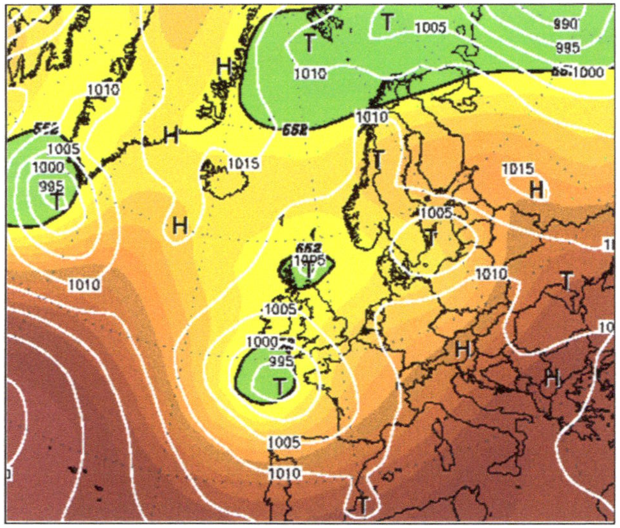

RAIN POTENTIAL

FROST/SNOW

AUGUST **3** SUNDAY

As the depression drifts away to the east of Ireland, a shower embedded low pressure trough moves over the country increasing cloud throughout, and bringing light generally westerly breezes, fresh and gusty at times around the coasts. Mostly dry at first before rain within the trough moves into the west of the country at first, and spreads throughout as the day progresses. Rain may be heavy at times in some places.

Belfast: Cloudy, occasional showers, warmer day cooler night, odd breezy spell
Dublin: Mostly cloudy, brief showers, mild day cold night
Cork: Changeable, overnight rain, sunny day, cooler, fresh breezes at times
Galway: Cloudy, occasional showers, cooler, fresh breezes

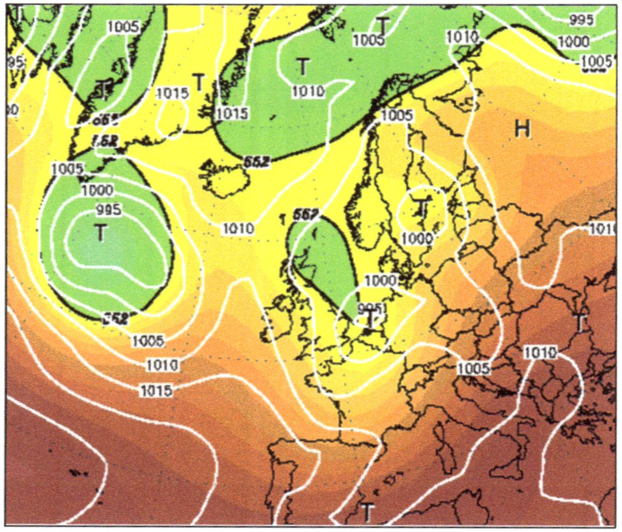

RAIN POTENTIAL

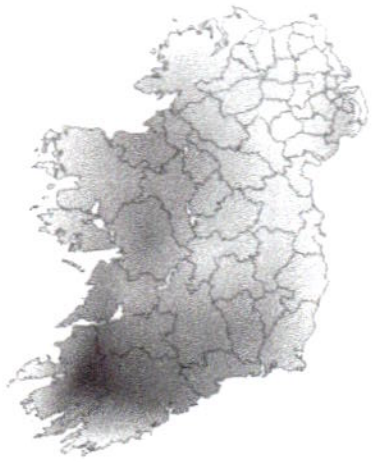

FROST/SNOW

AUGUST **4** MONDAY

A large depression to the northwest of Ireland directs a freshening southwesterly flow with occasional gusts over the country. Despite some good sunny spells, bands of showers and thunderstorms can be expected and include heavy rain and squally winds with the rain. Rain likely to be heaviest in the northwest and in parts of the midlands.

Belfast: Partly cloudy, outbreaks of rain, windy at times, chance of thunderstorms
Dublin: Unsettled and blustery at times, sunny, brief morning showers
Cork: Mostly sunny, brief showers, strong gusty winds
Galway: Strong winds, rain, some sunny spells, chance of gales

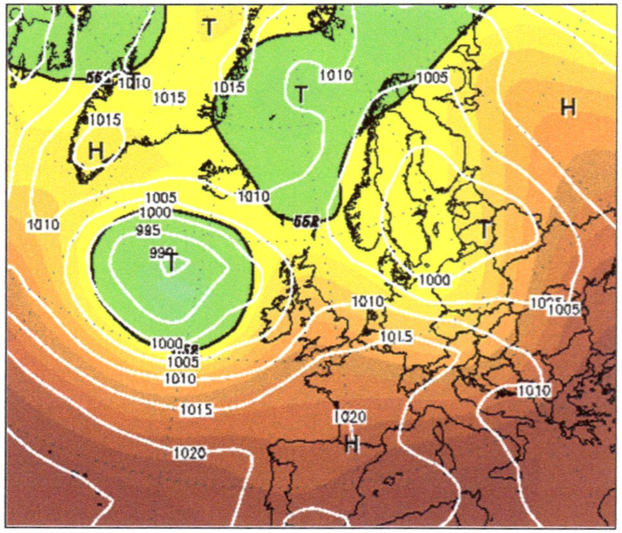

RAIN POTENTIAL

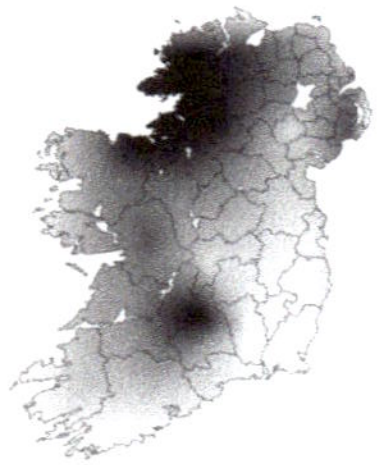

FROST/SNOW

AUGUST **5** TUESDAY

The depression over Ireland begins to weaken while it remains almost stalled over the country. Light to moderate west to southwest winds prevail throughout, fresher at times with occasional gusts. Another mixed day of sunny spells and light showers everywhere, as welll as further thunderstorms likely, and including the chance of some showers of hail, particularly in central districts.

Belfast: Partly cloudy, light showers, unsettled winds, chance of thunderstorms
Dublin: Partly cloudy, brief showers, blustery winds, chance of thunderstorms
Cork: Morning showers clearing to sunny day, strong winds at times
Galway: Sunny, brief squally showers, strong blustery winds

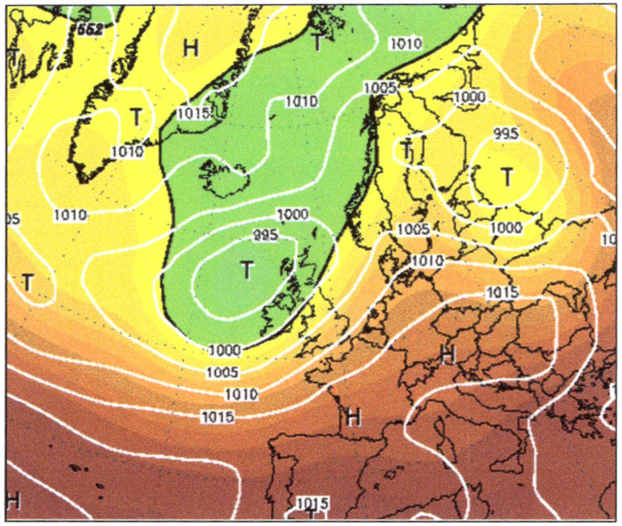

RAIN POTENTIAL

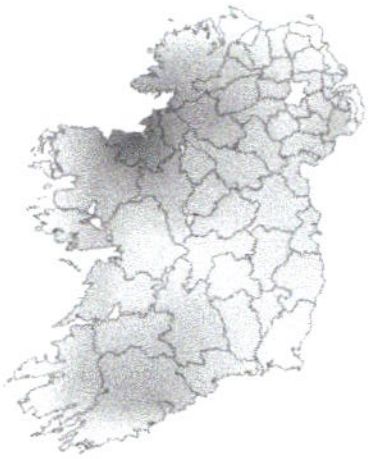

FROST/SNOW

AUGUST **6** WEDNESDAY

Low pressure continues to bring unsettled winds across the country, generally in a westerly flow. Mostly sunny at first, especially around the coastal fringes. Showers likely to develop later in the day once more, mostly in the east and north.

Belfast: Mostly sunny and dry, unsettled winds
Dublin: Scattered cloudy spells, brief showers, unsettled and blustery winds at times
Cork: Sunny, mostly dry, slightly warmer, unsettled winds, chance of showers
Galway: Sunny, mostly dry, strong winds at times

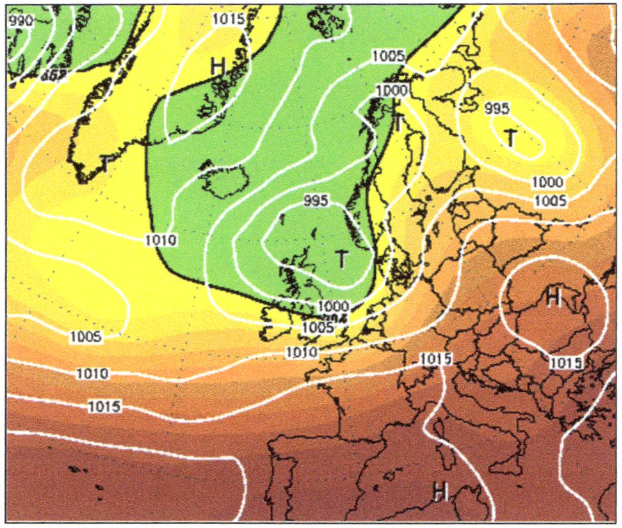

RAIN POTENTIAL

FROST/SNOW

AUGUST **7** THURSDAY

A complex of depressions near Ireland pass a low pressure trough over the country in a light to moderate northerly air, backing westerly at times. Mostly sunny in the north and east with overnight patches of fog likely, while cloudier in the south and west. Mostly dry in the far northwest of the country with some drizzle patches and occasional showers elsewhere.

Belfast: Mostly calm air with chance of patches of fog, partly cloudy, dry
Dublin: Mostly sunny, cooler, brief showers, winds easing to light
Cork: Cloudy, cooler, scattered showers, moderate breezes
Galway: Cloudy, brief showers and drizzle patches, mild day cold night, light breezes

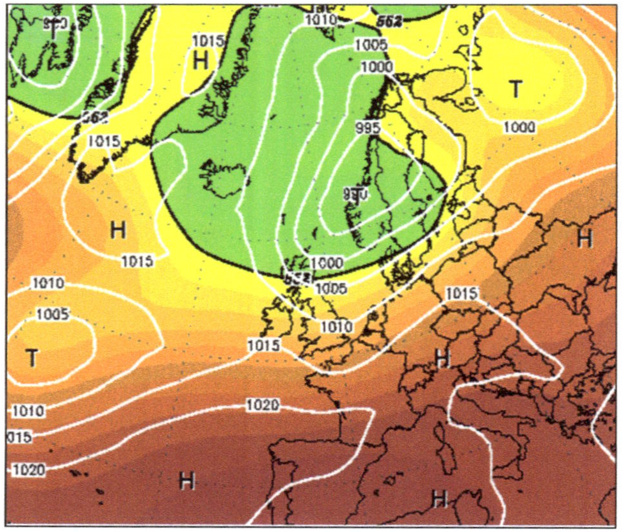

RAIN POTENTIAL

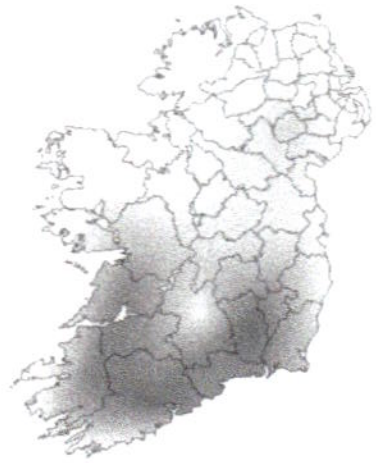

FROST/SNOW

AUGUST **8** FRIDAY

A weak high pressure system to the northwes tof Ireland ridges over the country directing a light northwesterly flow throughout. Very sunny and dry almost everywhere with patches of fog mostly in the east and south. Chance of some isolated showers possible mainly in the far northwestern districts.

Belfast: Very sunny, brief morning showers, mostly calm air
Dublin: Sunny, light breezes, mostly dry with chance of misty patches, cold overnight
Cork: Warmer, dry, sunny, mostly calm and pleasant air, chance of fog patches
Galway: Very sunny and dry, warm day cool night, mostly calm and pleasant air

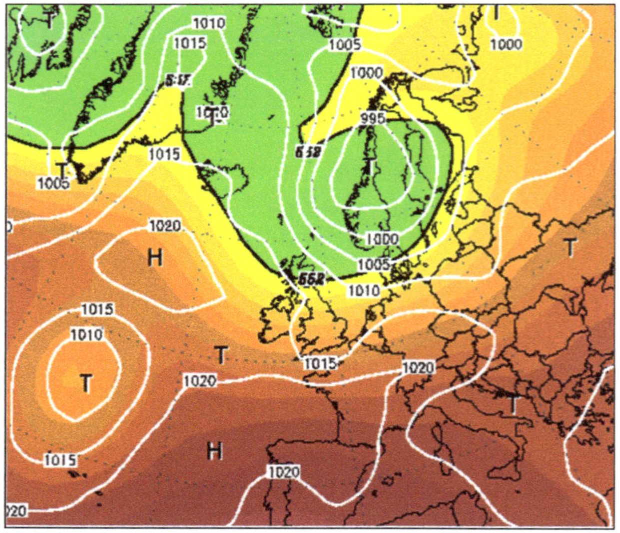

RAIN POTENTIAL

FROST/SNOW

AUGUST **9** SATURDAY

High pressure builds as it crosses the country bringing a very pleasant, sunny, and dry day throughout. Mostly calm to gentle breezes prevail generally in a northwest flow turning northerly then northeasterly in the south. Patches of fog overnight mostly along the coastal fringes.

Belfast: Sunny, dry, light but cold northwesterly flow
Dublin: Sunny, warmer day cold night, gentle breezes to mostly calm air
Cork: Pleasant and dry air, mild temperatures, very bright and sunny
Galway: Very sunny, dry and pleasant air, warmer day cooler night, gentle breezes

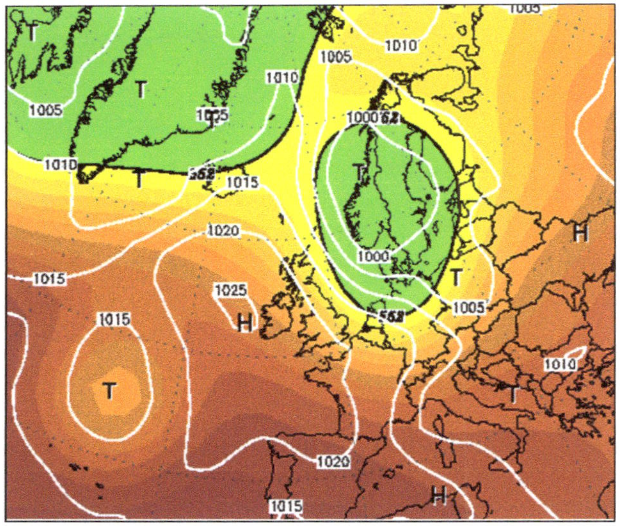

RAIN POTENTIAL

FROST/SNOW

AUGUST **10** SUNDAY

High pressure sits over Ireland directing a light air over the country, mostly in a south to southwest flow. Mostly sunny and dry with patches of overnight fog in many districts, except for increased cloud in the west and some isolated showers possible in the northwest to the midlands. Warmer than normal.

Belfast: Sunny, dry, mild day cold night, shifting breezy spells
Dublin: Warm day cool night, very bright and sunny, dry, gentle breezes
Cork: Very sunny, dry, mild temperatures, gentle breezes, patches of shallow fog
Galway: Gentle shifting breezes, light showers, occasional sunny spells

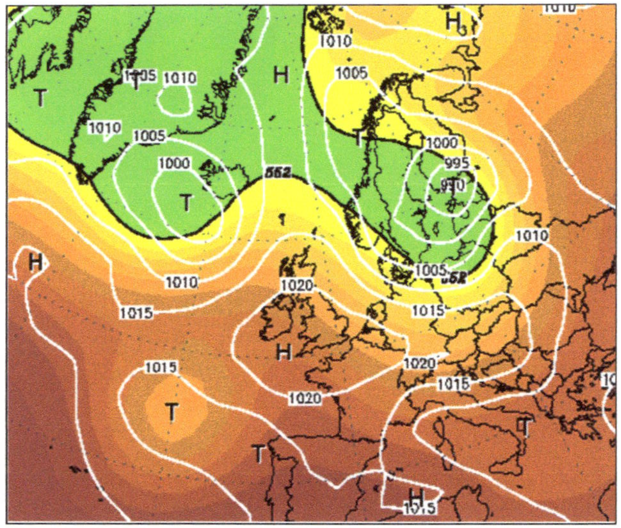

RAIN POTENTIAL

FROST/SNOW

AUGUST **11** MONDAY

A depression to the north of Ireland pushes the high pressure influence off the country in a gentle westerly flow. Cloud builds throughout with patches of fog or mists in sheltered districts. Some drizzle patches to light showers mostly in the north and west, while mostly dry in the south and east with rain threatening at times.

Belfast: Warm, cloudy, isolated showers, gentle breezes freshening later, fog
Dublin: Mostly calm and serene air, dull and gloomy, brief drizzle patches, cooler
Cork: Cloudy, warmer, brief misty showers, calm atmosphere, fog possible
Galway: Partly cloudy, light showers, gentle breezes

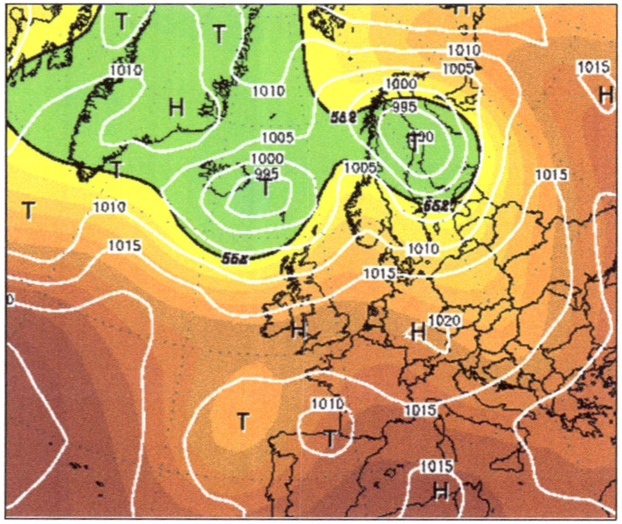

RAIN POTENTIAL

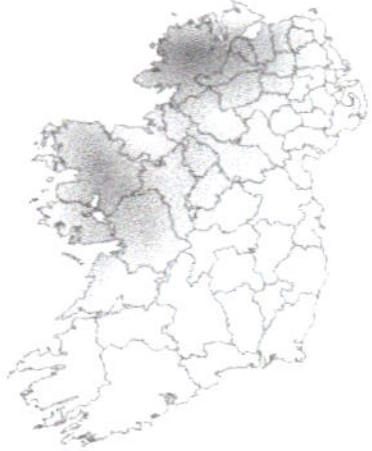

FROST/SNOW

AUGUST **12** TUESDAY

A low pressure trough flows over Ireland in a light, mostly westerly flow. A changeable day with a mix of long sunny spells and brief showers in most districts, with patches of drizzle becoming more persistent in the western fringes later in the day. Warmer and humid everywhere.

Belfast: Partly cloudy, dry and cooler, light breezes
Dublin: Partly cloudy, dry, gentle breezes, warm and humid air
Cork: Partly cloudy, isolated showers, gentle breezes
Galway: Sunny, brief showers, gentle breezes

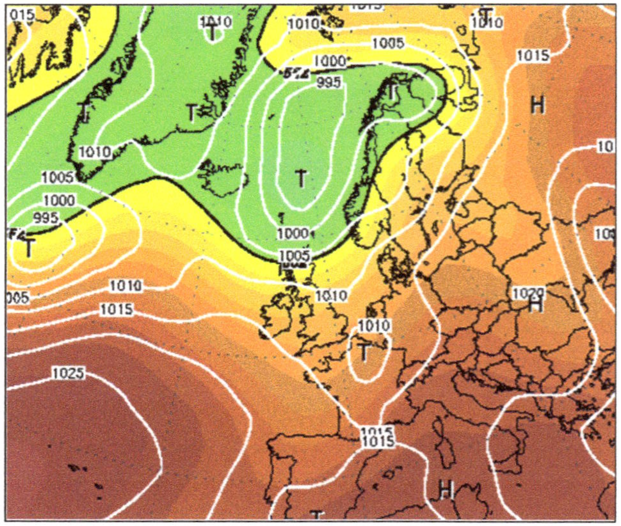

RAIN POTENTIAL

FROST/SNOW

AUGUST **13** WEDNESDAY

High pressure ridges near the south of Ireland while a low pressure trough maintains a generally westerly flow becoming stronger later in the day with a southwest change as a depression to the northwest of the country moves near to the north of Ireland. Cloudy everywhere with scattered rain activity moving into the western half of the country at first, spreading eastward as the day progresses, but remaining mostly dry along the eastern coastal fringes.

Belfast: Cloudy, warm southwest breezes freshening later in the day, brief showers
Dublin: Partly cloudy, warmer day cooler night, dry, light winds strengthening later
Cork: Overcast and dull, light showers, light breezes strengthening later in the day
Galway: Freshening onshore breezes, cloudy, cooler, rain

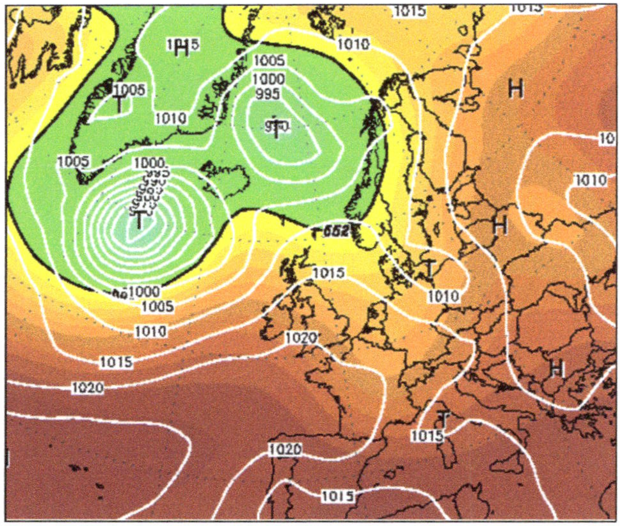

RAIN POTENTIAL

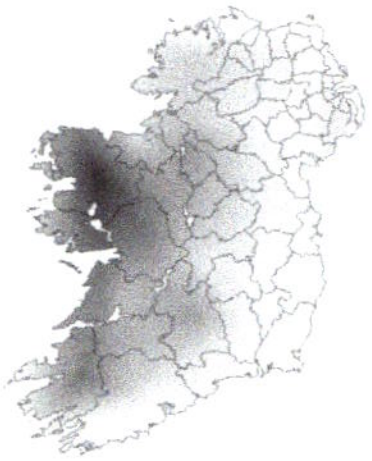

FROST/SNOW

AUGUST **14** THURSDAY

An intense depression moves onto Ireland in an unsettled and very strong southwest flow, with winds occasionally very gusty and possibly reaching gale force at times in some parts of the north and west. Stormy conditions build everywhere, as heavy rain moves into the west and spreads to the east and south. Rain activity likely to be lighter in the far north.

Belfast: Stormy, cloudy, heavy rain, warm and humid, winds to gale force at times
Dublin: Stormy air, rain, some sunny breakthroughs, chance of gales at times
Cork: Stormy air, windy, heavy rain, brief sunny outbreaks
Galway: Stormy air, cloudy, rain, chance of gales for a time

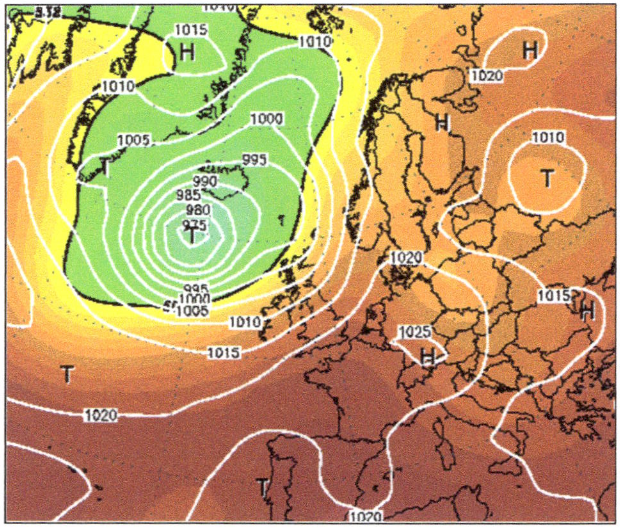

RAIN POTENTIAL

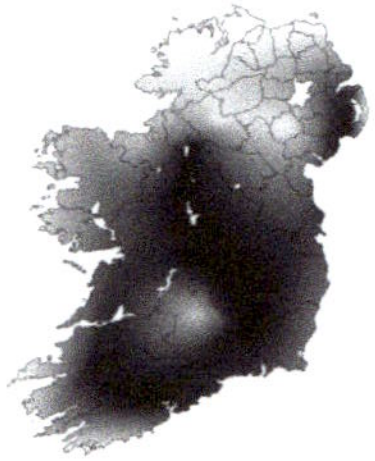

FROST/SNOW

AUGUST 15 FRIDAY

A depression to the north of Ireland directs a generally westerly flow over the country with winds easing throughout. Widespread cloud and rain almost everywhere except for mostly dry along the northern coastal fringes. Rain likely to be heavy at times in many areas over the southern half of the country, as well as in the northeast just south of Ulster.

Belfast: Overcast, scattered showers, blustery winds easing, slightly cooler
Dublin: Overcast, winds easing to light, rain, cooler
Cork: Partly cloudy, heavy rain, light breezes to mostly calm air
Galway: Partly cloudy, occasional showers, winds easing to light

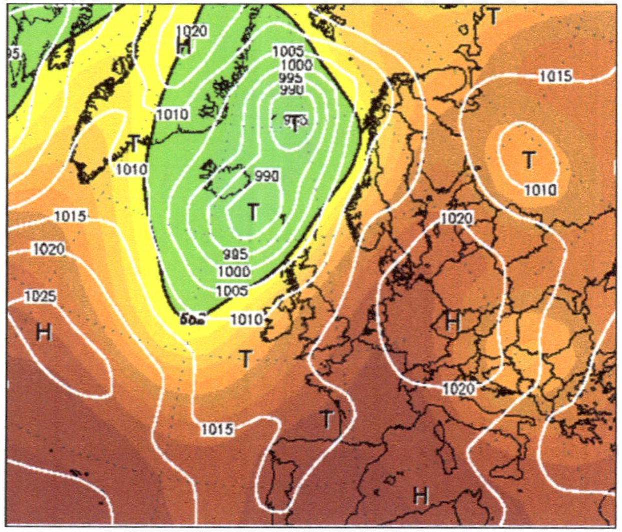

RAIN POTENTIAL

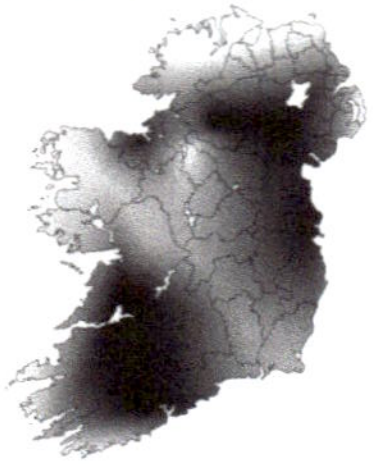

FROST/SNOW

AUGUST 16 SATURDAY

A weak depression drifts sits to the west of Ireland while high pressure pushes a ridge near the west of the country, with a lighter north to northwest flow prevailing throughout, fresher for a time later in the day. A changeable day with a mix of sunny spells and occasional shower sin most districts. Some showers may accompany thunderstorms and accompany brief hail showers mostly in the east to central districts, with cooler than normal temperatures almost everywhere.

Belfast: Changeable, mix of showers and sunny spells, cooler, fluctuating breezes
Dublin: Fluctuating winds, rain, some sunny spells, thunderstorms possible
Cork: Sunny, moderate breezes, occasional showers
Galway: Fresh breezes, sunny, light showers, mild day cool night

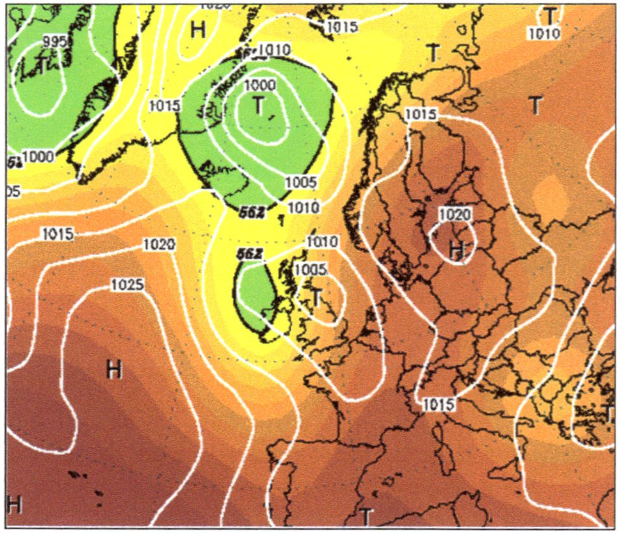

RAIN POTENTIAL

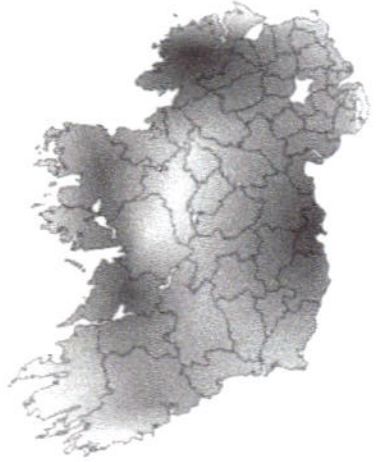

FROST/SNOW

AUGUST **17** SUNDAY

A weak depression to the northeast of Ireland directs a light northerly air over the country while an Atlantic high pressure ridge sits to the southwest. Mostly dry and sunny everywhere, with patches of fog likely mostly in the north. Chance of some isolated showers briefly in the northeast.

Belfast: Sunny, dry, mild temperatures, fresh breezes at times, chance of fog patches
Dublin: Partly cloudy, isolated showers, winds easing with occasional gusts
Cork: Sunny and dry, mild temperatures, light breezes
Galway: Sunny, dry, mild day cool night, moderate breezes

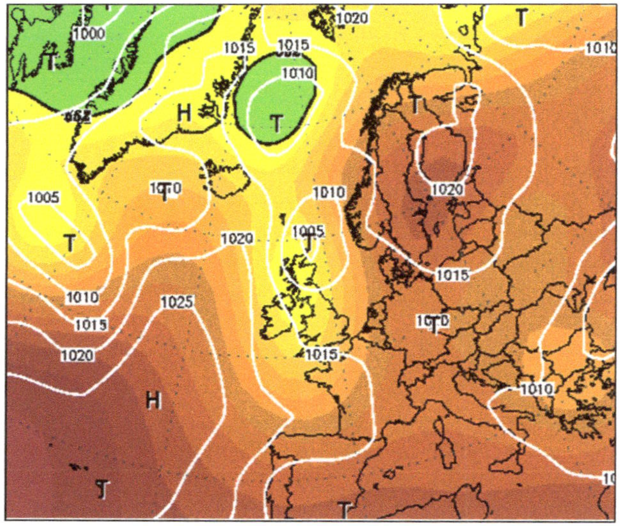

RAIN POTENTIAL

FROST/SNOW

AUGUST **18** MONDAY

A large anticyclonic system to the southwest of Ireland extends its high pressure ridge over the country in a very gentle west to southwest flow. Mostly dry and pleasant air throughout, with good sunny spells in most districts, although cloudier in the south. Chance of some isolated showers in the west to the midlands.

Belfast: Warmer day colder night, partly cloudy, dry, gentle breezes
Dublin: Partly cloudy and dry, mild day cold night, light breezes to mostly calm air
Cork: Mostly calm and pleasant air, sunny, dry
Galway: Mostly cloudy, occasional showers, cooler, moderate breezes

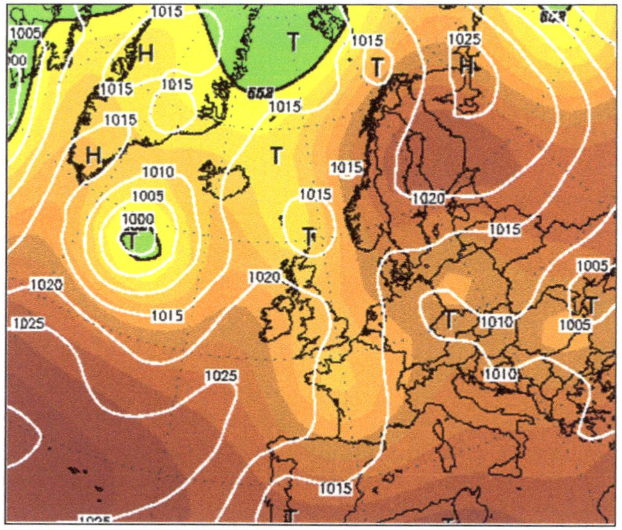

RAIN POTENTIAL

FROST/SNOW

AUGUST **19** TUESDAY

The high pressure ridge builds over Ireland with a light southerly flow prevailing. Increased cloud everywhere, with chance of patches of fog overnight in the east. Mostly dry everywhere at first, before spotty showers move into the western half of the country, while remaining dry in the east.

Belfast: Cooler, partly cloudy, isolated showers, moderate breezes
Dublin: Warm day cool night, partly cloudy, dry, calm air, chance of shallow fog
Cork: Dull and gloomy, light showers, cooler, light breezes
Galway: Mix of showers and brief sunny spells, fresh breezes at times

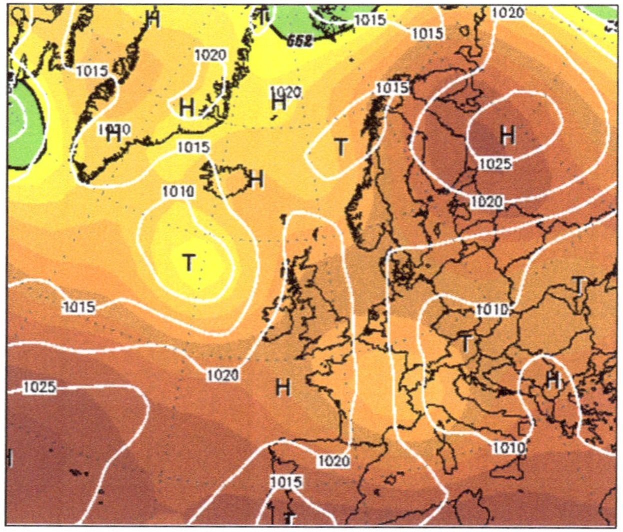

RAIN POTENTIAL

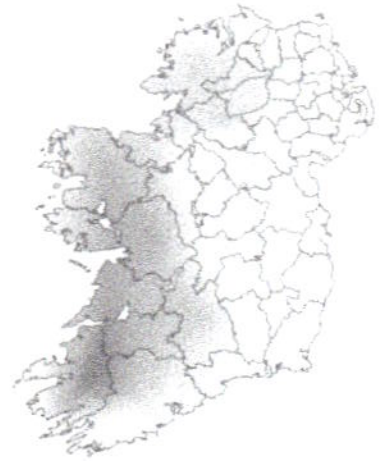

FROST/SNOW

AUGUST **20** WEDNESDAY

A weak depression to the northwest of Ireland moves onto the country in a gentle southerly flow, backing southeasterly as the low advances closer. Dull and gloomy everywhere with few sunny breakthroughs. A band of rain crosses into the country from the southwest and spreads northward during the day to bring showers to most districts by evening, although few likely to reach the far northwestern coasts. Rain may be heavy for a time in the southwest.

Belfast: Cloudy, mild temperatures, occasional showers, moderate breezes
Dublin: Cloudy, scattered showers, moderate breezes
Cork: Gentle southwesterly breezes, overcast, occasional showers
Galway: Cloudy, moderate breezes, rain

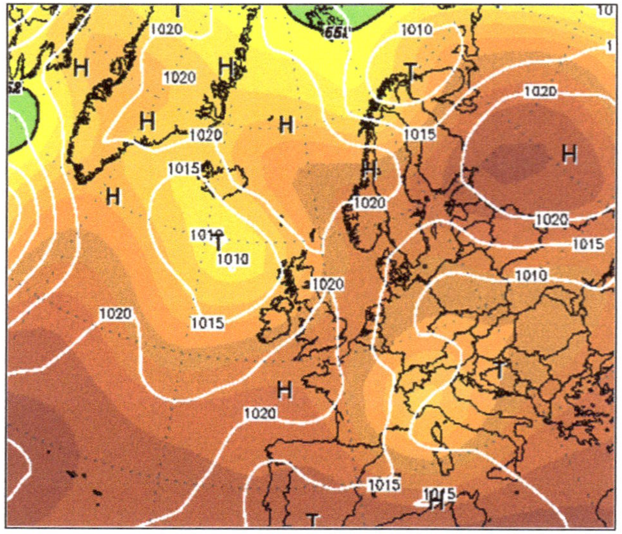

RAIN POTENTIAL

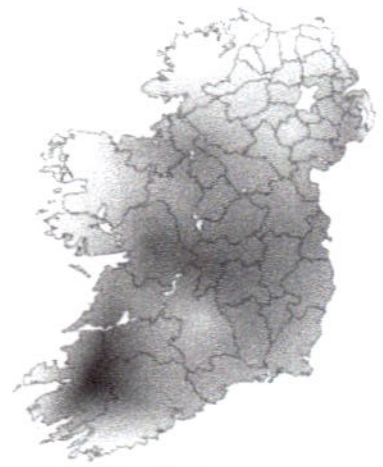

FROST/SNOW

AUGUST 21 THURSDAY

A weak depression continues to bring widespread rain activity throughout the country with heavy falls and scattered thunderstorms particularly in the northwest, and chance of hail showers in central parts. Cooler than normal and very dull and gloomy everywhere. Winds generally light and southerly at first, backing northwesterly by the end of the day.

Belfast: Overcast, heavy rain, fluctuating breezes
Dublin: Overcast, some showers and drizzle, mild temperatures, mostly calm air
Cork: Overcast, warmer, occasional showers, mostly calm air, chance of fog
Galway: Cloudy, passing showers, shifting breezy spells

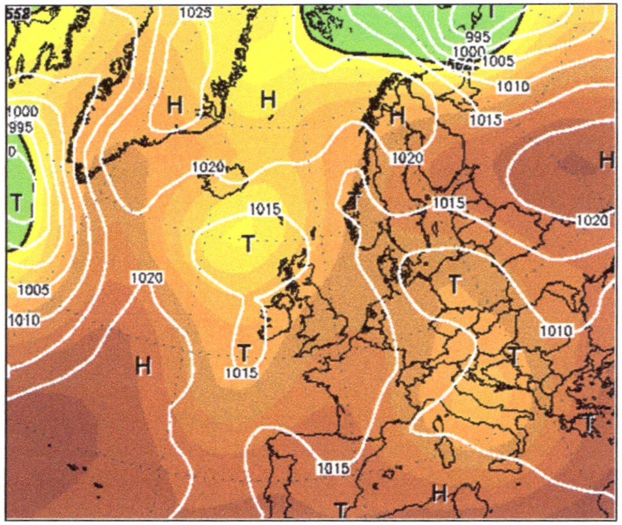

RAIN POTENTIAL

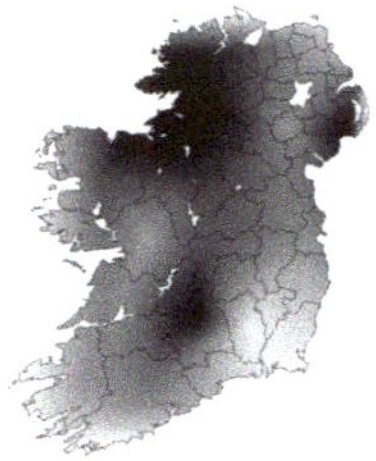

FROST/SNOW

AUGUST 22 FRIDAY

A weak depression remains almost stalled over the country bringing light to moderate northwest breezes throughout. Mostly sunny and dry day, with outbreaks of passing showers overnight in many districts, heavier at times in the northeast.

Belfast: Sunny, colder, light overnight showers, fresh breezes at times
Dublin: Sunny, dry day, brief evening showers, gentle shifting breezes
Cork: Mostly sunny, warmer day cooler night, light showers, light shifting breezes
Galway: Sunny, spells of light rain, warmer day colder night, fresh breezes at times

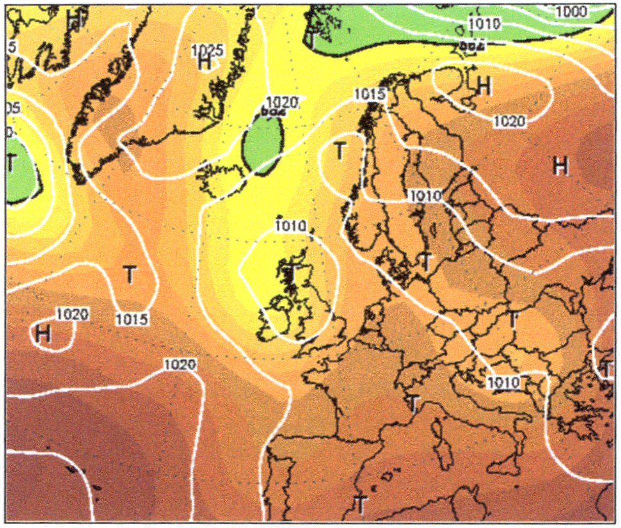

RAIN POTENTIAL

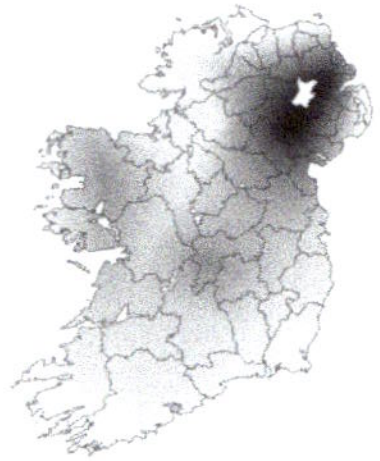

FROST/SNOW

AUGUST 23 SATURDAY

The depression over Ireland strengthens as it remains almost stalled over the country. Strong and unstable northwesterly winds prevail, with gusts around the northern coastal fringes, later easing with a easterly change. Frequent scattered cloud everywhere, more often in the south. Light showers at first in most districts at first, followed by long dry spells, before another rain band approaches the country later in the day.

Belfast: Partly cloudy, mostly dry, warmer, blustery at times
Dublin: Partly cloudy, light showers, unsettled breezy spells
Cork: Unsettled breezy spells, mostly cloudy, brief showers, chance of fog
Galway: Partly cloudy, dry, cold overnight, unsettled and strengthening winds

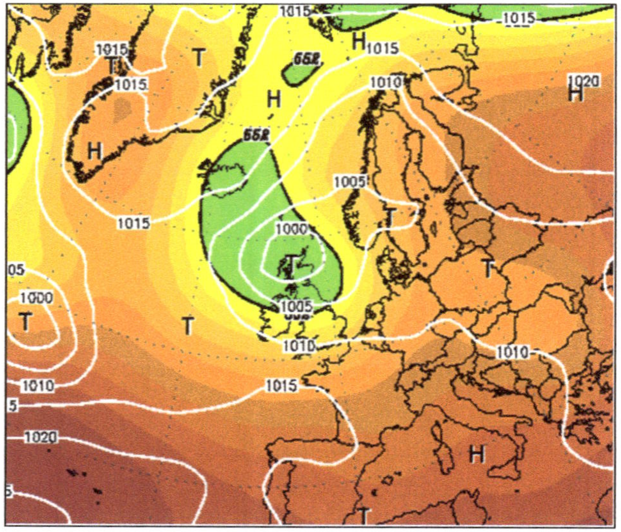

RAIN POTENTIAL

FROST/SNOW

AUGUST 24 SUNDAY

A rain embedded trough between two depressions crosses the country in a fresh to strong east to southeasterly flow. Widespread cloud and a stormy air prevails in most districts as a band of rain moves into the southwest at first and spreads up through the country during the day, easing to only a few showers by the time it reaches the far north. Rain likely to be particularly heavy for a time through the midlands. Winds may reach gale force for a time around the western coasts.

Belfast: Cloudy, mild, isolated showers, blustery at times, chance of misty drizzles
Dublin: Cloudy, rain, windy at times with occasional gusts
Cork: Unsettled windy spells, cloudy, occasional showers
Galway: Stormy air, overcast, heavy rain, colder, gales possible

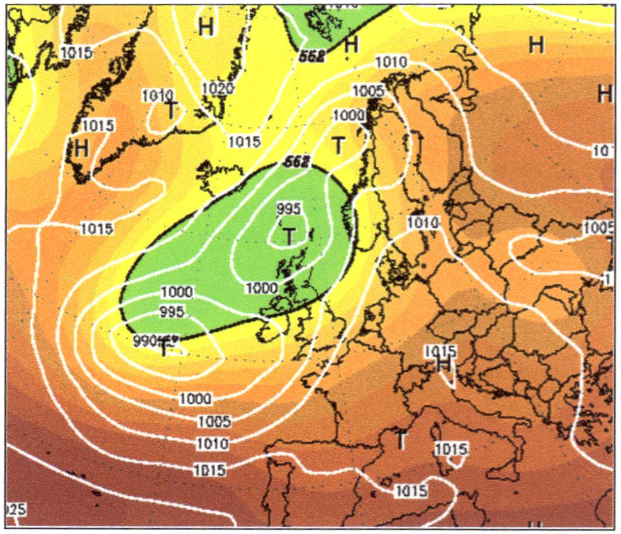

RAIN POTENTIAL

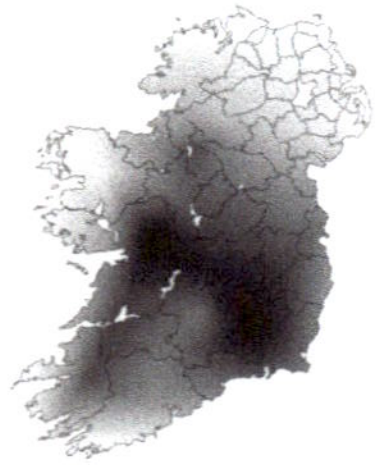

FROST/SNOW

AUGUST **25** MONDAY

A large depression fills over Ireland directing a fresh north to northwest flow over the country, strong around the northern and western coasts. An unsettled day everywhere, with widespread overnight showers clearing to mostly dry and sunny for much of the day, before another band of rain crosses the country once more later in the afternoon.

Belfast: Sunny, cooler, brief showers, strong breezes, chance of gales at times
Dublin: Unsettled and blustery air at times, mix of showers and sunny spells
Cork: Sunny, blustery spells, brief overnight showers
Galway: Unsettled blustery winds, sunny, light overnight showers

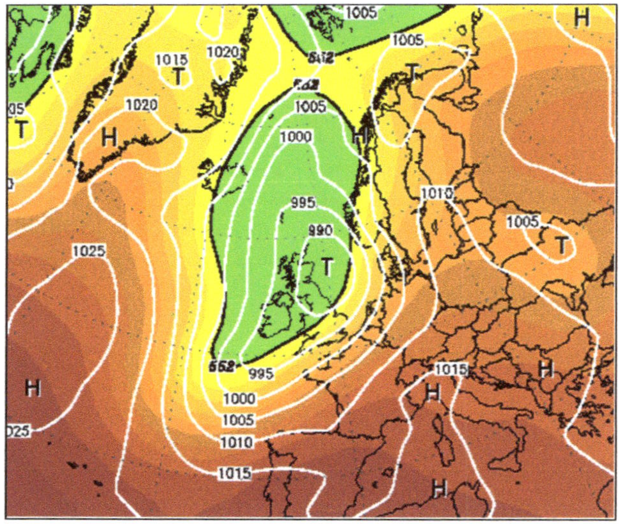

RAIN POTENTIAL

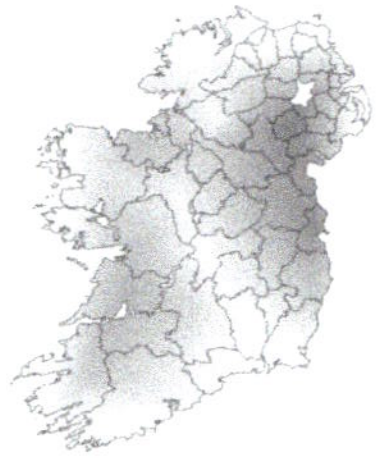

FROST/SNOW

AUGUST **26** TUESDAY

A depression drifts fruther north away from Ireland while a high pressure system in the Atlantic ridges to the southeast of the country. Fresh to strong northwest winds prevail for much of the day, later turning westerly. Some isolated overnight showers in most districts, clearing to a generally dry and mostly sunny day everywhere.

Belfast: Sunny, dry, slightly warmer day, cool overnight, strong winds, gales possible
Dublin: Strong winds, chance of gales, sunny day, isolated overnight showers
Cork: Partly cloudy, colder, mostly dry, threats of rain, gusty winds at times
Galway: Partly cloudy, isolated showers, strong blustery winds, chance of gales

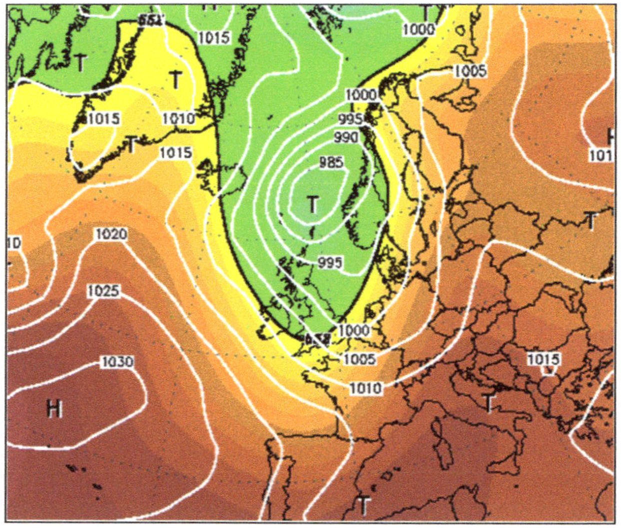

RAIN POTENTIAL

FROST/SNOW

AUGUST **27** WEDNESDAY

Low pressure systems to the north and west of Ireland direct a moderate west to southwest flow at first, while the high pressure to the southeast later brings a south to southeast change. Widespread cloud throughout with patches of light rain move across the western half of the country and spreads to the east as the day progresses. Chance of some patches of heavier rain in the southwest for a time.

Belfast: Overcast, rain, windy spells
Dublin: Cloudy, cooler, light showers, blustery at times
Cork: Overcast, scattered showers, cool, windy spells at times
Galway: Unsettled winds, overcast, rain, colder

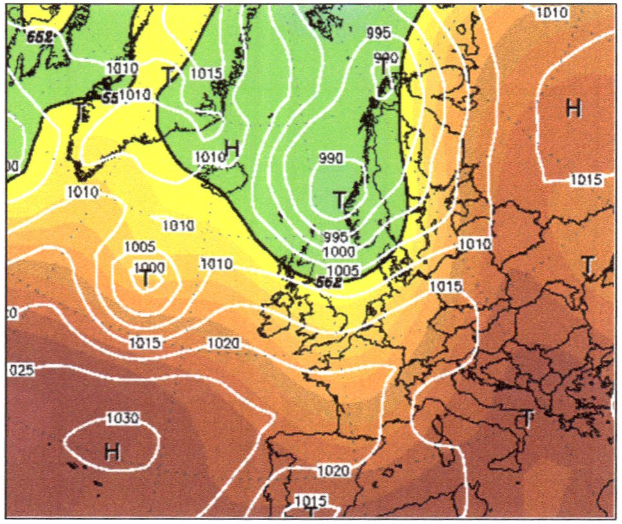

RAIN POTENTIAL

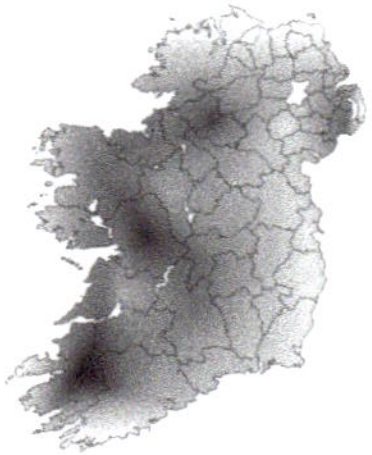

FROST/SNOW

AUGUST **28** THURSDAY

A depression consolidates over Ireland bringing a stormy air and fresh and gusty winds at times of variable directions, stablilsing into a northwesterly flow later in the day. Dull and gloomy everywhere with drizzly showers throughout.

Belfast: Stormy air, overcast, light drizzly showers, fluctuating winds
Dublin: Stormy air, overcast, scattered showers, unsettled winds with gusts
Cork: Unsettled and blustery winds, cloudy, light showers, warmer
Galway: Stormy winds, cloudy, light showers and drizzly spells

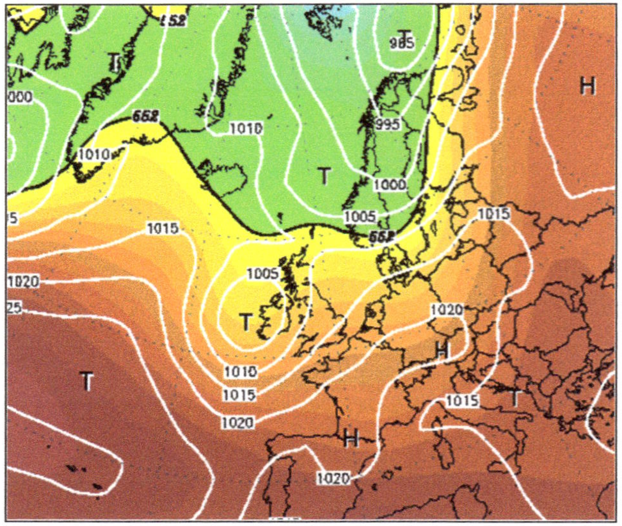

RAIN POTENTIAL

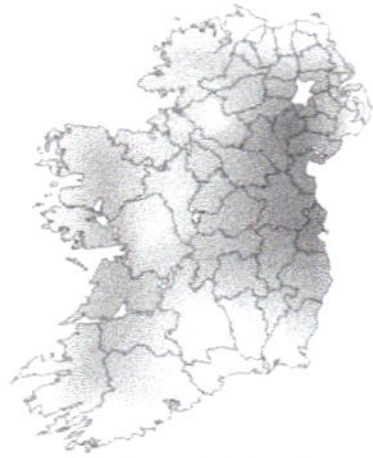

FROST/SNOW

AUGUST 29 FRIDAY

A weak depression lingers over Ireland bringing a stormy air everywhere with moderate to strong northerly winds with occasional gusts. Mostly cloudy throughout, remaining dry in the far south and northwest regions, while overnight showers affect elsewhere at first, and again at the end of the day. Rain may be heavy for a time in the east, while lighter in the west.

Belfast: Mostly cloudy, warmer, scattered showers, strong winds, possible gales
Dublin: Partly cloudy, dry, cold overnight, strong and stormy winds at times
Cork: Strong and blustery winds, mostly cloudy and dry, colder northerly air
Galway: Strong blustery winds, cloudy, rain, cooler

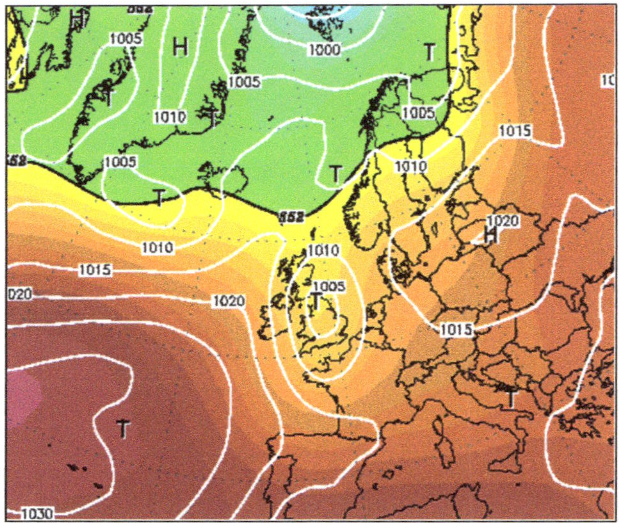

RAIN POTENTIAL

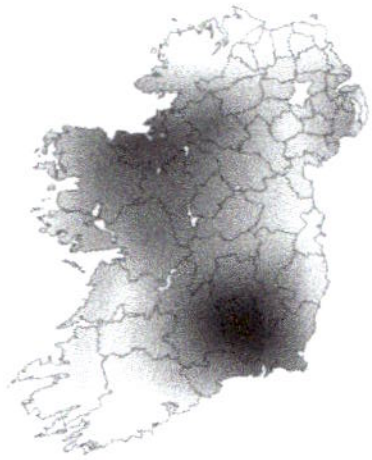

FROST/SNOW

AUGUST 30 SATURDAY

A large Atlantic high pressure ridge to the southwest of Ireland nears the southwest, but low pressure continues to prevail over the country with a north to northwest flow prevailing, easing from strong to moderate, but still gusty at times particularly around the coasts. A wet front brings heavy rain in the east and north overnight, with showers more isolated in the west and south. Mostly dry and sunny day after the rain clears, but showers remaining more persistent in the north and east.

Belfast: Partly cloudy, dry, blustery with possible gale gusts at times
Dublin: Partly cloudy, heavy rain, winds easing with lingering gusts
Cork: Sunny, warmer, light showers, unsettled and fluctuating winds
Galway: Sunny, warmer, mostly dry with chance of isolated showers, gusty winds

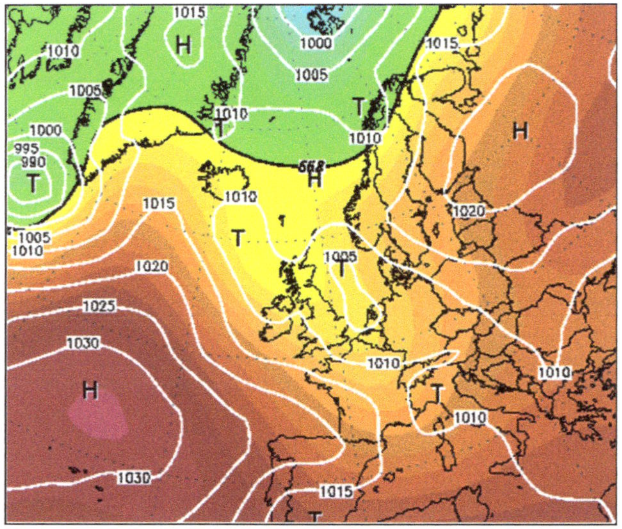

RAIN POTENTIAL

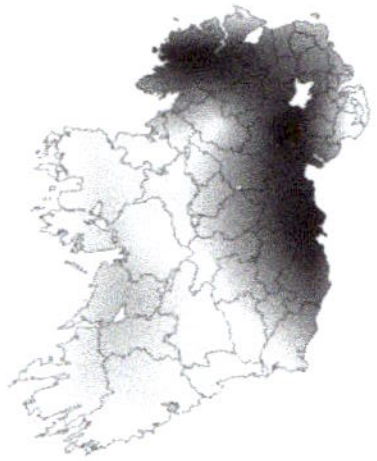

FROST/SNOW

AUGUST 31 SUNDAY

A large anticyclonic system in the Atlantic Ocean extends its high pressure ridge onto the country with fresh to blustery northerly winds prevailing, easing to lighter later in the day. Mostly dry and sunny everywhere, except for the chance of some lingering light showers overnight at first. Warmer day cooler night.

Belfast: Partly cloudy, dry, strong and cool northerly winds
Dublin: Partly cloudy, dry, mild day cool night, strong gusty winds
Cork: Partly cloudy, colder, dry, fresh winds with occasional gusts possible
Galway: Strong and blustery northerly winds, cooler, partly cloudy and dry

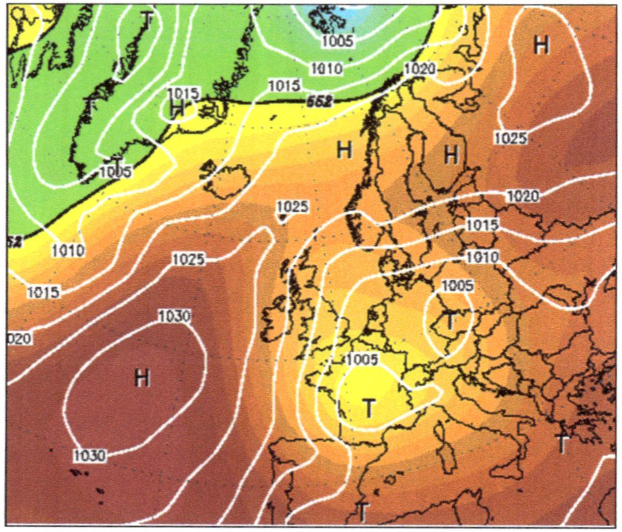

RAIN POTENTIAL

FROST/SNOW

SEPTEMBER

PHASES OF THE MOON

1st	Southern Declination
7th	Full Moon
9th	Crossing Equator
10th	Perigee #9
14th	Third Quarter
14th	Northern Declination
21st	New Moon
22nd	Crossing Equator
26th	Apogee
29th	First Quarter
29th	Southern Declination

1	2	3	4	5	6	7	8	9	10	11	12	13	14	15	16	17	18	19	20	21	22	23	24	25	26	27	28	29	30
Sun	Mon	Tue	Wed	Thur	Fri	Sat	Sun	Mon	Tue	Wed	Thur	Fri	Sat	Sun	Mon	Tue	Wed	Thur	Fri	Sat	Sun	Mon	Tue	Wed	Thur	Fri	Sat	Sun	Mon
V						F		XhN	P9				3Q ^							N	XhS				A			1Q V	

MONTHLY SUMMARY

1st-10th A large high pressure system centred to the west of Ireland brings a mostly dry and settled spell with light to moderate winds west to northwest breezes throughout. Some passing weak fronts during this spell are likely to bring some light rain or drizzle patches now and then, but good sunny spells can be expected daily, especially in the south and southeast. Warmer overnight temperatures than of recent, while daytime maximas will be close to the norm for this time of year.

1st Very sunny day.

1st-4th Very hot spell. Scattered fog patches overnight in many parts of the country.

3rd-6th Chance of snow grains in central parts

5th Very high pressure brings a very sunny day in Ireland.

9th-12th Chance of snow grains in the southern half of the country

Over the next four weeks high pressure will prevail for much of this outlook period, especially over the first half. Anticyclonic systems anchored to the southwest of Ireland or over the country will bring mostly settled conditions everywhere for the first couple of weeks. Following this, an active cold front arrives early in the third week bringing a wetter and colder period for a few days, before anticyclonic weather settles in again in the last week of this outlook period early in October. As a result, much of the next four weeks ahead will be drier than normal everywhere, as well as warm, with lighter than normal winds, and with plenty of sun. Many districts will only see around half their usual amount of rain that they typically receive at this time of year, and most of the rain is expected over the third week and into the fourth week. The east of Ireland can expect average sunshine for this time of year, while all other parts of the country can anticipate above average sunshine, especially in the southwest, with the first week being the sunniest spell over this outlook period. Maximum temperatures will be around half to one degree warmer than average everywhere, with the first couple of weeks particularly warmer than average for both maximas and minimas. The second half of this outlook period will see temperatures falling sharply and with the arrival of regular frosts by the early October. During this outlook period, there will be around 7 to 11 days when rain will fall, thunderstorms and hail on 1 day, fog on 7 days, frosts on 4-5 days, but no snow.

11th-25th Anticyclonic conditions prevail as a large high pressure system remains almost stalled to the southwest of the country bringing mostly settled conditions throughout. Some light showers at first over the northern half of the country before clearing to mainly dry everywhere. Light changeable breezes prevail, southwest at times, backing northwest. Warmer and sunnier than of recent.

12th-18th Frequent scattered fogs overnight.

13th Very bright and sunny day.

14th-15th Chance of wintry showers of snow grains overnight in central parts.

15th-18th High maximum daytime temperatures everywhere, with most districts over 20°C.

16th-20th Very sunny spell.

17th-19th Widespread fog. Very bright and sunny day with many parts of the country likely to receive over 12 hours of recorded sunshine, particularly in the southwest.

25th-30th Wet spell with some heavy falls in places as an active cold front crosses the country.

27th-28th Widespread frosts overnight likely, especially in the eastern to central areas. Windy at times.

30th Heavy rain likely in a number of districts, as well as the chance of hail showers.

IRELAND RAINFALL ESTIMATES

September 2025

d=mainly dry, n= nonrecordable, l=light shrs, s=significant shrs, r=rain, h=hvy falls

		ULSTER PROVINCE					CONNAUGHT PROVINCE						LEINSTER PROVINCE				MUNSTER PROVINCE						
		Alderg	Hillsbo	Armag	Loughl	Malin H	Sligo	Ardtarı	Belmul	Drums	Galway	Roscoı	Markre	Edende	Kilkenı	Dublin	Shann	Newpo	Tralee	Killarne	Cork	All	Moon
1st	Mon	d	d	d	d	d	d	d	d	d	d	d	d	d	d	d	d	d	d	d	d	0	V
2nd	Tues	d	d	d	l	d	d	l	d	d	d	d	d	d	d	d	d	d	d	d	d	2	
3rd	Wed	d	d	d	l	l	d	l	n	d	d	l	d	d	d	d	d	l	d	d	d	3	
4th	Thur	d	d	d	d	l	l	d	l	d	d	d	l	d	d	d	l	l	l	d	d	7	
5th	Fri	d	d	d	d	d	d	d	d	d	d	d	d	d	d	d	d	d	d	d	d	0	
6th	Sat	d	d	d	d	d	d	d	d	d	d	d	d	d	d	d	d	d	d	d	d	0	
7th	Sun	d	d	d	d	d	l	d	d	d	d	d	d	d	d	d	d	n	d	d	d	1	F
8th	Mon	l	d	l	d	l	n	d	d	d	d	d	d	d	d	d	d	d	d	d	d	5	
9th	Tues	d	d	d	l	l	l	l	d	l	d	d	n	d	d	d	d	l	d	d	d	8	XhN
10th	Wed	d	d	d	n	s	l	n	n	l	d	d	d	d	d	d	n	d	n	d	d	9	P9
11th	Thur	d	d	d	l	l	l	l	l	l	l	n	d	n	d	d	l	s	l	l	n	28	
12th	Fri	l	d	l	l	s	l	l	l	l	d	n	s	l	d	l	l	l	l	l	d	35	
13th	Sat	l	d	l	n	d	d	n	d	d	d	d	n	d	d	d	d	l	d	d	d	4	
14th	Sun	d	d	d	l	l	l	l	l	l	l	n	l	n	d	d	d	l	d	d	d	12	3Q ^
15th	Mon	d	d	d	n	l	d	n	d	d	d	d	d	d	d	l	d	d	d	d	d	3	
16th	Tues	d	d	d	d	l	d	d	n	d	d	d	d	d	d	d	d	d	d	d	d	2	
17th	Wed	d	d	d	d	d	d	d	d	d	d	d	d	d	d	d	d	d	d	d	d	0	
18th	Thur	d	d	d	d	l	d	d	d	d	d	d	d	d	d	d	d	d	d	d	d	2	
19th	Fri	d	d	d	l	n	l	l	d	d	d	l	d	d	d	d	d	n	d	d	d	5	
20th	Sat	d	d	d	d	d	n	d	l	n	n	n	l	l	d	d	d	n	d	d	d	4	
21st	Sun	d	d	d	d	l	d	d	d	d	d	d	d	d	d	d	d	d	d	d	d	1	N
22nd	Mon	d	d	d	d	d	d	d	d	d	d	d	d	d	d	d	d	d	d	d	d	0	XhS
23rd	Tues	n	l	d	l	s	l	l	s	l	l	l	l	n	d	d	d	s	d	l	d	35	
24th	Wed	d	d	d	d	l	l	d	d	d	d	d	s	d	d	l	l	d	l	d	l	10	
25th	Thur	n	l	d	h	l	h	h	l	r	s	h	h	d	d	d	d	h	d	l	d	220	
26th	Fri	l	l	l	l	h	s	l	h	r	l	s	l	l	n	l	s	s	l	s	s	121	A
27th	Sat	l	l	s	n	l	l	n	l	l	n	l	l	d	d	d	d	l	l	l	d	21	
28th	Sun	l	l	l	s	s	s	s	s	s	r	l	l	l	l	d	l	r	d	l	d	84	
29th	Mon	l	l	s	l	s	l	l	l	n	s	l	l	s	l	r	r	l	s	s	l	71	1Q V
30th	Tues	l	l	d	l	s	l	l	l	l	s	l	d	r	h	l	l	s	h	h	l	143	
	Estimate:	12	7	17	66	77	76	66	47	40	38	35	45	25	27	15	28	101	36	73	8	838	
	Average:	76	81	68	99	96	104	106	102	79	100	105	104	78	70	60	76	132	128	87	95	1846	
	Trend:	drr	drr	drr	drr	drr	drr	drr	drr	drr	drr	drr	drr	drr	drr	drr	drr	drr	drr	drr	drr	drr	

IRELAND SUNSHINE ESTIMATES

September 2025

F=fine (8-12 hours of sunshine), pc= partly cloudy (4-7 hours), c=cloudy (1-3 hours), o=overcast (0 hours)

		ULSTER PROVINCE					CONNAUGHT PROVINCE						LEINSTER PROVINCE				MUNSTER PROVINCE						
		Alderg	Hillsbo	Armag	Loughl	Malin H	Sligo	Ardtari	Belmul	Drums	Galway	Roscoi	Markre	Edende	Kilkenı	Dublin	Shann	Newpo	Tralee	Killarne	Cork	All	Moon
1st	Mon	F	F	F	F	F	pc	F	o	F	c	c	F	F	F	F	pc	F	F	F	F	172	V
2nd	Tues	c	c	c	c	c	c	c	o	c	c	c	c	c	pc	c	pc	pc	pc	pc	pc	68	
3rd	Wed	F	F	F	F	F	F	F	c	F	c	c	F	F	F	pc	c	pc	c	c	F	134	
4th	Thur	c	c	c	c	c	c	c	c	pc	c	c	c	pc	pc	c	c	pc	c	c	F	64	
5th	Fri	pc	pc	pc	pc	pc	pc	pc	pc	F	pc	pc	pc	F	F	F	pc	pc	F	F	F	154	
6th	Sat	pc	pc	pc	pc	pc	pc	pc	pc	c	pc	pc	pc	c	c	c	pc	c	pc	pc	c	85	
7th	Sun	pc	pc	pc	pc	pc	c	pc	pc	pc	pc	pc	pc	pc	pc	pc	c	pc	F	F	pc	108	F
8th	Mon	pc	pc	pc	pc	pc	c	pc	c	pc	c	c	pc	pc	pc	pc	c	c	F	F	c	85	
9th	Tues	c	c	c	c	c	c	c	pc	c	pc	pc	c	c	pc	c	pc	pc	pc	pc	F	75	XhN
10th	Wed	o	o	o	o	o	o	o	o	c	c	c	o	c	c	c	c	c	c	c	c	23	P9
11th	Thur	c	c	c	c	c	c	c	c	c	c	c	c	c	c	c	c	c	c	c	c	46	
12th	Fri	o	o	o	o	o	o	o	c	o	o	c	o	o	o	o	o	o	o	o	o	10	
13th	Sat	F	F	F	F	F	F	F	F	F	F	F	F	F	F	F	F	F	F	F	F	200	
14th	Sun	c	c	c	c	c	o	c	o	c	o	o	c	c	c	c	o	c	pc	pc	c	42	3Q ^
15th	Mon	c	c	c	c	c	o	c	pc	c	pc	pc	c	c	c	pc	c	c	pc	pc	o	67	
16th	Tues	c	c	c	c	c	o	c	F	c	F	F	c	c	pc	c	F	pc	F	F	pc	96	
17th	Wed	F	F	F	F	F	F	F	pc	F	F	F	F	F	F	F	F	F	F	F	pc	203	
18th	Thur	c	c	c	c	c	pc	c	pc	pc	F	pc	c	pc	pc	o	F	F	F	F	F	120	
19th	Fri	c	c	c	c	c	c	c	c	c	pc	c	c	c	pc	o	pc	pc	F	F	F	79	
20th	Sat	pc	pc	pc	pc	pc	c	pc	o	pc	c	o	pc	pc	pc	F	c	c	o	o	c	78	
21st	Sun	o	o	o	o	o	o	o	o	o	o	o	o	o	o	o	o	o	o	o	o	5	N
22nd	Mon	o	o	o	o	o	o	o	o	o	o	o	o	o	o	o	o	o	o	o	o	7	XhS
23rd	Tues	o	o	o	o	o	o	o	c	c	c	c	o	c	c	o	o	c	pc	pc	pc	34	
24th	Wed	pc	pc	pc	pc	pc	pc	pc	pc	pc	pc	pc	pc	pc	pc	pc	pc	pc	F	F	pc	110	
25th	Thur	pc	pc	pc	pc	pc	c	pc	o	F	c	c	pc	F	F	F	pc	F	c	c	F	124	
26th	Fri	pc	pc	pc	pc	pc	pc	pc	pc	pc	pc	pc	pc	pc	pc	c	pc	pc	pc	pc	pc	104	A
27th	Sat	F	F	F	F	F	pc	F	pc	F	F	F	F	F	F	F	F	F	pc	pc	F	174	
28th	Sun	o	o	o	o	o	o	o	o	c	o	o	o	c	c	c	c	c	o	o	pc	25	
29th	Mon	c	c	c	c	c	pc	c	pc	c	c	pc	c	c	c	c	c	c	o	o	o	59	1Q V
30th	Tues	c	c	c	c	c	pc	c	c	pc	c	c	c	pc	pc	c	c	c	c	c	pc	66	
Estimate hours:		118	118	118	118	118	102	118	108	136	119	114	118	136	155	134	131	143	170	170	175	2617	
Average hours:		118	118	111	111	105	99	101	110	96	111	120	111	119	115	129	111	95	109	108	124	2219	
Trend:		av	av	av	av	more	av	more	av	more	av	av	av	more	more	av	more	more	more	more	more	more	

SEPTEMBER 1 MONDAY

A large anticyclonic system to the southwest of Ireland ridges over the country directing a very light and pleasant northwesterly flow. Dry and sunny everywhere, with patches of fog in sheltered areas, especially in inland parts. Chance of some brief drizzle patches in the far north overnight.

Belfast: Warmer, sunny, dry, mostly calm air, chance of fog patches
Dublin: Warm day cool night, dry, very sunny, fresh breezes at times
Cork: Sunny, dry, warm, moderate breezes
Galway: Warm, dry, occasional sunny spells, gentle breezes to mostly calm air

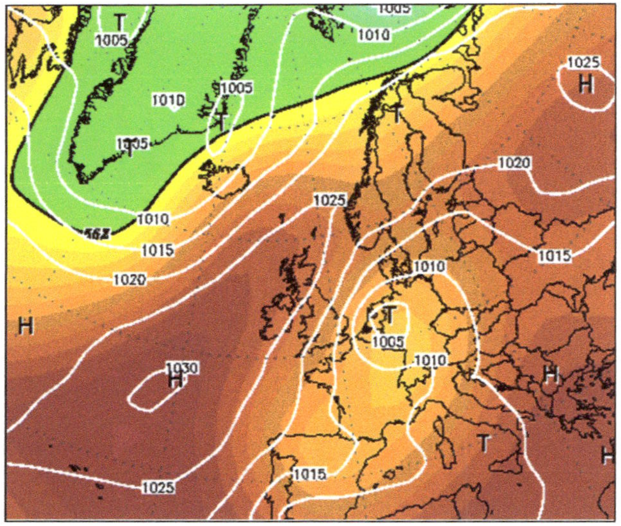

RAIN POTENTIAL

FROST/SNOW

SEPTEMBER 2 TUESDAY

Anticyclonic conditions prevail over Ireland. Warm and dry everywhere, mostly cloudy in the north and east, while sunnier in the west and south. Patches of mist and fog overnight in many places, with some misty patches briefly turning to light drizzle for a time overnight. A pleasant mostly calm air to light northwest breezes prevail.

Belfast: Warmer, cloudy, dry with odd threat of rain, gentle breezes
Dublin: Warm and cloudy, dry, moderate northerly breezes
Cork: Partly cloudy, warm and dry, light breezes
Galway: Dry, warmer, occasional sunny spells, gentle breezes, chance of fog

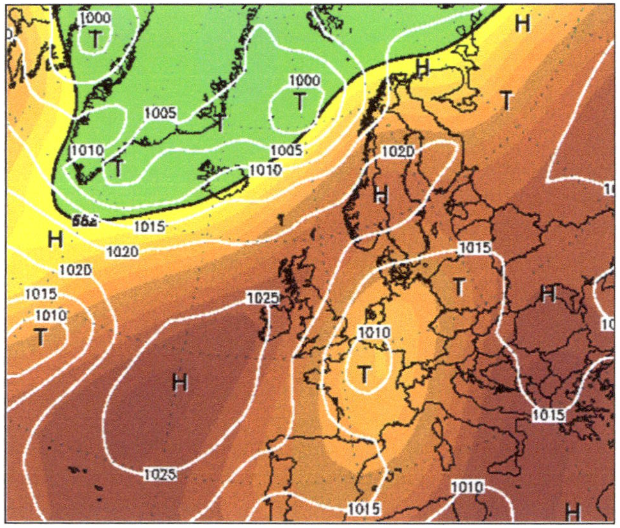

RAIN POTENTIAL

FROST/SNOW

SEPTEMBER **3** WEDNESDAY

Anticyclonic. Warm, mostly sunny and dry throughout with more frequent cloudy spells mainly in the south. Some overnight drizzle patches likely overnight along the western coastal fringes, and patches of fog mostly in western parts. Gentle breezes prevail, generally southwest in the south of the country and northwest in the north.

Belfast: Warm and sunny, dry, gentle breezes, chance of overnight misty patches
Dublin: Mostly sunny, warm and dry, moderate breezes
Cork: Very warm and sunny, dry, light breezes
Galway: Cloudy, dry, gentle breezes to mostly calm air, fog patches possible

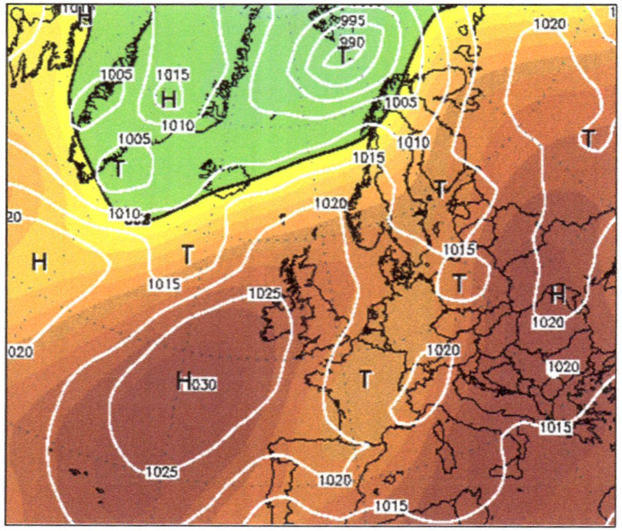

RAIN POTENTIAL

FROST/SNOW

SEPTEMBER **4** THURSDAY

A large high pressure system sits over Ireland and directs a light westerly flow over the country, with more moderate breezes around the southern and northern coasts. Mostly sunny in the east, with cloudier spells elsewhere. Patches of overnight mist in many districts, some turning to light drizzle at times, particularly in the north.

Belfast: Cloudy and dry, very warm, gentle breezes, chance of misty patches
Dublin: Warm, mostly cloudy, dry, gentle shifting breezes
Cork: Sunny, warm, dry, light shifting breezes, chance of patches of mist
Galway: Gentle westerly breezes, cloudy, mostly dry with odd threat of rain

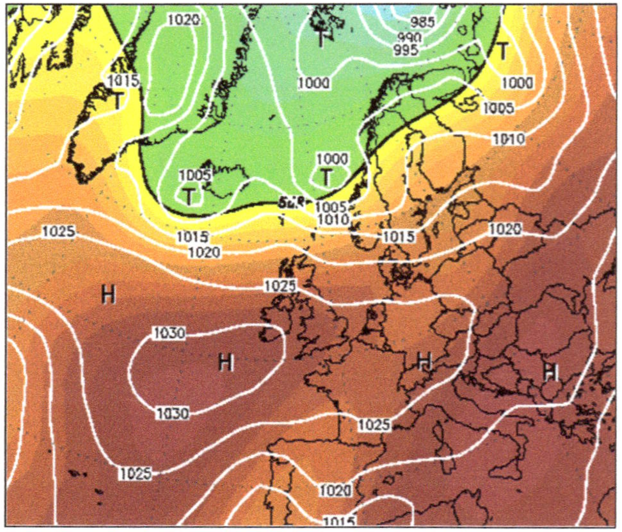

RAIN POTENTIAL

FROST/SNOW

SEPTEMBER **5** FRIDAY

A large high pressure system to the west of Ireland directs a gentle northwest change over the country. Mostly dry and sunny everywhere, except for the chance of some isolated showers mostly in the north.

Belfast: Partly cloudy, dry, cooler, light northwesterly breezes
Dublin: Gentle northerly breezes, dry, sunny, cooler, pleasant air
Cork: Light northerly breezes, sunny, dry, cooler
Galway: Partly cloudy, dry, gentle breezes, mild day cold night

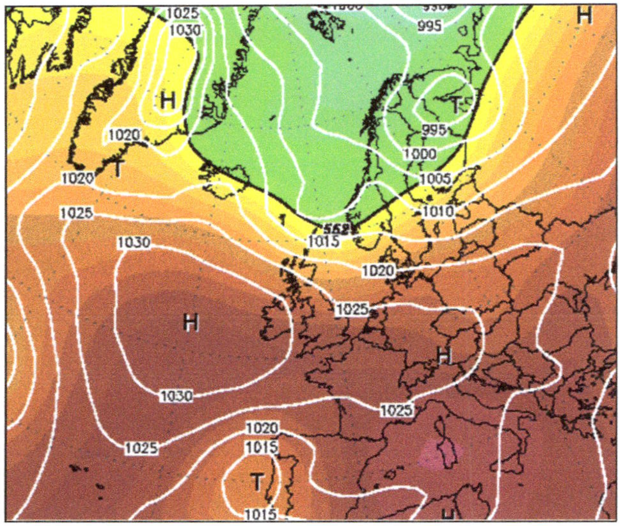

RAIN POTENTIAL

FROST/SNOW

SEPTEMBER **6** SATURDAY

Anticyclonic conditions continue to prevail over the country with light northwest breezes prevailing. Partly cloudy and dry almost everywhere, except for the chance of some isolated overnight showers to the north. Sunniest in the north and west, with increased scattered cloud elsewhere.

Belfast: Partly cloudy, dry, moderate breezes
Dublin: Cloudy, dry, mild temperatures, pleasant air with gentle breezes
Cork: Cooler, dry, cloudy, gentle shifting breezes
Galway: Partly cloudy and dry, gentle breezes

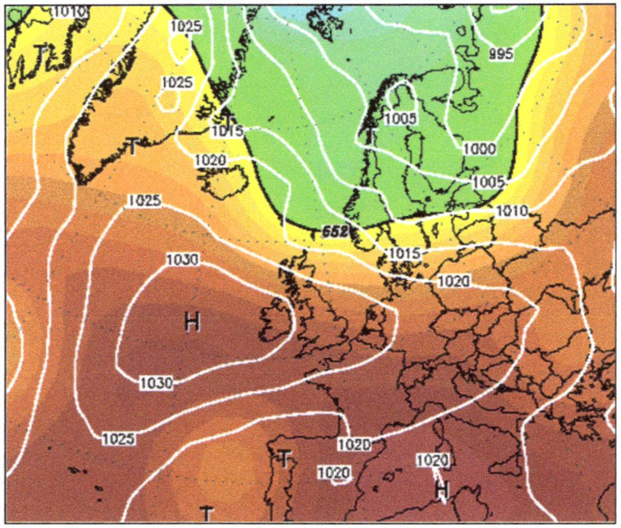

RAIN POTENTIAL

FROST/SNOW

SEPTEMBER 7 SUNDAY

A large high pressure system centred to the west of Ireland continues to bring pleasant conditions across the country. Winds generally light north to northwesterly, allowing for patches of overnight fog possible. Mostly dry with good sunny spells throughout, except for the chance of some isolated overnight showers briefly around the northern coastal fringes.

Belfast: Pleasant and calm air, dry, occasional sunny spells
Dublin: Partly cloudy, mostly dry, serene air, chance of brief misty patches
Cork: Partly cloudy, dry, mostly calm air
Galway: Occasional sunny spells, dry, mild day cool night, calm air, possible fog

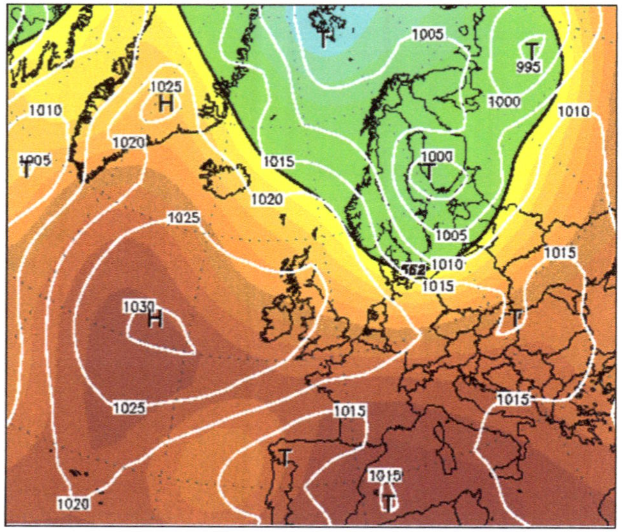

RAIN POTENTIAL

FROST/SNOW

SEPTEMBER 8 MONDAY

The high pressure system to the west of Ireland strengthens with light northerly breezes prevailing, becoming fresher with a northwest change later in the day. Mostly dry and cloudy in the south and west, while sunnier and mostly dry elsewhere, except for some brief overnight drizzle patches in the north,

Belfast: Partly cloudy, dry, moderate northerly breezes
Dublin: Partly cloudy, dry, mild day cool night, gentle breezes
Cork: Cloudy and dry, mild temperatures, light breezes
Galway: Cloudy and dry, mild day cool night, moderate breezes

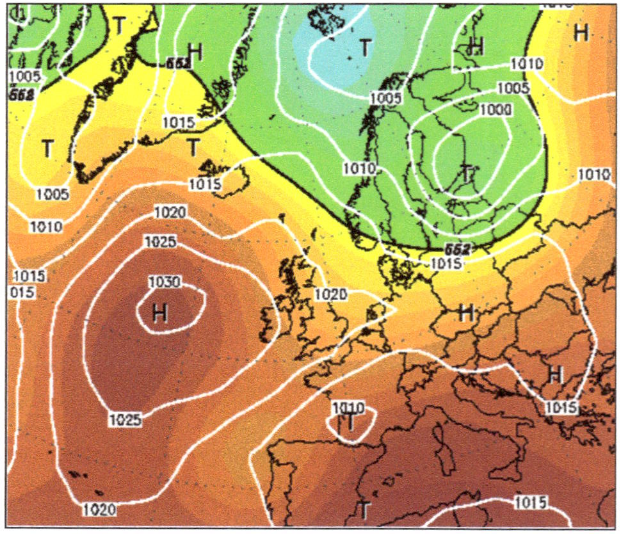

RAIN POTENTIAL

FROST/SNOW

SEPTEMBER 9 TUESDAY

The high pressure ridge maintains a light northwesterly flow over the country, with occasional fresher spells now and then. Generally cloudy in the north and east, with the odd mist or drizzle patch overnight. Sunnier and dry over the southern half of the country.

Belfast: Cloudy, fresh northwesterly breezes, dry, chance of hazy patches and drizzle
Dublin: Cloudy, mostly dry, light breezes, chance of brief misty patches
Cork: Warmer, sunny, dry, fresh breezes at times
Galway: Unsettled breezy spells, dry, mild temperatures, passing sunny outbreaks

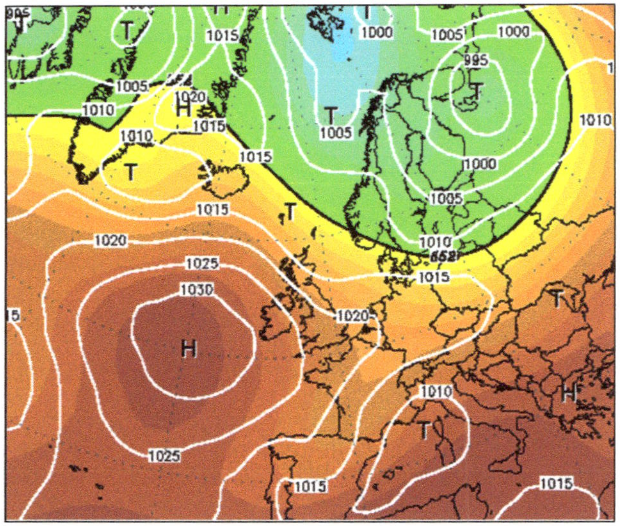

RAIN POTENTIAL

FROST/SNOW

SEPTEMBER 10 WEDNESDAY

The large anticyclonic system that has lingered to the west of Ireland over the past few days slowly drifts southward but maintains a light to moderate northwest flow throughout, sometimes fresher around the coasts with a westerly change. Increased cloud everywhere, with a mostly dry day, except for some drizzle patches in the north and outbreaks of light rain in the northwest.

Belfast: Overcast and dull, dry with passing threats of rain, windy at times
Dublin: Cloudy, dry, moderate breezes
Cork: Cloudy and dry, mild temperatures, fresh breezes
Galway: Cloudy, dry, fresh winds

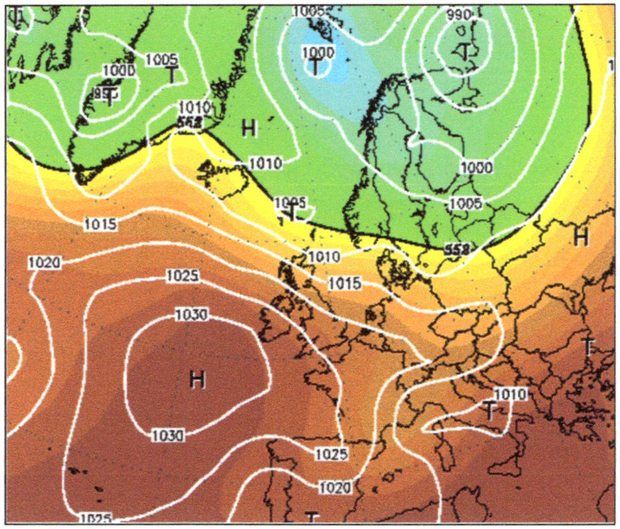

RAIN POTENTIAL

FROST/SNOW

SEPTEMBER **11** THURSDAY

The high pressure system continues to slowly drift south of Ireland, maintaining a mostly gentle westerly flow, but later freshening with a southwest change. Cloudy everywhere with patches of light rain or drizzle in many areas, particularly in the western half of the country.

Belfast: Cloudy, dry, cooler, fresh breezes
Dublin: Cloudy and dry, milder, moderate breezes freshening for a time
Cork: Cloudy, isolated showers, mild temperatures, fresh breezes
Galway: Mostly cloudy, windy at times, light showers

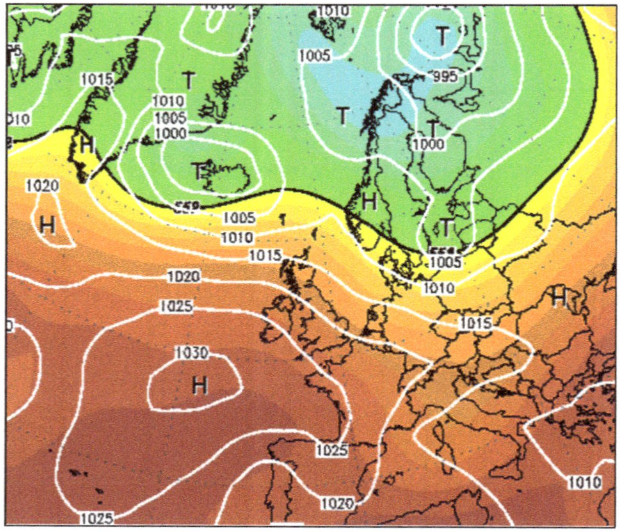

RAIN POTENTIAL

FROST/SNOW

SEPTEMBER **12** FRIDAY

The high pressure ridge slips further south of Ireland while a depression to the north of the country gets nearer, with freshening breezes throughout in a west to northwest flow. Very cloudy and dull everywhere, with patches of drizzle in most districts, and brief outbreaks of rain later in the day, with showers later clearing in the north and contracting to the south.

Belfast: Dry, dull and gloomy air, warmer, moderate westerly winds
Dublin: Unsettled and windy at times, overcast, brief showers
Cork: Overcast and dull, dry, warmer, moderate westerly breezes
Galway: Overcast, dry with odd threat of rain, blustery at times

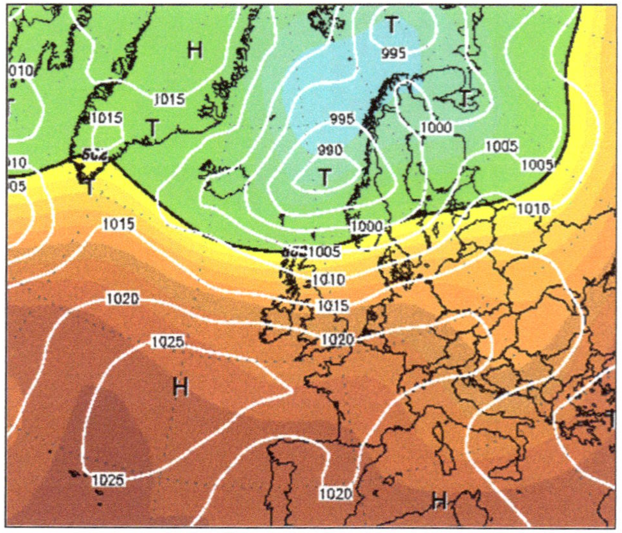

RAIN POTENTIAL

FROST/SNOW

SEPTEMBER 13 SATURDAY

The large high pressure system that has lingered to the west of Ireland for much of this month once more drifts northwards to ridge over the country directing a light to moderate northwest flow throughout. Mostly dry and sunny everywhere, except for some light showers overnight in the north of the country.

Belfast: Sunny and dry, cooler, moderate northwesterly breezes
Dublin: Sunny and dry, mild day cold night, light northwesterly breezes
Cork: Sunny and dry, moderate breezes
Galway: Sunny, dry, mild day cool night, moderate northwesterly breezes

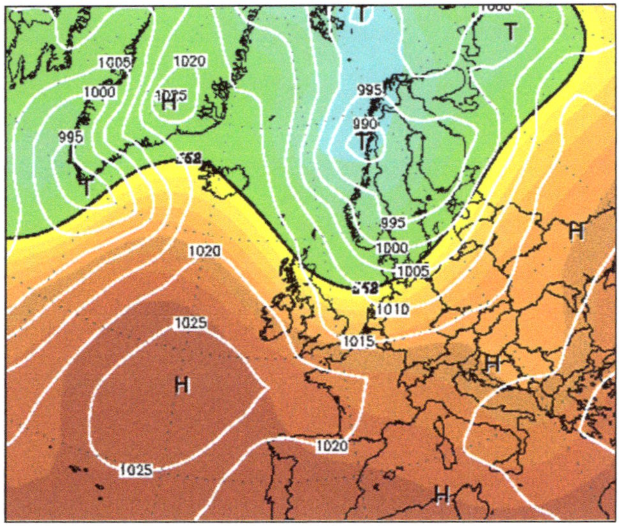

RAIN POTENTIAL

FROST/SNOW

SEPTEMBER 14 SUNDAY

The anticyclonic system finally moves onto the country building cloud throughout. Generally winds will be light to mostly calm, with patches of fog and mist overnight in many districts. Some spotty showers or drizzle patches likely in the west and midlands, while mainly dry elsewhere.

Belfast: Cloudy, dry, warmer day cooler night, calm and misty air
Dublin: Mostly cloudy and dry, mild day cold night, calm and serene air
Cork: Mostly cloudy, dry, warm day cool night, calm and pleasant air
Galway: Overcast, drizzly showers, cooler, calm atmosphere, fog patches possible

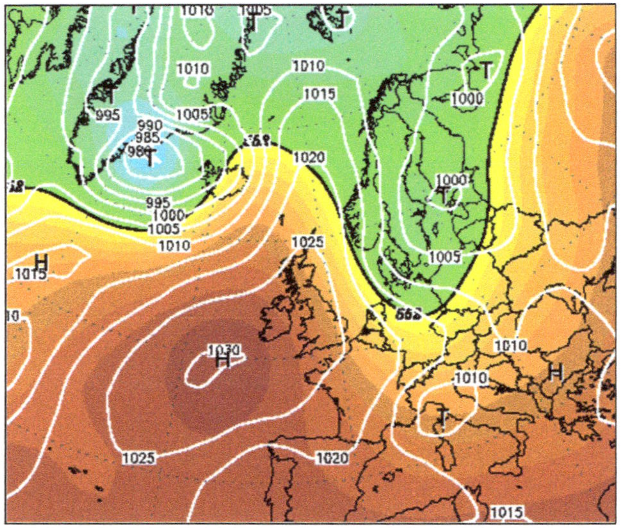

RAIN POTENTIAL

FROST/SNOW

SEPTEMBER 15 MONDAY

Anticyclonic. Light to moderate westerly winds everywhere. Generally cloudy at first, with some drizzle patches in the north and northwest, while dry elsewhere and increasing sunny spells as the day advances.

Belfast: Dry, cloudy, moderate westerly breezes
Dublin: Warmer, partly cloudy, isolated evening showers, gentle breezes
Cork: Overcast, dry, cooler, mostly calm and serene air
Galway: Partly cloudy, mild and dry, fresh breezes

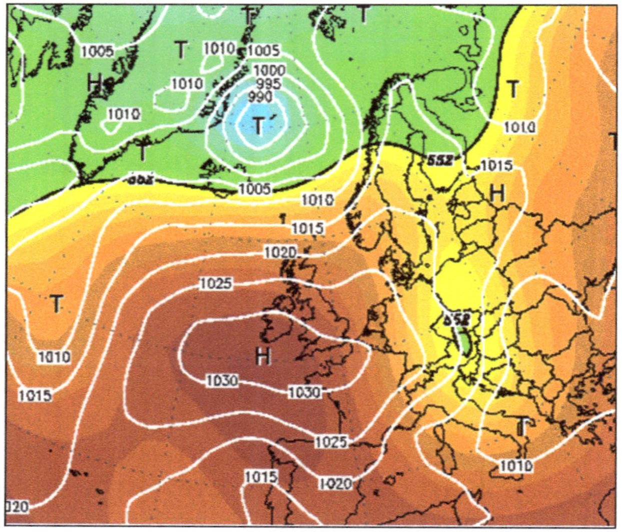

RAIN POTENTIAL

FROST/SNOW

SEPTEMBER 16 TUESDAY

Anticyclonic conditions continue. Light variable breezes to mostly calm atmosphere everywhere, with patches of fog and mist overnight here and there. Mostly cloudy in the north and east, and sunny in the west and south. Chance of some outbreaks of rain or drizzle around the northwest and northern coastal friges, but dry elsewhere.

Belfast: Cloudy, warmer, dry, calm and pleasant air, chance of overnight fog
Dublin: Mostly cloudy, dry, gentle breezes to mostly serene atmosphere
Cork: Warm, dry, occasional sunny spells, calm and pleasant air
Galway: Sunny, dry, calm atmosphere, fog patches possible

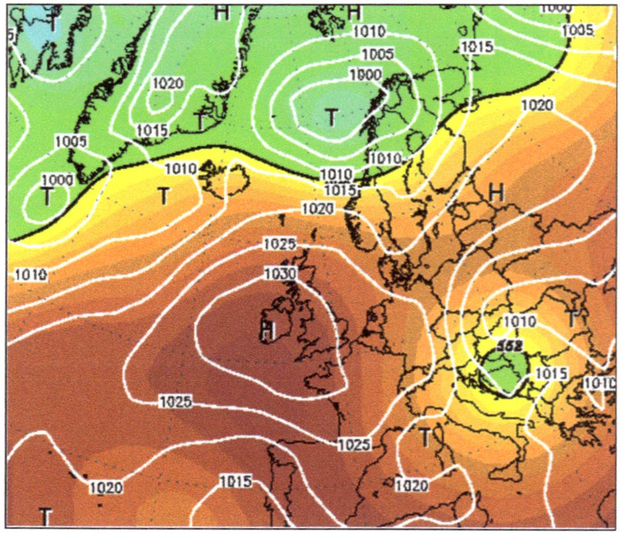

RAIN POTENTIAL

FROST/SNOW

SEPTEMBER **17** WEDNESDAY

A large anticyclonic system sits over Ireland bringing a very pleasant and mostly calm air throughout. Patches of fog in many districts over night, clearing everywhere to a dry and sunny day, slightly cloudier in the east and southeast. Warmer than normal for this time of year.

Belfast: Dry, sunny, cooler, calm and serene air, fog likely
Dublin: Sunny, dry, calm and pleasant air, cool overnight
Cork: Partly cloudy, dry, calm atmosphere, fog likely
Galway: Calm and serene air, sunny and warm day, cold night, dry, shallow fog likely

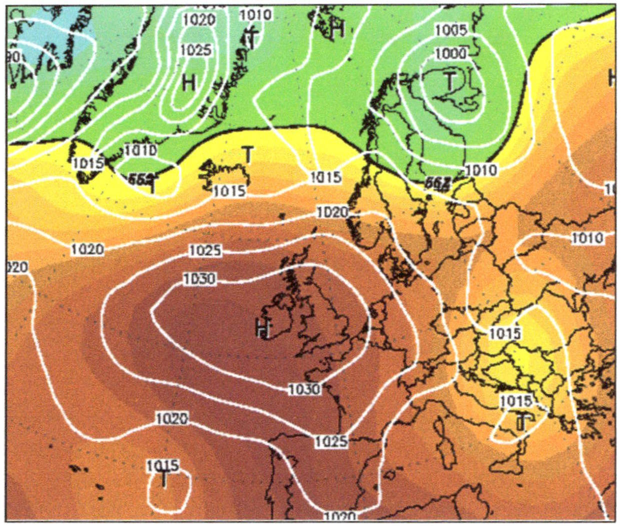

RAIN POTENTIAL

FROST/SNOW

SEPTEMBER **18** THURSDAY

High pressure continues to prevail over Ireland bringing another mostly dry and pleasant day with light winds throughout. Increased cloud in the north and east, while sunnier in the west and south. Fog patches in most districts overnight, as well as the chance of some isolated patches of drizzle briefly along the northern coasts.

Belfast: Pleasant and mostly calm air, cloudy, warm, dry, chance of misty patches
Dublin: Overcast and cooler, dry, calm atmosphere, chance of misty patches
Cork: Sunny, dry, warm, calm atmosphere, fog patches likely
Galway: Sunny, dry, pleasant and serene air, fog likely

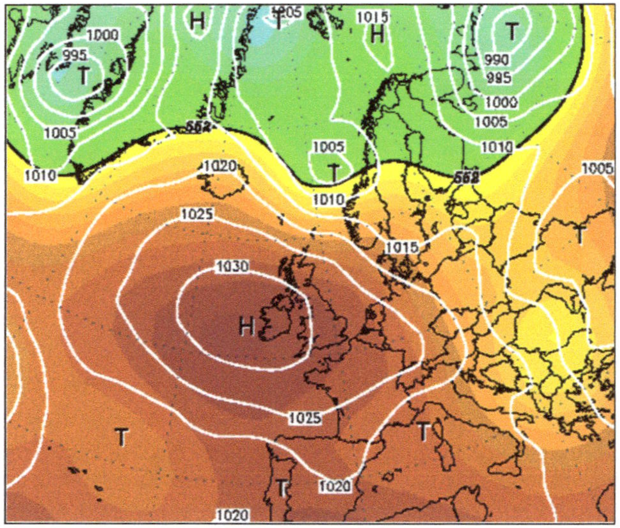

RAIN POTENTIAL

FROST/SNOW

SEPTEMBER **19** FRIDAY

High pressure remains almost stalled over Ireland bringing another day of gentle breezes throughout, but building and becoming fresher at times mostly around the northwestern coasts, with a westerly change. Sunny and dry in the south, while cloudier and mostly dry elsewhere, with the occasional threat of rain in the north. Fog patches likely in many districts overnight once again.

Belfast: Cloudy, mostly dry with odd threat of rain, cooler, gentle shifting breezes
Dublin: Overcast, dry, odd threat of rain, gentle shifting breezes, misty patches likely
Cork: Sunny and warm, dry, light shifting breezes, chance of fog
Galway: Partly cloudy, dry, mild day cold night, light shifting breezes, chance of fog

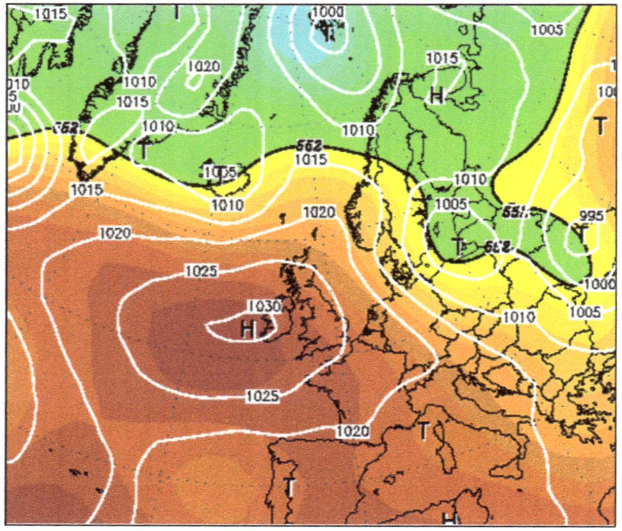

RAIN POTENTIAL

FROST/SNOW

SEPTEMBER **20** SATURDAY

The large anticyclone that has brought stable conditions over the past week slowly slips to the southwest of Ireland while a low pressure system to the northeast nears the northern coasts, with a moderate northwest flow prevailing, fresher and occasionally gusty around the northern coasts. Mostly sunny and dry in the east, while cloudier and mostly dry elsewhere. Chance of some drizzle patches overnight mainly in the north. Warmer.

Belfast: Partly cloudy, dry, mild temperatures, light northwesterly breezes
Dublin: Sunny, light northwesterly breezes, dry, warmer day colder night
Cork: Cloudy, dry, mild temperatures, light breezes
Galway: Cloudy, isolated showers, moderate breezes

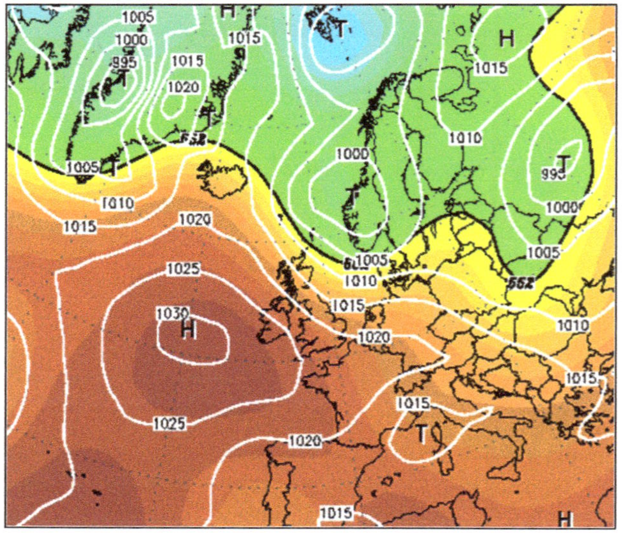

RAIN POTENTIAL

FROST/SNOW

SEPTEMBER 21 SUNDAY

The high pressure ridge continues over Ireland with light to moderate west to northwest breezes prevailing. Overnight mist or fog patches in some inland districts, while drizzle patches likely in the west. Generally dry throghout the day but with with little sunshine as widespread cloud develops everywhere.

Belfast: Overcast, dry with threats of rain, warm day cool night, moderate breezes
Dublin: Dull and gloomy, dry, mild day cold night, light northerly air
Cork: Dull and gloomy, mostly calm air, dry, some threats of rain
Galway: Overcast, dry with odd threat of rain, mild temperatures, light breezes

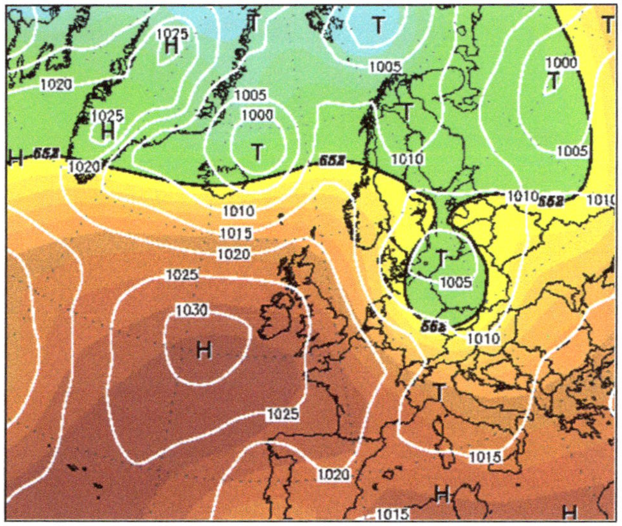

RAIN POTENTIAL

FROST/SNOW

SEPTEMBER 22 MONDAY

The high pressure system swells back up over Ireland in a light westerly air once more. Generally dull and gloomy, warm, humid, and mostly dry everywhere widespread cloud prevailing. Some patches of mist or fog likely overnight.

Belfast: Overcast, cooler, dry, passing threats of rain, gentle shifting breezes
Dublin: Overcast, dry, gentle shifting breezes
Cork: Overcast and dull, dry, mild temperatures, gentle breezes
Galway: Light shifting breezy spells, dull and gloomy air, dry

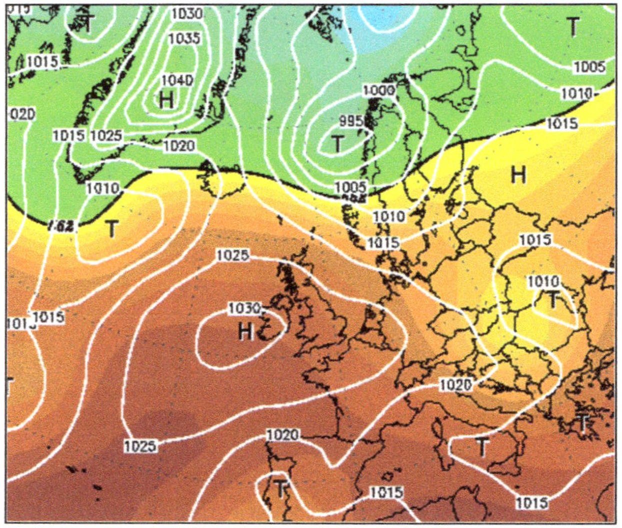

RAIN POTENTIAL

FROST/SNOW

SEPTEMBER **23** TUESDAY

The large anticyclonic system that has been stalled over Ireland for a while now, weakens, as depressions to the north of the country edge closer. Light variable winds throughout, becoming fresher with a southwest change later in the day. Widespread cloud prevails, slightly sunnier in the south. Some light overnight showers in most districts, except around the east and southern coasts.

Belfast: Overcast, mostly dry day with brief overnight showers, calm atmosphere
Dublin: Overcast and dry, odd threat of rain, mild temperatures, serene atmosphere
Cork: Partly cloudy, dry, calm atmosphere
Galway: Cloudy, calm air, isolated showers

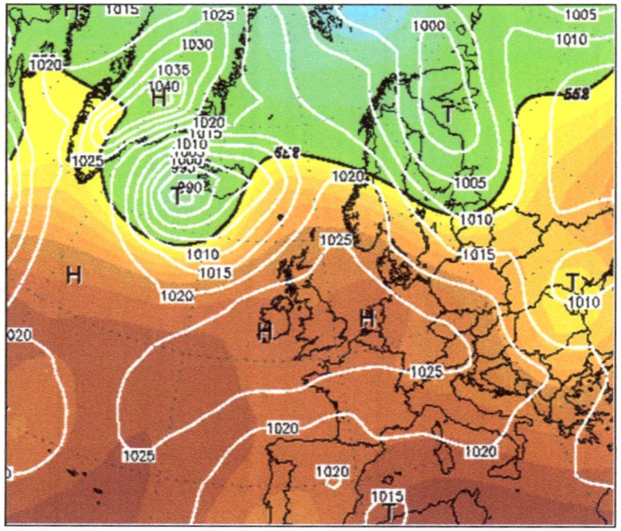

RAIN POTENTIAL

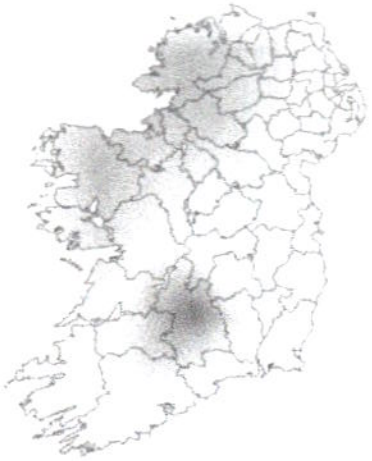

FROST/SNOW

SEPTEMBER **24** WEDNESDAY

The high pressure ridge departs to the south as a depression to the north moves onto the country for a time, with fresh to strong northwest winds prevailing over the northern half of the country, and lighter to the south. A changeable day with a mix of sunny spells and isolated showers and passing drizzle patches in most districts, although mostly dry inthe southwest. Cooler.

Belfast: Partly cloudy, dry, freshening breezes
Dublin: Freshening winds, partly cloudy, isolated showers
Cork: Changeable, some sunny spells, isolated showers, cooler, odd breezy spell
Galway: Partly cloudy, fresh breezy spells, dry, chance of overnight drizzle patches

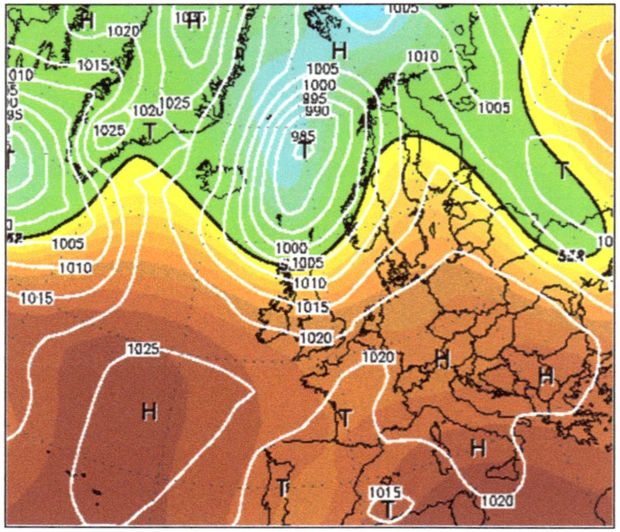

RAIN POTENTIAL

FROST/SNOW

SEPTEMBER **25** THURSDAY

High pressure moves back onto Ireland in a light to moderate south to southwest flow. Mostly dry and sunny at first with patches of fog mostly in central parts. A depression to the north of Ireland moves onto the country later in the day building stronger winds in northern and western parts, and increasing threats of rain in the north and west, with showers developing overnight in many districts, heavier in the northwest.

Belfast: Partly cloudy, light overnight showers, gentle breezes strengthening later
Dublin: Sunny, dry, mild day cold night, mostly calm and pleasant air
Cork: Calm and pleasant air, sunny, dry
Galway: Mostly cloudy, occasional showers, cooler, gentle breezes freshening later

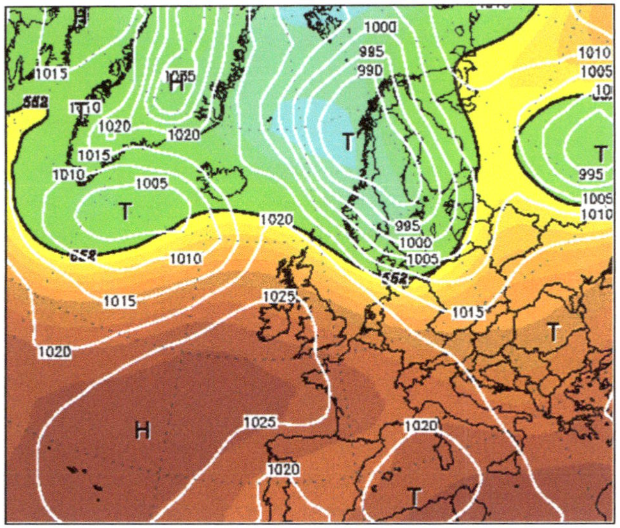

RAIN POTENTIAL

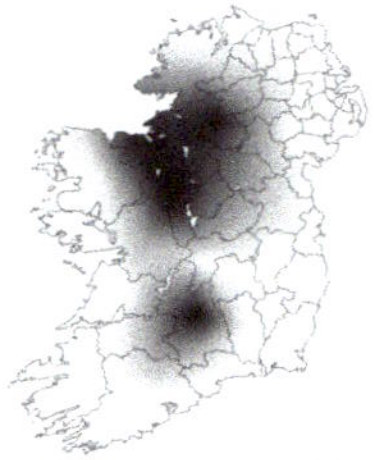

FROST/SNOW

SEPTEMBER **26** FRIDAY

A depression sits to the north of Ireland and extends a blustery westerly flow over the country. A band of rain crosses the northwest at first, and spreads in a southeasterly tract across the country, clearing the south by the end of the day. A changeable day with a mix of sunny spells and scattered showers in most districts, heavier in the north and west.

Belfast: Mostly cloudy, brief showers and drizzle patches, windy, chance of gales
Dublin: Increased cloudy spells, brief showers, strong blustery winds
Cork: Partly cloudy, light showers, strengthening winds with occasional gusts
Galway: Partly cloudy, light showers, blustery spells

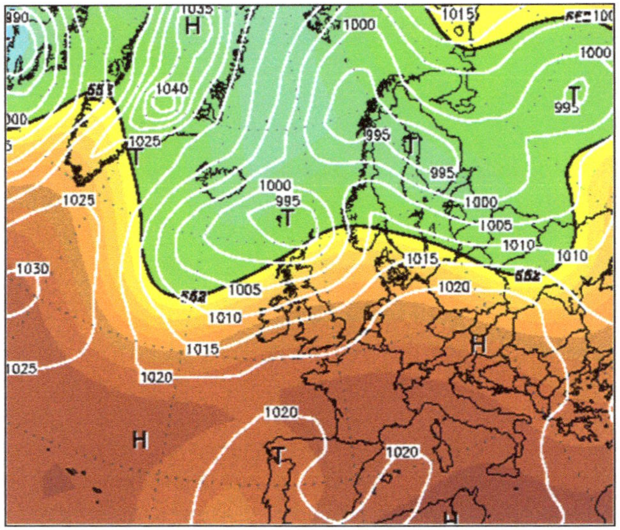

RAIN POTENTIAL

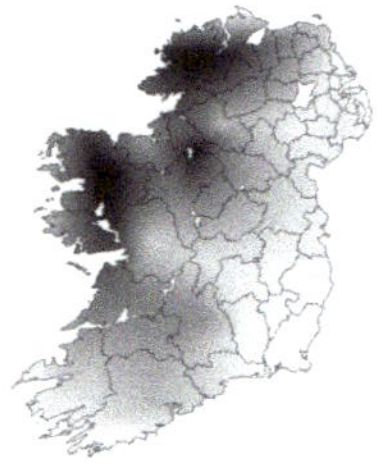

FROST/SNOW

SEPTEMBER **27** SATURDAY

Low pressure directs fresh to strong northerly winds over the country, with light overnight showes along the western coastal fringes at first, and spreading elsewhere later before clearing everywhere and bringing a mostly dry and sunny day. Showers likely to return again in the evening in the north.

Belfast: Sunny, colder, isolated showers, fresh winds easing
Dublin: Sunny, colder, dry, strong winds at times
Cork: Sunny, colder, dry, breezy, chance of misty drizzle spells
Galway: Sunny, colder, isolated showers, strong and unsettled winds

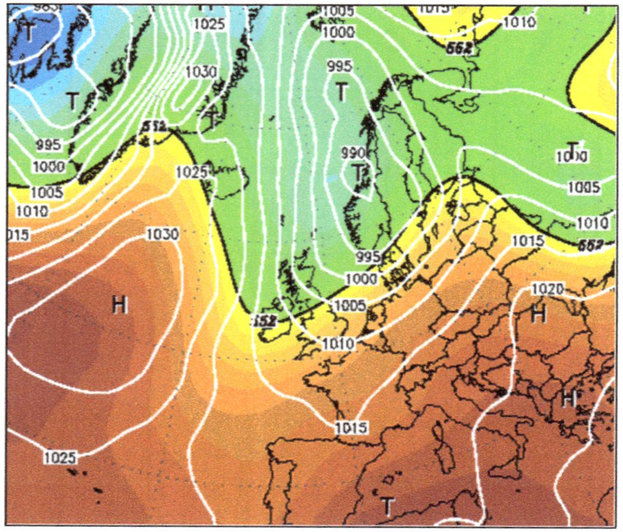

RAIN POTENTIAL

FROST/SNOW

SEPTEMBER **28** SUNDAY

A high pressure ridge to the west extends a moderate north to northwest flow over the country with occasional gusts now and then. Mostly dry with some sunny spells at first, before cloud builds and outbreaks of rain develop in the north and west and spread eastward as the day progresses, remaining mostly dry along the eastern coastal fringes and in the far south.

Belfast: Overcast, light showers, colder, fresh to strong winds at times
Dublin: Cloudy, dry, strong and cold northerly winds
Cork: Partly cloudy and dry, cool, fresh northerly breezes at times
Galway: Overcast, rain, mild day cold night, windy spells

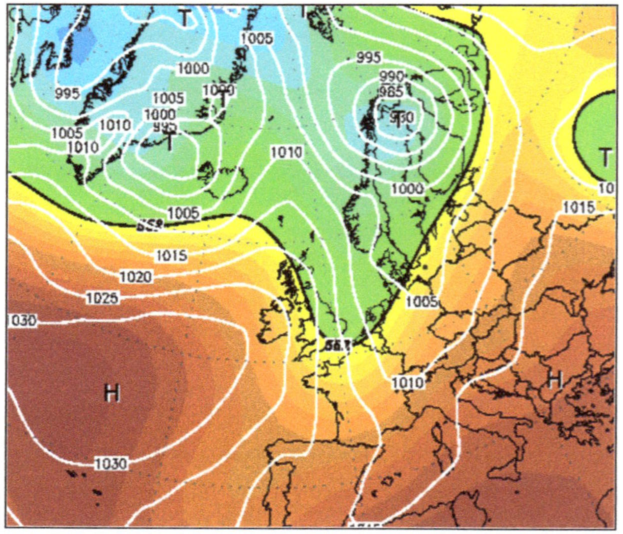

RAIN POTENTIAL

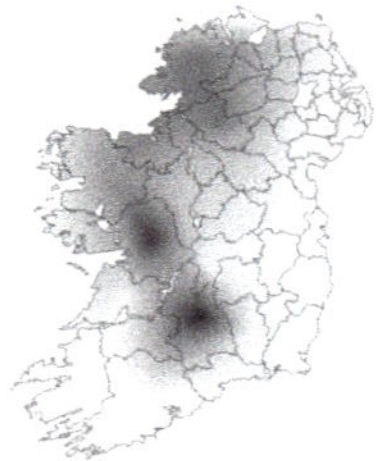

FROST/SNOW

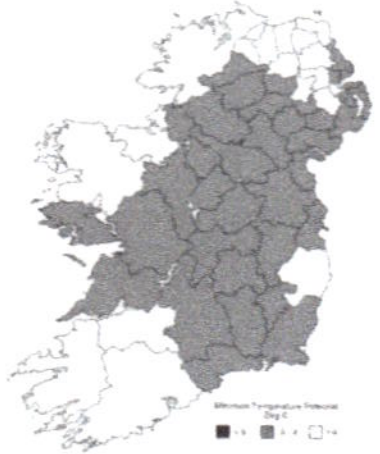

SEPTEMBER **29** MONDAY

Low pressure to the north of Ireland directs a stronger westerly flow over the country, with chance of winds reaching gale force for a time along the coastal fringes. Outbreaks of rain and drizzle overnight in most districts, with more persistent rain through central parts, clearing during the day before returning later. Some brief sunny spells in the north during the day, while dull and gloomy almost everywhere else.

Belfast: Partly cloudy and cool, isolated showers, fresh to strong westerlies
Dublin: Westerly change, cloudy, rain, warmer
Cork: Unsettled and blustery winds at times, overcast, light showers
Galway: Mostly cloudy, some showers, slightly warmer, strong winds, chance of gales

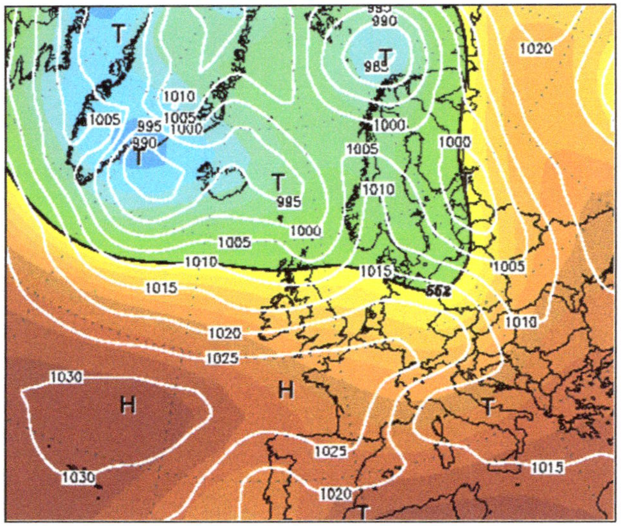

RAIN POTENTIAL

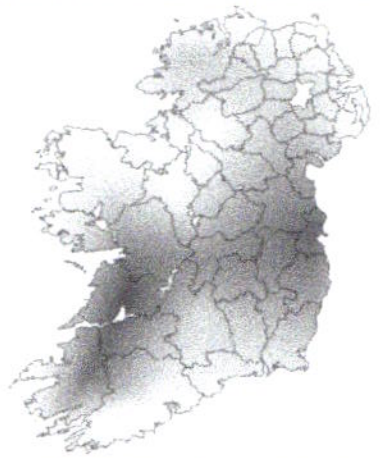

FROST/SNOW

SEPTEMBER **30** TUESDAY

A depression to the north of Ireland directs a strong westerly air over the country, with gusts up to gale force at times especially around the coasts. Mostly cloudy and with showers everywhere, with a band of rain bringing heavier falls at times in the south and southwest, while some sunnier spells in the far southeast.

Belfast: Cloudy, light showers, strong and blustery westerly winds, chance of gales
Dublin: Strong winds with chance of gales, cloudy, brief squalls and drizzle patches
Cork: Mostly sunny, brief showers, strong gusty winds, chance of gales at times
Galway: Stormy air, cloudy, scattered showers, strong winds with gale gusts possible

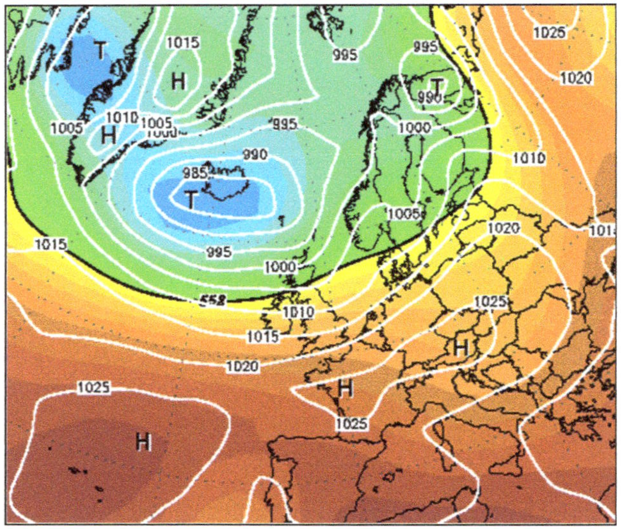

RAIN POTENTIAL

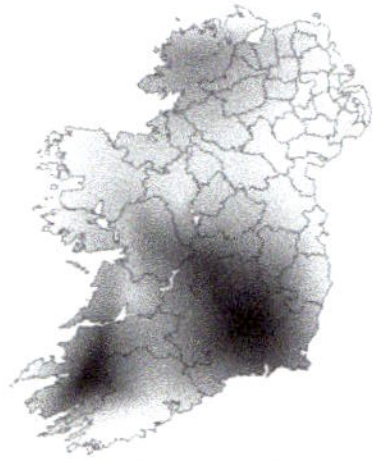

FROST/SNOW

OCTOBER

PHASES OF THE MOON

7th	Full Moon
7th	Crossing Equator
8th	Perigee #6
12th	Northern Declination
13th	Third Quarter
20th	Crossing Equator
21st	New Moon
23rd	Apogee
26th	Southern Declination
29th	First Quarter

1	2	3	4	5	6	7	8	9	10	11	12	13	14	15	16	17	18	19	20	21	22	23	24	25	26	27	28	29	30	31
Tue	Wed	Thur	Fri	Sat	Sun	Mon	Tue	Wed	Thur	Fri	Sat	Sun	Mon	Tue	Wed	Thur	Fri	Sat	Sun	Mon	Tue	Wed	Thur	Fri	Sat	Sun	Mon	Tue	Wed	Thur
						F XhN	P6				^	3Q							XhS	N		A			V			1Q		

MONTHLY SUMMARY

1st-4th A deep depression passes near the northwest of the country bringing an unsettled and stormy air. Strong west to southwest winds and frequent showers prevail. Cloudier everywhere with milder temperatures.
1st Heavy rain likely in a number of districts.
3rd Chance of wintry overnight showers of snow grains in some central districts.
4th Chance of thunderstorms.
5th-10th An anticyclonic system centred to the north of Ireland directs a cool north to northeast airflow over the country, later backing southeasterly. Active troughs bring bands of light showers now and then, slightly heavier in eastern parts.
6th-8th Widespread frosts likely especially in central parts to the eastern districts.
10th snow grains central parts

Over the next four weeks high pressure will prevail the most, with only a low pressure period expected around the third week. This will bring a fairly stable time with mostly dry and sunny weather and light breezes predominately southerly in flow. This will result in mostly warmer than normal temperatures everywhere, especially along the western coastal fringes during the first couple of weeks, before dropping closer to the norms for this time of year. It will be drier than normal throughout, with only be short spells of wet weather on a few days, the heaviest falls of which will be in the west to southwestern districts, and some eastern districts may see less than half their usual rainfall for this time of year. Sunshine hours will be above normal almost everywhere during the next four weeks, except in the south where it will be relatively dull. Maximum and minimum temperatures will be mostly above average in everywhere, by as much as 1-2°C above the norm for this time of year in some districts, and frosts will be mostly confined to inland districts with coastal regions remaining frost free. During this outlook period, there is likely to be between 8 and 12 days with rain, thunderstorms on 1 day, and hail on 2 days. Frosts can be expected on 4-10 days in inland districts, but none around the coasts, while fogs are likely to be observed on up to 14 days. Strong winds to gale force may occur on 3 days but no days of snow are expected.

11th-13th A band of rain crosses the country bringing showers into the southeast at first before spreading and becoming heavier as it moves into the northwest. Slightly warmer than in the past few days.
11th-21st Frequent fog during this period, sometimes extensive and slow to clear.
12th Very warm day.
14th-31st High pressure prevails and brings a mostly dry period with light winds throughout broken only by the passage of a couple of Atlantic fronts. Warm south to southeast winds mostly prevail bringing milder than normal temperatures including milder overnight.
14th-16th Very sunny spell, especially in the west, with overnight frosts likely in inland eastern parts.
16th-17th Widespread fog.
18th-19th An Atlantic fronts brings a brief wet spell to Ireland, heaviest in western parts.
18th-25th Mild south to southeast winds bring a spell of very mild weather including unusually mild overnight temperatures.
20th Widespread fog likely which may be slow to clear.
21st-22nd Chance of snow grains being observed in central areas.
24th-27th Unusually sunny spell.
25th-26th A wet front associated with a depression to the west of Ireland brings a brief wet spell to Ireland with heaviest falls in the west of the country and chance of some showers of snow grains in some southern parts.
26th-28th Cold spell with some frosts in inland parts.

IRELAND RAINFALL ESTIMATES

October 2025

d=mainly dry, n= nonrecordable, l=light shrs, s=significant shrs, r=rain, h=hvy falls

		ULSTER PROVINCE					CONNAUGHT PROVINCE						LEINSTER PROVINCE				MUNSTER PROVINCE						
		Alderg	Hillsbo	Armag	Loughl	Malin H	Sligo	Ardtari	Belmul	Drums	Galway	Roscoi	Markre	Edende	Kilkeni	Dublin	Shannc	Newpo	Tralee	Killarne	Cork	All	Moon
1st	Wed	l	l	l	d	l	d	d	s	l	l	d	d	d	l	s	s	n	h	n	h	67	
2nd	Thur	n	l	d	l	d	l	l	d	d	l	l	s	l	d	d	d	l	s	l	l	23	
3rd	Fri	l	l	l	l	l	s	l	s	s	h	s	d	l	s	l	s	r	r	r	s	103	
4th	Sat	s	s	s	s	s	r	s	s	s	s	l	r	l	l	s	r	r	s	s	d	113	
5th	Sun	l	d	l	d	l	l	d	l	d	d	n	l	d	d	l	l	l	l	l	d	14	
6th	Mon	d	d	d	d	d	d	d	d	d	d	d	d	d	d	d	d	d	d	d	d	0	
7th	Tues	d	d	d	d	d	d	d	d	d	d	d	d	d	d	d	d	d	d	d	d	0	F XhN
8th	Wed	d	d	d	d	d	d	d	d	d	d	n	d	l	l	n	n	d	l	d	d	7	P6
9th	Thur	l	l	d	h	n	s	h	d	s	l	l	h	l	l	l	l	l	d	d	n	79	
10th	Fri	l	d	l	s	l	l	s	n	l	d	d	l	l	d	s	d	d	d	d	d	22	
11th	Sat	d	d	n	d	d	d	d	d	d	d	d	d	d	n	d	d	d	d	d	n	1	
12th	Sun	l	s	d	s	d	r	s	l	s	r	r	s	l	s	l	n	r	s	r	d	120	^
13th	Mon	l	l	s	l	r	s	l	r	l	l	l	l	s	l	s	s	l	r	d	s	88	3Q
14th	Tues	l	d	s	d	d	d	d	d	d	d	d	l	d	d	d	d	d	d	d	d	11	
15th	Wed	d	d	d	d	d	d	d	d	d	d	d	d	d	d	d	d	d	n	d	d	1	
16th	Thur	d	d	d	d	d	d	d	l	d	d	d	n	d	d	d	d	d	d	d	d	2	
17th	Fri	d	d	d	d	d	d	d	d	d	d	d	n	d	l	d	d	d	n	d	d	2	
18th	Sat	l	l	n	r	s	r	r	r	s	l	s	r	l	l	l	s	r	s	s	s	116	
19th	Sun	l	d	l	d	l	d	d	d	d	l	d	n	l	d	d	d	l	h	s	s	33	
20th	Mon	l	d	l	d	d	d	d	d	d	d	d	d	d	d	d	d	s	d	d	d	13	XhS
21st	Tues	d	d	d	d	s	d	d	l	d	d	d	d	d	n	d	d	l	d	d	d	8	N
22nd	Wed	d	d	d	d	d	d	d	d	d	d	d	d	d	l	d	d	d	d	d	d	1	
23rd	Thur	d	d	d	d	d	d	d	s	d	n	d	d	d	d	d	d	l	d	d	d	12	A
24th	Fri	n	l	d	l	d	s	l	l	s	d	l	l	d	d	d	d	s	s	l	d	36	
25th	Sat	l	l	l	d	s	l	d	s	l	s	l	l	l	s	l	l	l	s	s	l	59	
26th	Sun	l	d	l	l	l	l	l	l	d	n	d	l	d	d	d	d	l	s	d	r	26	V
27th	Mon	d	d	d	d	l	d	d	n	d	d	l	d	d	d	d	d	d	d	d	d	3	
28th	Tues	d	d	d	d	d	d	d	d	d	d	l	d	d	d	d	d	n	d	d	d	2	
29th	Wed	d	d	d	d	d	d	d	l	d	n	d	d	d	d	d	d	d	d	d	d	1	1Q
30th	Thur	d	d	d	d	d	d	d	d	d	d	d	d	d	d	d	d	d	d	d	d	0	
31st	Fri	d	d	d	r	d	s	r	l	l	h	s	s	d	n	d	d	s	d	s	d	74	
	Estimate:	22	19	25	67	47	64	67	64	50	75	48	67	21	28	32	39	77	102	62	60	1035	
	Average:	90	100	86	129	121	134	136	146	112	129	140	134	104	93	79	105	176	183	133	138	2467	
	Trend:	drr	drr	drr	drr	drr	drr	drr	drr	drr	drr	drr	drr	drr	drr	drr	drr	drr	drr	drr	drr	drr	

IRELAND SUNSHINE ESTIMATES

October 2025

F=fine (8-12 hours of sunshine), pc= partly cloudy (4-7 hours), c=cloudy (1-3 hours), o=overcast (0 hours)

		ULSTER PROVINCE					CONNAUGHT PROVINCE						LEINSTER PROVINCE				MUNSTER PROVINCE						
		Alderg	Hillsbo	Armag	Loughl	Malin I	Sligo	Ardtari	Belm ul	Drums	Galway	Roscoi	Markre	Edende	Kilkenr	Dublin	Shannc	Newpo	Tralee	Killarne	Cork	All	Moon
1st	Wed	c	c	c	c	c	pc	c	pc	c	pc	pc	c	c	c	c	pc	pc	pc	pc	c	81	
2nd	Thur	pc	pc	pc	pc	pc	c	pc	pc	pc	pc	pc	pc	pc	F	F	pc	F	pc	pc	F	132	
3rd	Fri	c	c	c	c	c	c	c	c	c	c	c	c	c	c	c	c	c	c	c	c	62	
4th	Sat	c	c	c	c	c	c	c	pc	pc	pc	pc	c	pc	pc	c	pc	pc	pc	pc	pc	97	
5th	Sun	pc	pc	pc	pc	pc	pc	pc	pc	pc	F	F	pc	pc	F	pc	F	F	pc	pc	F	156	
6th	Mon	pc	pc	pc	pc	pc	pc	pc	c	pc	c	c	pc	pc	pc	pc	c	pc	pc	pc	pc	103	
7th	Tues	F	F	F	F	F	F	F	F	F	F	F	F	F	pc	pc	F	F	F	F	pc	181	F XhN
8th	Wed	c	c	c	c	c	pc	c	pc	c	pc	pc	c	c	c	c	c	c	c	c	c	68	P6
9th	Thur	o	o	o	o	o	o	o	pc	o	c	c	o	o	c	o	o	o	pc	pc	c	30	
10th	Fri	pc	pc	pc	pc	pc	c	pc	c	c	pc	pc	pc	c	c	pc	pc	pc	c	c	o	70	
11th	Sat	c	c	c	c	c	c	c	F	c	pc	F	c	c	c	pc	pc	pc	c	c	o	80	
12th	Sun	c	c	c	c	c	c	c	c	o	c	c	c	o	o	o	o	o	o	o	o	22	^
13th	Mon	pc	pc	pc	pc	pc	pc	pc	pc	pc	pc	pc	pc	pc	pc	c	pc	pc	pc	pc	pc	101	3Q
14th	Tues	pc	pc	pc	pc	pc	pc	pc	pc	F	pc	pc	pc	F	F	F	pc	F	F	F	F	149	
15th	Wed	pc	pc	pc	pc	pc	c	pc	o	pc	c	c	pc	pc	pc	pc	c	pc	o	o	c	70	
16th	Thur	pc	pc	pc	pc	pc	c	pc	o	pc	c	c	pc	pc	pc	F	pc	pc	F	F	pc	104	
17th	Fri	c	c	c	c	c	pc	c	pc	c	pc	pc	c	c	c	c	c	c	c	c	c	70	
18th	Sat	o	o	o	o	o	o	o	o	o	o	o	o	o	o	o	o	o	o	o	o	7	
19th	Sun	pc	pc	pc	pc	pc	pc	pc	pc	pc	pc	pc	pc	pc	c	pc	c	c	o	o	o	87	
20th	Mon	c	c	c	c	c	o	c	pc	c	c	c	c	c	c	pc	c	c	o	o	o	45	XhS
21st	Tues	o	o	o	o	o	o	o	o	o	o	o	o	o	c	o	o	o	o	o	c	7	N
22nd	Wed	o	o	o	o	o	o	o	o	o	o	o	o	o	o	o	o	o	o	o	o	1	
23rd	Thur	o	o	o	o	o	o	o	c	o	c	c	o	o	o	o	pc	c	o	o	o	19	A
24th	Fri	c	c	c	c	c	o	c	pc	pc	c	pc	c	pc	pc	pc	c	c	o	o	c	69	
25th	Sat	c	c	c	c	c	pc	c	pc	c	pc	pc	c	c	c	c	pc	pc	pc	pc	pc	80	
26th	Sun	pc	pc	pc	pc	pc	pc	pc	F	pc	pc	pc	pc	pc	c	pc	pc	pc	pc	pc	c	119	V
27th	Mon	pc	pc	pc	pc	pc	pc	pc	pc	pc	pc	pc	pc	pc	pc	pc	c	pc	c	c	c	105	
28th	Tues	pc	pc	pc	pc	pc	c	pc	o	c	c	c	pc	c	c	pc	c	c	c	c	o	60	
29th	Wed	c	c	c	c	c	c	c	c	c	c	c	c	c	c	c	o	c	o	o	c	27	1Q
30th	Thur	c	c	c	c	c	o	c	o	c	o	o	c	c	c	c	o	c	o	o	c	26	
31st	Fri	c	c	c	c	c	o	c	o	c	o	o	c	c	c	c	o	o	o	o	o	20	
Estimate hours:		114	114	114	114	114	102	114	123	113	119	121	114	113	112	126	116	114	97	97	98	2244	
Average hours:		91	91	87	82	74	74	76	82	75	85	88	85	86	84	96	82	71	80	78	87	1653	
Trend:		more	more	more	more	more	more	more	more	more	more	more	more	more	more	more	more	more	more	more	more	more	

OCTOBER **1** WEDNESDAY

A depression to the north of Ireland extends a strong southwest flow over the country along with a band of heavy rain through southern districts, weakening to lighter showers over the northern half. Winds may reach gale force at times in the north and west. Rain activity like to ease by afternoon allowing for some sunny breakthroughs to appear.

Belfast: Cloudy, isolated showers, unsettled and blustery winds
Dublin: Stormy air, cloudy, wet and windy, chance of gales
Cork: Heavy rain, some sunny spells, mild temperatures, windy
Galway: Partly cloudy, brief showers, mild day cool night, windy, chance of gales

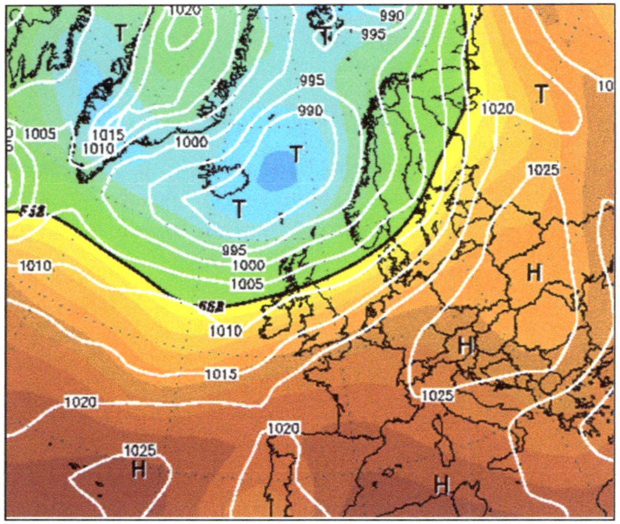

RAIN POTENTIAL

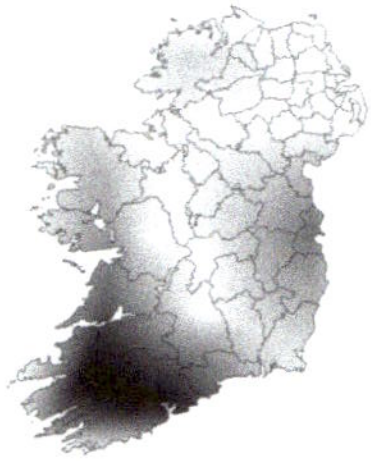

FROST/SNOW

OCTOBER **2** THURSDAY

Depressions to the north of Ireland maintain a fresh and gusty southwest flow over the country. A mixed day of sunny spells and isolated showers. Chance of overnight mist or fog patches here and there.

Belfast: Partly cloudy, isolated showers, windy spells
Dublin: Sunny, dry, warmer day cooler night, moderate winds
Cork: Winds easing to light, sunny, brief morning showers, mild day cool night
Galway: Partly cloudy, brief showers, fresh breezes, mild day cool night

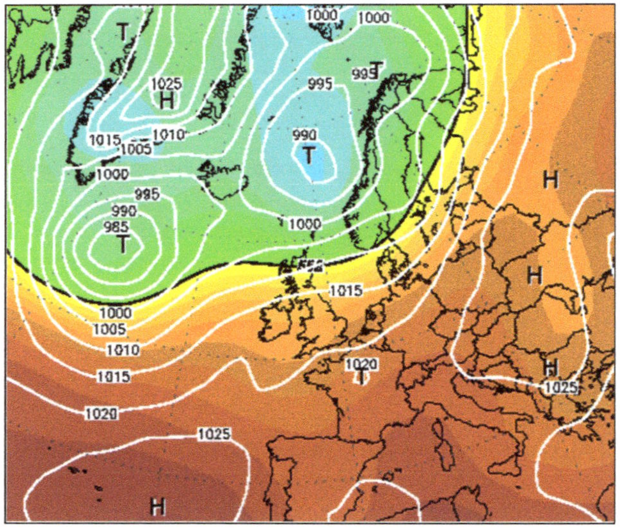

RAIN POTENTIAL

FROST/SNOW

OCTOBER **3** FRIDAY

A depression to the north of Ireland intensifies with a strong southwesterly flow throughout, with very strong and gusty winds, occasionally to gale force. A stormy air prevails with widespread cloud and spells of rain and drizzle throughout. Rain activity likely to be more persistent in the south and west, while lighter in the far north.

Belfast: Mostly cloudy, warmer, light showers, strong winds, chance of gales
Dublin: Mostly cloudy, light showers, squally and stormy winds
Cork: Cloudy, occasional showers, warmer, strong and unsettled winds
Galway: Stormy air, cloudy, heavy rain, gusts and gales

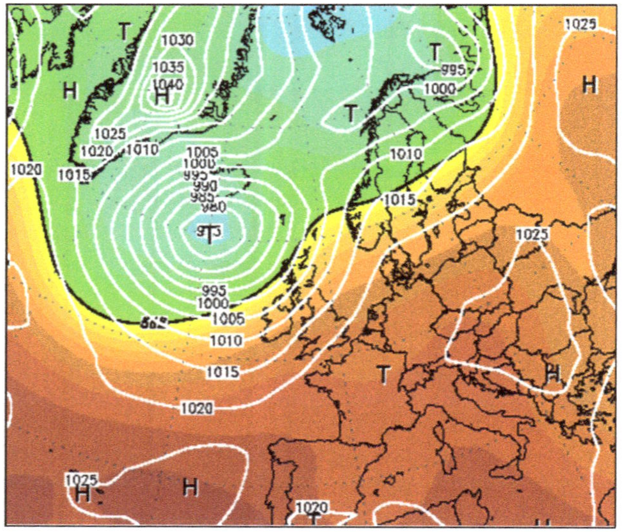

RAIN POTENTIAL

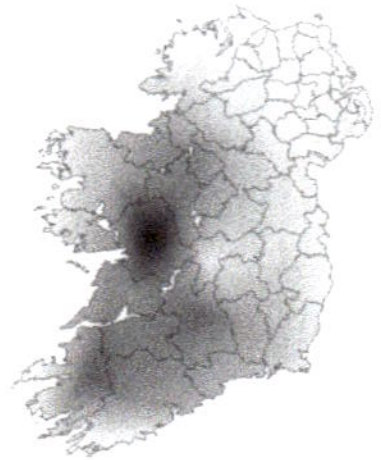

FROST/SNOW

OCTOBER **4** SATURDAY

Stormy conditions continue throughout the country as the depression to the north of Ireland persists. Strong west to southwest winds prevail, occasionally up to gale force at times and with gusts possibly stronger. Scattered squally showers in most districts. Cloudy in the north and more frequent sunny breakthroughs elsewhere.

Belfast: Stormy air, mostly cloudy, squally showers
Dublin: Scattered showers, some sunny spells, cooler, strong and unsettled winds
Cork: Mostly sunny and dry, cooler, strong and unsettled winds
Galway: Stormy air, squally showers, some sunny spells, cooler

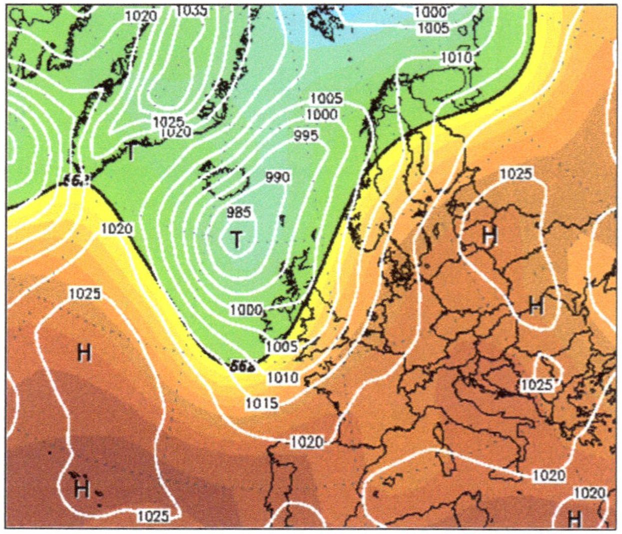

RAIN POTENTIAL

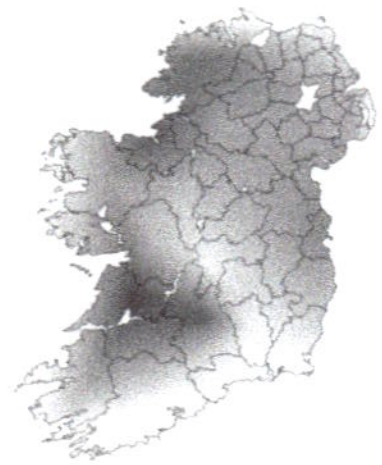

FROST/SNOW

OCTOBER 5 SUNDAY

The depression to the north of Ireland moves away to the northeast allowing a high pressure system to ridge over the south of the country. Winds strong and gusty at first in a strong northwesterly flow, before easing later ahead of the high. Some isolated showers linger overnight, mostly in the west and north, clearing to a mostly sunny and dry day throughout as the high establishes its influence.

Belfast: Mostly sunny, dry, colder, strong winds, possible gusts to gale force at times
Dublin: Mostly sunny, brief showers, cold overnight, strong blustery winds
Cork: Sunny and dry, strong blustery winds, mild day cool night
Galway: Strong and blustery winds, sunny, dry, mild day cool night, chance of gales

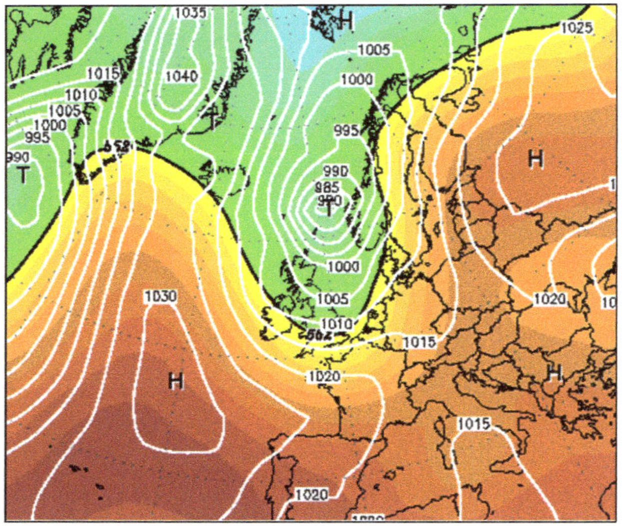

RAIN POTENTIAL

FROST/SNOW

OCTOBER 6 MONDAY

A high pressure system ridges over the country extending light northerly breezes and colder temperatures with the chance of some patches of frost overnight. Mostly dry day with occasional sunny spells throughout. Chance of some isolated showers briefly mostly in northern parts.

Belfast: Partly cloudy, dry, cold northerly breezes
Dublin: Partly cloudy, dry, cold overnight, moderate northerly breezes
Cork: Partly cloudy, cooler, dry, winds easing to moderate
Galway: Cloudy, dry, cooler, winds easing with occasional gusts

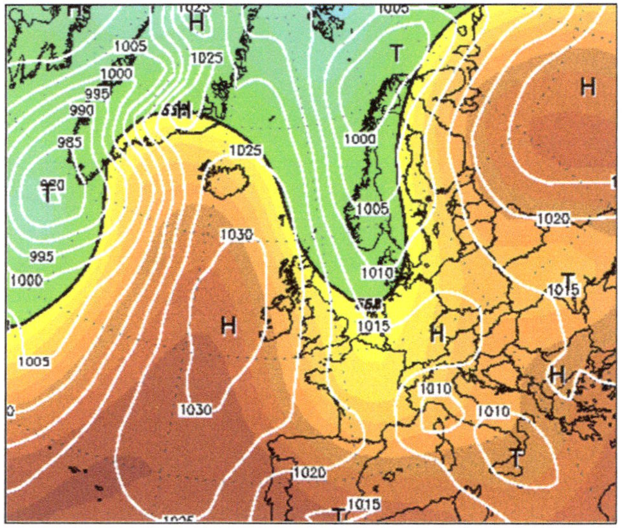

RAIN POTENTIAL

FROST/SNOW

OCTOBER 7 TUESDAY

High pressure brings a mostly calm atmosphere over the country in a variable to northeasterly flow. Mostly sunny and dry everywhere, with patches of fog and frost overnight in many districts, especially in the north and west, and particularly cold in central districts.

Belfast: Sunny, dry, cool, chance of frost in sheltered parts, fluctuating breezes
Dublin: Cold northerly winds, partly cloudy, dry
Cork: Mostly sunny, cool and dry, light breezes
Galway: Sunny, dry, mild day cold night, chance of frost, chance of fog patches

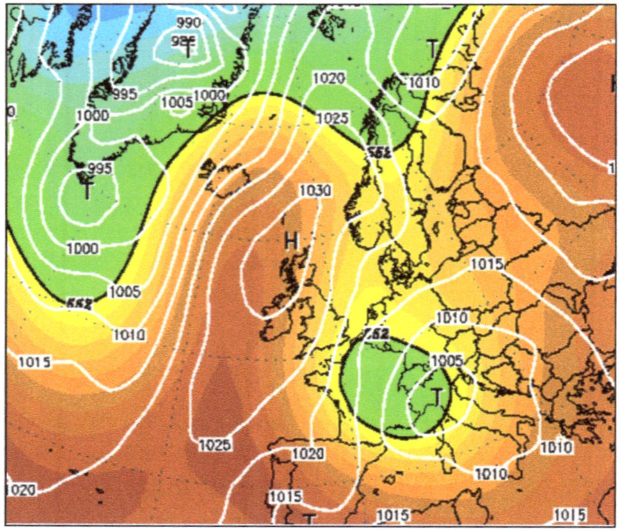

RAIN POTENTIAL

FROST/SNOW

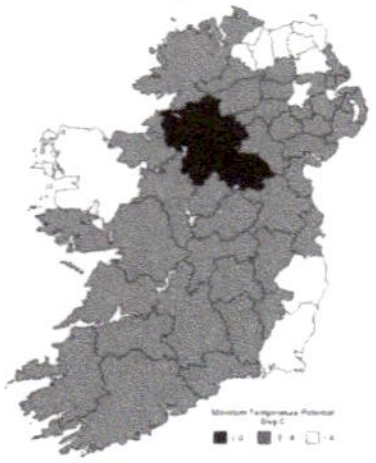

OCTOBER 8 WEDNESDAY

High pressure continues to prevail over Ireland directing a gentle, variable to mostly northeasterly air throughout. Cold overnight with patches of frost and fog particularly in the north and west once more, and some misty patches in the north turning more drizzly at times. Some brief showers in the east and southwest, while mainly dry elsewhere. Cloudy in the north and east, sunnier in the west and south.

Belfast: Cloudy, dry, light shifting breezes, chance of fog overnight
Dublin: Cloudy, isolated showers, moderate breezes
Cork: Light shifting breezes, dry, warmer, occasional sunny spells
Galway: Dry, some sunny outbreaks, milder day, overnight frost, fluctuating breezes

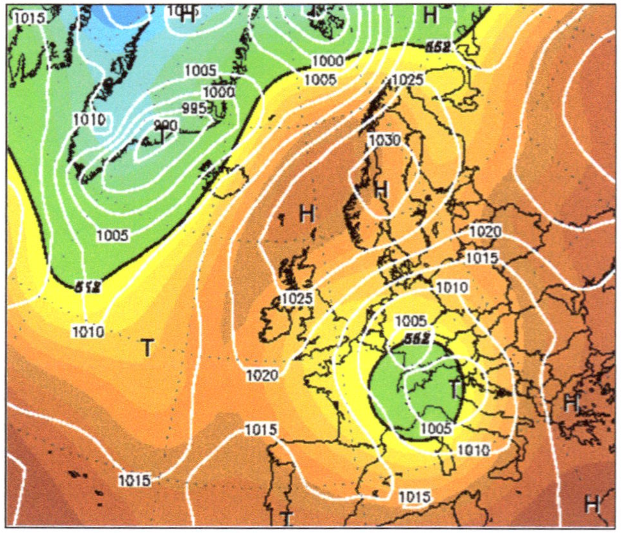

RAIN POTENTIAL

FROST/SNOW

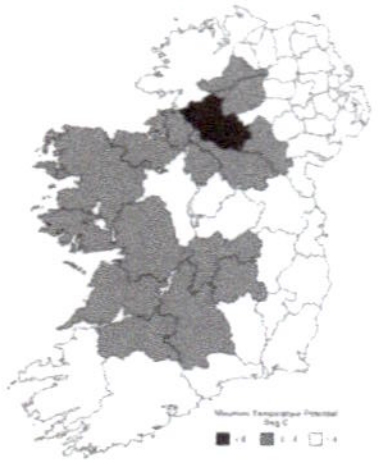

OCTOBER 9 THURSDAY

A low pressure trough forms over Ireland with light east to northeast winds prevailing. Widespread cloud throughout, with patches of fog overnight once more in the north, and drizzly showers at times in most districts, although remaining mostly dry in the southern and western coastal fringes. Some showers likely to be heavier for a time in the northwest.

Belfast: Overcast, spells of light showers, gentle breezes to mostly calm air, fog likely
Dublin: Overcast, brief showers, milder temperatures, light winds
Cork: Cloudy, brief showers of light rain, occasional breezy spells
Galway: Cloudy, light showers, cool, moderate breezes

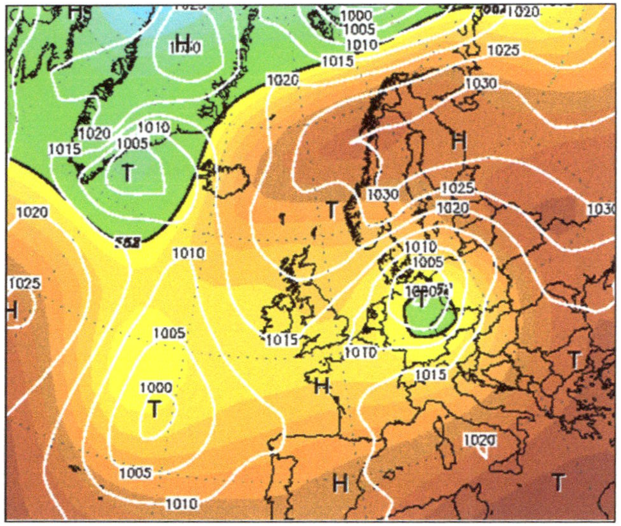

RAIN POTENTIAL

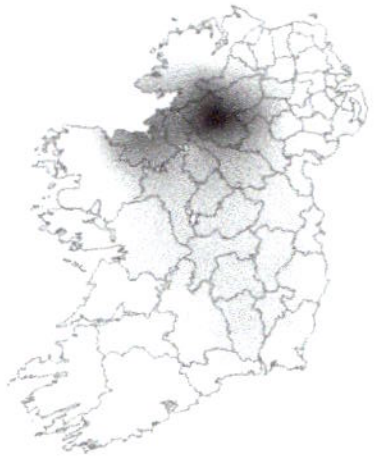

FROST/SNOW

OCTOBER 10 FRIDAY

A trough between two low pressure systems crosses over Ireland directing a very light to mostly calm air throughout, variable and shifting from southeasterly to northeasterly. Patches of drizzle overnight in the east, fog and mist form in many central districts, and mostly dry in the west and south. A changeable day in the east and north with some sunny spells and light showers, while mainly dry and cloudy elsewhere. Milder temperatures.

Belfast: Partly cloudy and mostly dry, calm and cool air
Dublin: Changeable, mix of sunny spells and occasional showers, mostly calm air
Cork: Overcast, dry, coolish, mostly calm air, possible scattered fog patches
Galway: Partly cloudy and dry, gentle breezes to calm air, fog likely, warmer

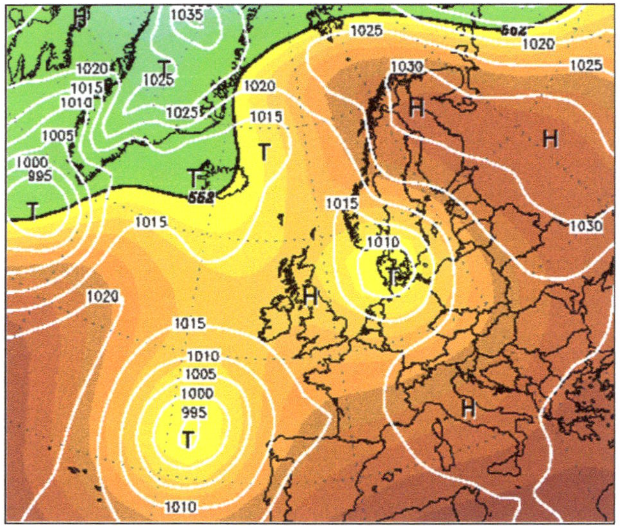

RAIN POTENTIAL

FROST/SNOW

OCTOBER **11** SATURDAY

A high pressure ridge prevails over Ireland with low pressure to the south. Winds will be variable in direction and gentle to mostly calm. Mostly dry throughout with good sunny spells in the west and cloudier elsewhere. Overnight fog likely to be widespread and slow to clear. Warmer than normal for this time of year.

Belfast: Mostly cloudy, dry, warmer day cooler night, mostly calm atmosphere
Dublin: Partly cloudy, dry, mild day cool night, gentle breezes, fog patches possible
Cork: Overcast, gentle breezes, outbreaks of drizzle, fog in places
Galway: Mostly sunny, dry, mild day cool night, shifting breezes, chance of light fog

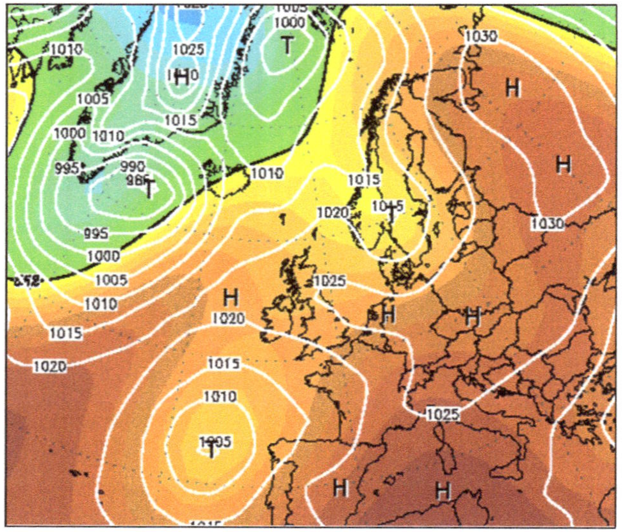

RAIN POTENTIAL

FROST/SNOW

OCTOBER **12** SUNDAY

A high pressure centre to the east of Ireland ridges over the country while a low pressure trough sits to the west. Variable winds prevail, predominately easterly, in a light flow, fresher at times around the northwest coasts. Dull and gloomy throughout with patches of rain and drizzle in many districts, with some heavier spots in the west. Chance of fog patches overnight in the south before turning to drizzly showers.

Belfast: Cloudy, mild day cool night, occasional showers, moderate breezes
Dublin: Overcast and dull, brief showers, gentle shifting breezes
Cork: Overcast, mostly dry with brief misty patches, gentle breezes, fog likely
Galway: Moderate shifting breezes, cloudy, rain

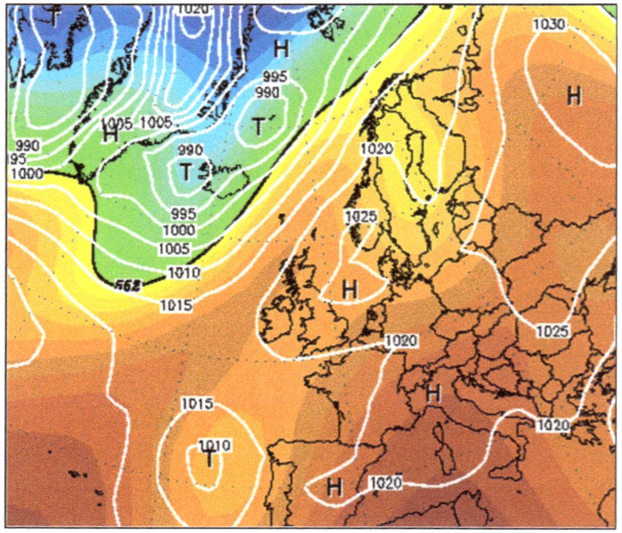

RAIN POTENTIAL

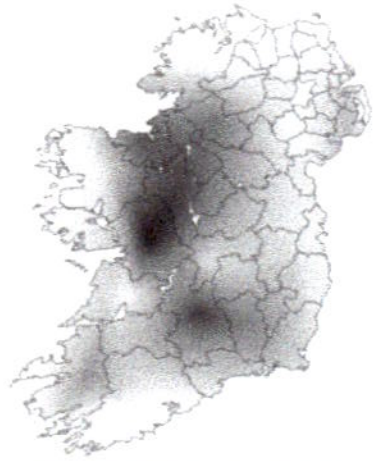

FROST/SNOW

OCTOBER 13 MONDAY

A complex of high and low pressure centres gather around Ireland, with a low pressure trough prevailing. Winds generally light and in a southerly flow, freshening with a westerly change. A mixed day with a few sunny spells, and widespread light showers and drizzle patches moving up through the country, lingering in the northeast while clearing elsewhere. Chance of some patches of fog overnight mostly in the south.

Belfast: Partly cloudy, light showers, gentle shifting breezes
Dublin: Cloudy, occasional showers, mostly calm air
Cork: Changeable, mix of rain and sunny spells, mostly calm air, fog possible
Galway: Partly cloudy, light drizzly showers, gentle shifting breezes

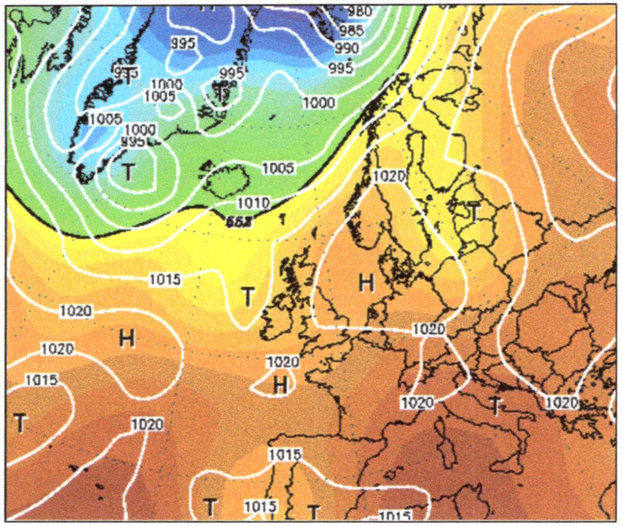

RAIN POTENTIAL

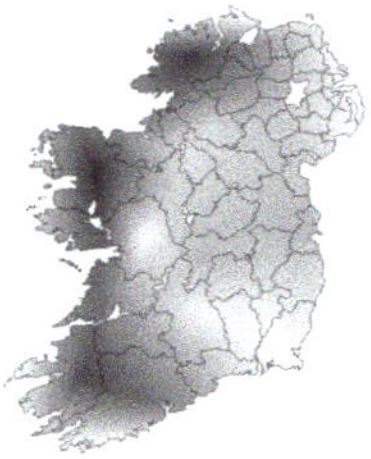

FROST/SNOW

OCTOBER 14 TUESDAY

A high pressure ridge directs a westerly flow over the country with light breezes throughout. Some lingering showers overnight, mostly in the northeast, and frost and fog patches elsewhere, clearing quickly to a mostly dry and sunny day. Sunniest in the east with increased cloudy spells to the west.

Belfast: Mostly sunny and dry, mild day cool night, light shifting breezes
Dublin: Gentle breezes, sunny and dry, mild day cool night
Cork: Light breezes, sunny, dry, mild day cool night
Galway: Partly cloudy, dry, mild day cold night, light shifting breezes

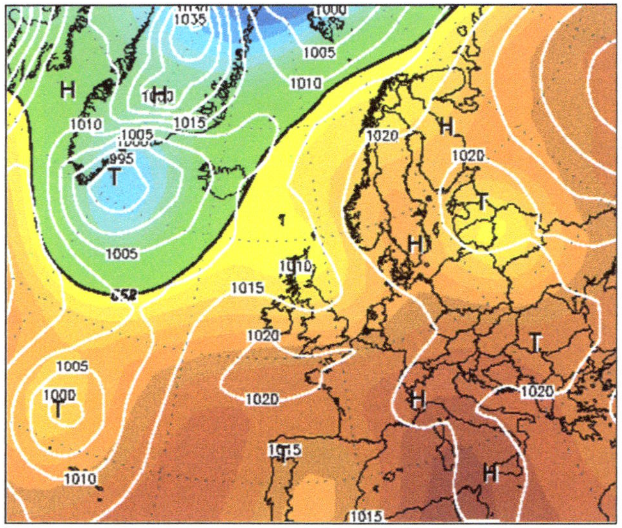

RAIN POTENTIAL

FROST/SNOW

OCTOBER **15** WEDNESDAY

High pressure prevails. Winds generally light throughout in a south to southeast flow. Some fog overnight in sheltered districts. Milder everywhere, mostly dry, and with good sunny spells during the day, slightly cloudier in the west.

Belfast: Partly cloudy and dry, mostly calm air
Dublin: Partly cloudy, dry, mild day cold night, gentle breezes to mostly calm air
Cork: Partly cloudy, dry, mild day cool night, gentle southerly flow
Galway: Cloudy and dry, mild day cool night, gentle breezes

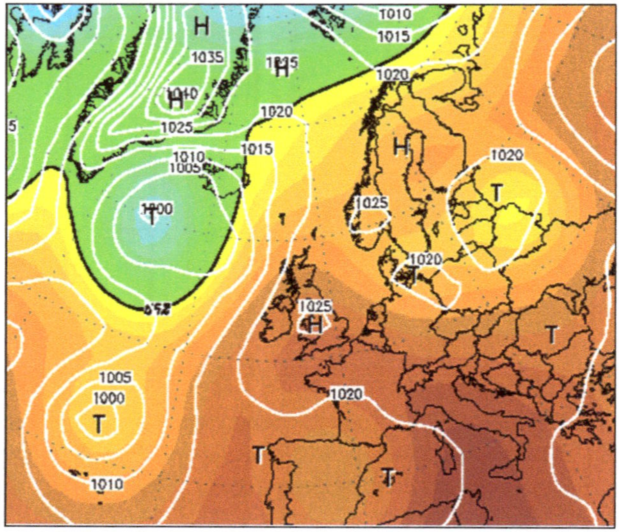

RAIN POTENTIAL

FROST/SNOW

OCTOBER **16** THURSDAY

An anticyclonic system lingers over Ireland in a gentle to mostly calm air throughout. Widespread fog overnight, and although mostly dry and sunny in most districts after the fog clears, there may be some isolated showers briefly around the northwest coasts. Sunniest in the east while increased scattered cloud elsewhere, cloudiest in the west.

Belfast: Partly cloudy, dry, warmer, mostly calm air, chance of patches of fog
Dublin: Sunny, dry, mild day cold night, calm and serene atmosphere, fog likely
Cork: Partly cloudy and mild, dry, calm atmosphere, fog
Galway: Cloudy, warmer, dry, calm and serene air, fog likely

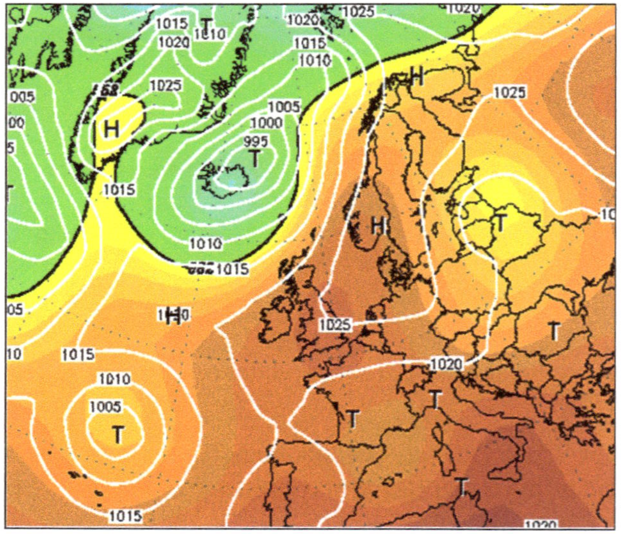

RAIN POTENTIAL

FROST/SNOW

OCTOBER **17** FRIDAY

A large anticyclonic system continues to linger over Ireland bringing a mostly calm air throughout, with overnight fog in many districts, often slow to clear. A dry day everywhere, with some sunny spells after the fog lifts.

Belfast: Calm and cloudy air, dry, mild day cool night, fog likely
Dublin: Cloudy, calm and pleasant air, cool, mostly dry, patches of fog
Cork: Cloudy, dry, increasing threats of rain, calm atmosphere with patches of fog
Galway: Serene air, mild day cold night, some sunny spells, dry, shallow fog

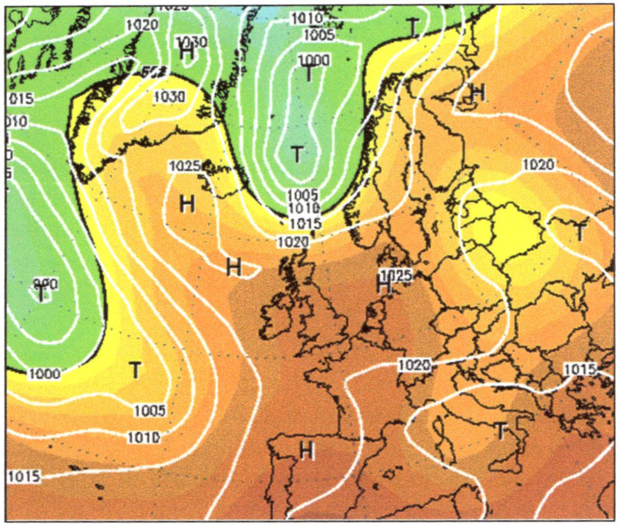

RAIN POTENTIAL

FROST/SNOW

OCTOBER **18** SATURDAY

High pressure prevails over Ireland in a light to moderate southeast flow, freshening later in the day with a southwest change. Widespread cloud throughout, with overnight fog patches mostly in the north and east, and misty patches of drizzle and occasional showers developing almost everywhere. A band of rain likely to cross the western half of the country later in the day bringing some pockets of heavier falls.

Belfast: Gloomy and calm air, brief showers, colder, overnight fog patches likely
Dublin: Overcast, light showers, milder temperatures, calm air
Cork: Overcast and dull, spells of light rain, warmer, calm air, fog likely
Galway: Calm air, overcast, drizzle patches and showers, possible fog

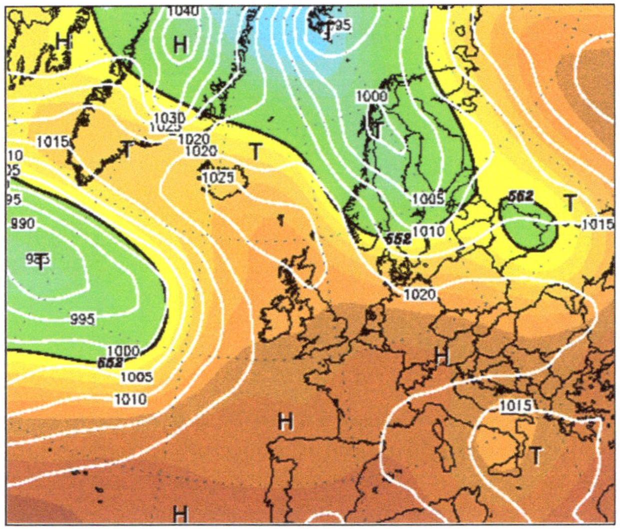

RAIN POTENTIAL

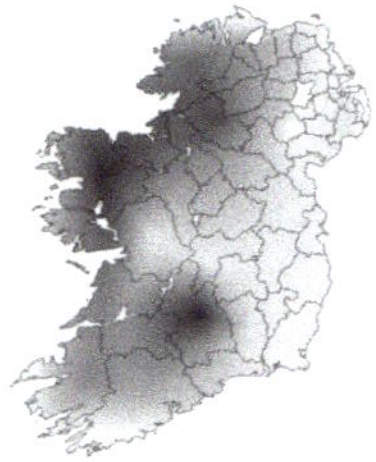

FROST/SNOW

OCTOBER **19** SUNDAY

A deep depression to the west of Ireland pushes the high pressure ridge eastwards across the country, in a light to moderate southwest flow, backing northwesterly for a time later in the day. Some patches of showers and drizzle at first in most districts, contracting to the south of the country as the day progresses, allowing for some short sunny breakthroughs and drier conditions elsewhere.

Belfast: Partly cloudy, slightly warmer, dry, freshening breezes later in the day
Dublin: Partly cloudy, dry, moderate breezes
Cork: Overcast, passing showers, mild temperatures, moderate breezes
Galway: Partly cloudy, warmer, light showers, freshening breezes

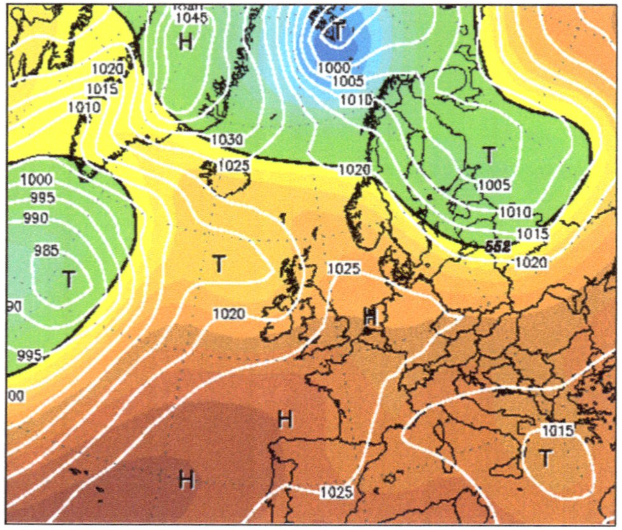

RAIN POTENTIAL

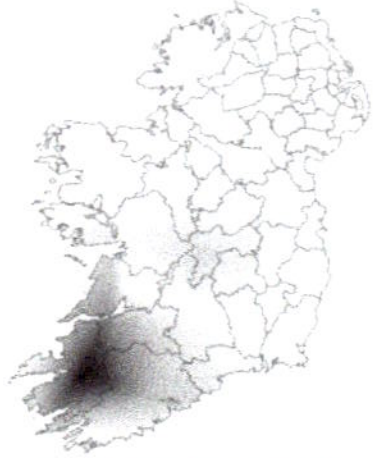

FROST/SNOW

OCTOBER **20** MONDAY

.A low pressure trough brings a light southwesterly flow over the country, fresher and gustier at times around the northern coasts. Mostly cloudy throughout with patches of fog and mist overnight in many places, slow to clear. Mostly dry day with some sunny breakthroughs, before some evening drizzle showers likely to arrive, mostly in the northeast and southwestern Leinster areas.

Belfast: Cloudy, mostly calm, dry, mild day cool night, shallow fog in places
Dublin: Mostly calm and dry air, occasional sunny breakthroughs, mild day cold night
Cork: Overcast, dry, odd threat of rain, mild temperatures, fog likely, calm air
Galway: Cloudy, calm air, dry, mild day cool night, fog likely

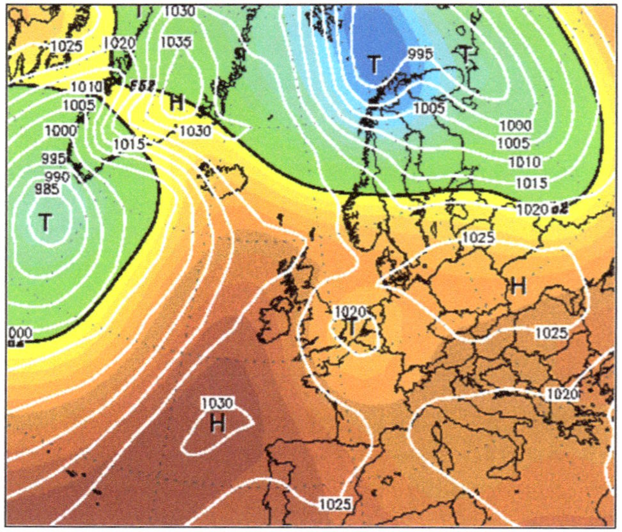

RAIN POTENTIAL

FROST/SNOW

OCTOBER **21** TUESDAY

The high pressure ridges over the east of Ireland while low pressure lingers to the west, with light southwesterly breezes prevailing, fresher at times along the western coastal fringes. Mostly cloudy everywhere, with patches of drizzle in the north, more isolated in the west and around the northeast of Munster. Dry elsewhere. Longer sunny breakthroughs likely in the south and southeast.

Belfast: Dull and gloomy air, mild temperatures, dry, moderate breezes
Dublin: Overcast, mostly calm air, dry and mild, odd threat of rain, chance of drizzle
Cork: Cloudy and mostly calm air, dry, misty patches, warmer
Galway: Mostly calm air, dry, overcast with some threats of rain, warmer

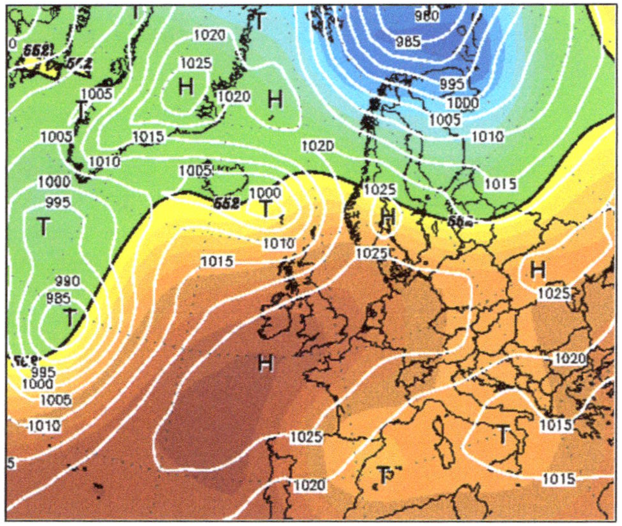

RAIN POTENTIAL

FROST/SNOW

OCTOBER **22** WEDNESDAY

High pressure continues to prevail over Ireland in a moderate south to southwest flow throughout, with occasional gusts around the western coasts. Widespread cloud everywhere, with isolated misty drizzle patches overnight, mostly in the west and south. Generally dry but gloomy elsewhere.

Belfast: Overcast, dry, threats of rain and drizzle, moderate breezes
Dublin: Gentle breezes, overcast, dry, mild temperatures, light drizzle patches
Cork: Overcast, light shifting breezes, dry, cooler
Galway: Gentle southerly breezes, dull and gloomy air, dry and cooler

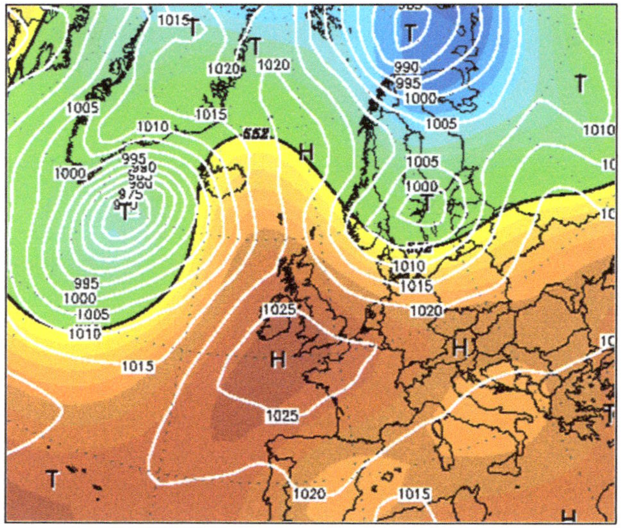

RAIN POTENTIAL

FROST/SNOW

OCTOBER 23 THURSDAY

The high pressure ridge maintains a light southwest flow over the country, while a depression to the west of Ireland nears. Dull and gloomy and mostly dry+DL21 everywhere, except for some drizzle patches mostly in the west later in the day. Some sunny spells likely in central districts.

Belfast: Overcast, mostly dry with some misty patches, moderate breezes
Dublin: Overcast, dry with odd threat of rian, moderate breezes
Cork: Overcast and dull, moderate southerly breezes, dry
Galway: Cloudy, isolated showers, mild temperatures, gentle breezes

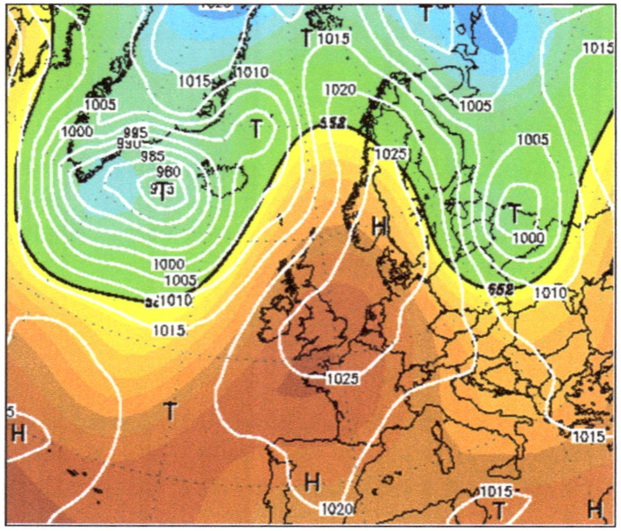

RAIN POTENTIAL

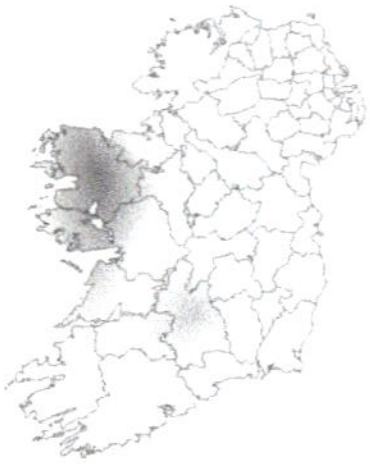

FROST/SNOW

OCTOBER 24 FRIDAY

A depression to the northwest of Ireland moves onto the country with a light south to southwesterly flow continuing throughout. Mostly dry and with the occasional sunny spell during the day after pockets of light rain or drizzle clear mostly in the western half fo the country. Warmer than normal. Chance of fog in inland parts.

Belfast: Mostly cloudy, isolated showers, light breezes
Dublin: Mostly sunny and dry, warmer day cooler night, moderate breezes
Cork: Cloudy, dry, threats of rain, mild day cool night, chance of misty patches
Galway: Occasional sunny spells, dry, mild temperatures, moderate breezes

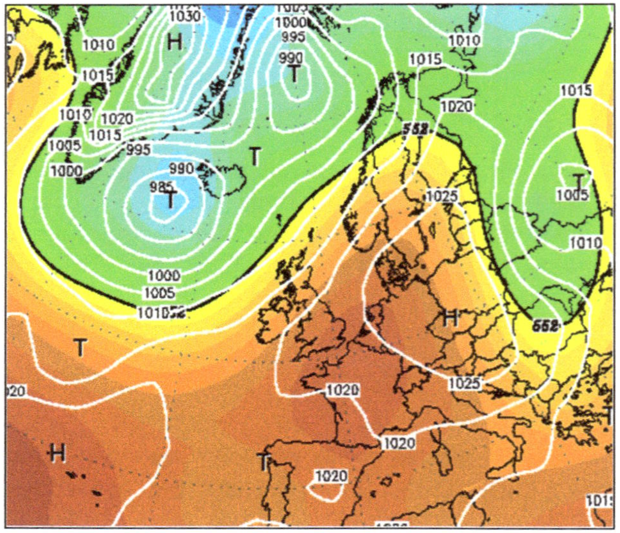

RAIN POTENTIAL

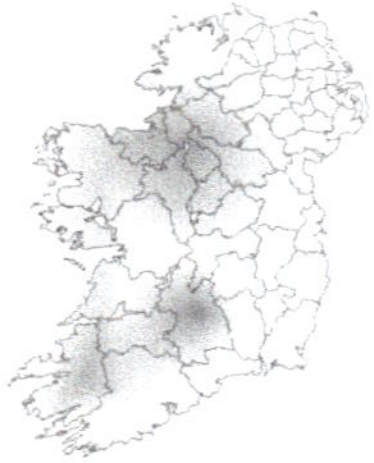

FROST/SNOW

OCTOBER **25** SATURDAY

A depression to the north of Ireland extends a stronger south to southwest flow over the country with chance of gale gusts at times around the west and northern coasts. A changeable day with a mix of some sunny spells and occasional showers in most districts as a band of rain and drizzle crosses the country from the west at first to the east later in the day. Some falls likely to be heavier at times in western parts.

Belfast: Cloudy, cooler, light showers, windy at times
Dublin: Cloudy, colder, isolated showers, fresh breezes at times
Cork: Changeable, light showers and sunny spells, warmer day cool night
Galway: Changeable, mix of showers and sunny spells, windy at times

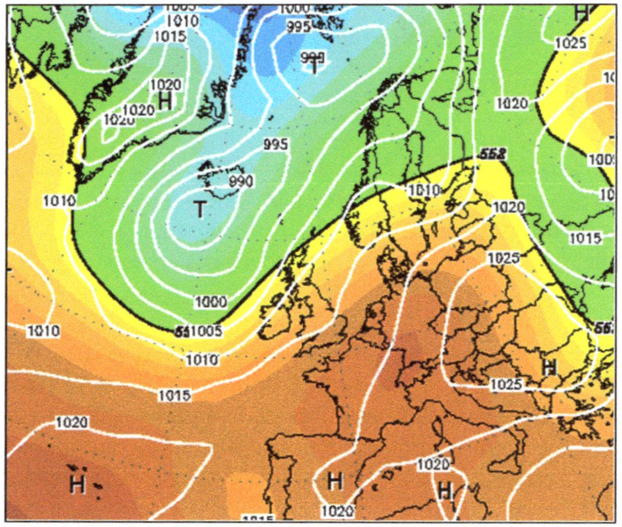

RAIN POTENTIAL

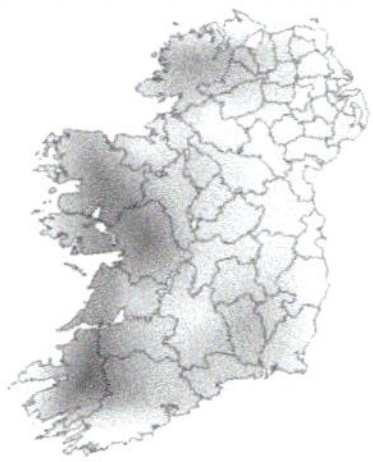

FROST/SNOW

OCTOBER **26** SUNDAY

Strong and unsettled southwest winds prevail, with a northwest change later in the day. Further changeable conditions throughout with a mix of sunny spells and some isolated showers in most districts, with more frequent showers and the occasional heavier outbreaks of rain activity in the south of the country.

Belfast: Partly cloudy, cool, dry, blustery, chance of gales at times
Dublin: Partly cloudy, mostly dry, threat of rain, mild day cool night, blustery at times
Cork: Cloudy, wet and windy, cooler
Galway: Partly cloudy, isolated showers, mild day cold night, blustery, near gales

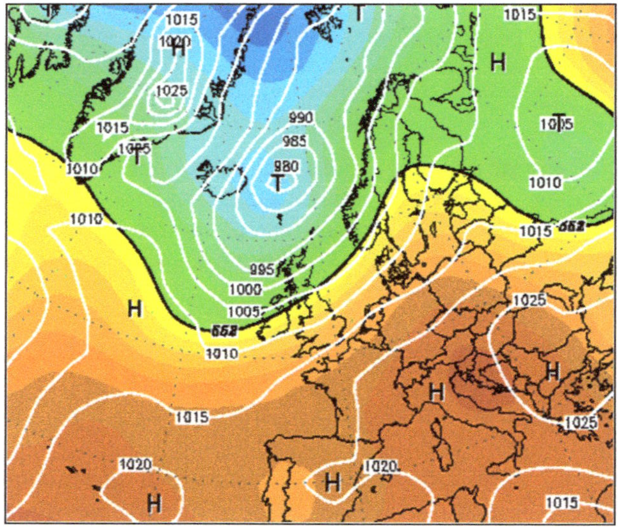

RAIN POTENTIAL

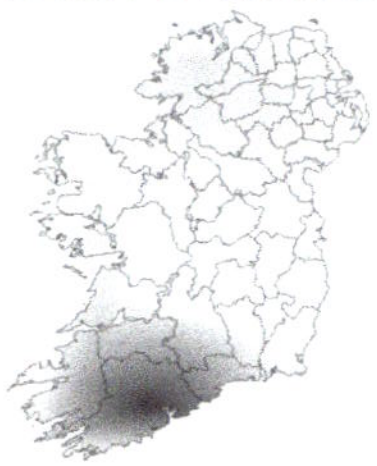

FROST/SNOW

OCTOBER **27** MONDAY

Changeable conditions prevail across the country as a high pressure ridge nears the south of Ireland and the low pressure trough departs to the north. A northwest flow prevails for a time, veering westerly later in the day as the high pressure ridge moves up over the country. Isolated overnight showers likely in some central parts, while dry elsewhere. Scattered cloudy spells and mostly dry day everywhere. Chance of fog patches overnight, mostly in the west.

Belfast: Partly cloudy and dry, mild day cold night, light shifting breezes
Dublin: Fluctuating breezes, dry, occasional sunny spells, cold overnight
Cork: Partly cloudy, dry, cool, moderate shifting breezes
Galway: Partly cloudy, dry, cooler, chance of overnight fog patches

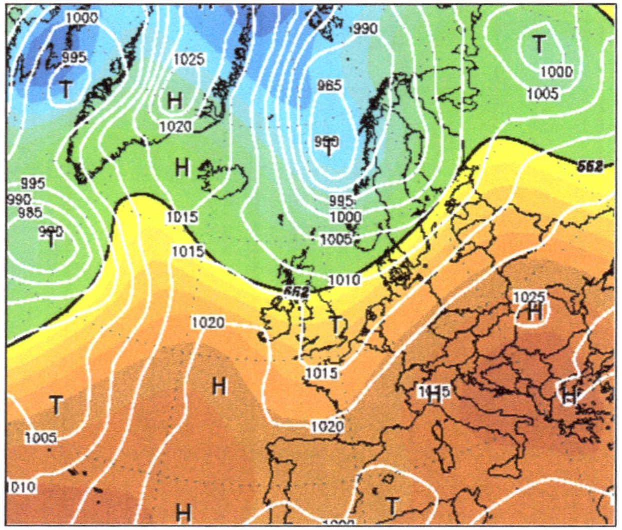

RAIN POTENTIAL

FROST/SNOW

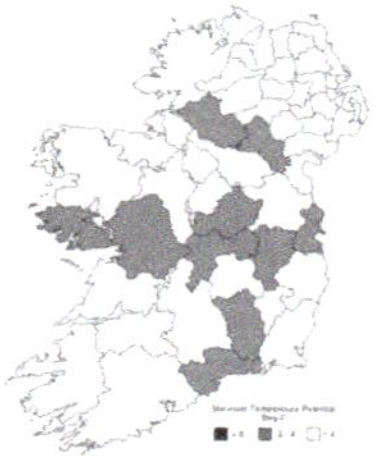

OCTOBER **28** TUESDAY

An anticyclonic system crosses the country with generally light southerly breezes prevailing. Patches of fog and frosts overnight and dry throughout the day with good sunny spells in the east, and cloudier elsewhere.

Belfast: Mostly cloudy, dry, colder, fluctuating breezes
Dublin: Partly cloudy, dry, fluctuating breezes, chance of overnight frost pockets
Cork: Fluctuating breezes, dull and gloomy air, dry with passing threats of rain
Galway: Cloudy, dry, cold overnight with frost and fog patches possible

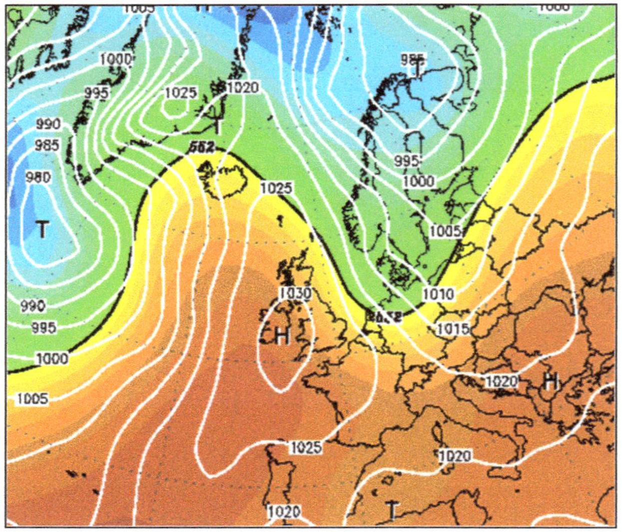

RAIN POTENTIAL

FROST/SNOW

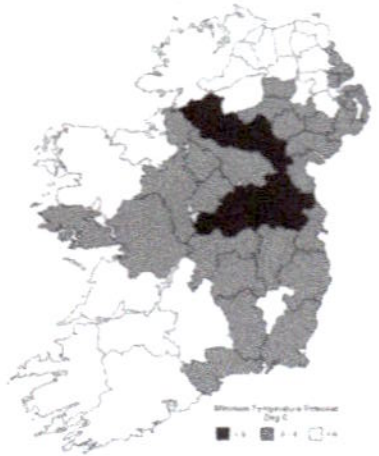

OCTOBER **29** WEDNESDAY

Anticyclonic to the east, with low pressure to the west, and a mostly light south to southwest flow prevailing, freshening around the western coasts. Cloudy and dry everywhere, except for the chance of some isolated showers in the far northwest.

Belfast: Cloudy, dry, moderate breezes
Dublin: Cloudy, warmer, dry, shifting windy spells
Cork: Cloudy, dry, freshening breezes later in the day
Galway: Fresh and unsettled winds, cloudy, isolated showers, warmer overnight

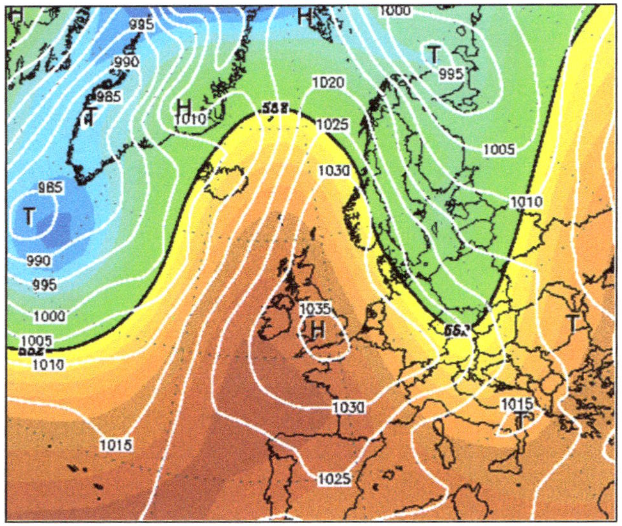

RAIN POTENTIAL

FROST/SNOW

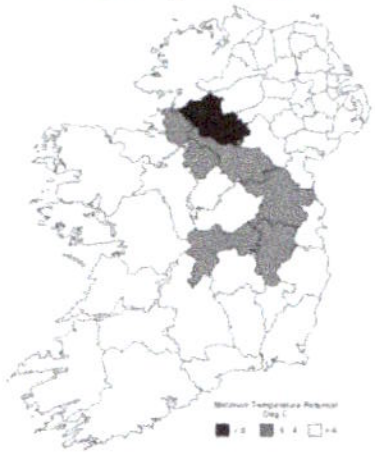

OCTOBER **30** THURSDAY

A large high pressure system maintains a southerly flow throughout, with patches of mist and fog likely overnight in many districts, and some frost patches in the east. Mostly cloudy and dry day, but very dull and gloomy in the west with increasing threats of rain.

Belfast: Cloudy, dry, mild day cool overnight, moderate winds
Dublin: Cloudy and dry, mild day cold night, moderate breezes
Cork: Cloudy and mild, dry, breezy spells
Galway: Overcast, dry with increasing threats of rain, windy

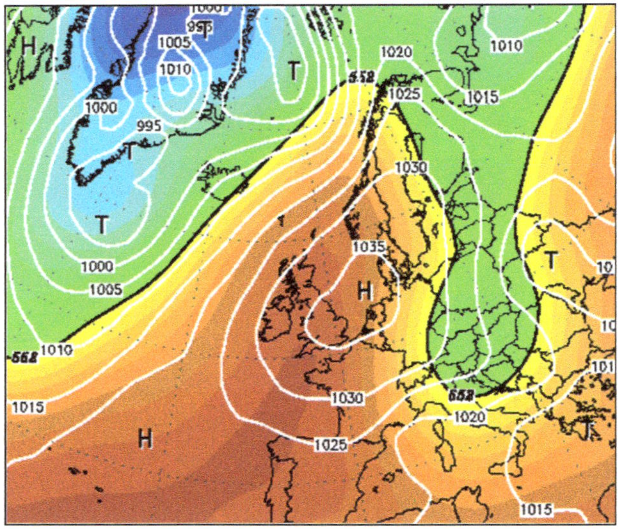

RAIN POTENTIAL

FROST/SNOW

OCTOBER **31** FRIDAY

High pressure continues to linger over the east of Ireland while a depression to the west nears the country. Generaly light to moderate south to southeasterly winds prevail throughout, but fresher with occasional gusts developing around the western and northern coasts. Dull and gloomy day everywhere, with outbreaks of rain in many inland districts, but remaining mostly dry around the coasts. Rain may be more persistent in the southwest of Connaught.

Belfast: Cloudy and cool, dry, fresh breezes at times
Dublin: Mostly cloudy, dry, mild temperatures, windy at times
Cork: Overcast and dry with increasing threats of rain, fresh breezes at times
Galway: Overcast, heavy rain, strong winds at times, mild temperatures

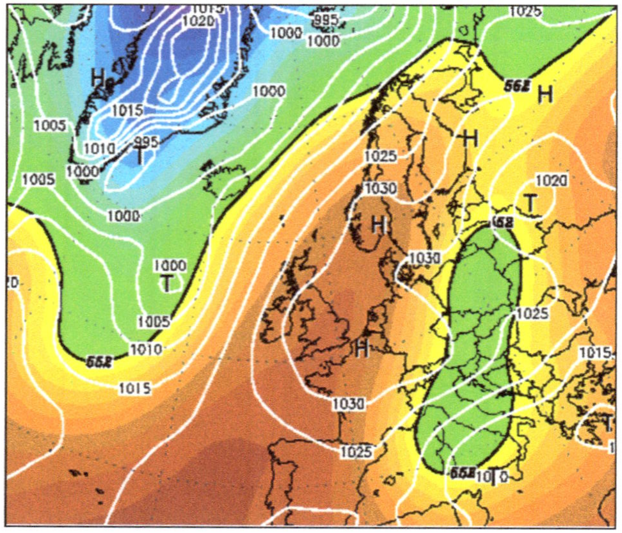

RAIN POTENTIAL

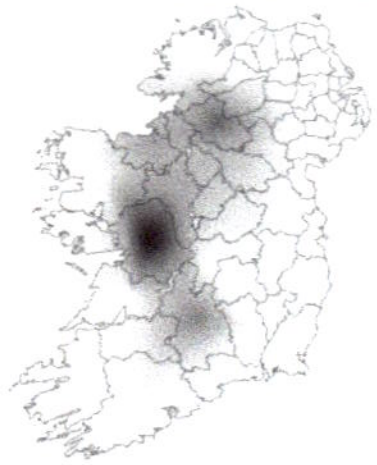

FROST/SNOW

NOVEMBER

PHASES OF THE MOON

3rd	Crossing Equator
5th	Full Moon
5th	Perigee #1
8th	Northern Declination
12th	Third Quarter
16th	Crossing Equator
20th	New Moon
20th	Apogee
22nd	Southern Declination
28th	First Quarter
30th	Crossing Equator

1	2	3	4	5	6	7	8	9	10	11	12	13	14	15	16	17	18	19	20	21	22	23	24	25	26	27	28	29	30
Fri	Sat	Sun	Mon	Tue	Wed	Thur	Fri	Sat	Sun	Mon	Tue	Wed	Thur	Fri	Sat	Sun	Mon	Tue	Wed	Thur	Fri	Sat	Sun	Mon	Tue	Wed	Thur	Fri	Sat
		XhN		F P1			^				3Q				XhS				N A		V						1Q		XhN

MONTHLY SUMMARY

1st-4th High pressure prevails and brings a mostly dry period with light winds throughout broken only by the passage of a passing Atlantic front. The warmer south to southeast winds at the end of last month drop to below average temperatures during this period.

1st Brief wet spell as an Atlantic front brushes up through the country.

2nd-4th Very sunny spell.

3rd-4th Scattered frosts mostly in inland parts but some into eastern areas. Some pockets of frost may see very low overnight temperatures.

3rd-6th Chance of widespread fogs.

5th-10th A weak frontal system brings a showery spell across the country.

5th Chance of wintry showers of snow grains in central areas.

6th-7th Chance of heavy rain mostly in the north and northwest of the country. Strong and blustery southwest winds likely, possibly up to gale force in some places. Temperatures likely to be unusually warmer than normal.

7th-8th Chance of thunderstorms in the west of the country and hail showers.

9th-10th Isolated showers likely but some may fall as snow grains overnight in some central areas.

Over the next four week areas of high pressure near to or over Ireland will prevail for much of this outlook period, bringing mostly dry conditions with light winds over the first couple of weeks, and unusually warmer temperatures over the first week. The subsequent couple of weeks will be a mix of short spells of wet and windy weather and dry and calmer conditions, ending this outlook period under the influence of low pressure. Rain amounts will be below normal for this outlook period almost everywhere, with most districts in the southern half of the country seeing around half their usual rainfall. Very little rain will be observed during the first couple of weeks, but some heavy rain can be expected at times during the second half of this outlook period. Sunshine hours per day will vary considerably across the country as a whole, with much of the southern districts up to 10% sunnier, the northwest of Ireland may only receive around 40% of their normal sunshine hours for this time of year. Temperatures may average well above normal everywhere during this outlook period, with many districts possibly up to two degrees warmer than average. Temperatures, while starting this outlook period much warmer than average, will gradually get cooler, but still remain about the norm throughout. Overall, winds will be lighter than is typical for this time of year. During this outlook period, there will be around 7-8 days with rain in the east and southeast, but up to 20 wet days in the north and northwest. Thunderstorms may be heard on 1 day and hail on 4 days. Fogs can be expected on 5 days and frosts on 7 days mostly in inland parts, but no frosts are expected around the coastal fringes. Gale force winds are only expected on a couple of days, but no days are expected to see snow.

11th-26th An anticyclonic system stalls near the southwest of the country bringing a mostly dry spell and warmer spell with light west to southwest winds throughout. Only the occasional weak frontal system will bring brief spells of rain now and then, mostly around the north and northwestern coastal fringes.

11th-17th Chance of warmer than normal daytime temperatures throughout.

12th-23rd Frosts in inland to eastern parts

16th High pressure may reach its highest value around this time.

18th-19th An intense depression to the north of Ireland brings a stormy air across the country and a band of heavy rain into many districts. Very cold overnight with the chance of wintry precipitation falling as snow grains in some central areas.

21st-24th Chance of thunderstorms and light snow in the west of the country, and snow grains may be observed in central parts.

25th Widespread fog likely.

27th-30th A depression crossing to the north of Ireland brings widespread cloud the occasional band of rain mostly to northern parts of the country. Some falls may be heavy briefly in the west and southwest, as well as wintry and falling as hail. During this spell, there is a chance of snow in the west and north. Frosts are likely in eastern regions to the central districts.

30th Chance of widespread fog.

IRELAND RAINFALL ESTIMATES

November 2025

d=mainly dry, n= nonrecordable, l=light shrs, s=significant shrs, r=rain, h=hvy falls

		ULSTER PROVINCE					CONNAUGHT PROVINCE						LEINSTER PROVINCE				MUNSTER PROVINCE						
		Alderg	Hillsbo	Armag	Loughl	Malin H	Sligo	Ardtarı	Belmul	Drums	Galway	Roscoı	Markre	Edendı	Kilkenı	Dublin	Shann	Newpo	Tralee	Killarnı	Cork	All	Moon
1st	Sat	n	l	d	s	l	s	s	r	r	r	l	r	l	s	l	h	l	h	s	r	152	
2nd	Sun	d	d	d	d	d	d	d	l	d	d	d	d	d	d	d	l	l	l	d	l	7	
3rd	Mon	d	d	d	d	d	d	d	d	d	d	d	d	d	d	d	d	d	d	d	d	0	XhN
4th	Tues	d	d	d	d	d	n	d	d	d	d	d	l	d	d	d	d	l	d	d	d	3	
5th	Wed	l	l	d	l	l	l	l	s	l	l	l	n	l	l	l	l	l	s	l	l	40	F P1
6th	Thur	s	s	l	h	h	h	h	h	h	r	h	r	s	s	d	l	h	s	r	s	260	
7th	Fri	s	l	s	s	s	s	s	h	s	s	s	s	l	l	s	s	r	s	r	s	138	
8th	Sat	l	d	l	l	l	l	l	l	d	d	l	s	l	d	n	l	l	l	l	n	31	^
9th	Sun	d	d	d	l	l	n	l	d	l	n	d	d	d	d	d	d	d	l	d	d	5	
10th	Mon	d	d	d	d	d	d	d	d	d	d	d	d	d	d	d	d	d	d	d	d	0	
11th	Tues	d	d	d	d	s	d	d	l	d	d	d	d	d	d	d	d	d	d	d	d	6	
12th	Wed	d	d	d	l	l	d	l	d	d	d	d	d	d	d	d	d	l	d	d	d	6	3Q
13th	Thur	d	d	d	n	d	d	n	l	d	d	d	d	d	d	l	d	d	d	d	d	4	
14th	Fri	d	d	d	l	d	l	l	l	d	l	n	l	d	d	n	d	s	d	d	d	12	
15th	Sat	l	d	l	d	l	d	d	l	d	d	d	d	d	d	l	l	d	l	l	n	10	
16th	Sun	n	d	l	d	d	n	d	l	d	n	d	d	d	d	d	d	l	d	n	d	4	XhS
17th	Mon	d	d	d	s	l	s	s	l	s	l	l	s	d	l	d	d	s	d	l	l	40	
18th	Tues	l	l	l	d	s	l	d	l	n	n	l	l	l	l	l	l	l	l	l	l	25	
19th	Wed	d	d	d	d	n	d	d	d	d	d	d	d	d	d	d	d	d	d	d	d	1	
20th	Thur	l	l	d	l	l	l	l	d	l	l	l	l	n	d	d	d	l	d	d	d	15	N A
21st	Fri	n	d	l	d	l	n	d	l	d	d	d	d	d	d	n	d	d	n	l	d	6	
22nd	Sat	d	n	d	s	l	l	s	l	d	l	l	s	l	d	n	l	s	l	l	d	33	V
23rd	Sun	l	l	n	l	d	l	l	n	s	d	l	l	d	d	d	d	l	d	l	l	22	
24th	Mon	l	l	l	d	s	l	d	l	l	n	l	l	l	l	s	d	l	d	n	d	26	
25th	Tues	l	d	s	d	d	n	d	n	l	d	l	l	l	l	l	n	l	l	d	l	20	
26th	Wed	d	d	d	n	d	l	n	l	n	l	l	l	d	l	d	d	s	l	d	d	17	
27th	Thur	s	r	d	h	r	r	h	r	h	r	h	s	h	h	r	h	h	h	h	h	331	
28th	Fri	r	s	h	l	s	s	l	h	l	l	l	l	s	s	r	l	h	r	s	l	160	1Q
29th	Sat	r	r	s	l	l	n	l	s	l	l	l	n	l	s	s	l	l	l	l	l	79	
30th	Sun	s	s	s	d	l	d	d	l	d	l	d	d	l	d	d	d	d	d	d	l	22	XhN
	Estimate:	57	58	55	85	91	83	85	125	83	63	67	62	48	56	51	62	100	101	83	63	1477	
	Average:	80	88	75	119	109	128	130	134	102	120	129	128	88	86	73	94	170	168	121	120	2263	
	Trend:	drr	drr	drr	drr	drr	drr	drr	av	drr	drr	drr	drr	drr	drr	drr	drr	drr	drr	drr	drr	drr	

IRELAND SUNSHINE ESTIMATES

November 2025

F=fine (8-12 hours of sunshine), pc= partly cloudy (4-7 hours), c=cloudy (1-3 hours), o=overcast (0 hours)

		ULSTER PROVINCE					CONNAUGHT PROVINCE						LEINSTER PROVINCE				MUNSTER PROVINCE						
		Alderg	Hillsbo	Armag	Loughl	Malin I	Sligo	Ardtar	Belmul	Drums	Galway	Roscoı	Markre	Edende	Kilkenı	Dublin	Shann	Newpo	Tralee	Killarne	Cork	All	Moon
1st	Sat	o	o	o	o	o	o	o	c	o	c	c	o	o	o	o	o	o	c	c	o	11	
2nd	Sun	pc	pc	pc	pc	pc	c	pc	c	pc	pc	pc	pc	pc	pc	F	pc	pc	pc	pc	pc	108	
3rd	Mon	pc	pc	pc	pc	pc	pc	pc	pc	pc	pc	pc	pc	pc	pc	pc	F	pc	F	F	pc	150	XhN
4th	Tues	pc	pc	pc	pc	pc	pc	pc	c	pc	pc	c	pc	pc	pc	c	F	pc	c	c	pc	92	
5th	Wed	c	c	c	c	c	c	c	c	o	c	c	c	o	o	o	c	c	c	c	o	29	F P1
6th	Thur	o	o	o	o	o	o	o	o	o	o	o	o	o	o	c	o	o	o	o	o	10	
7th	Fri	pc	pc	pc	pc	pc	pc	pc	c	pc	c	c	pc	pc	pc	pc	pc	pc	c	c	pc	89	
8th	Sat	pc	pc	pc	pc	pc	pc	pc	c	pc	c	c	pc	pc	pc	pc	c	c	c	c	c	73	^
9th	Sun	c	c	c	c	c	o	c	o	c	o	o	c	c	c	c	c	c	o	o	c	20	
10th	Mon	o	o	o	o	o	o	o	o	o	c	o	o	o	o	o	c	c	o	o	o	12	
11th	Tues	o	o	o	o	o	o	o	o	c	o	o	o	c	c	c	o	c	c	c	c	21	
12th	Wed	c	c	c	c	c	c	c	o	c	o	o	c	c	c	pc	o	c	o	o	o	41	3Q
13th	Thur	o	o	o	o	o	o	o	o	o	o	o	o	o	o	o	o	o	o	o	o	1	
14th	Fri	o	o	o	o	o	o	o	o	o	o	o	o	o	o	o	o	o	o	o	o	0	
15th	Sat	c	c	c	c	c	c	c	o	c	o	o	c	c	o	c	c	c	o	o	o	18	
16th	Sun	c	c	c	c	c	c	c	c	pc	c	c	c	pc	pc	pc	pc	pc	pc	pc	pc	84	XhS
17th	Mon	c	c	c	c	c	o	c	c	c	c	c	c	c	c	c	c	c	c	c	c	35	
18th	Tues	c	c	c	c	c	c	c	c	pc	c	c	c	pc	pc	c	c	c	c	c	pc	75	
19th	Wed	c	c	c	c	c	o	c	o	c	c	o	c	c	c	c	c	c	c	c	pc	34	
20th	Thur	c	c	c	c	c	c	c	c	c	c	c	c	c	c	c	o	c	o	o	c	28	N A
21st	Fri	c	c	c	c	c	o	c	o	c	c	o	c	c	c	pc	c	c	o	o	c	39	
22nd	Sat	c	c	c	c	c	o	c	o	pc	c	c	c	pc	pc	pc	c	pc	c	c	pc	61	V
23rd	Sun	o	o	o	o	o	o	o	o	o	o	o	o	o	c	o	o	o	c	c	c	12	
24th	Mon	o	o	o	o	o	o	o	o	o	c	o	o	o	c	o	c	c	o	o	c	16	
25th	Tues	o	o	o	o	o	o	o	c	o	o	c	o	o	o	o	o	o	o	o	o	5	
26th	Wed	o	o	o	o	o	o	o	o	o	o	o	o	o	o	o	o	o	o	o	o	3	
27th	Thur	o	o	o	o	o	o	o	o	o	o	o	o	o	o	o	o	o	o	o	o	1	
28th	Fri	o	o	o	o	o	c	o	o	o	o	o	o	o	c	o	c	c	o	o	c	15	1Q
29th	Sat	o	o	o	o	o	o	o	o	o	o	o	o	o	o	o	o	o	o	o	o	2	
30th	Sun	o	o	o	o	o	o	o	c	c	c	c	o	c	c	c	c	c	c	c	c	38	XhN
Estimate hours:		54	54	54	54	54	42	54	40	61	54	47	54	61	68	66	67	67	51	51	70	1122	
Average hours:		59	59	57	52	45	51	49	52	55	60	72	65	70	67	72	58	50	53	52	65	1161	
Trend:		av	av	av	av	av	av	av	less	av	av	less	less	av	av	av	av	more	av	av	av	less	

NOVEMBER **1** SATURDAY

A complex of depressions sit to the north and northwest of Ireland directing a very fresh southwest flow over the country, stronger around the coasts. Stormy air everywhere, with widespread cloud and rain most districts. Rain will be heavy in the south and west, while more isolated in the far northeast.

Belfast: Overcast, isolated showers, odd windy spell
Dublin: Gloomy and grey, light showers, mild temperatures, moderate winds
Cork: Overcast, light breezes, rain
Galway: Cloudy, rain, some windy spells

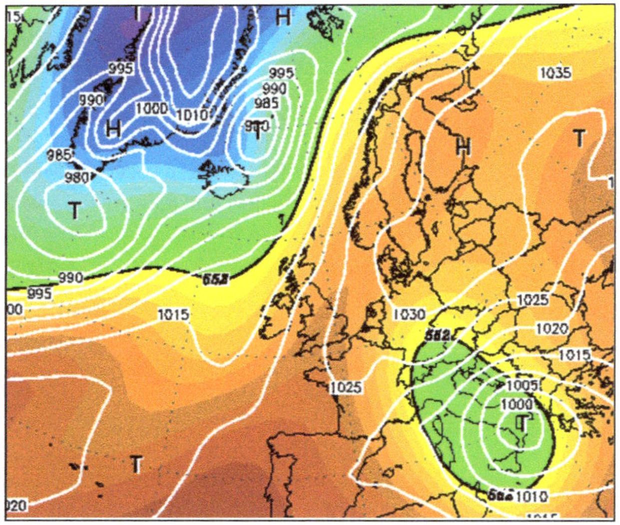

RAIN POTENTIAL

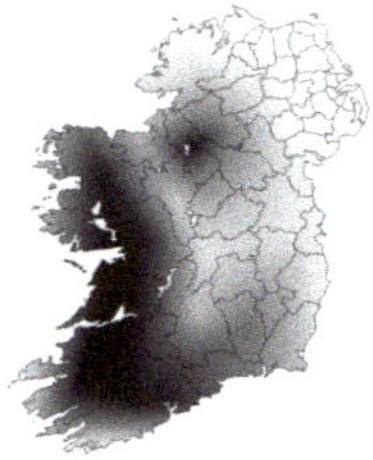

FROST/SNOW

NOVEMBER **2** SUNDAY

A trough associated with a depression to the northwest of Ireland drifts away to the north as a high pressure ridge moves onto the north of the country. Mostly sunny and dry throughout, except for some brief showers mostly overnight in the south and some misty patches in the west and south. Winds generally in a gentle southwest flow throughout later shifting to southeasterly.

Belfast: Partly cloudy, dry, moderate breezes
Dublin: Sunny, dry, mild day cold night, moderate breezes
Cork: Mostly sunny, brief showers, gentle breezes, chance of misty patches
Galway: Partly cloudy, dry, light breezes, chance of misty patches

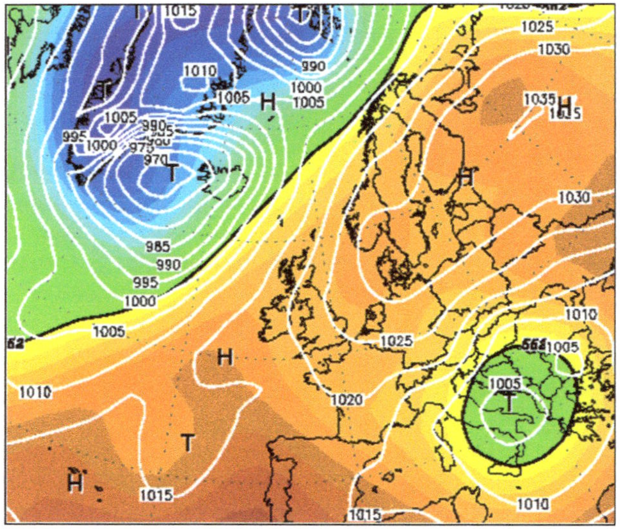

RAIN POTENTIAL

FROST/SNOW

NOVEMBER **3** MONDAY

A large high pressure system lingers over Ireland bringing a gentle southeast to southerly flow throughout. Overnight patches of frost and fog in numerous districts, with fog patches very dense in some parts. Mostly sunny and dry day after the fog lifts.

Belfast: Mostly sunny, dry, mild day cold night with chance of frost and misty patches
Dublin: Mostly sunny, dry, cooler, frost patches overnight, light shifting breezes
Cork: Partly cloudy, colder, dry, occasional breezy spells
Galway: Mostly sunny, dry, cooler, frost and fog patches likely, moderate breezes

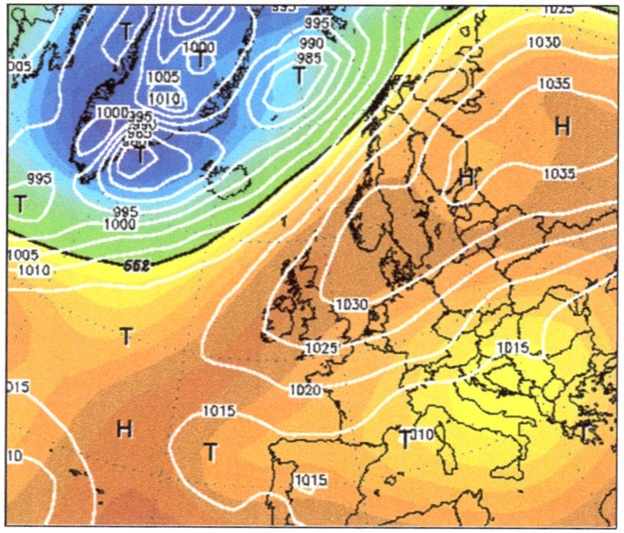

RAIN POTENTIAL

FROST/SNOW

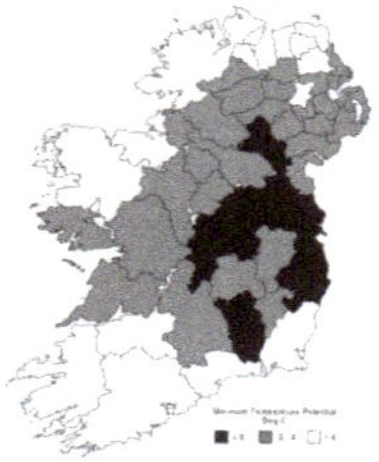

NOVEMBER **4** TUESDAY

High pressure continues to prevail over Ireland with overnight patches of frost and fog and gentle breezes of variable direction. Mostly sunny and dry in the west and south, and dry with increased cloudy spells and longer lingering misty patches in the north and east. Chance of some isolated showers around the southeast overnight.

Belfast: Light breezes to mostly calm air, sunny spells, dry, frost and fog patches
Dublin: Mostly cloudy, dry, serene air, chance of overnight frost and fog patches
Cork: Partly cloudy, dry, mostly calm and pleasant air, chance of fog overnight
Galway: Partly cloudy, dry and pleasant air, overnight frost and fog patches likely

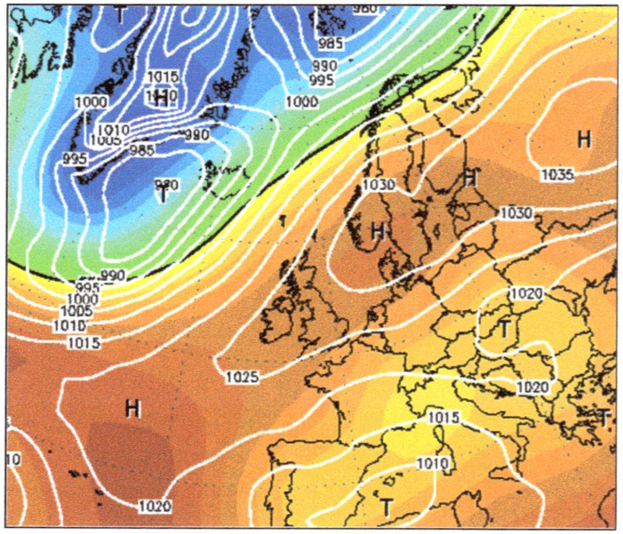

RAIN POTENTIAL

FROST/SNOW

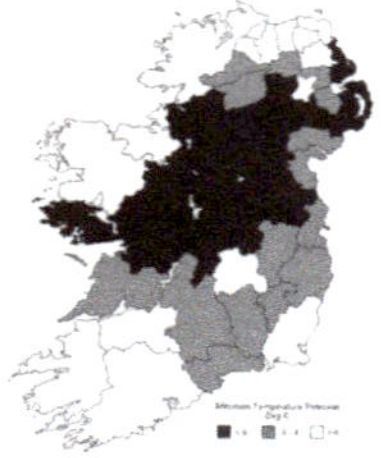

NOVEMBER **5** WEDNESDAY

An intense depression to the north of Ireland moves down onto the country bringing a strengthening south to southwest flow, with winds very gusty at times around the north and western coasts. Widespread cloud builds throughout the country bringing a dull and gloomy day to most dfistricts and after fog clears, outbreaks of brief showers and drizzle patches spread almost everywhere, before evening fog sets back in.

Belfast: Cloudy, cooler, brief showers, blustery winds with chance of gales at times
Dublin: Overcast, slightly milder, brief showers, strong winds at times
Cork: Overcast, warmer, brief showers and drizzle, freshening winds, chance of fog
Galway: Cloudy, mild day cold night, light showers developing

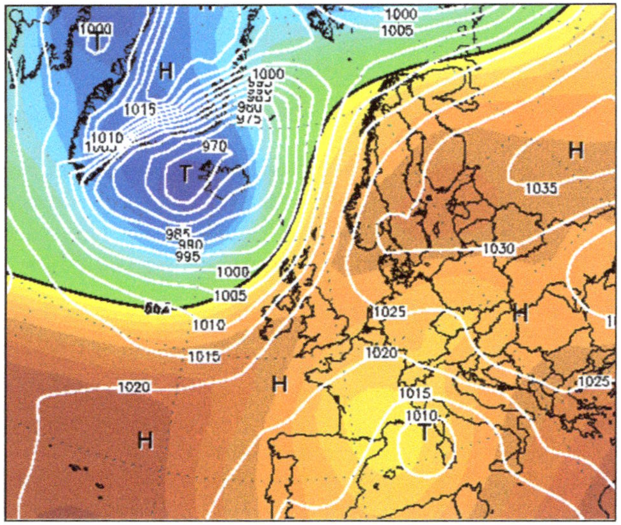

RAIN POTENTIAL

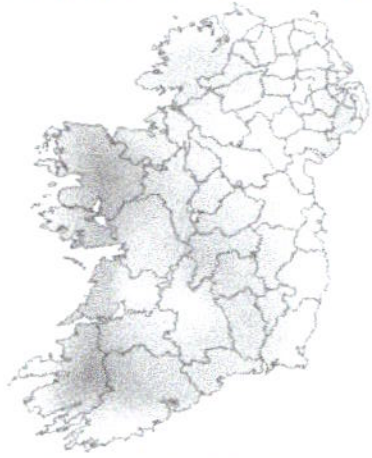

FROST/SNOW

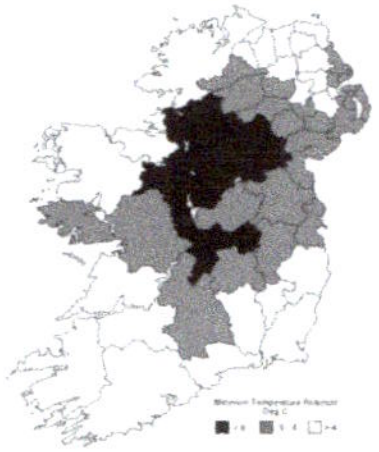

NOVEMBER **6** THURSDAY

A stormy air associated with a large depression north of Ireland prevails over the country with moderate to strong southwest winds prevailing, with some gale force gusts likely around the northern and western coasts for a time. Widespread cloud and bands of rain cross the country bringing a wet day everywhere, with heavy falls in particular in the north and west.

Belfast: Overcast, windy at times, rain
Dublin: Cloudy, warmer, dry, winds easing to moderate
Cork: Overcast, light showers, mild day cold night, moderate breezes, fog possible
Galway: Stormy, overcast, rain, chance of overnight fog, winds strengthening later

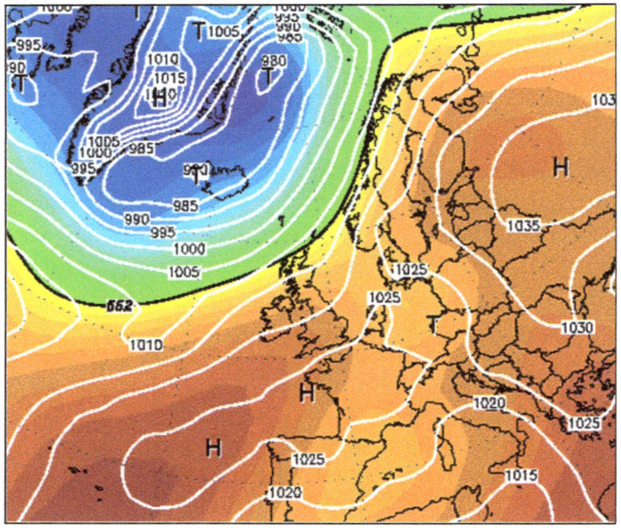

RAIN POTENTIAL

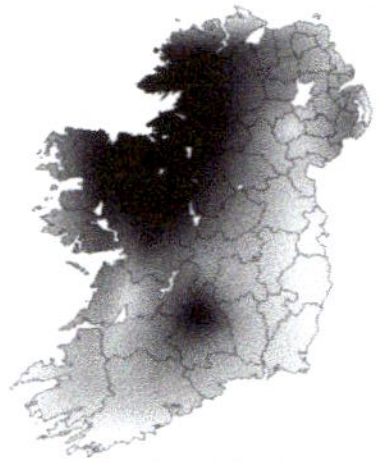

FROST/SNOW

NOVEMBER 7 FRIDAY

Stormy conditions continue throughout the country as a northern depression maintains a very strong and gusty westerly flow throughout. A changeable day with overnight rain in most districts, easing during the day and allowing some sunny breakthroughs. Chance of thunderstorms and winds to gale force at times mosty in northern parts.

Belfast: Stormy air, warmer, some sunny spells, brief squally showers
Dublin: Changeable, sunny spells and showers, unsettled winds, chance of gales
Cork: Changeable, mix of sunny spells and outbreaks of rain, stormy winds
Galway: Changeable, mx of showers and sunshine, stormy, gales and thunderstorms

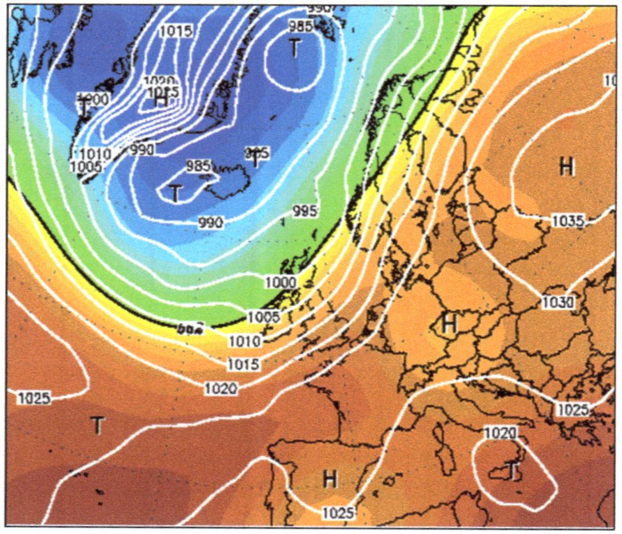

RAIN POTENTIAL

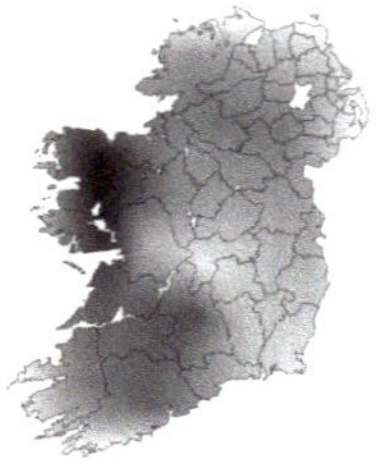

FROST/SNOW

NOVEMBER 8 SATURDAY

Low pressure troughs continue to bring fresh to strong west to northwest winds over the country, easing later in the day. A changeable day with a mix of sunny spells and bands of showers in most districts. Chance of thunderstorms and hail possible around the northern coasts.

Belfast: Partly cloudy, mostly dry with odd threat of rain, cooler, blustery
Dublin: Unsettled, blustery, changeable, some sunny spells, isolated showers, cooler
Cork: Cloudy, cooler, isolated showers, winds easing with fresh gusts at times
Galway: Cloudy, dry, mild day cold night, strong winds easing later in the day

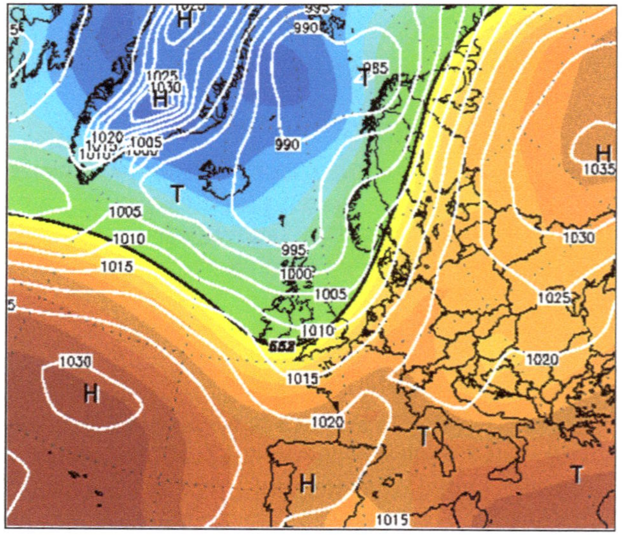

RAIN POTENTIAL

FROST/SNOW

NOVEMBER 9 SUNDAY

A high pressure ridge settles over Ireland directing a light to moderate westerly flow throughout, becoming stronger and gustier later in the day over the northern half of the country. Cloudy and mostly dry everywhere, except for the chance of some isolated drizzle patches in the north and west. Chance of overnight fog patches briefly in some sheltered northern areas.

Belfast: Cooler, unsettled winds, cloudy, mostly dry with odd drizzle patch possible
Dublin: Cloudy, dry, windy at times
Cork: Cloudy, dry, slightly warmer, moderate shifting breezes
Galway: Overcast, isolated showers, fresh and blustery winds at times

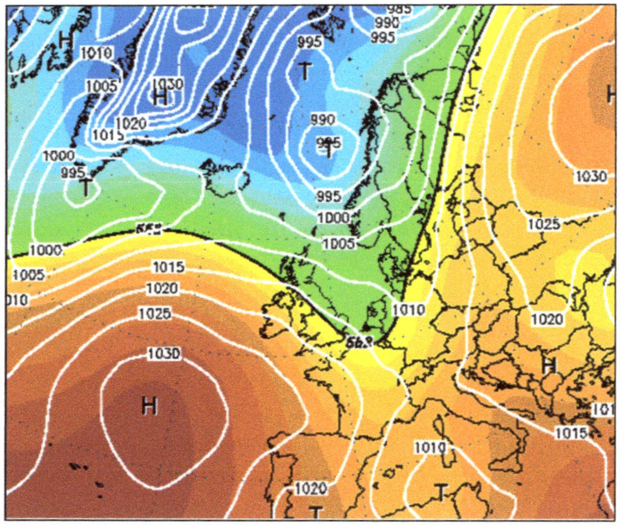

RAIN POTENTIAL

FROST/SNOW

NOVEMBER 10 MONDAY

Low pressure to the north of Ireland intensifies the winds along the edge of the high pressure ridge over Ireland, with west to southwest winds prevailing and up to gale force at times over the northern half of the country. Dull and gloomy throughout and although mostly dry, rain is likely to threaten in a number of districts, and chance of some possible drizzly spells building in the northwest.

Belfast: Overcast and dull, milder temperatures, dry with odd threat of rain, blustery
Dublin: Overcast, mild temperatures, dry, threats of rain, strong winds with gusts
Cork: Overcast and dull, mostly dry with odd threat of rain, moderate breezes
Galway: Cloudy, dry, mild temperatures, strong winds to near gales likely

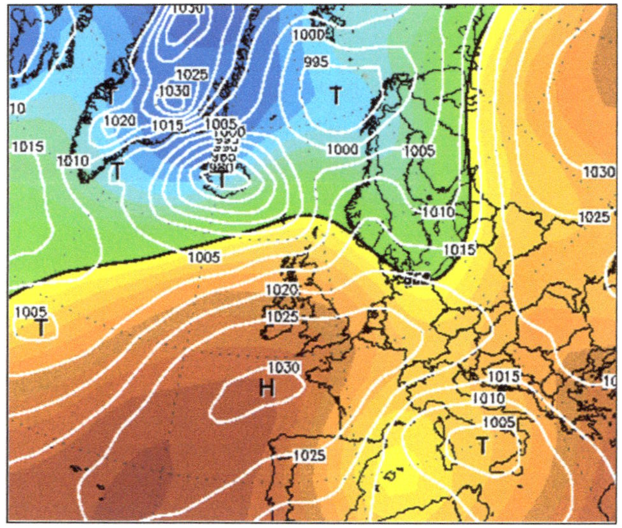

RAIN POTENTIAL

FROST/SNOW

NOVEMBER 11 TUESDAY

The high pressure ridge maintains a southwest flow over the country, with fresh winds prevailing, gusty at times especially in northern parts and along the western coastal fringes. Mostly cloudy with some drizzle patches or spotty showers, mostly in the north and west, while mostly dry with brief sunny breakthroughs elsewhere. Warmer than normal.

Belfast: Strong winds to near gales, overcast, warmer, dry with threats of rain
Dublin: Cloudy, dry, strong and unsettled winds
Cork: Cloudy, dry, mild day cool night, windy for a time, patches of fog likely
Galway: Overcast, dry, warmer, strong and unsettled winds, gales possible

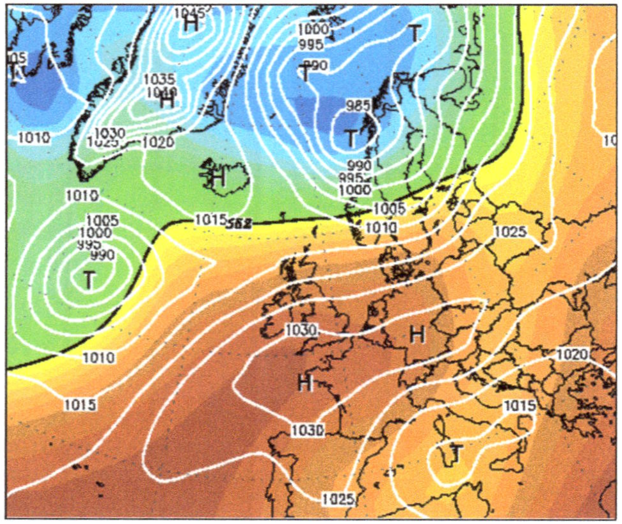

RAIN POTENTIAL

FROST/SNOW

NOVEMBER 12 WEDNESDAY

A large anticylonic system continues to linger over Ireland directing a light southwest to westerly flow with chance of patches of fog overnight, particularly in some northern and southern reaches. Some passing sunny spells in the north and east, while dull and very gloomy in the west and south. Mostly dry day, while some drizzle patches likely overnight in the northwest, and possibly around northern parts of the Munster province.

Belfast: Winds easing to light, cloudy, mild temperatures, dry, chance of fog
Dublin: Partly cloudy, dry, slightly cooler, light shifting breezes
Cork: Overcast, cooler, dry, gentle shifting breezes, fog likely
Galway: Overcast, dry, mild temperatures, winds easing to light

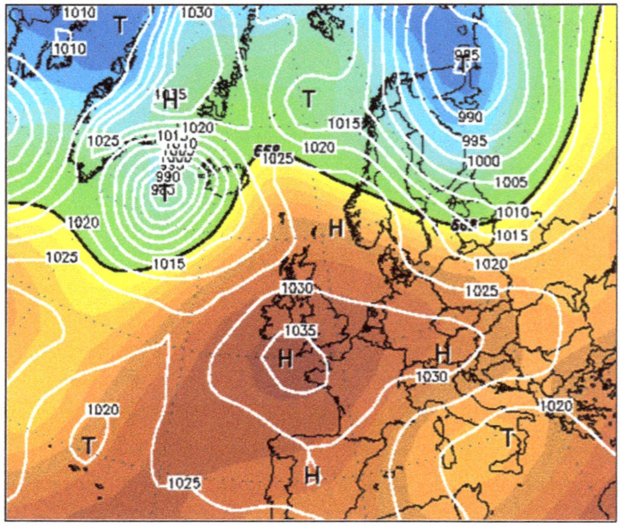

RAIN POTENTIAL

FROST/SNOW

NOVEMBER **13** THURSDAY

Light, variable breezes to a mostly calm air prevails under the influence of a large high pressure system that remains almost stalled over Ireland. Mostly dry everywhere with widespread cloud prevailing, and the likelihood of overnight fog mostly in the north and east. Chance of some isolated drizzle patches in the west, briefly spreading to the east for a time. Some patches of frost overnight.

Belfast: Calm atmosphere, dry, overcast, mild temperatures, fog likely
Dublin: Overcast and dull, patches of light drizzle, mostly calm air
Cork: Overcast and dull, dry, gentle breezes to mostly calm air, fog possible
Galway: Mostly calm and serene air, widespread cloud, dry

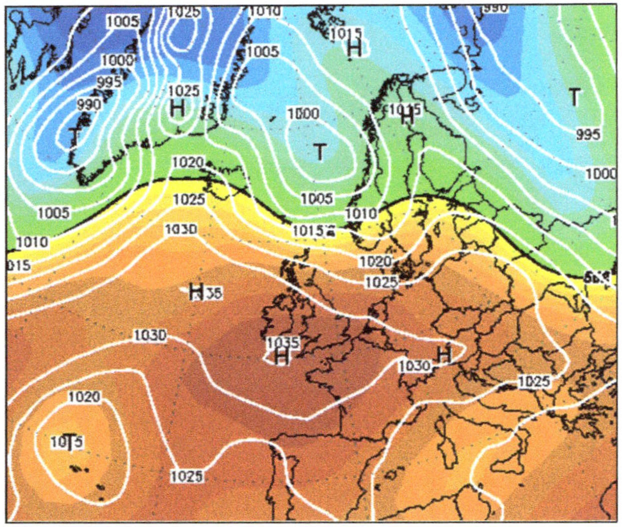

RAIN POTENTIAL

FROST/SNOW

NOVEMBER **14** FRIDAY

High pressure continues to dominate the country with light variable breezes to a mostly calm air at first, but a little more fresher with a southwest change. Dull and gloomy everywhere with brief drizzle patches mostly in the west to inland parts, while remaining mostly dry elsewhere. Chance of fog patches mostly in the west to central districts.

Belfast: Overcast and dull, dry, gentle shifting breezes
Dublin: Overcast, brief drizzly showers, gentle shifting breezes
Cork: Overcast, light shifting breezes, dry, cooler
Galway: Overcast, brief showers, moderate shifting breezes, chance of fog patches

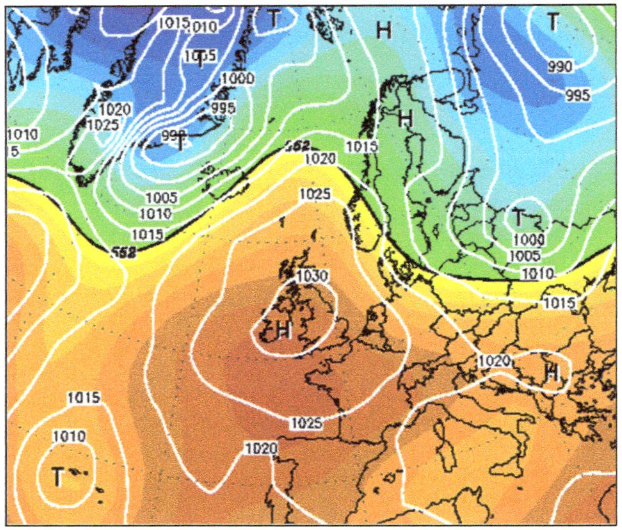

RAIN POTENTIAL

FROST/SNOW

NOVEMBER **15** SATURDAY

High pressure fills as it lingers over Ireland bringing another day with light variable breezes to a mostly calm air, with some fresher breezes mostly along the northern coasts in a west to northwest flow. Mostly cloudy to overcast throughout. A band of light showers crosses the country from the northwest at first and moving southwards bringing some isolated showers into most districts, although mostly dry in the east and southeast. Inland frost pockets likely.

Belfast: Cloudy, mostly dry with odd threat of rain, odd windy spell
Dublin: Calm and serene air, cold, cloudy, isolated drizzly showers
Cork: Slightly warmer, widespread cloud, brief outbreaks of drizzle, calm air
Galway: Overcast, dry, odd threats of rain, calm and serene air, chance of drizzle

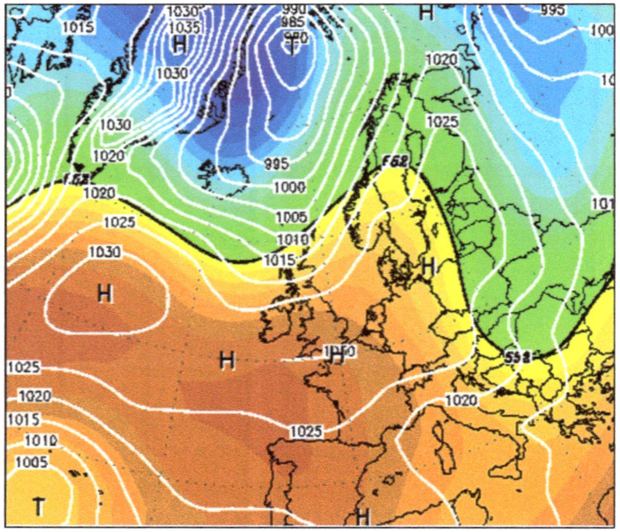

RAIN POTENTIAL

FROST/SNOW

NOVEMBER **16** SUNDAY

The large anticyclonic system slowly drifts to the west of Ireland but maintains its ridge over the country, while depressions intensify to the north of the country. Mostly dry with some sunny spells throughout, and the chance of patches of fog mostly in the west. Light west to northwest winds prevail, with the occasional gust around the northern coasts. Some threats of rain overnight.

Belfast: Mostly cloudy and dry, mild day cold night, freshening winds
Dublin: Gentle breezes, dry, occasional sunny spells, cool
Cork: Dry, cool, gentle onshore breezes to mostly calm air, some sunny spells
Galway: Light breezes, cloudy, isolated showers, cold overnight, shallow fog likely

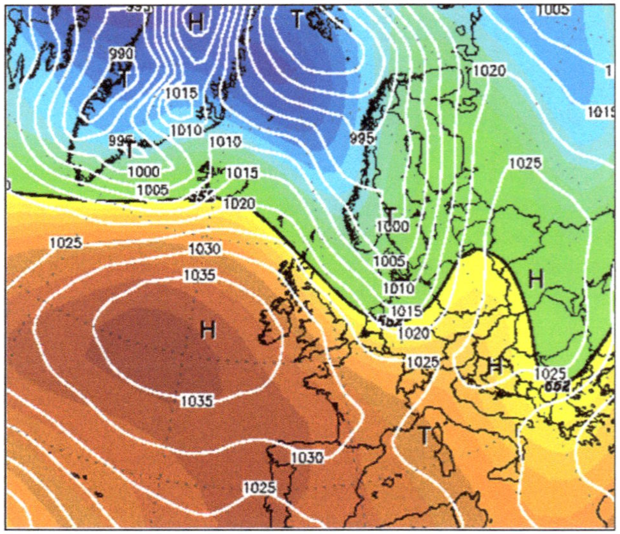

RAIN POTENTIAL

FROST/SNOW

NOVEMBER **17** MONDAY

A depression in the North Atlantic north of Ireland moves down onto the northern half of ireland, whie the high pressure ridge drifts to the south of the country. Cloudy everywhere, with strong west to northwest winds over the northern half of Ireland, and lighter winds to the south. A changeable day with a mix of isolated showers in most districts, and brief sunny breakthroughs.

Belfast: Cloudy, dry, strong blustery winds, possible gales at times, odd threat of rain
Dublin: Cloudy and cool, dry, strong northwest winds
Cork: Cloudy, isolated showers, fresh breezes
Galway: Cloudy, freshening winds, occasional gusts, light showers, warmer

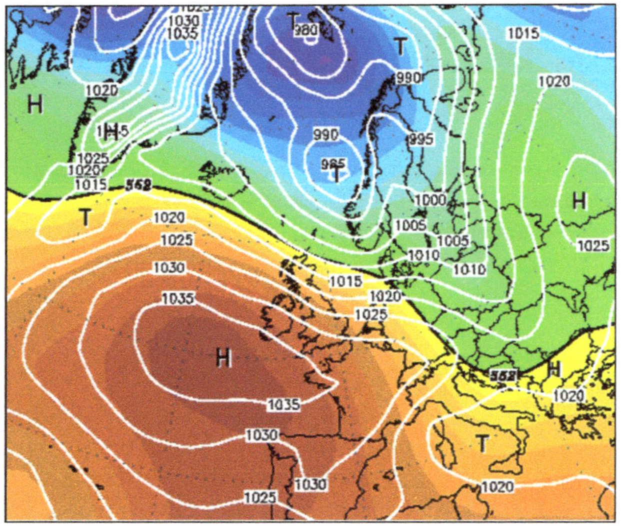

RAIN POTENTIAL

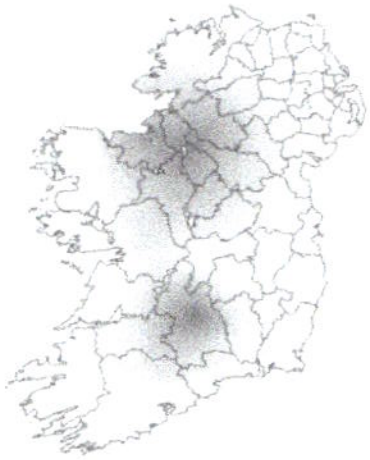

FROST/SNOW

NOVEMBER **18** TUESDAY

A depression to the north of Ireland directs a strong and blustery west to northwest flow over most of the country, while a high pressure ridge maintains a lighter northwest flow over southern parts. Mostly cloudy throughout. A band of rain crosses the northern coasts and treks southeasterly bringing some showers to most of the country, clearing later in the day. Colder everywhere with chance of overnight patches of frost. Chance of gale gusts developing around the northern coasts.

Belfast: Mostly cloudy, light showers, strong and gusty winds at times
Dublin: Cloudy, light showers, unsettled and changeable winds
Cork: Partly cloudy, light showers, mild day cold night, moderate northwesterlies
Galway: Unsettled winds for a time, cloudy, isolated showers, slightly cooler

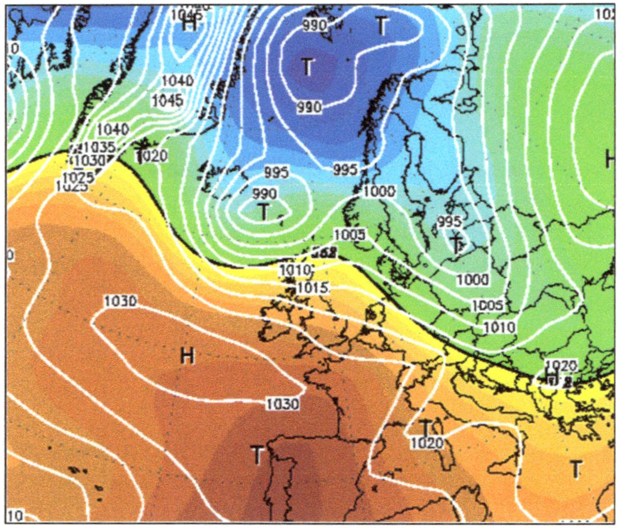

RAIN POTENTIAL

FROST/SNOW

NOVEMBER 19 WEDNESDAY

A large high strengthens to the west of Ireland while an intense depression stretches down to the east of the country, directing a moderate west to northwest flow through the western half of the country and stronger west to northwest winds to the east. Some isolated overnight showers likely in the far north. Mostly dry and cloudy day throughout except for some longer sunny spells in the south.

Belfast: Unsettled and blustery winds at times, cloudy, dry, cooler
Dublin: Cloudy, dry, cold northwesterly winds
Cork: Partly cloudy, cooler, dry, fresh to strong northwest winds
Galway: Cloudy, dry, mild day cold night, strong winds at times

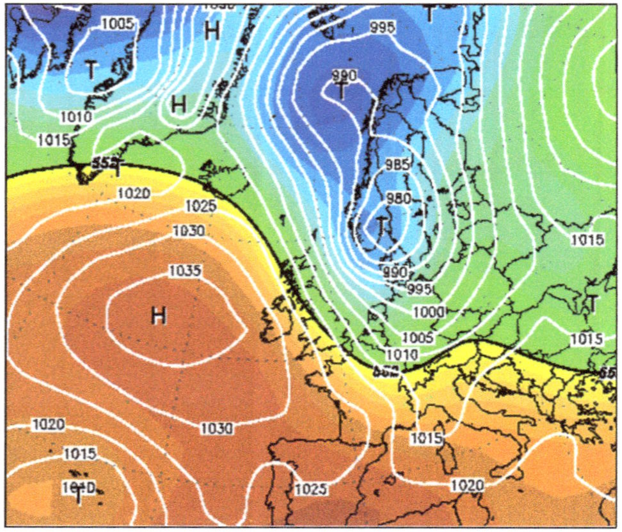

RAIN POTENTIAL

FROST/SNOW

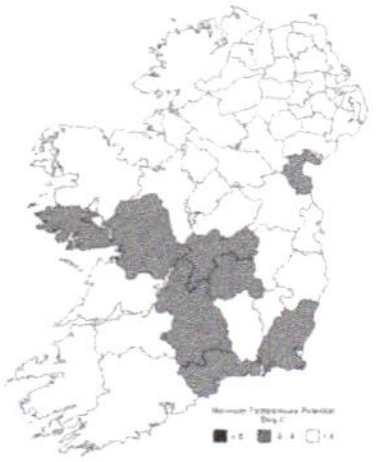

NOVEMBER 20 THURSDAY

The high pressure ridge moves back up onto Ireland directing a light to moderate west to northwest flow, fresher at times in the north. Cloudy everywhere, with some isolated showers in the north to the midlands, while remaining mostly dry with the occasional threat of rain elsewhere. Sunniest in the southeast.

Belfast: Unsettled and windy, cloudy, brief showers
Dublin: Cloudy, dry, milder temperatures, fresh and cold winds
Cork: Cloudy, dry, fresh windy spells
Galway: Cloudy, isolated showers, mild temperatures, strong winds to near gales

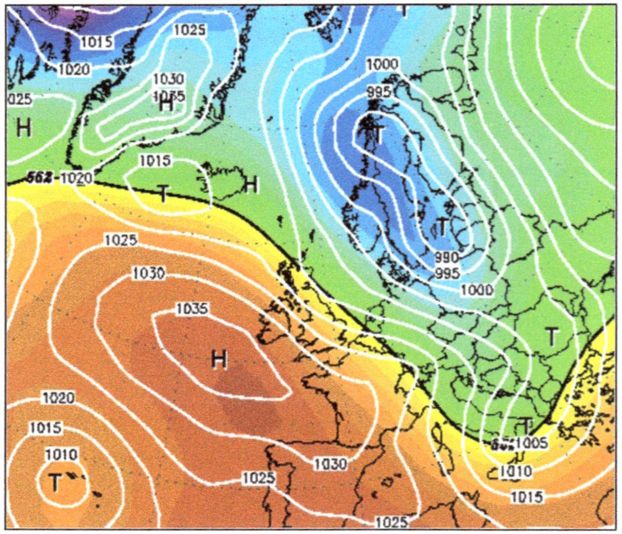

RAIN POTENTIAL

FROST/SNOW

NOVEMBER 21 FRIDAY

The high pressure ridge contracts to the southeast of the country while low pressure pushes near the northeast. Moderate to fresh winds prevail over the northern half of Ireland, while lighter to the south, in a northwest flow. Mostly dry at first, with some patchy light rain developing for a time, before clearing and allowing for some sunny breakthroughs. Chance of thunderstorms and light snow in the west overnight.

Belfast: Cloudy, dry, mild temperatures, breezy at times
Dublin: Strong winds, partly cloudy, colder, overnight frosts, isolated showers
Cork: Cloudy and cool, dry with passing threats of rain, strong winds at times
Galway: Cloudy, dry, strong winds, chance of thunderstorms and overnight snow

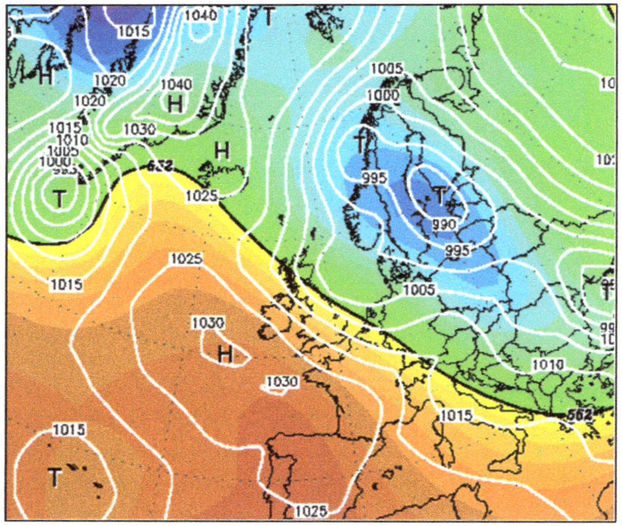

RAIN POTENTIAL

FROST/SNOW

NOVEMBER 22 SATURDAY

A weakening anticyclonic system moves back up over the country bringing a very light air of variable direction throughout. Partly cloudy in the east and south, while mostly cloudy in the north and west. Some light showers and drizzle patches over the western half of the country, and mostly dry in the coastal fringes of the east and south. Chance of overnight thunderstorms and possible light snowfall in the west. Frosty air.

Belfast: Cloudy, mostly calm air and cooler, isolated showers, frost and fog patches
Dublin: Mostly calm, partly cloudy, frosts, isolated wintry showers, chance of snow
Cork: Calm and pleasant air, dry, occasional sunny spells, colder
Galway: Mostly calm air, frosts, cloudy, overnight light snow and thunderstorms

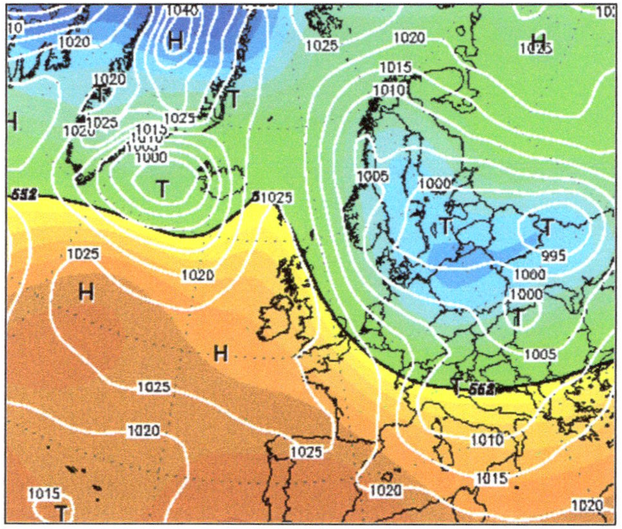

RAIN POTENTIAL

FROST/SNOW

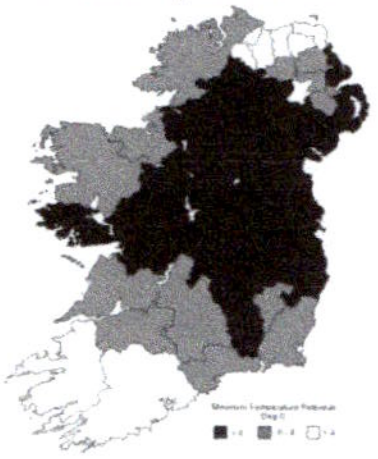

NOVEMBER **23** SUNDAY

Dull and gloomy everywhere with a low pressure trough to the northeast, and the high pressure ridge remaining fairly stable to the west. Mostly overcast everywhere with some patches of drizzle or rain mostly in the north and more isolated into central parts to the southeast. Dry with passing threats of rain elsewhere. Winds generally in a northwest flow, light in most districts, but fresher with gusts in the north.

Belfast: Overcast, light showers, cold, overnight frosts, moderate breezes
Dublin: Overcast, milder, dry with increasing threats of rain, moderate breezes
Cork: Cloudy, breezy, light showers
Galway: Overcast, dry, moderate breezes

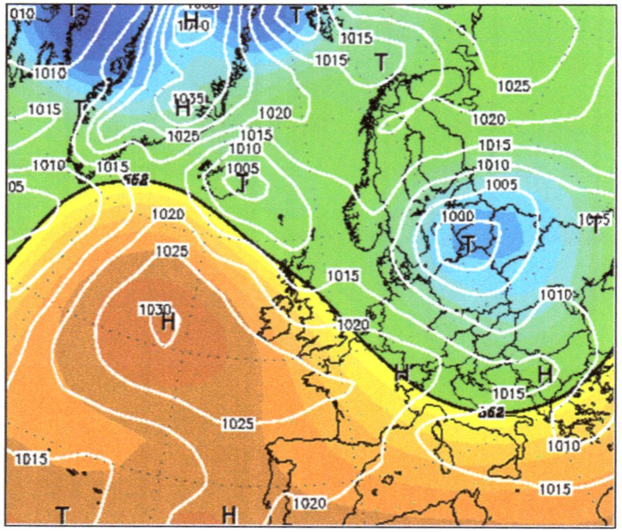

RAIN POTENTIAL

FROST/SNOW

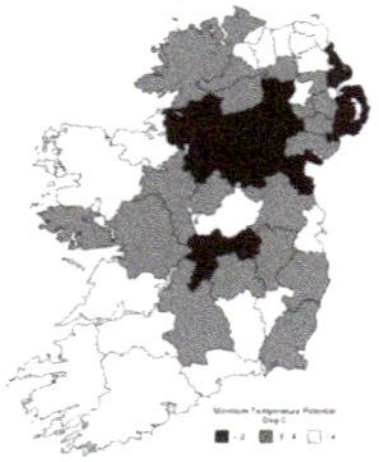

NOVEMBER **24** MONDAY

Fresh to strong north to northwest winds prevail over Ireland with high pressure to the west and low pressure to the east. Dull and gloomy in the north and east with light showers, while mostly dry with some brief sunny breakthroughs in the west and south. Chance of some thunderstorms and possible light snow for a time mostly in the west.

Belfast: Overcast, milder, light showers, possible drizzle and shallow fog
Dublin: Overcast, scattered showers, light winds, chance of shallow fog overnight
Cork: Cloudy, dry, moderate northerly breezes
Galway: Cloudy, isolated showers, warmer, chance of thunderstorms and snow

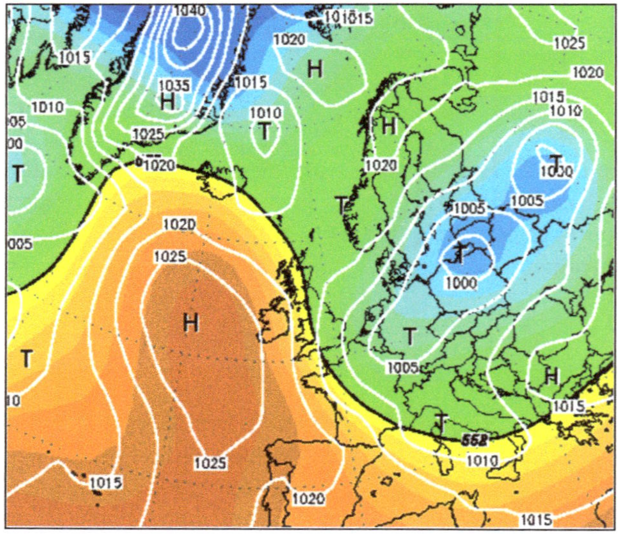

RAIN POTENTIAL

FROST/SNOW

NOVEMBER **25** TUESDAY

A anticyclonic system decays as it lingers over Ireland bringing light generally southerly breezes to most districts. Overcast everywhere with increasing threats of rain, overnight patches of mist and fog in many districts, with some fog patches lasting well into the late morning and shifting to light showers in many districts.

Belfast: Overcast, mostly dry, passing threats of rain, milder, pockets of fog
Dublin: Overcast and dull, isolated showers, gentle shifting breezes
Cork: Overcast, light showers, cooler, shifting breezes, chance of fog overnight
Galway: Overcast, fluctuating breezes, mostly dry with threats of rain, drizzly spells

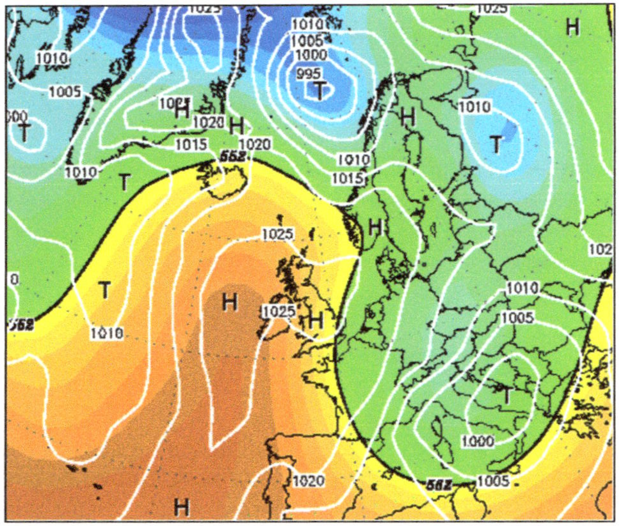

RAIN POTENTIAL

FROST/SNOW

NOVEMBER **26** WEDNESDAY

The high pressure system that had remained almost stalled to the west of Ireland now crosses the country to the east allowing for a weak depression to slip to the southwest of Ireland. A light south to southwest flow prevails, with widespread cloud everywhere. Some spotty showers in the west to inland parts for a time, while mostly dry with increasing threats of rain elsewhere. Chance of fog patches in the north.

Belfast: Overcast, cooler, dry, odd threat of rain, gentle breezes, chance of fog
Dublin: Gentle southerly breezes, overcast, dry, milder temperatures
Cork: Dull and gloomy, mostly dry, possible light drizzle, gentle southerly breezes
Galway: Overcast, light showers, mild temperatures, breezy at times

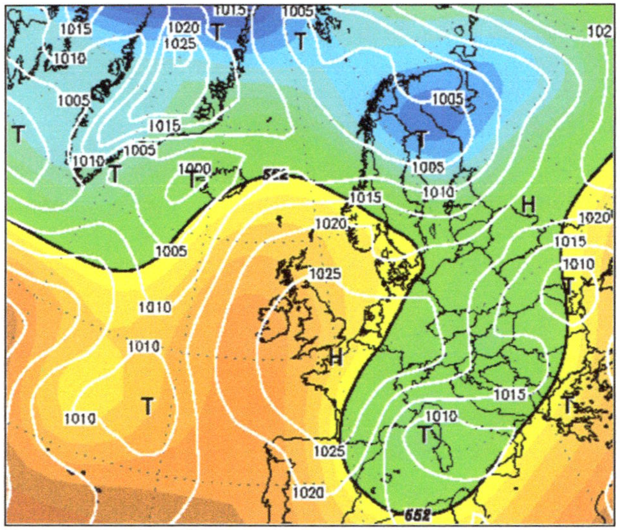

RAIN POTENTIAL

FROST/SNOW

NOVEMBER 27 THURSDAY

Low pressure brings a fresh to strong south to southwest flow over Ireland with a rain embedded trough moving onto the country. Rain spreads from the northwest at first and down into the south during the day bringing wet conditions everywhere. Rain likely to be very heavy in many parts of the country, especially in the west to the south.

Belfast: Overcast, rain, freshening winds for a time
Dublin: Overcast, rain, breezy spells
Cork: Overcast, heavy rain, cooler, moderate breezes
Galway: Overcast, outbreaks of rain, mild temperatures, windy

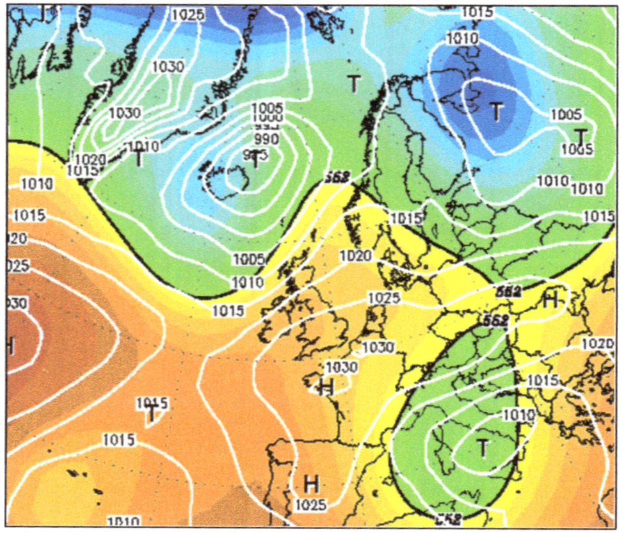

RAIN POTENTIAL

FROST/SNOW

NOVEMBER 28 FRIDAY

An intense North Sea depression moves down through the country bringing very cold temperatures but light variable winds throughout, with some gusty spells briefly around the coasts. Mostly overcast with wintry showers in all districts, although chance of mostly dry around the southeast coastal fringes. Heavy rain in pockets in the northeast, west and south, and snow likely in the west and north.

Belfast: Overcast, wintry showers, chance of fog and possible overnight snow
Dublin: Overcast, colder, rain, light to moderate breezes, chance of misty spells
Cork: Cloudy and colder, isolated showers, some breezy spells
Galway: Overcast, colder, light wintry showers, breezy spells, chance of snow flurries

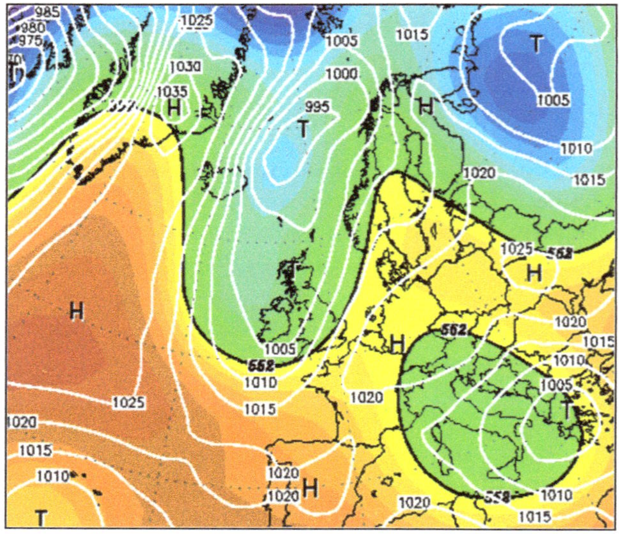

RAIN POTENTIAL

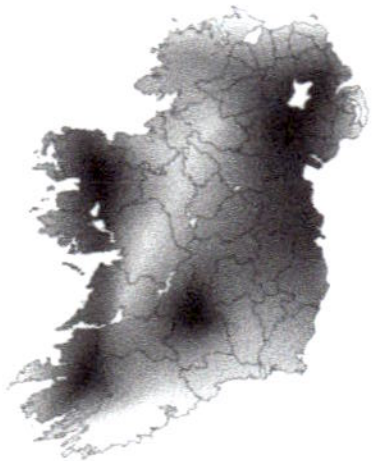

FROST/SNOW

NOVEMBER 29 SATURDAY

A depression consolidates over Ireland with winds strengthening to strong with gale gusts at times, particularly in the north and west, of variable directions. Overcast and dull with scattered wintry showers throughout, as well as the likelihood of thunderstorms mostly in the west, and some heavy falls still likely in the far northeast. Chance of overnight snow mostly in the north and west.

Belfast: Stormy, overcast, heavy rain, colder, gale winds and overnight snow possible
Dublin: Overcast, stormy, cold, rain, fluctuating winds with gusts, chance of gales
Cork: Colder, stormy, overcast, squally showers, fluctuating and unsettled winds
Galway: Overcast, cold, some showers, strong and blustery winds, gales for a time

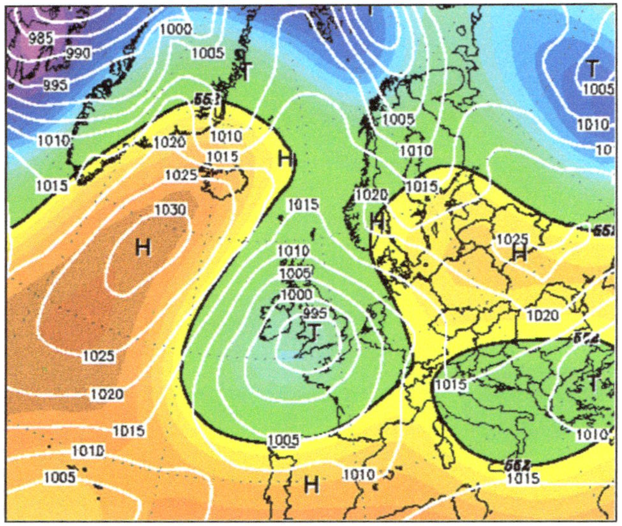

RAIN POTENTIAL

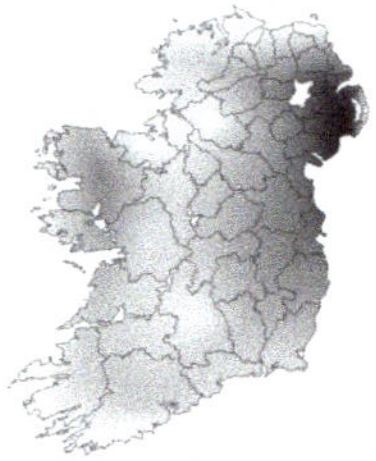

FROST/SNOW

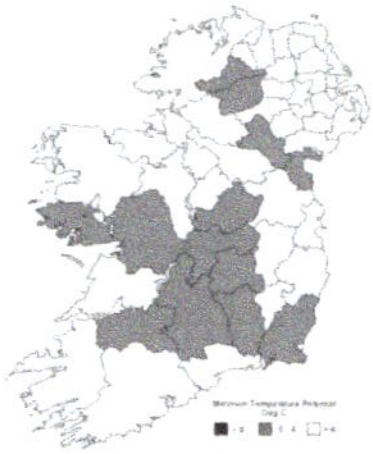

NOVEMBER 30 SUNDAY

Low pressure slips southwards to be centred to the south of Ireland directing fresh to strong east to southeast winds throughout. Cloudy everywhere with patches of frost in inland parts overnight. Showers contract to the northeast and inland districts, while mostly dry around the eastern and western coastal areas.

Belfast: Cold and overcast, occasional showers, strong winds
Dublin: Cloudy, cold, mostly dry, threats of rain, unsettled and changeable winds
Cork: Cloudy, slightly warmer, brief showers, unsettled and changeable winds
Galway: Mostly cloudy, light showers, cold, unsettled and blustery winds, gales likely

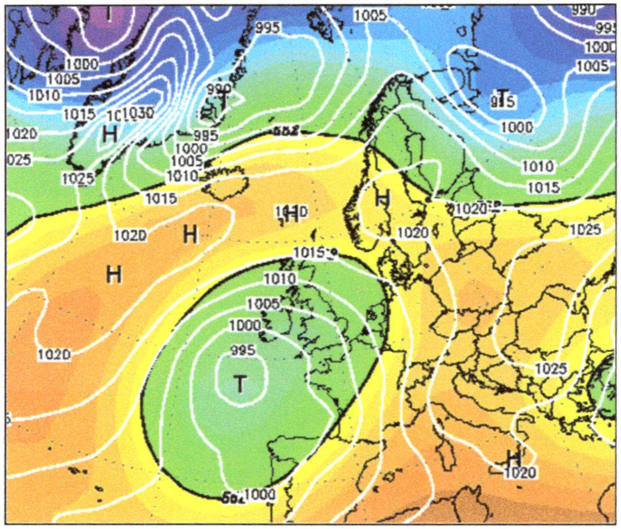

RAIN POTENTIAL

FROST/SNOW

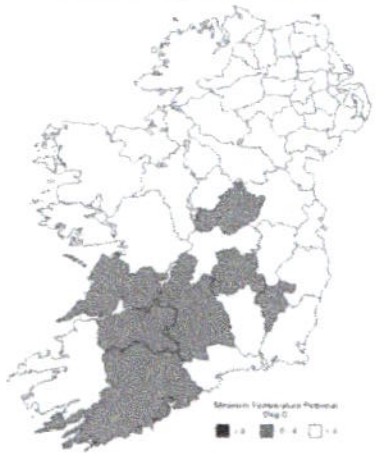

DECEMBER

PHASES OF THE MOON

4th	Full Moon
4th	Perigee #2
5th	Northern Declination
11th	Third Quarter
13th	Crossing Equator
17th	Apogee
19th	Southern Declination
20th	New Moon
27th	First Quarter
27th	Crossing Equator

1	2	3	4	5	6	7	8	9	10	11	12	13	14	15	16	17	18	19	20	21	22	23	24	25	26	27	28	29	30	31
Sun	Mon	Tue	Wed	Thur	Fri	Sat	Sun	Mon	Tue	Wed	Thur	Fri	Sat	Sun	Mon	Tue	Wed	Thur	Fri	Sat	Sun	Mon	Tue	Wed	Thur	Fri	Sat	Sun	Mon	Tue
			F P2	^						3Q		XhS				A		V	N							1Q XhN				

MONTHLY SUMMARY

1st-4th widespread frosts, severe in central parts

1st-7th High pressure returns bringing a return to mostly dry and settled weather throughout the country, although a brief showery spells may interrupt the flow midway through the week. Warmer than normal on most days.

2nd-4th A brief cold snap and some wintry showers including hail possible, with the chance of light snow in the west. Cold northerly winds prevail.

5th-7th Milder spell with no frosts. Widespread patches of overnight fog likely.

8th-10th Wet spell with unsettled and blustery winds and heavy rain at times. It may be especially wet and windy in the north and northwest. Winds generally southerly at first but later back westerly as they strengthen.

10th A depression may bring a stormy air across the country. Chance of thunderstorms and wintry rain bringing snow grains or hail in central districts.

Over the next four weeks, conditions may be milder than is typical for this time of year, especially in the southwest of the country, as well as wetter but also sunnier than normal. Depressions centred to the northwest of Ireland will bring an unsettled start to this outlook period and over the first couple of weeks and again in the last week, with a high pressure interval and milder weather early in the new year. Rainfall amounts overall will vary greatly across the country, but average out overall, with very wet weather in the western half of the country, but below average rainfall in the east, especially drier than average along the eastern and southeastern coasts. Much of the middle couple of weeks of this outlook period, especially over the third week, will be dry with very few areas seeing any precipitation of any significance while some rainfall can be expected on every other day. The next four weeks will also be sunnier than normal in most parts of the country, with eastern parts possibly receiving up to twice as much sunshine hours than the north and west. Overall, both maximum and minimum temperatures will be unusually milder than is normal for this time of year, especially in the first week ahead, but minimum temperatures will be colder than normal in the third week, when cold and mostly dry weather brings widespread ground frosts almost everywhere except around the coastal fringes. Many coastal areas may not see any frosts over the next four weeks, while inland parts may see up to 10-12 frost days. Winds will be stronger than normal for this time of year overall, with strong winds prevailing over the first week, before easing to moderate to mostly calm conditions for the next couple of weeks, then returning to blustery at times in the last week. During this outlook period, the is likely to be around 15-20 rain days, which is about average for this time of year, but with more wet days in the west and less in the east. Thunderstorms and hail can be expected on 5 days, fogs on 4 days, frosts on 6-10 days mostly in inland districts, snow on 1 day, and winds up to gale force at times on 11 days.

11th-19th A series of deep depressions to the west of the country pass close to the northwest, bringing a series of frontal systems across Ireland with some rain or showers daily. Some falls will be heavy falls at times and possibly accompanied by thunderstorms, particularly in the western half of the country where localised pockets of flooding may be noted. Strong, mostly southwest winds will prevail during this spell with gales likely at times along the western and northern coasts, possibly generating huge sea swells. Despite the stormy air, temperatures will be well above normal with no air frosts likely.

11th-12th Very heavy rain likely, mostly in the west. Chance of thunderstorms and hail.

11th-13th Chance of ground frosts in sheltered parts in the east.

14th-15th Chance of thunderstorms and wintry showers of snow grains in central areas. Very strong winds particularly around the western coasts.

16th-18th Scattered outbreaks of heavy rain occasionally. Possible wintry precipitation of hail or snow likely. Chance of thunderstorms.

17th-19th An intense depression brings a stormy air with chance of scattered thunderstorms, hail showers, and gale force winds.

17th- 21st Frosts in many parts, severe patches in the east.

20th-31st An anticyclonic system south of Ireland slowly moves northeastwards up through the country in a light, mostly southeasterly flow, to become centred in the North Sea during this spell. Mostly dry with very few areas receiving any noticeable rain activity except for some light rain at first in western parts, and a brief outbreak of heavy rain in the far southwest early in the transit. Colder than average, with ground frosts becoming widespread, and severe at times in inland and eastern parts.

22nd-23rd Milder with strong southerly winds and chance of heavy rain briefly in the extreme southwest.

23rd-31st Mostly dry spell throughout the country.

24th-31st Widespread frosts overnight, severe at times in some eastern and inland parts

27th-31st Chance of frosts being severe especially in the midlands to the eastern regions.

IRELAND RAINFALL ESTIMATES

December 2025

d=mainly dry, n= nonrecordable, l=light shrs, s=significant shrs, r=rain, h=hvy falls

		ULSTER PROVINCE					CONNAUGHT PROVINCE						LEINSTER PROVINCE				MUNSTER PROVINCE						
		Alderg	Hillsbo	Armag	Loughl	Malin H	Sligo	Ardtarı	Belmul	Drums	Galway	Roscoı	Markre	Edende	Kilkenı	Dublin	Shannı	Newpo	Tralee	Killarne	Cork	All	Moon
1st	Mon	l	d	l	l	l	l	l	s	d	d	d	n	d	d	l	d	s	d	l	d	24	
2nd	Tues	l	l	l	l	l	l	l	s	d	n	d	d	d	d	d	d	l	l	l	l	21	
3rd	Wed	l	l	d	l	l	l	l	l	l	s	s	s	l	d	d	d	r	d	l	d	47	
4th	Thur	l	l	l	l	s	l	l	s	l	d	l	l	n	d	n	l	l	s	d	l	31	F P2
5th	Fri	d	n	d	l	l	l	l	d	n	d	n	l	n	d	d	d	l	l	l	d	17	^
6th	Sat	d	n	d	d	d	l	d	l	n	l	l	n	l	d	d	n	l	l	d	n	9	
7th	Sun	l	s	l	h	l	h	h	l	r	h	r	h	s	s	d	l	h	l	r	l	209	
8th	Mon	l	l	s	l	r	l	l	r	s	l	l	l	l	l	l	s	l	h	s	r	95	
9th	Tues	l	l	n	r	s	r	r	l	s	h	r	s	l	l	l	l	r	l	l	l	112	
10th	Wed	s	r	l	s	h	s	s	r	s	r	s	s	r	l	s	r	s	s	r	r	167	
11th	Thur	s	r	l	s	r	s	s	s	r	h	r	r	h	s	l	h	h	s	h	s	234	3Q
12th	Fri	l	l	s	r	s	s	r	r	r	l	s	s	l	l	s	r	r	h	r	r	157	
13th	Sat	l	l	l	l	l	l	l	s	l	l	l	r	n	d	d	l	s	l	l	n	48	XhS
14th	Sun	s	l	s	s	l	s	s	s	l	l	r	l	l	s	l	l	l	s	h	r	109	
15th	Mon	l	l	l	s	s	s	s	s	s	r	r	l	l	l	d	s	r	s	r	l	109	
16th	Tues	s	r	s	h	s	h	h	h	r	h	r	h	r	r	r	h	h	r	h	r	370	
17th	Wed	h	h	r	r	l	h	r	s	h	h	h	h	h	r	l	s	h	l	h	s	301	A
18th	Thur	r	h	s	r	h	h	r	h	h	s	h	s	r	l	r	h	h	h	h	h	322	
19th	Fri	s	l	s	s	l	s	s	s	s	n	l	l	l	l	l	s	s	s	r	r	90	V
20th	Sat	l	d	l	n	d	l	n	n	d	d	n	d	d	d	d	d	d	l	l	d	8	N
21st	Sun	d	d	d	n	d	s	n	d	d	n	n	l	n	d	d	d	n	n	r	d	19	
22nd	Mon	d	d	d	l	d	l	l	s	d	d	n	d	d	d	d	d	d	h	l	s	47	
23rd	Tues	d	d	d	l	l	l	l	l	d	l	d	l	d	d	d	d	l	l	d	d	11	
24th	Wed	d	d	d	d	d	d	d	l	d	d	d	d	d	d	d	d	d	d	d	d	1	
25th	Thur	d	d	d	d	d	d	d	d	d	d	d	d	d	d	d	d	d	d	d	d	0	
26th	Fri	d	d	d	d	d	d	d	d	d	d	d	d	d	d	d	d	d	d	d	d	0	
27th	Sat	d	d	d	d	d	d	d	d	d	d	d	d	d	d	d	d	d	d	d	d	0	1Q XhN
28th	Sun	d	d	d	d	d	d	d	d	d	d	d	d	d	d	d	d	d	d	d	d	0	
29th	Mon	d	d	d	d	d	d	d	d	d	d	d	d	d	d	d	d	d	d	d	d	0	
30th	Tues	d	d	d	d	d	d	d	d	d	d	d	d	d	d	d	d	d	d	d	d	0	
31st	Wed	l	l	d	l	d	s	l	d	l	l	l	s	l	l	d	d	s	n	s	d	42	
	Estimate:	77	93	61	137	123	171	137	138	137	170	144	162	92	65	52	125	218	155	236	109	2600	
	Average:	79	90	77	124	116	126	128	137	110	123	138	126	95	84	73	104	180	155	112	133	2309	
	Trend:	av	av	drr	wtr	av	wtr	av	av	wtr	wtr	av	wtr	av	drr	drr	wtr	wtr	av	wtr	drr	wtr	

IRELAND SUNSHINE ESTIMATES

December 2025

F=fine (8-12 hours of sunshine), pc= partly cloudy (4-7 hours), c=cloudy (1-3 hours), o=overcast (0 hours)

		ULSTER PROVINCE					CONNAUGHT PROVINCE						LEINSTER PROVINCE				MUNSTER PROVINCE						
		Alderg	Hillsbo	Armag	Loughl	Malin H	Sligo	Ardtari	Belmul	Drums	Galway	Roscoi	Markre	Edende	Kilkenı	Dublin	Shannc	Newpo	Tralee	Killarne	Cork	All	Moon
1st	Mon	c	c	c	c	c	pc	c	c	c	c	c	c	c	pc	c	pc	pc	pc	pc	pc	79	
2nd	Tues	c	c	c	c	c	c	c	c	c	c	c	c	c	c	c	c	c	c	c	pc	44	
3rd	Wed	o	o	o	o	o	o	o	o	c	pc	c	o	c	pc	c	pc	pc	pc	pc	pc	59	
4th	Thur	o	o	o	o	o	o	o	o	o	o	o	o	o	o	o	o	o	c	c	o	6	F P2
5th	Fri	o	o	o	o	o	o	o	o	c	o	o	o	c	c	c	o	o	o	o	c	13	^
6th	Sat	o	o	o	o	o	o	o	o	o	o	o	o	o	o	o	o	o	o	o	o	6	
7th	Sun	o	o	o	o	o	o	o	o	o	o	o	o	o	o	o	o	o	o	o	o	7	
8th	Mon	c	c	c	c	c	c	c	c	c	c	c	c	c	c	c	c	c	c	c	pc	56	
9th	Tues	c	c	c	c	c	c	c	o	pc	c	o	c	pc	pc	pc	c	c	c	c	pc	64	
10th	Wed	o	o	o	o	o	o	o	o	o	o	o	o	o	o	o	o	o	o	o	o	10	
11th	Thur	c	c	c	c	c	o	c	o	c	o	o	c	c	c	c	o	c	c	c	c	29	3Q
12th	Fri	c	c	c	c	c	o	c	o	c	c	c	c	c	c	c	c	c	c	c	c	39	
13th	Sat	c	c	c	c	c	c	c	c	pc	c	c	c	pc	pc	pc	pc	pc	c	c	pc	69	XhS
14th	Sun	o	o	o	o	o	o	o	o	o	o	o	o	o	o	o	o	o	o	o	o	6	
15th	Mon	c	c	c	c	c	o	c	c	c	c	c	c	c	c	c	c	c	c	c	pc	40	
16th	Tues	o	o	o	o	o	o	o	o	o	o	o	o	o	o	o	o	o	o	o	o	0	
17th	Wed	c	c	c	c	c	c	c	c	c	c	c	c	c	c	c	c	c	c	c	c	40	A
18th	Thur	o	o	o	o	o	o	o	o	o	o	o	o	o	c	o	o	c	o	o	c	12	
19th	Fri	c	c	c	c	c	o	c	o	c	o	o	c	c	c	pc	o	c	c	c	o	33	V
20th	Sat	pc	pc	pc	pc	pc	pc	pc	c	pc	pc	pc	pc	pc	pc	pc	pc	pc	pc	pc	pc	113	N
21st	Sun	o	o	o	o	o	o	o	o	o	o	o	o	o	o	c	o	o	o	o	o	9	
22nd	Mon	o	o	o	o	o	o	o	o	o	o	o	o	o	o	o	o	o	o	o	o	0	
23rd	Tues	o	o	o	o	o	o	o	o	o	o	o	o	o	o	o	o	o	o	o	o	3	
24th	Wed	o	o	o	o	o	o	o	o	o	o	o	o	o	o	o	o	o	o	o	o	1	
25th	Thur	o	o	o	o	o	o	o	o	o	o	o	o	o	o	o	o	o	o	o	o	2	
26th	Fri	o	o	o	o	o	o	o	o	o	o	o	o	o	o	o	o	o	o	o	o	2	
27th	Sat	pc	pc	pc	pc	pc	pc	pc	pc	pc	pc	pc	pc	pc	pc	pc	pc	pc	pc	pc	pc	123	1Q XhN
28th	Sun	o	o	o	o	o	o	o	pc	o	c	c	o	o	o	o	o	o	o	o	o	17	
29th	Mon	c	c	c	c	c	o	c	pc	c	pc	pc	c	c	c	pc	pc	c	c	c	o	64	
30th	Tues	o	o	o	o	o	o	o	o	c	o	o	o	c	c	o	o	c	o	o	c	10	
31st	Wed	c	c	c	c	c	c	c	c	c	c	c	c	c	c	pc	o	c	c	c	o	59	
Estimate hours:		48	48	48	48	48	33	48	41	57	45	43	48	57	66	63	49	57	51	51	69	1014	
Average hours:		39	39	34	34	28	34	34	33	33	36	42	47	45	46	53	44	30	36	36	48	774	
Trend:		av	av	more	more	more	av	more	av	more	av	av	av	more	more	av	av	more	more	more	more	more	

DECEMBER **1** MONDAY

A depression centred to the south of Ireland directs a moderate north to northeasterly flow over the country with fresher spells around the north and eastern coasts. Some patches of mist overnight mostly around the coasts. A changeable day of some sunny spells and occasional showers in most districts. Chance of frost in sheltered parts overnight.

Belfast: Mostly cloudy and dry, windy at times
Dublin: Unsettled breezy spells, mostly cloudy, brief showers
Cork: Partly cloudy and cold, dry, chance of fog patches
Galway: Cloudy, dry, moderate breezes, chance of patches of fog overnight

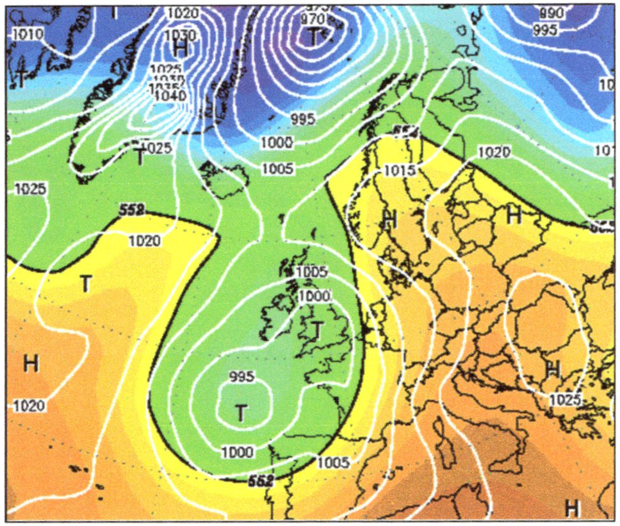

RAIN POTENTIAL

FROST/SNOW

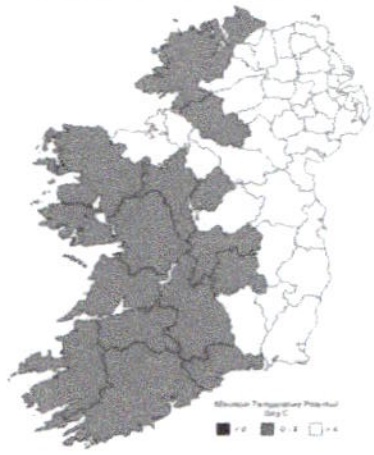

DECEMBER **2** TUESDAY

The depression over Ireland contracts to the immediate north of the country bringing further cold north to northeast breezes throughout, blustery at times around the coasts. Mostly cloudy everywhere and dry in the east. Some wintry showers elsewhere, including the likelihood of snow flurries overnight, mostly in the west.

Belfast: Cloudy, light drizzly showers, cold, unsettled winds
Dublin: Cloudy, cold, dry, moderate to light breezes
Cork: Partly cloudy, brief showers, gentle breezes
Galway: Cloudy, isolated wintry showers, moderate winds, chance of snow flurries

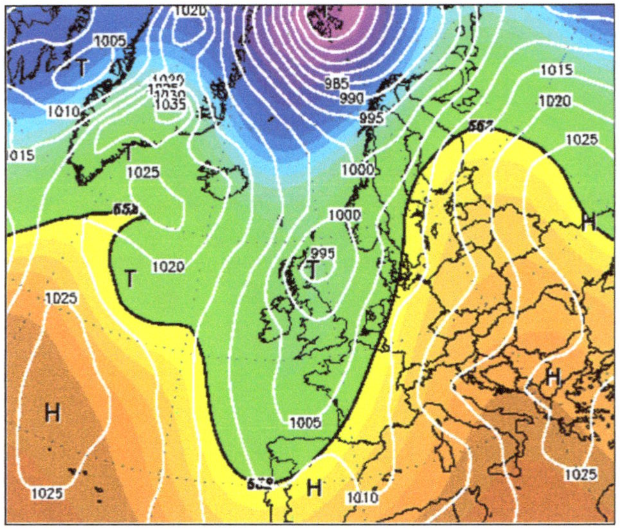

RAIN POTENTIAL

FROST/SNOW

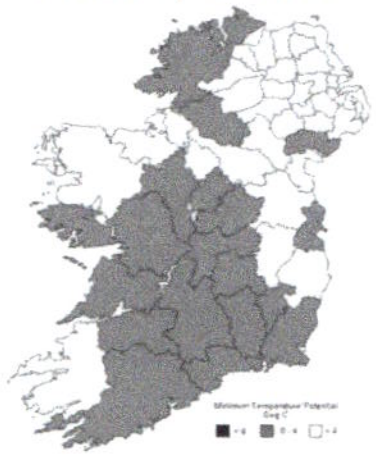

DECEMBER 3 WEDNESDAY

An Atlantic high pressure system ridges back onto the country, with north to northeast breezes prevailing throughout, more variable later in the day. Mostly suny and dry around the southern and eastern coastal regions, while cloudier with passing showers elsewhere. Chance of frosts forming after dark.

Belfast: Overcast, colder, brief showers, cold northerly breezes, overnight frosts
Dublin: Cloudy, cold and crisp air with frosts, dry, light to moderate breezes
Cork: Mostly sunny and dry, cold with severe frosts in places, gentle breezes
Galway: Changeable, sunny spells, rain, cold with frosts overnight, chance of snow

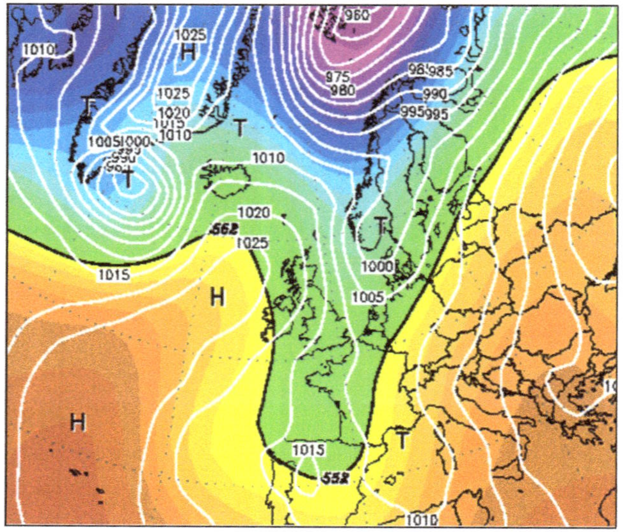

RAIN POTENTIAL

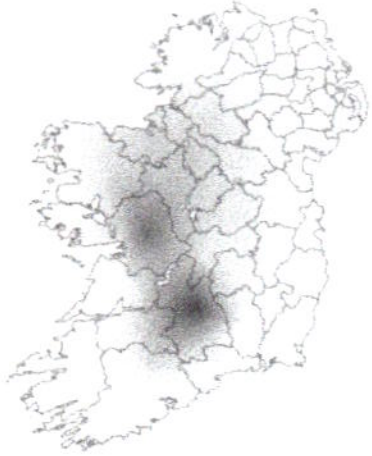

FROST/SNOW

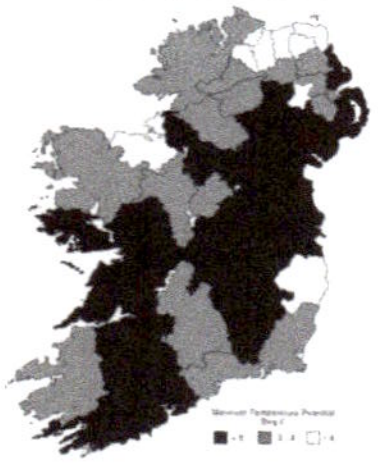

DECEMBER 4 THURSDAY

The high pressure ridge contacts in a southerly trek down through the country while an intense depression to the north nears the northern half of Ireland. Cloud builds to a dull and gloomy day everywhere with passing wintry showers throughout. Winds lighter in a westerly flow in the south, while strong and gusty in the north.

Belfast: Overcast, isolated showers, cold with frosts, strong gusty winds at times
Dublin: Overcast, slightly warmer, isolated showers, fluctuating winds
Cork: Overcast, light rain, mild day cold overnight, frost, light shifting breezes
Galway: Overcast, dry, windy at times, cold overnight with frost patches

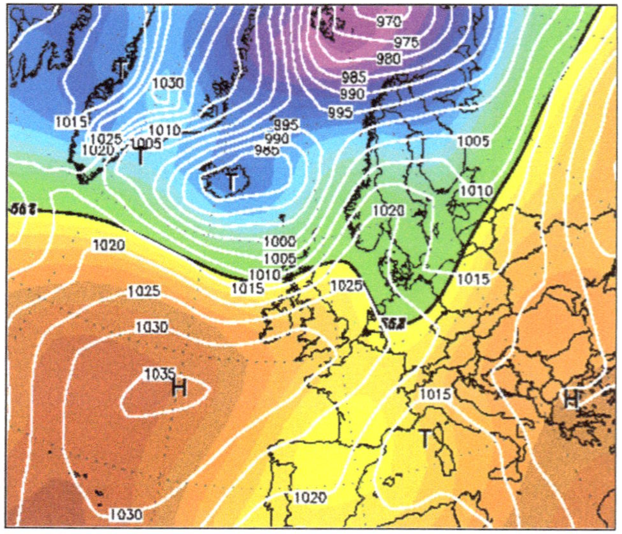

RAIN POTENTIAL

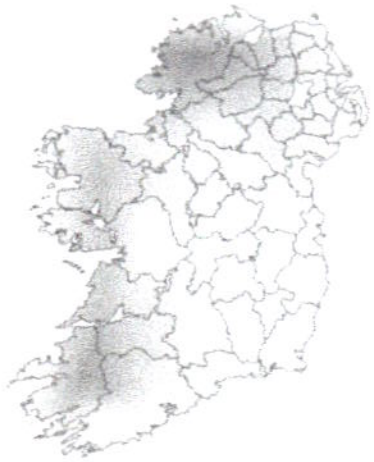

FROST/SNOW

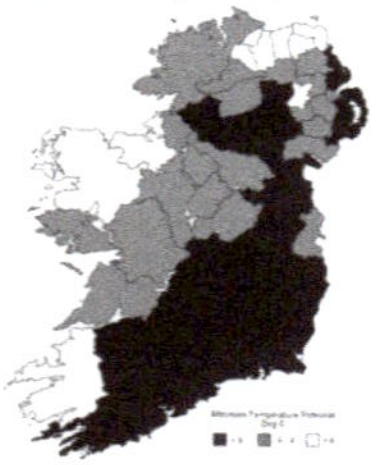

DECEMBER 5 FRIDAY

The high pressure ridge holds back the depression to the north of Ireland, allowing for a moderate to fresh northwesterly breezes to prevail. Mostly cloudy in central parts while sunnier around the coasts with patches of misty and drizzle. Some isolated showers in the north west, and far southwest, while mostly dry with passing threats of rain elsewhere.

Belfast: Strong and blustery winds easing later, widespread cloud, isolated showers
Dublin: Cloudy, cool, dry, windy spells
Cork: Cloudy and dry, milder temperatures, fresh breezes at times
Galway: Unsettled winds, overcast, mostly dry, threatsof rain, chance of mists

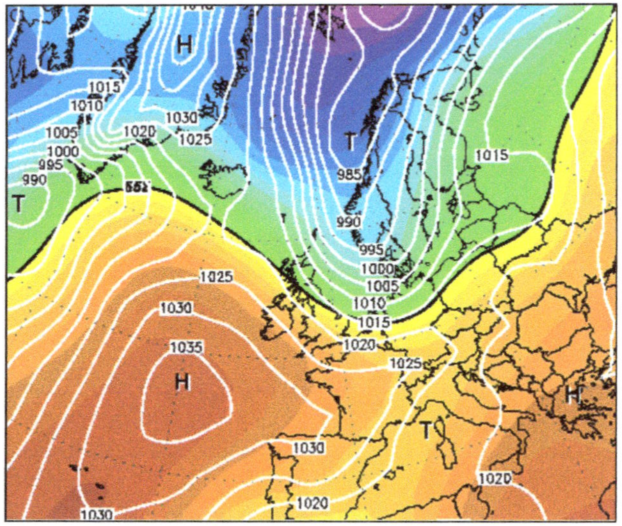

RAIN POTENTIAL

FROST/SNOW

DECEMBER 6 SATURDAY

High pressure builds over Ireland with a light northwesterly air prevailing. Patches of fog overnight in many district but no frosts likely. Widespread cloud and mostly dry everywhere with light drizzle patches developing later in the day and building threats of rain, mostly in central parts.

Belfast: Overcast, milder temperatures, isolated showers, chance of fog overnight
Dublin: Overcast and dry with odd threat of rain, gentle breezes, chance of mists
Cork: Overcast, isolated showers, gentle breezes to mostly calm air, fog possible
Galway: Overcast, light showers, light winds, chance of fog

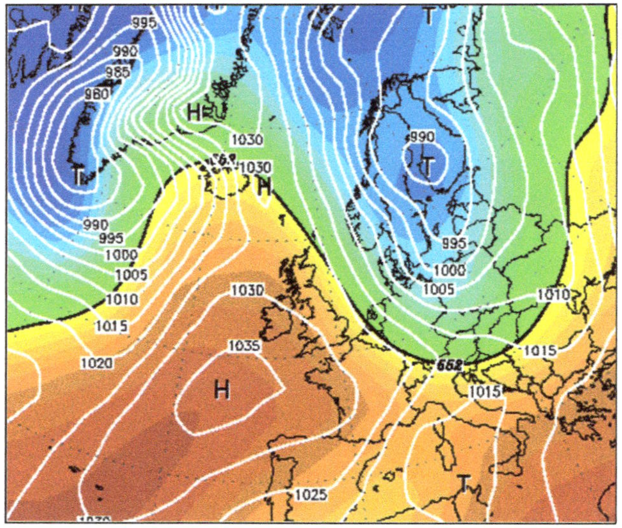

RAIN POTENTIAL

FROST/SNOW

DECEMBER **7** SUNDAY

A large anticyclonic system brings widespread dull and gloomy conditions throughout the country, and with a light to moderate south to southwest flow prevailing. Rain and wintry drizzle develops in the north and northwest and first before spreading almost everywhere, except for remaining mostly dry on the southern coasts. Rain heavy for a time in the west to midlands.

Belfast: Gentle breezes, overcast, occasional showers, fog likely in places
Dublin: Overcast, dry with increasing threats of rain, light breezes
Cork: Overcast, light showers, mostly calm and serene air, fog likely
Galway: Overcast, light breezes to mostly calm air, heavy rain

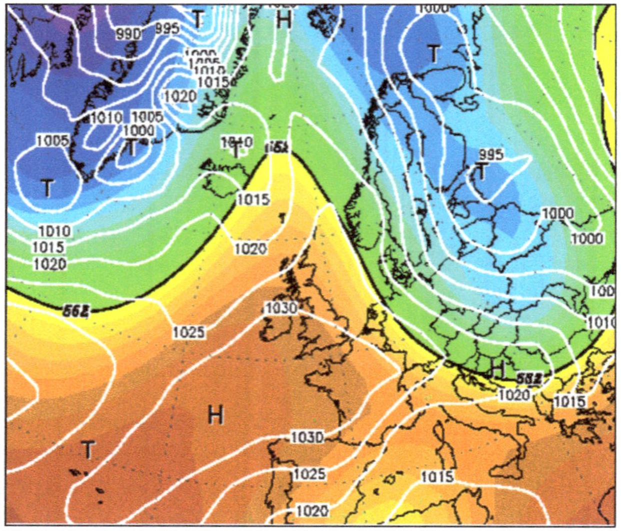

RAIN POTENTIAL

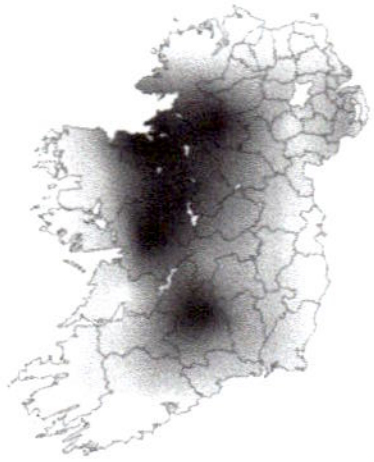

FROST/SNOW

DECEMBER **8** MONDAY

A low pressure trough moves onto Ireland with light to moderate southerly winds prevailing at first, but freshening and becoming more gusty with a southwest change. Widespread cloud and showers almost everywhere, more isolated in the east, and heavier in the northwest and south.

Belfast: Cloudy, light showers, mild temperatures, gentle breezes, fog in places
Dublin: Light shifting breezes, cloudy, mild temperatures, brief showers
Cork: Changeable, rain, some sunny spells, gentle shifting breezes
Galway: Changeable, mix of showers and sunny spells, moderate breezes

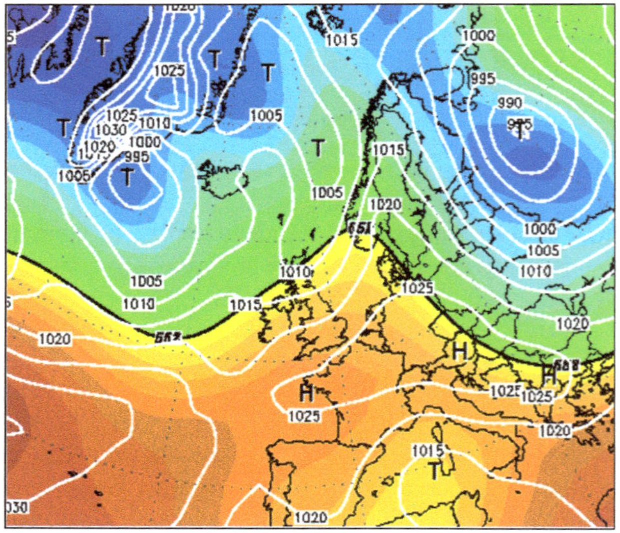

RAIN POTENTIAL

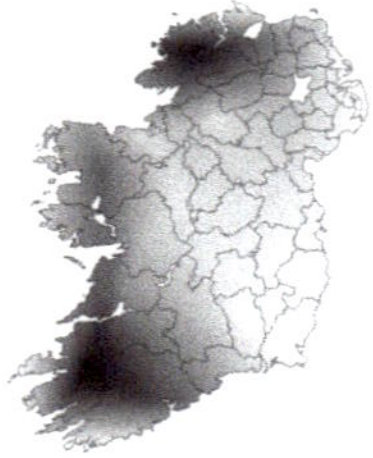

FROST/SNOW

DECEMBER **9** TUESDAY

A depression centred to the northeast of Ireland directs a fresh to strong west to northwest flow over the country. Mostly cloudy and with rain everywhere except for mostly dry conditions and more frequent sunny breakthroughs likely in the east and south coastal regions. Wintry rain activity heavier and snow possible in parts of the west to the midlands.

Belfast: Mostly cloudy, light showers and drizzles, strong and unsettled winds
Dublin: Unsettled and changeable winds, occasional sunny spells, isolated showers
Cork: Partly cloudy, milder temperatures, light showers, unsettled winds at times
Galway: Cloudy, heavy rain, blustery winds to gales, cold with frost, chance of snow

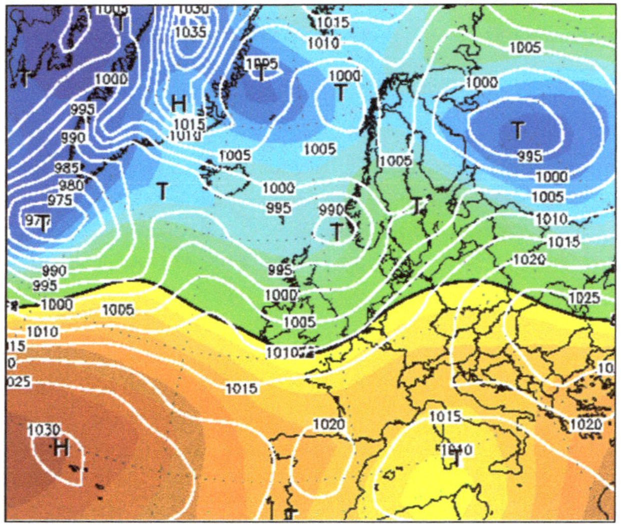

RAIN POTENTIAL

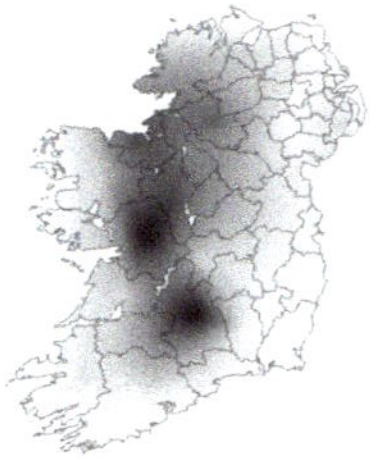

FROST/SNOW

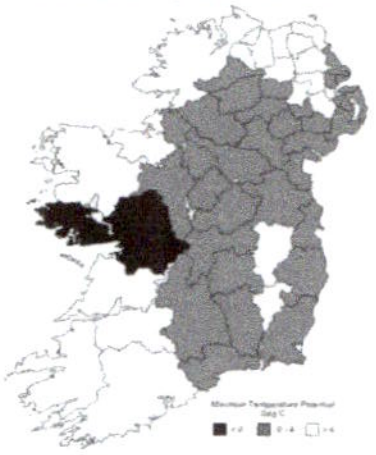

DECEMBER **10** WEDNESDAY

An intense depression to the west of Ireland brings strong and blustery westerly winds throughout with gale gusts likely in around the coasts. A stormy air prevails, with widespread cloud and frequent rain everywhere, including heavy falls at times in the western half of Ireland, where some precipitation will also be wintry with snow and hail showers likely.

Belfast: Overcast and gloomy, rain, strong and blustery winds
Dublin: Overcast, occasional showers, slightly warmer, strong and blustery winds
Cork: Unsettled and blustery winds, overcast, rain
Galway: Stormy air, overcast, rain, colder, chance of overnight snow or hail

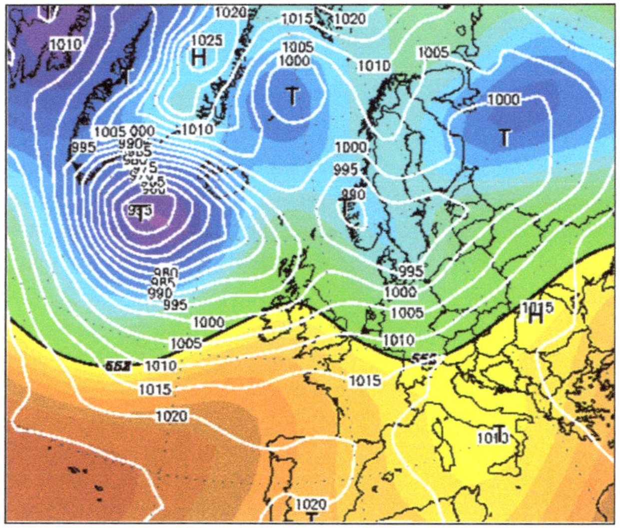

RAIN POTENTIAL

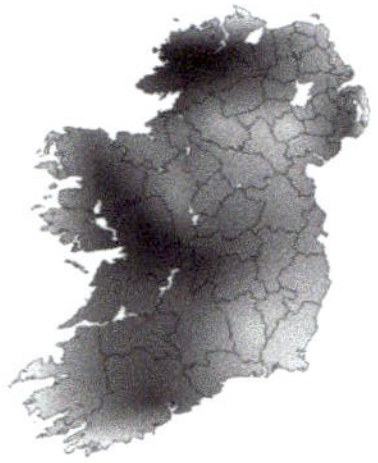

FROST/SNOW

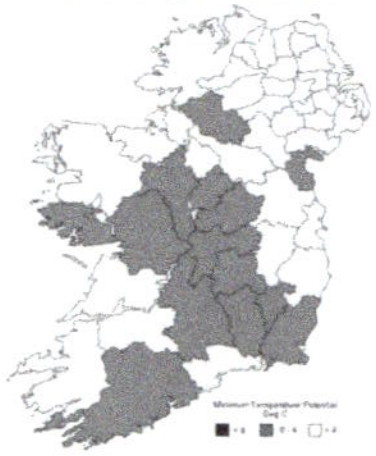

DECEMBER 11 THURSDAY

A depression centred to the northwest of Ireland maintains a stormy air throughout the country, with fresh and gusty westerly winds prevailing, with gale gusts likely at times around the west and northern coasts. A mixed day with brief sunny spells and bands of rain and wintry showers everywhere. Thunderstorms likely and chance of snow, particularly in the north. Rain heavy at times in many western parts of the country.

Belfast: Stormy air, cloudy, rain, gusts and gales with winds to storm force possible
Dublin: Stormy air, cloudy, cold, brief showers, winds to storm force at times likely
Cork: Strong and stormy winds, cloudy, cooler with squally showers
Galway: Overcast, heavy rain, strong winds, thunderstorms, snow, violent wind gusts

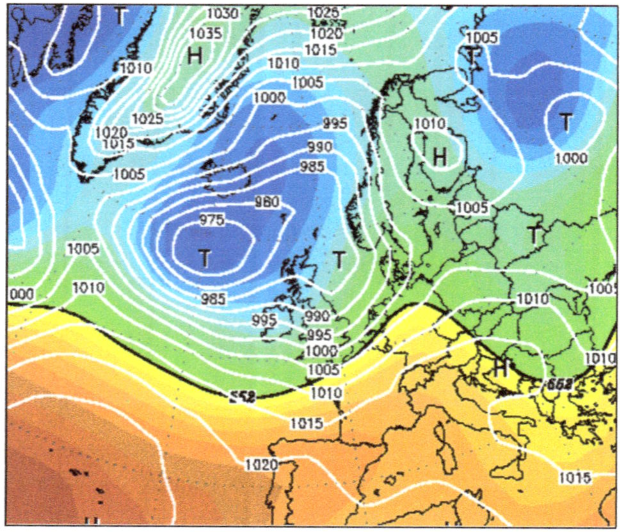

RAIN POTENTIAL

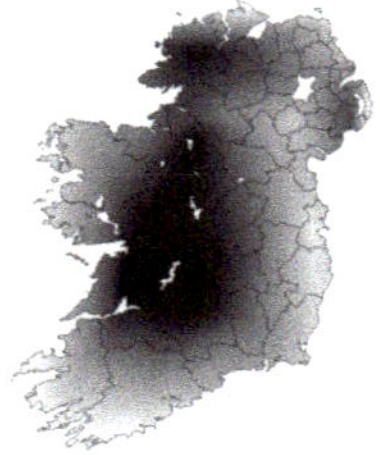

FROST/SNOW

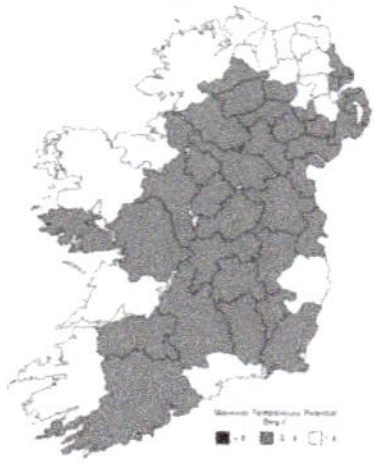

DECEMBER 12 FRIDAY

The depression to the north of Ireland decays slightly as the low pressure centre nears the country. Strong westerlies prevail throughout, backing southwesterly at times and up to gale force with strong gusts especially in the west. Cloudy with outbreaks of rain everywhere, with heavy falls in the west and south. Rain activity easing later in the day.

Belfast: Cloudy, cooler, light showers, strong and unsettled winds
Dublin: Mostly cloudy, outbreaks of rain, strong winds with gusts and gales at times
Cork: Cloudy, rain, milder temperatures, blustery winds with gales at times
Galway: Blustery, cloudy, squally showers

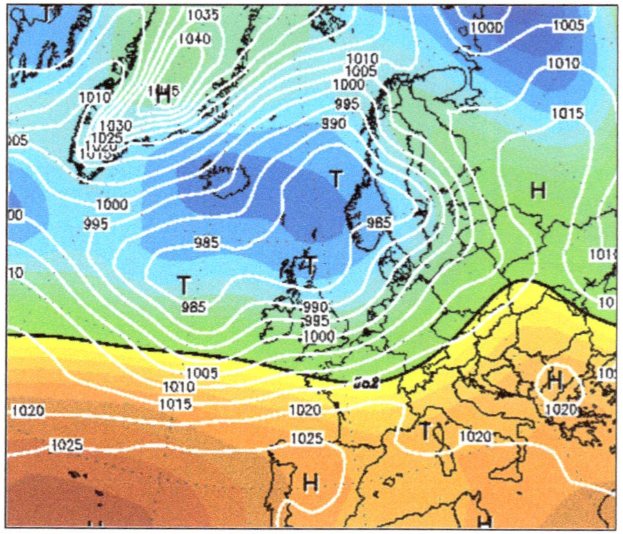

RAIN POTENTIAL

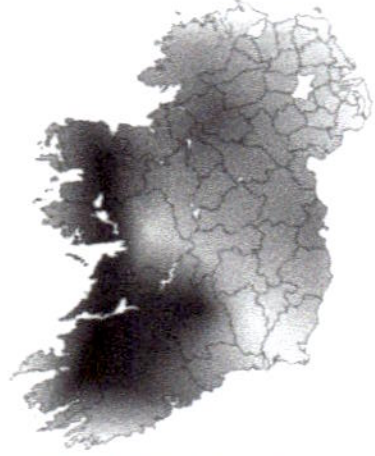

FROST/SNOW

DECEMBER 13 SATURDAY

The large low pressure system that has brought stormy conditions over Ireland the past couple of days drifts to the east of the country, while high pressure sits to the south. Winds remain fresh to strong throughout in a northwesterly flow, reaching gale force at times in many places. A mixed day of brief sunny spells and scattered wintry showers almost everywhere, although clearing to mostly dry in the eastern coastal fringes.

Belfast: Cloudy, cold, squally showers, strong and gusty northwest winds
Dublin: Stormy air, brief sunny spells, mostly dry, threats of rain, gales and gusts
Cork: Partly cloudy, isolated showers, cooler, strong winds with gusts
Galway: Unsettled and blustery winds, cloudy, some showers, warmer day cold night

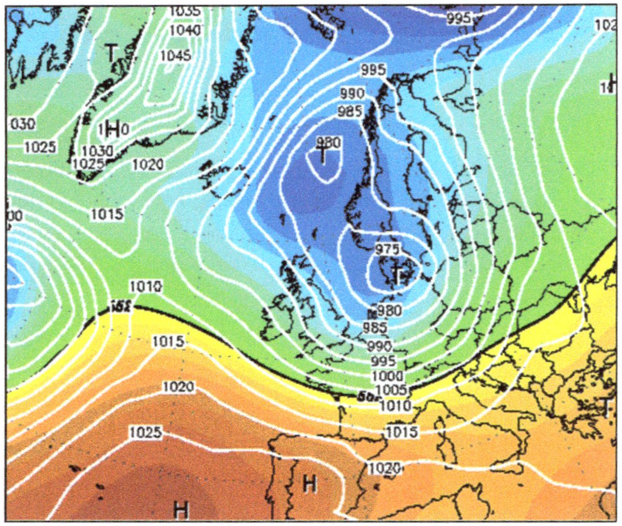

RAIN POTENTIAL

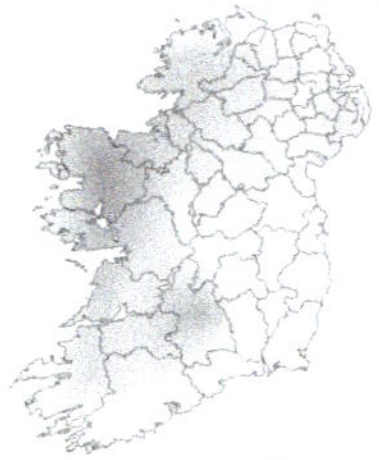

FROST/SNOW

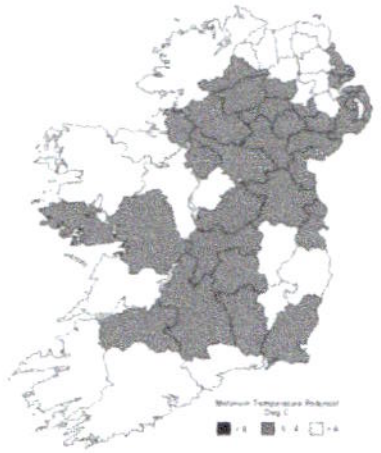

DECEMBER 14 SUNDAY

Another depression to the west of Ireland moves onto the country behind the depression departing to the east. Strong and gusty southwest winds and widespread cloud prevails throughout. Some rain in the west at first, spreading drizzle and patchy showers into all other districts as the day progresses. Rain may be heavy for a time in inland western parts.

Belfast: Overcast, passing showers, milder temperatures, strong unsettled winds
Dublin: Overcast, warmer, drizzly showers, strong unsettled winds, gale gusts likely
Cork: Strong and unsettled gusty winds, overcast, rain
Galway: Overcast, light showers, milder, strong winds to storm force possible

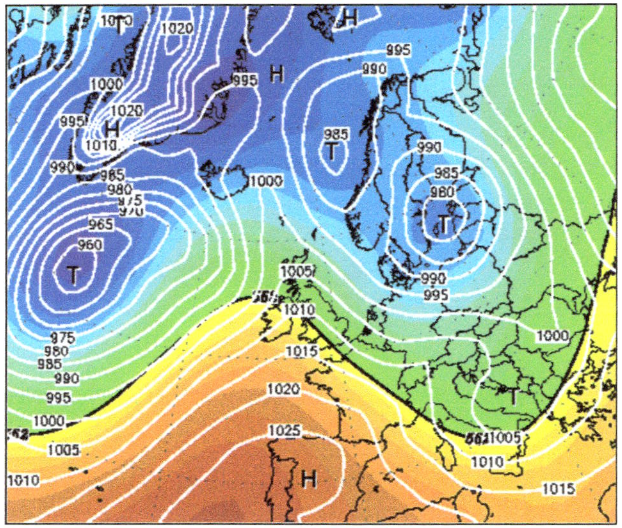

RAIN POTENTIAL

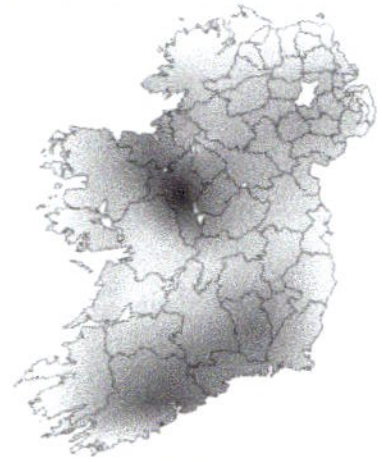

FROST/SNOW

DECEMBER **15** MONDAY

A stormy air builds over Ireland as a depression to the northwest of the country directs a very strong and blustery southwest flow throughout, with winds reaching near gale force at times in many parts. Mostly cloudy with widespread showers in most districts, although more isolated in the east and mostly dry around the coastal fringes of Leinster. Longer sunny spells in the south likely.

Belfast: Cloudy, warmer, scattered showers, chance of storm force winds for a time
Dublin: Storm winds, cloudy, mostly dry with increasing threats of rain
Cork: Partly cloudy, brief squalls, storm winds
Galway: Storm winds, cloudy, rain

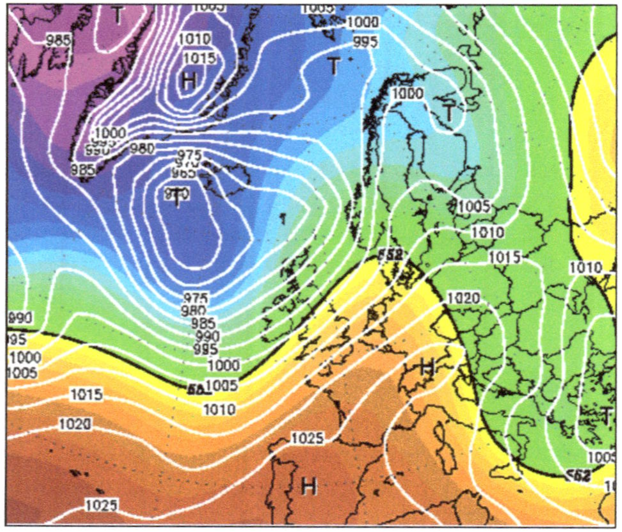

RAIN POTENTIAL

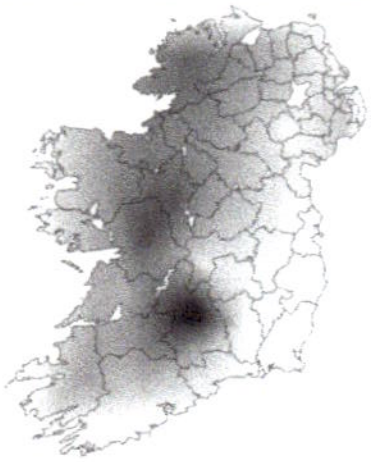

FROST/SNOW

DECEMBER **16** TUESDAY

An intense depression to the north of Ireland directs a stormy air over the country, with strong and gusty west to southwest winds prevailing. Dull and gloomy and with widespread rain everywhere. Rain likely to be heavy at times in many districts.

Belfast: Violent storm, overcast, rain, gales
Dublin: Overcast, rain, strong winds with gales gusts at times
Cork: Stormy air, overcast, rain, milder temperatures, gale winds
Galway: Strong winds to gale force, overcast, heavy rain

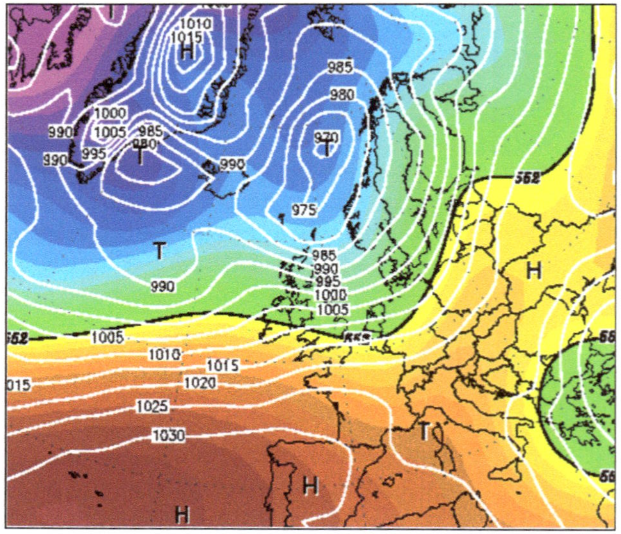

RAIN POTENTIAL

FROST/SNOW

DECEMBER **17** WEDNESDAY

A stormy air persists over Ireland as a depression centred to the north of the country maintains a very strong and blustery northwesterly flow over the country, with winds up to gale force at times. Cloudy and wet everywhere, with heavy rain in many districts, but lighter around the coasts. Colder everywhere, with chance of snow in some parts, especially in western districts. Chance of frost overnight.

Belfast: Storm, cloudy, heavy rain, gusts and gales
Dublin: Cloudy, colder, isolated showers, strong winds to storm force at times
Cork: Cloudy, colder, scattered showers, strong winds to storm force likely
Galway: Cloudy, cooler, heavy wintry rain, chance of snow and storm force winds

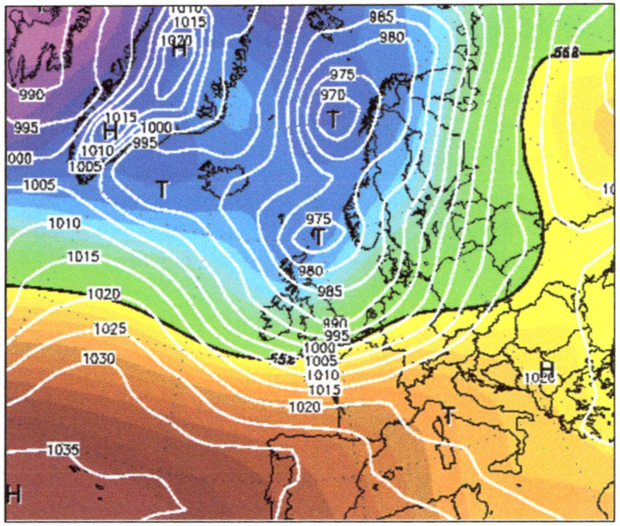

RAIN POTENTIAL

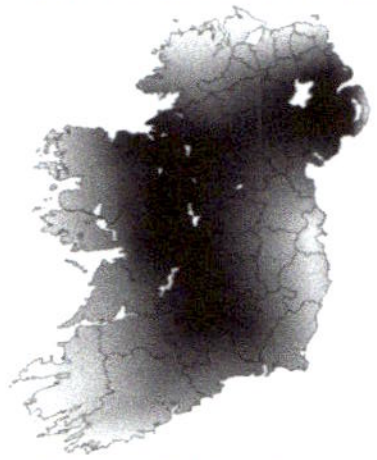

FROST/SNOW

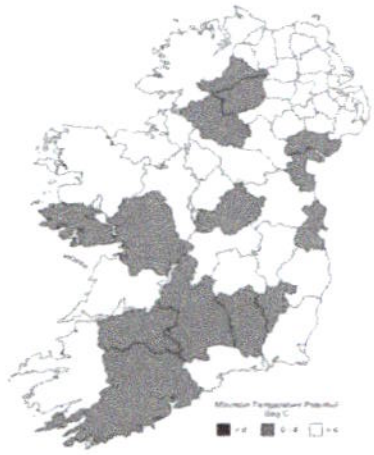

DECEMBER **18** THURSDAY

Low pressure trough between two depressions bring a strong and unsettled southerly flow at first over the country, backing west to southwest later with chance of gale gusts at times. Storm conditions continue everywhere with heavy rain in almost all districts, as well as the chance of snow and hail at times mostly in the west.

Belfast: Overcast, colder, heavy rain, strong and unsettled winds
Dublin: Stormy air, overcast, rain, cold overnight
Cork: Cloudy, heavy rain, strong and unsettled winds
Galway: Overcast, wintry rain, strong and unsettled winds, snow and hail likely

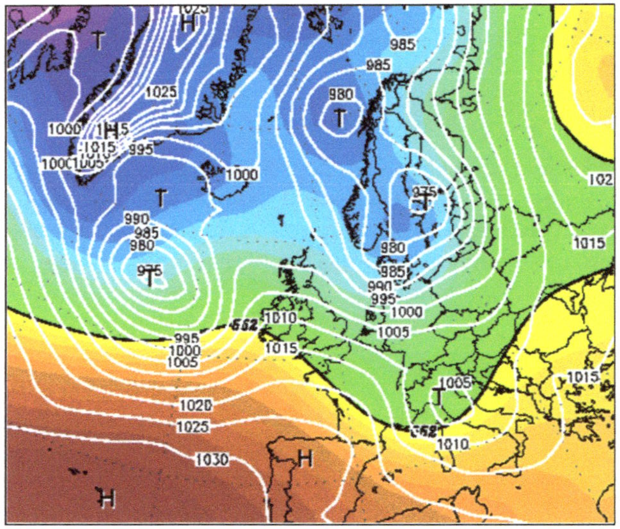

RAIN POTENTIAL

FROST/SNOW

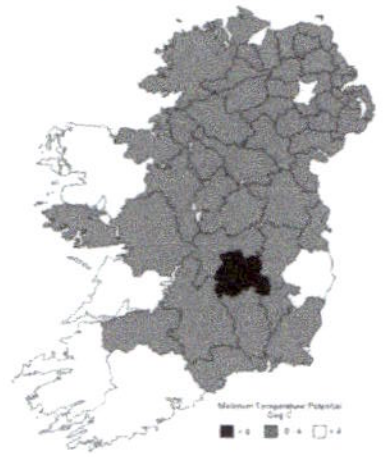

DECEMBER 19 FRIDAY

An intense depression again moves onto Ireland bringing a stormy air once more across the country with strong to gale force west to northwest winds prevailing. Widespread cloud and wintry rain activity everywhere, with the chance of snow again mostly in western parts, but lighter shower activity in the east. Some brief sunny spells mostly in the north and east.

Belfast: Stormy air, cloudy, mild day cold night, brief squally showers
Dublin: Colder and stormy air, some sunny spells, isolated squalls, gales
Cork: Overcast, colder, rain, stormy air with winds to storm force for a time
Galway: Stormy air, overcast, brief squalls, chance of snow

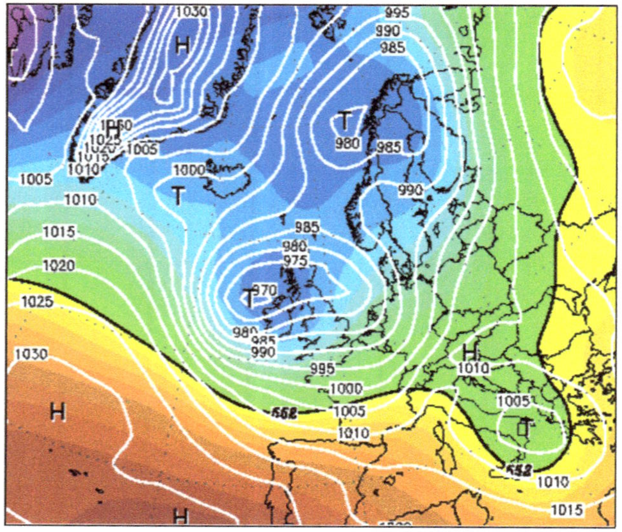

RAIN POTENTIAL

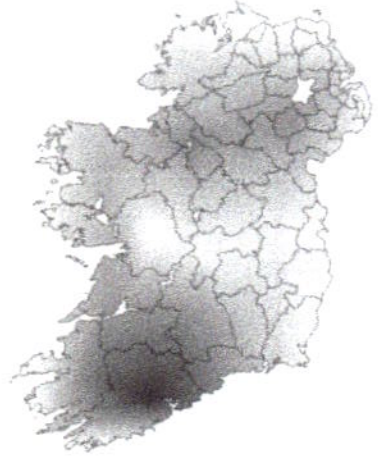

FROST/SNOW

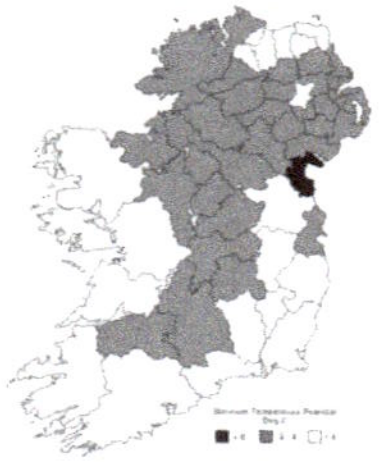

DECEMBER 20 SATURDAY

The depression that has brought stormy conditions over the past few days drifts away to the east, allowing for a high pressure ridge to near the southwest coasts. Winds remain fresh to strong in a northwest flow at first, but ease with a northerly change ahead of the high. Increased sunny spells and mostly dry everywhere, except for some lingering isolated showers mostly in the southwest and in the northeast. Overnight frost pockets likely.

Belfast: Partly cloudy, dry, cold, winds easing with fresh gusts at times
Dublin: Partly cloudy, dry, cold, winds easing with occasional gusts
Cork: Partly cloudy, cold and dry, winds easing with occasional gusts
Galway: Strong winds easing to near gales at times, partly cloudy, dry

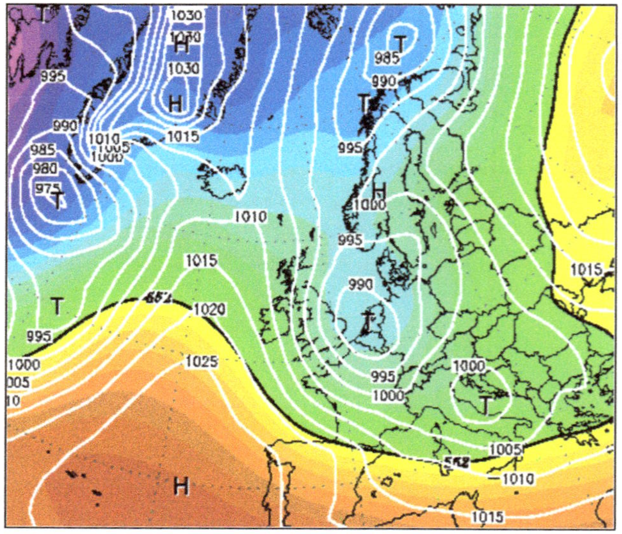

RAIN POTENTIAL

FROST/SNOW

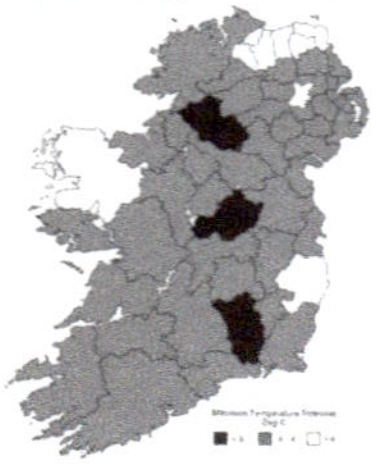

DECEMBER **21** SUNDAY

High pressure expands up through Ireland in an light southerly flow in the east, stronger southwesterlies in the west and, freshening later in the day. Widespread cloud throughout brings a dull and gloomy day everywhere and some showers around the northwest. Mainly dry elsewhere. Very cold overnight with frosts in many parts.

Belfast: Overcast and dull, dry, cold, odd threat of rain, windy spells
Dublin: Cloudy, dry, cold, moderate southerly winds
Cork: Overcast, mostly dry with odd threat of rain, cold overnight, winds easing
Galway: Strong winds, overcast, isolated showers, cold overnight with frosts likely

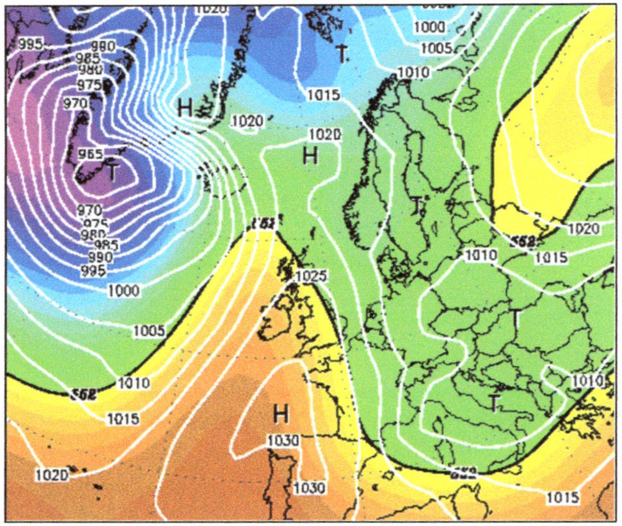

RAIN POTENTIAL

FROST/SNOW

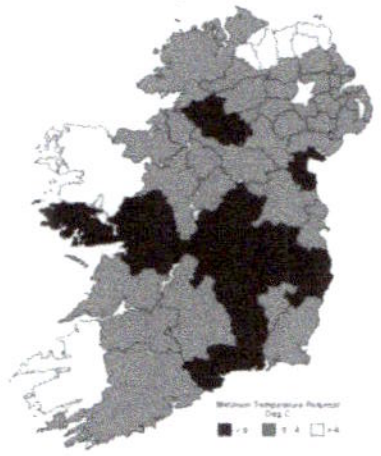

DECEMBER **22** MONDAY

Fresh to strong south to southwest breezes prevail across Ireland as the high pressure ridge moves to the east of the country and a low pressure trough nears the western coasts. Overcast and gloomy everywhere with a wet front bringing occasional rain activity in the south and west of the country, with heavy falls at times in the southwest. Mainly dry with the occasional threat of rain in the north and east.

Belfast: Strong southerly winds, overcast, dry with odd threat of rain, cool
Dublin: Overcast and dull, mostly dry with passing threats of rain, blustery at times
Cork: Overcast, scattered showers, fresh onshore winds with occasional gusts
Galway: Overcast, dry with odd threat of rain, milder temperatures, windy at times

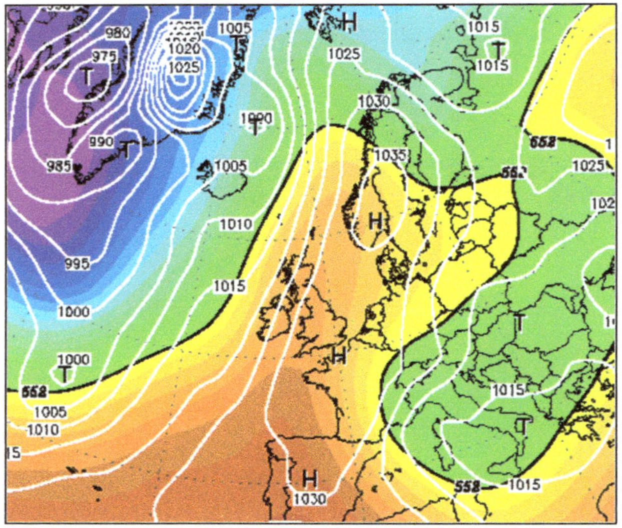

RAIN POTENTIAL

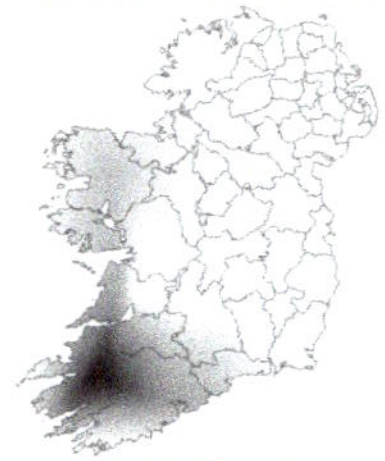

FROST/SNOW

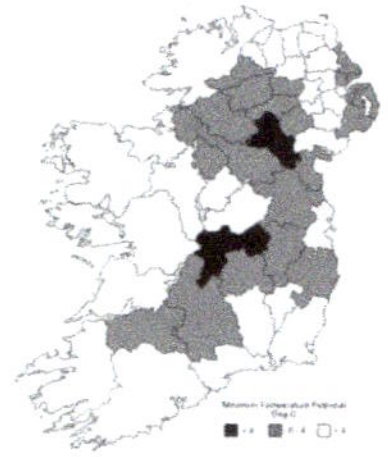

DECEMBER 23 TUESDAY

High pressure prevails to the east, low pressure to the west, and a high pressure ridge sits over the country directing a south to southwest flow throughout, lighter in the east while stronger and gusty with winds possibly to gale force for a time in the west. Mostly dry and mild with widespread cloud over much of the country, except for some spotty showery spells moving into the western half of Ireland for a time.

Belfast: Unsettled blustery winds, overcast, milder, dry with threats of rain
Dublin: Unsettled and strong winds, dull and gloomy air, dry with odd threat of rain
Cork: Overcast, dry, strong and blustery winds, gales possible
Galway: Dull and gloomy, widespread cloud, isolated showers, strong gusty winds

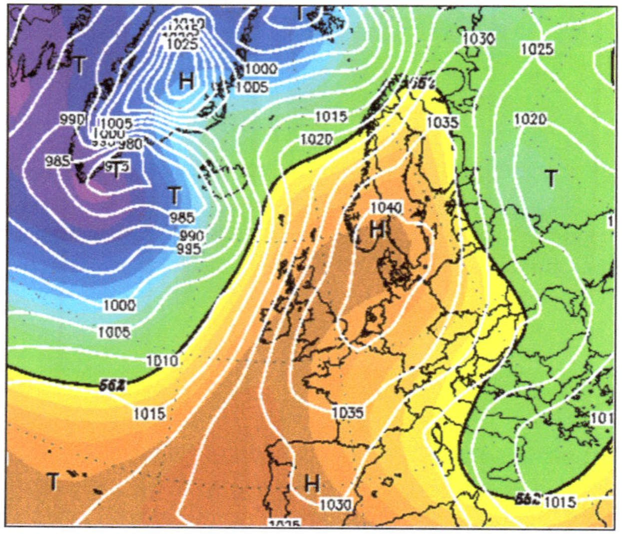

RAIN POTENTIAL

FROST/SNOW

DECEMBER 24 WEDNESDAY

A large anticyclonic system remains almost stalled to the east of Ireland while a depression intensifies to the west. Mainly southerly winds prevail throughout, lighter in the east while stronger and with gusts in the west. Mostly dry and dull everywhere. Very cold overnight with patches of frost likely especially in the east of the country.

Belfast: Overcast, dry, strong southerly winds at times
Dublin: Overcast, fresh breezes, dry, passing threats of rain, chance of frost
Cork: Overcast, windy, dry
Galway: Overcast and blustery air, dry with passing threats of rain

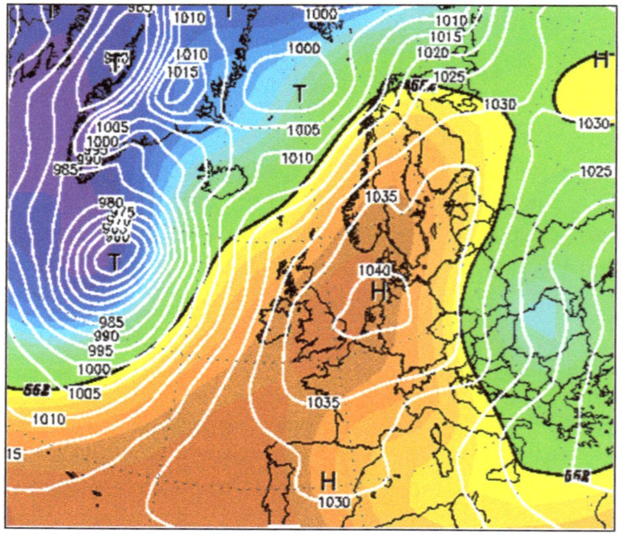

RAIN POTENTIAL

FROST/SNOW

DECEMBER **25** THURSDAY

A fresh to strong south to southwest flow continues to prevail throughout the country as the high pressure ridge lingers over the country and the depression in the west contracts away to the northwest. Dull and gloomy conditions persist everywhere but with a mostly dry day prevailing. Very cold in inland parts with frost patches likely, particularly in the east. Chance of some pockets of mist mosty in central parts.

Belfast: Overcast, cool and dry with passing threats of rain, fresh windy spells
Dublin: Cold northerly winds, gloomy air with widespread cloudy, dry
Cork: Fresh and cold northerly winds, dry, widespread cloud and gloomy air
Galway: Overcast, colder, dry, cold northerly winds

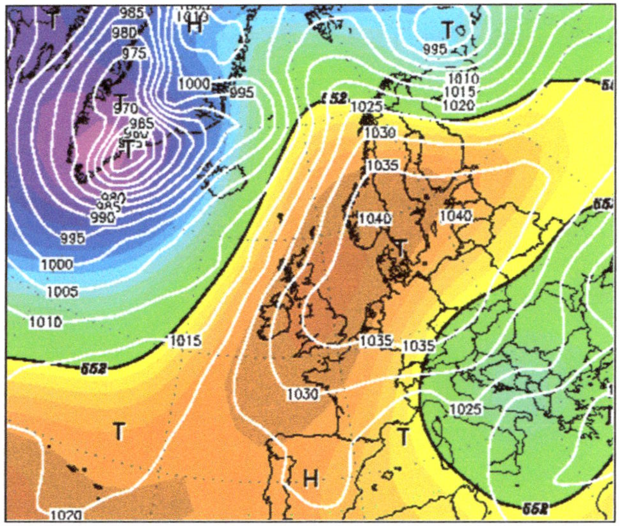

RAIN POTENTIAL

FROST/SNOW

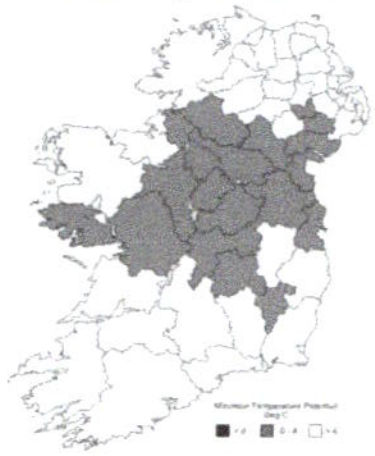

DECEMBER **26** FRIDAY

High pressure continues to bring a dry, very cold, and mostly overcast day throughout the country. Severe frosts likely, mostly in eastern parts. Fresh to strong southerly winds prevail in most parts, but more southeasterly and milder in the south.

Belfast: Overcast, colder, dry, strong winds, possible gales at times
Dublin: Overcast, cold, dry, threats of rain, moderate winds with occasional gusts
Cork: Overcast and dull, dry, cold, fresh windy spells
Galway: Cold, dry, overcast, strong winds at times

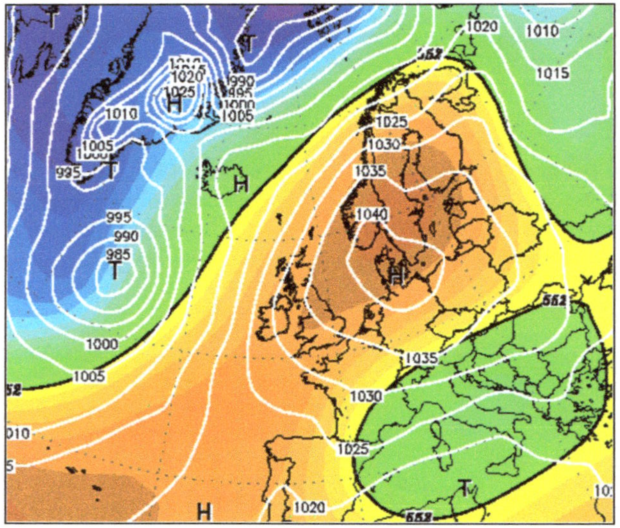

RAIN POTENTIAL

FROST/SNOW

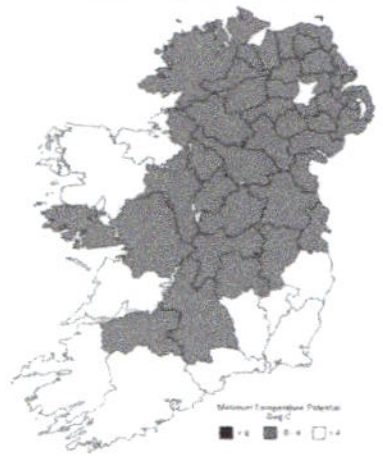

DECEMBER **27** SATURDAY

High pressure continues to prevail over Ireland with moderate to fresh south to southeasterly breezes prevailing, strong and with gusts at times in the west and northern coasts. Very cold overnight with severe ground frosts in many areas. Dry and with occasional sunny spells everywhere.

Belfast: Colder and dry, blustery at times, occasional sunny spells, frost patches likely
Dublin: Partly cloudy, dry, cold with frosts, fresh breezes
Cork: Partly cloudy, cold, dry, fresh and unsettled winds
Galway: Occasional sunny spells, dry, cold, chance of frost patches, windy spells

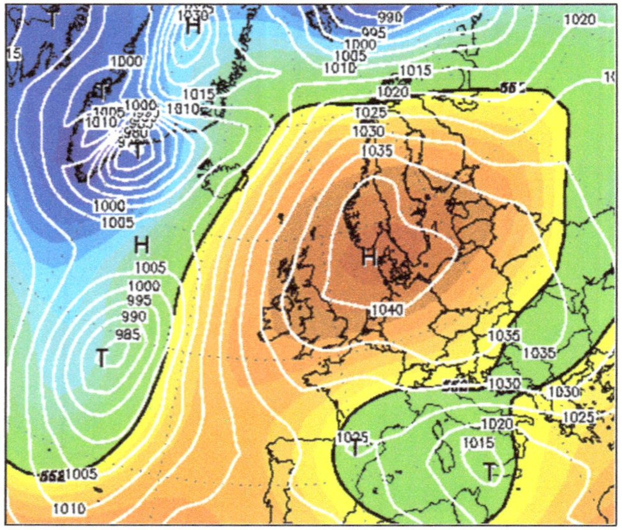

RAIN POTENTIAL

FROST/SNOW

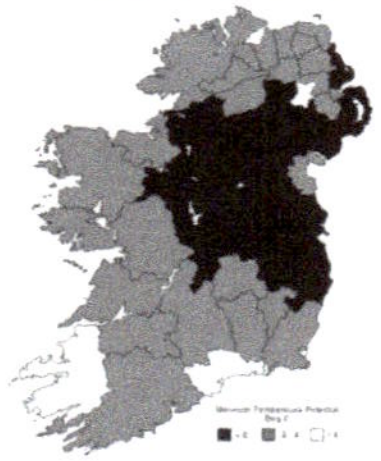

DECEMBER **28** SUNDAY

A large high pressure system situated to the east of the UK expands and brings a light to moderate southeast flow over Ireland. Patches of mist and scattered frosts in many districts overnight. Dry with widespread cloud everywhere, except for some brief sunny breakthroughs in the northwest.

Belfast: Overcast, dry, threats of rain, cold, chance of frost pockets, winds easing
Dublin: Dull and gloomy, dry, cold, moderate breezes with odd gusty spell
Cork: Overcast, dull and gloomy, cold, dry, some windy spells
Galway: Cloudy and cold, frosts in sheltered places, dry, fresh breezes at times

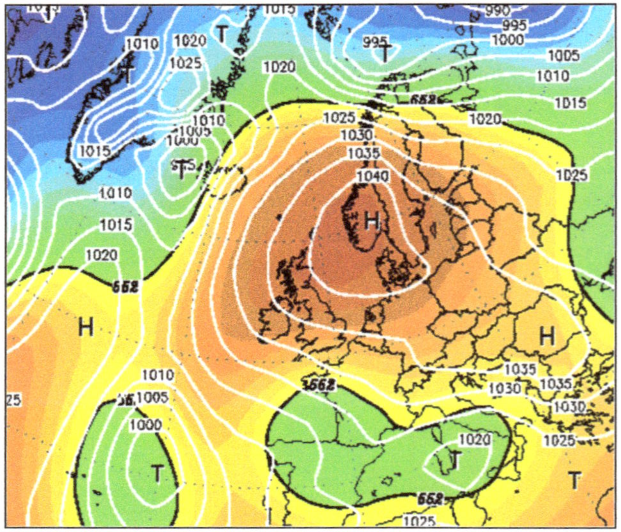

RAIN POTENTIAL

FROST/SNOW

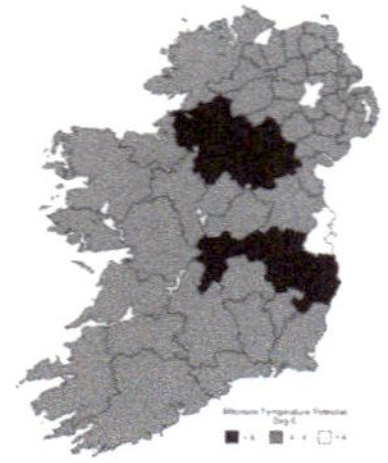

DECEMBER 29 MONDAY

A very large anticyclonic system fills over Europe and brings a mostly calm and cold air over Ireland, with very gentle easterly breezes at times. Cold, dry, and scattered cloudy spells throughout, sunniest through central parts of the country from the west coast to the east coast, while thicker cloud to the north and south. Overnight misty patches in many places.

Belfast: Mostly calm and foggy air, cloudy, dry, cold, frosts in sheltered parts
Dublin: Partly cloudy, dry, cold with frost overnight, light breezes to mostly calm air
Cork: Dull and gloomy with widespread cloud, dry, cold, moderate breezes
Galway: Partly cloudy, dry and cold, light breezes, chance of misty patches

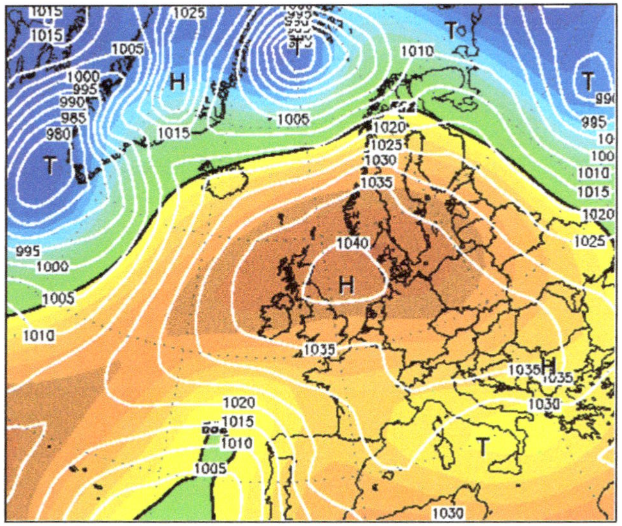

RAIN POTENTIAL

FROST/SNOW

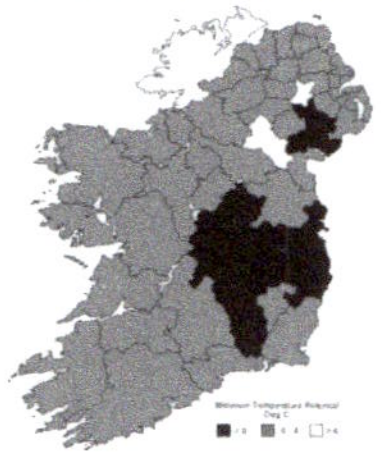

DECEMBER 30 TUESDAY

A large high pressure system over Europe continues to bring a dry and mostly cloudy day throughout Ireland, with patches of frost and fog in many places. Mostly overcast in the east and midlands, with occasional sunny spells elsewhere. Light to moderate southeasterly breezes prevail.

Belfast: Overcast, cold and dry, mostly light breezes, chance of frost and fog patches
Dublin: Cold, dull and gloomy atmosphere, dry, moderate breezes
Cork: Mostly cloudy and dry, warmer, moderate southeast breezes
Galway: Overcast and dull, dry, cold with frosts, moderate breezes

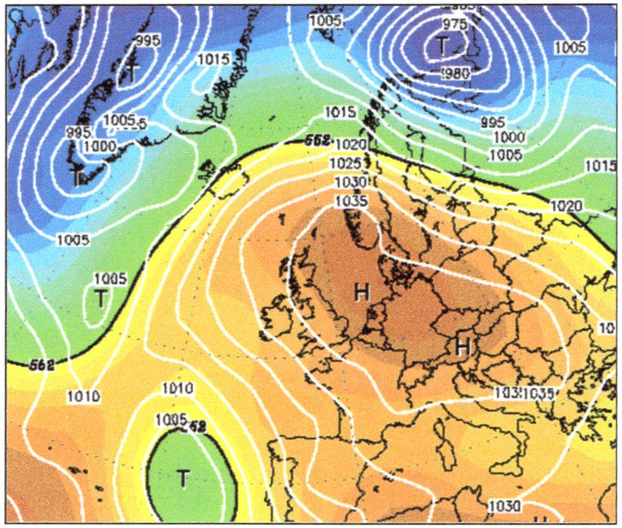

RAIN POTENTIAL

FROST/SNOW

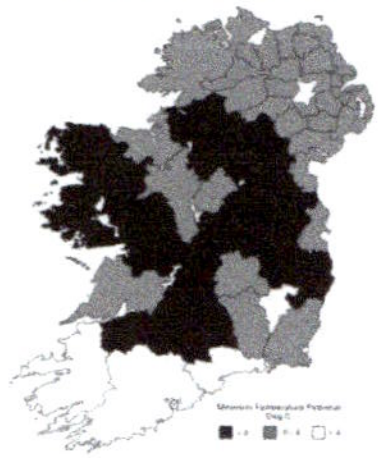

DECEMBER 31 WEDNESDAY

A light to moderate southeasterly flow continues to prevail over Ireland directed from a very large anticyclonic system slowly lingering across Europe. A low pressure trough associated with a depression to the northwest, nears the western coasts. A mostly dry day with scattered cloud prevails at first everywhere after fog patches clear, before cloud builds and some showers develop later in the day in most districts except for remaining mostly dry in the western and far northern coastal fringes.

Belfast: Mostly cloudy, very cold with severe frosts, light showers, light breezes
Dublin: Partly cloudy, dry, very cold, frosts and patches of fog overnight
Cork: Overcast, dry with some threats of rain, cool, moderate breezes
Galway: Cloudy, brief showers, cold with frosts overnight, moderate breezes

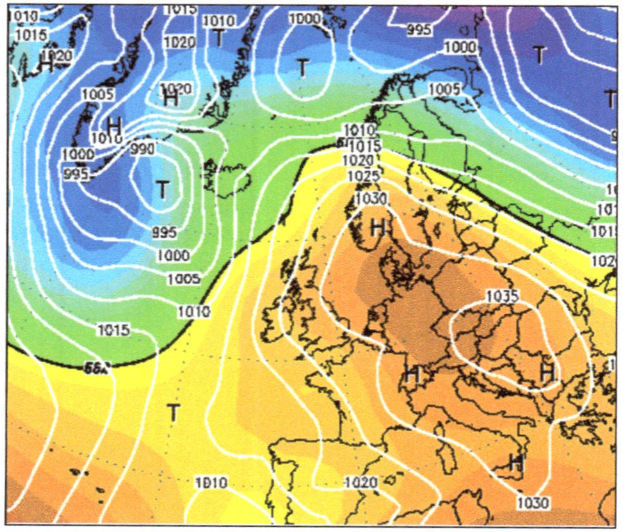

RAIN POTENTIAL

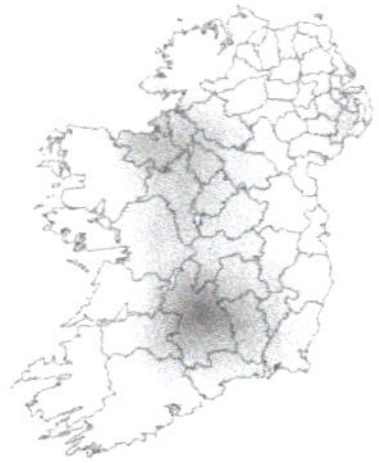

FROST/SNOW

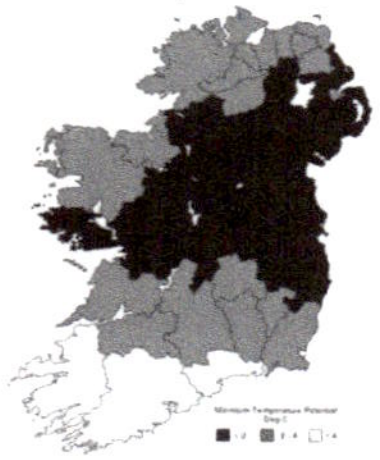

APPENDICES

RAINFALL EXPECTATIONS

WETTER OR DRIER BY 10MM THAN LONG-TERM AVERAGES

	Ulster Province					Connaught Province						Leinster Province				MUNSTER PROVINCE					
Date	Aldergrove	Hillsborough	Armagh	Lough Navar	Malin Head	Sligo	Ardtarmon	Belmullet	Drumsna	Galway	Roscommon	Markree Castle	Edenderry	Kilkenny	Dublin	Shannon	Newport	Tralee	Killarney	Cork	All
JAN	w tr	w tr	w tr	w tr	w tr	w tr	w tr	w tr	w tr	w tr	w tr	w tr	w tr	w tr	w tr	w tr	w tr	drr	w tr	w tr	w tr
FEB	drr	drr	drr	drr	drr	drr	drr	drr	drr	drr	drr	drr	drr	drr	av	drr	drr	drr	w tr	drr	drr
MAR	w tr	w tr	w tr	w tr	av	w tr	w tr	w tr	w tr	w tr	w tr	w tr	w tr	w tr	w tr	w tr	w tr	w tr	w tr	w tr	w tr
APR	drr	drr	drr	drr	drr	drr	drr	drr	drr	drr	drr	drr	drr	drr	drr	drr	drr	drr	drr	drr	drr
MAY	av	av	av	w tr	w tr	w tr	w tr	w tr	w tr	w tr	w tr	w tr	drr	drr	drr	av	w tr	av	w tr	drr	w tr
JUNE	w tr	w tr	w tr	av	drr	drr	drr	av	av	w tr	w tr	av	drr	w tr	w tr	w tr	w tr	w tr	w tr	w tr	w tr
JULY	w tr	w tr	w tr	w tr	w tr	w tr	w tr	w tr	w tr	w tr	w tr	w tr	w tr	w tr	w tr	w tr	w tr	drr	w tr	w tr	w tr
AUG	w tr	w tr	w tr	w tr	w tr	w tr	w tr	av	w tr	w tr	drr	w tr	w tr	w tr	w tr	w tr	drr	w tr	w tr	w tr	w tr
SEP	drr	drr	drr	drr	drr	drr	drr	drr	drr	drr	drr	drr	drr	drr	drr	drr	drr	drr	drr	drr	drr
OCT	drr	drr	drr	drr	drr	drr	drr	drr	drr	drr	drr	drr	drr	drr	drr	drr	drr	drr	drr	drr	drr
NOV	drr	drr	drr	drr	drr	drr	drr	av	drr	drr	drr	drr	drr	drr	drr	drr	drr	drr	drr	drr	drr
DEC	av	av	drr	w tr	av	w tr	av	av	w tr	w tr	av	w tr	av	drr	drr	w tr	w tr	av	w tr	drr	w tr
Overall	av	drr	av	av	av	av	drr	av	av	av	drr	av	drr	av	av	av	av	drr	w tr	drr	drr
year	861	944	812	1179	1107	1260	1284	1245	1047	1193	1289	1260	971	840	758	978	1607	1583	1131	1228	22578
estimates	**822**	**833**	**810**	**1106**	**1071**	**1355**	**1106**	**1170**	**1141**	**1257**	**1154**	**1181**	**821**	**854**	**765**	**891**	**1554**	**1328**	**1579**	**1062**	**21859**

	Ulster Province					Connaught Province						Leinster Province				MUNSTER PROVINCE					
Date	Aldergrove	Hillsborough	Armagh	Lough Navar	Malin Head	Sligo	Ardtarmon	Belmullet	Drumsna	Galway	Roscommon	Markree Castle	Edenderry	Kilkenny	Dublin	Shannon	Newport	Tralee	Killarney	Cork	All
WINTER	av	av	av	w tr	w tr	w tr	w tr	w tr	w tr	w tr	w tr	w tr	av	w tr	w tr	w tr	w tr	w tr	w tr	w tr	w tr
SPRING	drr	drr	drr	av	av	av	av	av	drr	av	drr	av	drr	av	drr	drr	av	av	w tr	drr	drr
SUMMER	w tr	w tr	w tr	w tr	av	w tr	av	av	w tr	w tr	w tr	w tr	w tr	w tr	w tr	w tr	w tr	av	w tr	w tr	w tr
AUTUMN	drr	drr	drr	drr	drr	drr	drr	drr	drr	drr	drr	drr	drr	drr	drr	drr	drr	drr	drr	drr	drr
Overall	av	drr	av	av	av	av	drr	av	av	av	drr	av	drr	av	av	av	av	drr	w tr	drr	drr
year	861	944	812	1179	1107	1260	1284	1245	1047	1193	1289	1260	971	840	758	978	1607	1583	1131	1228	22578
estimates	**822**	**833**	**810**	**1106**	**1071**	**1355**	**1106**	**1170**	**1141**	**1257**	**1154**	**1181**	**821**	**854**	**765**	**891**	**1554**	**1328**	**1579**	**1062**	**21859**

RAINFALL EXPECTATIONS

WETTER OR DRIER BY 25% THAN LONG-TERM AVERAGES

	Ulster Province					Connaught Province						Leinster Province				MUNSTER PROVINCE					
Date	Aldergrove	Hillsborough	Armagh	Lough Navar	Malin Head	Sligo	Ardtarmon	Belmullet	Drumsna	Galway	Roscommon	Markree Castle	Edenderry	Kilkenny	Dublin	Shannon	Newport	Tralee	Killarney	Cork	All
JAN	w tr	av	w tr	av	w tr	w tr	av	w tr	w tr	w tr	w tr	w tr	av	w tr	w tr	av	w tr	av	w tr	av	w tr
FEB	drr	drr	drr	drr	drr	av	drr	drr	drr	drr	drr	drr	drr	drr	av	drr	drr	drr	w tr	drr	drr
MAR	w tr	w tr	w tr	w tr	av	w tr	av	w tr	w tr	w tr	av	av	av	w tr	w tr	av	av	w tr	w tr	w tr	w tr
APR	drr	drr	drr	drr	drr	drr	drr	drr	drr	drr	drr	drr	drr	drr	drr	drr	drr	drr	drr	av	drr
MAY	av	av	av	w tr	w tr	w tr	av	w tr	w tr	w tr	av	av	av	av	drr	av	w tr	av	w tr	av	av
JUNE	w tr	w tr	w tr	av	drr	av	av	av	av	w tr	w tr	av	av	w tr	w tr	w tr	av	w tr	w tr	w tr	av
JULY	w tr	w tr	w tr	w tr	w tr	w tr	w tr	av	w tr	w tr	w tr	w tr	w tr	w tr	w tr	w tr	w tr	av	w tr	w tr	w tr
AUG	w tr	w tr	w tr	av	av	w tr	av	av	w tr	av	av	w tr	w tr	w tr	w tr	w tr	av	av	w tr	av	w tr
SEP	drr	drr	drr	drr	av	drr	drr	drr	drr	drr	drr	drr	drr	drr	drr	drr	av	drr	av	drr	drr
OCT	drr	drr	drr	drr	drr	drr	drr	drr	drr	drr	drr	drr	drr	drr	drr	drr	drr	drr	drr	drr	drr
NOV	drr	drr	drr	drr	av	drr	drr	av	av	drr	drr	drr	drr	drr	drr	drr	drr	drr	drr	drr	drr
DEC	av	av	av	av	av	w tr	av	av	av	w tr	av	w tr	av	av	drr	av	av	av	w tr	av	av
Overall	av	av	av	av	av	av	av	av	av	av	av	av	av	av	av	av	av	av	w tr	av	av
year	861	944	812	1179	1107	1260	1284	1245	1047	1193	1289	1260	971	840	758	978	1607	1583	1131	1228	22578
estimates	822	833	810	1106	1071	1355	1106	1170	1141	1257	1154	1181	821	854	765	891	1554	1328	1579	1062	21859

	Ulster Province					Connaught Province						Leinster Province				MUNSTER PROVINCE					
Date	Aldergrove	Hillsborough	Armagh	Lough Navar	Malin Head	Sligo	Ardtarmon	Belmullet	Drumsna	Galway	Roscommon	Markree Castle	Edenderry	Kilkenny	Dublin	Shannon	Newport	Tralee	Killarney	Cork	All
WINTER	av	av	av	av	drr	w tr	av	w tr	w tr	w tr	av	av	av	w tr	w tr	av	w tr	av	w tr	av	w tr
SPRING	av	av	av	av	av	av	av	av	av	av	av	av	drr	av	drr	drr	av	av	w tr	av	av
SUMMER	w tr	w tr	w tr	w tr	av	w tr	av	av	w tr	w tr	w tr	w tr	w tr	w tr	w tr	w tr	w tr	av	w tr	w tr	w tr
AUTUMN	drr	drr	drr	drr	drr	drr	drr	drr	drr	drr	drr	drr	drr	drr	drr	drr	drr	drr	drr	drr	drr
Overall	av	av	av	av	av	av	av	av	av	av	av	av	av	av	av	av	av	av	w tr	av	av
year	861	944	812	1179	1107	1260	1284	1245	1047	1193	1289	1260	971	840	758	978	1607	1583	1131	1228	22578
estimates	822	833	810	1106	1071	1355	1106	1170	1141	1257	1154	1181	821	854	765	891	1554	1328	1579	1062	21859

RAINFALL ESTIMATION SUMMARIES

% WETTER OR DRIER OF LONG-TERM AVERAGES

	Ulster Province					Connaught Province						Leinster Province				MUNSTER PROVINCE					
Date	Aldergrove	Hillsborough	Armagh	Lough Navar	Malin Head	Sligo	Ardtarmon	Belmullet	Drumsna	Galway	Roscommon	Markree Castle	Edenderry	Kilkenny	Dublin	Shannon	Newport	Tralee	Killarney	Cork	**All**
JANUARY	28%	17%	33%	22%	67%	61%	15%	64%	57%	68%	35%	42%	19%	29%	54%	22%	45%	-10%	130%	10%	40%
	w tr	w tr	w tr	w tr	w tr	w tr	w tr	w tr	w tr	w tr	w tr	w tr	w tr	w tr	w tr	w tr	w tr	drr	w tr	w tr	w tr
FEBRUARY	-48%	-48%	-56%	-56%	-59%	-24%	-57%	-48%	-37%	-25%	-41%	-44%	-58%	-27%	-20%	-44%	-42%	-25%	46%	-25%	-36%
	drr	drr	drr	drr	drr	drr	drr	drr	drr	drr	drr	drr	drr	drr	drr	drr	drr	drr	w tr	drr	drr
MARCH	36%	28%	27%	37%	11%	57%	19%	31%	69%	47%	17%	12%	21%	79%	50%	16%	17%	78%	176%	63%	43%
	w tr	w tr	w tr	w tr	w tr	w tr	w tr	w tr	w tr	w tr	w tr	w tr	w tr	w tr	w tr	w tr	w tr	w tr	w tr	w tr	w tr
APRIL	-90%	-94%	-84%	-88%	-87%	-91%	-89%	-89%	-95%	-99%	-98%	-90%	-93%	-84%	-96%	-98%	-97%	-95%	-97%	-99%	-93%
	drr	drr	drr	drr	drr	drr	drr	drr	drr	drr	drr	drr	drr	drr	drr	drr	drr	drr	drr	drr	drr
MAY	-4%	-4%	-8%	39%	108%	34%	13%	62%	46%	38%	15%	20%	-22%	-23%	-39%	-13%	64%	4%	44%	-22%	19%
	drr	drr	drr	w tr	w tr	w tr	w tr	w tr	w tr	w tr	w tr	w tr	drr	drr	drr	drr	w tr	w tr	w tr	drr	w tr
JUNE	54%	25%	83%	-9%	-34%	-20%	-23%	-11%	2%	44%	47%	-5%	-17%	72%	37%	34%	21%	36%	79%	88%	24%
	w tr	w tr	w tr	drr	drr	drr	drr	drr	w tr	w tr	w tr	drr	drr	w tr	w tr	w tr	w tr	w tr	w tr	w tr	w tr
JULY	116%	102%	145%	61%	41%	112%	46%	22%	133%	90%	123%	87%	124%	144%	181%	62%	91%	-11%	69%	49%	85%
	w tr	w tr	w tr	w tr	w tr	w tr	w tr	w tr	w tr	w tr	w tr	w tr	w tr	w tr	w tr	w tr	w tr	drr	w tr	w tr	w tr
AUGUST	44%	35%	63%	20%	14%	30%	16%	9%	32%	18%	-16%	27%	44%	63%	54%	47%	-10%	23%	40%	15%	25%
	w tr	w tr	w tr	w tr	w tr	w tr	w tr	w tr	w tr	w tr	drr	w tr	w tr	w tr	w tr	w tr	drr	w tr	w tr	w tr	w tr
SEPTEMBER	-84%	-91%	-76%	-34%	-20%	-27%	-38%	-54%	-49%	-62%	-67%	-57%	-68%	-61%	-74%	-63%	-23%	-72%	-16%	-92%	-55%
	drr	drr	drr	drr	drr	drr	drr	drr	drr	drr	drr	drr	drr	drr	drr	drr	drr	drr	drr	drr	drr
OCTOBER	-75%	-81%	-71%	-48%	-61%	-52%	-51%	-56%	-55%	-42%	-66%	-50%	-80%	-70%	-59%	-63%	-56%	-45%	-54%	-57%	-58%
	drr	drr	drr	drr	drr	drr	drr	drr	drr	drr	drr	drr	drr	drr	drr	drr	drr	drr	drr	drr	drr
NOVEMBER	-29%	-34%	-26%	-28%	-16%	-35%	-35%	-7%	-19%	-48%	-48%	-52%	-46%	-35%	-29%	-34%	-41%	-40%	-31%	-47%	-35%
	drr	drr	drr	drr	drr	drr	drr	drr	drr	drr	drr	drr	drr	drr	drr	drr	drr	drr	drr	drr	drr
DECEMBER	-3%	4%	-21%	11%	6%	36%	7%	1%	24%	38%	4%	29%	-3%	-23%	-29%	20%	21%	-1%	111%	-18%	13%
	drr	w tr	drr	w tr	w tr	w tr	w tr	w tr	w tr	w tr	w tr	w tr	drr	drr	drr	w tr	w tr	drr	w tr	drr	w tr
WINTER	11%	4%	8%	15%	40%	54%	10%	43%	43%	52%	14%	16%	3%	25%	25%	23%	25%	22%	154%	17%	32%
	w tr	w tr	w tr	w tr	w tr	w tr	w tr	w tr	w tr	w tr	w tr	w tr	w tr	w tr	w tr	w tr	w tr	w tr	w tr	w tr	w tr
SPRING	-17%	-19%	-19%	0%	8%	7%	-14%	4%	13%	1%	-18%	-15%	-29%	-6%	-30%	-27%	-3%	-4%	43%	-13%	-6%
	drr	drr	drr	**av**	w tr	w tr	drr	w tr	w tr	w tr	drr	drr	drr	drr	drr	drr	drr	drr	w tr	drr	drr
SUMMER	71%	54%	95%	26%	9%	43%	15%	7%	54%	48%	46%	38%	49%	91%	85%	47%	30%	16%	62%	48%	44%
	w tr	w tr	w tr	w tr	w tr	w tr	w tr	w tr	w tr	w tr	w tr	w tr	w tr	w tr	w tr	w tr	w tr	w tr	w tr	w tr	w tr
AUTUMN	-63%	-69%	-58%	-37%	-34%	-39%	-42%	-38%	-41%	-50%	-60%	-53%	-65%	-55%	-53%	-53%	-42%	-50%	-36%	-63%	-49%
	drr	drr	drr	drr	drr	drr	drr	drr	drr	drr	drr	drr	drr	drr	drr	drr	drr	drr	drr	drr	drr
OVERALL	-5%	-12%	0%	-6%	-3%	8%	-14%	-6%	9%	5%	-10%	-6%	-15%	2%	1%	-9%	-3%	-16%	40%	-14%	-3%
	drr	drr	**av**	drr	drr	w tr	drr	drr	w tr	w tr	drr	drr	drr	w tr	w tr	drr	drr	drr	w tr	drr	drr

ESTIMATED SUNSHINE HOURS TREND

TRENDS BY MONTHS FOR SELECTED TOWNS

	Ulster Province						Connaught Province							Leinster Province					MUNSTER PROVINCE						
Date	Aldergrove	Hillsborough	Armagh	Lough Navar	Malin Head	Ulster Overall	Sligo	Ardtarmon	Belmullet	Drumsna	Galway	Roscommon	Connaught Overall	Markree Castle	Edenderry	Kilkenny	Dublin	Leinster Overall	Shannon	Newport	Tralee	Killarney	Cork	Munster Overall	Ireland Overall
Dec 2023	snnr	snnr	snnr	snnr	snnr	snnr	av	snnr	av	snnr	av	av	snnr	av	snnr	snnr	snnr	snnr	av	snnr	snnr	snnr	av	snnr	snnr
Jan	av	av	av	av	av	av	av	av	av	av	av	cldr	cldr	av	av	av	av	av	av	av	av	av	av	av	cldr
Feb	av	av	av	av	av	snnr	av	av	av	av	av	av	snnr	av	av	av	av	snnr	av	av	av	av	av	av	snnr
Winter	snnr	snnr	snnr	snnr	snnr	snnr	av	snnr	av	snnr	av	av	av	av	snnr	snnr	snnr	snnr	av	snnr	av	av	av	av	snnr
Mar	av	av	av	av	av	av	av	av	av	av	av	av	cldr	av	av	av	av	av	av	av	av	av	av	cldr	cldr
Apr	snnr	snnr	snnr	snnr	snnr	snnr	snnr	snnr	snnr	snnr	snnr	snnr	snnr	snnr	snnr	snnr	snnr	snnr	snnr	snnr	snnr	snnr	snnr	snnr	snnr
May	av	av	snnr	av	av	snnr	av	av	snnr	snnr	snnr	snnr	snnr	snnr	snnr	snnr	snnr	snnr	av	snnr	av	av	av	snnr	snnr
Spring	snnr	snnr	snnr	snnr	snnr	snnr	snnr	snnr	snnr	snnr	snnr	snnr	snnr	snnr	snnr	snnr	snnr	snnr	snnr	snnr	snnr	snnr	snnr	snnr	snnr
June	snnr	snnr	snnr	av	av	snnr	snnr	snnr	snnr	snnr	snnr	snnr	snnr	snnr	snnr	snnr	av	snnr	snnr	snnr	av	av	snnr	snnr	snnr
July	cldr	cldr	av	cldr	av	cldr	av	av	snnr	av	av	av	av	cldr	cldr	av	cldr	cldr	av	av	av	av	av	av	cldr
Aug	av	av	snnr	snnr	snnr	snnr	snnr	snnr	snnr	snnr	snnr	snnr	snnr	snnr	snnr	snnr	av	snnr	snnr	snnr	snnr	snnr	snnr	snnr	snnr
Summer	av	av	snnr	av	snnr	snnr	snnr	snnr	snnr	snnr	snnr	snnr	snnr	snnr	av	snnr	av	snnr	snnr	snnr	snnr	snnr	snnr	snnr	snnr
Sep	av	av	av	av	av	av	av	av	av	snnr	av	av	snnr	av	av	snnr	av	snnr	av	snnr	snnr	snnr	snnr	snnr	snnr
Oct	snnr	snnr	snnr	snnr	snnr	snnr	snnr	snnr	snnr	snnr	snnr	snnr	snnr	snnr	snnr	snnr	snnr	snnr	snnr	snnr	av	av	av	snnr	snnr
Nov	av	av	av	av	av	av	av	av	av	av	av	cldr	cldr	av	av	av	av	cldr	av	av	av	av	av	av	av
Autumn	av	av	snnr	snnr	snnr	snnr	av	snnr	snnr	snnr	snnr	av	snnr	av	snnr	snnr	av	snnr	snnr	snnr	snnr	snnr	snnr	snnr	snnr
Dec	av	av	av	av	snnr	snnr	av	av	av	snnr	av	av	snnr	av	av	av	av	av	av	snnr	av	av	snnr	snnr	snnr
Year	snnr	snnr	snnr	snnr	snnr	snnr	snnr	snnr	snnr	snnr	snnr	snnr	snnr	snnr	snnr	snnr	snnr	snnr	snnr	snnr	snnr	snnr	snnr	snnr	snnr

ESTIMATED SUNSHINE HOURS

	Ulster Province					Connaught Province						Leinster Province				Munster Province							
Date	Aldergrove	Hillsborough	Armagh	Lough Navar	Malin Head	Sligo	Ardtarmon	Belmullet	Drumsna	Galway	Roscommon	Markree Castle	Edenderry	Kilkenny	Dublin	Shannon	Newport	Tralee	Killarney	Cork	Estimated Totals	Av Totals	Month Trend
Jan	46	46	46	46	46	29	46	27	50	31	29	46	50	54	64	35	44	31	31	44	844	946	less
Feb	78	78	78	78	78	74	78	71	79	70	70	78	79	80	81	69	75	63	63	79	1496	1268	more
Mar	94	94	94	94	94	79	94	90	97	89	89	94	97	100	110	88	94	78	78	89	1839	1987	less
Apr	214	214	214	214	214	202	214	218	225	223	221	214	225	235	225	227	231	238	238	246	4448	2957	more
May	203	203	203	203	203	194	203	215	206	203	209	203	206	209	212	192	201	176	176	206	4024	3611	more
Jun	188	188	188	188	188	199	188	235	187	210	222	188	187	186	176	184	185	168	168	196	3814	3173	more
Jul	119	119	119	119	119	107	119	169	134	148	159	119	134	149	131	128	138	164	164	168	2722	2841	less
Aug	163	163	163	163	163	162	163	171	169	175	173	163	169	175	164	179	177	165	165	186	3372	2793	more
Sep	118	118	118	118	118	102	118	108	136	119	114	118	136	155	134	131	143	170	170	175	2617	2219	more
Oct	114	114	114	114	114	102	114	123	113	119	121	114	113	112	126	116	114	97	97	98	2244	1653	more
Nov	54	54	54	54	54	42	54	40	61	54	47	54	61	68	66	67	67	51	51	70	1122	1161	average
Dec	48	48	48	48	48	33	48	41	57	45	43	48	57	66	63	49	57	51	51	69	1014	774	more
Estimate	1437	1437	1437	1437	1437	1325	1437	1508	1513	1486	1497	1437	1513	1588	1550	1464	1526	1451	1451	1626	29557	25381	more
Average	1314	1314	1192	1272	1230	1200	1204	1278	1134	1276	1355	1280	1327	1284	1438	1273	1140	1245	1237	1391	25381		
Trend	more	more	more	more	more	more	more	more	more	more	more	more	more	more	more	more	more	more	more	more	more		

ESTIMATED AVERAGE MONTHLY MAXIMUM TEMPERATURES

	Ulster Province					Connaught Province						Leinster Province				Munster Province					
	Aldergrove	Hillsborough	Armagh	Lough Navar	Malin Head	Sligo	Ardtarmon	Belmullet	Drumsna	Galway	Roscommon	Markree Castle	Edenderry	Kilkenny	Dublin	Shannon	Newport	Tralee	Killarney	Cork	All
Jan	10.0	10.0	10.0	9.6	9.4	9.3	9.8	10.3	9.4	9.9	9.4	9.3	9.6	10.2	10.0	10.5	10.8	9.6	11.7	9.7	**9.9**
Feb	8.2	8.2	8.2	8.3	8.5	8.1	8.0	9.3	8.1	8.4	8.1	8.1	7.4	8.5	7.8	8.6	8.8	7.4	9.1	7.9	**8.3**
Mar	10.7	10.7	10.7	10.5	10.0	10.0	11.1	10.9	10.2	10.2	10.2	10.0	10.3	11.0	10.6	11.0	10.9	10.3	11.4	10.1	**10.5**
Apr	14.6	14.6	14.6	14.2	12.2	12.8	15.8	13.8	14.7	14.1	14.7	12.8	14.5	15.5	13.3	14.8	14.7	14.5	15.1	13.9	**14.2**
May	16.6	16.6	16.6	16.3	13.9	14.8	16.9	15.5	16.0	15.3	16.0	14.8	16.1	17.3	15.5	16.5	16.1	16.1	16.8	15.4	**16.0**
Jun	18.2	18.2	18.2	18.0	15.1	16.4	19.2	17.7	18.1	16.4	18.1	16.4	17.3	18.7	16.6	18.2	18.4	17.3	18.0	16.6	**17.5**
Jul	18.2	18.2	18.2	17.7	15.5	16.5	18.5	17.1	17.5	17.7	17.5	16.5	17.7	18.7	17.6	17.7	17.8	17.7	17.8	17.3	**17.6**
Aug	18.7	18.7	18.7	18.1	16.3	17.1	19.4	17.3	18.2	17.7	18.2	17.1	19.1	19.4	18.4	18.7	18.6	19.1	18.7	17.7	**18.3**
Sep	18.8	18.8	18.8	18.1	16.2	17.3	19.5	17.5	18.2	18.4	18.2	17.3	19.0	19.7	18.6	18.9	18.7	19.0	19.4	19.0	**18.5**
Oct	16.1	16.1	16.1	15.7	14.9	15.6	16.5	16.0	15.6	15.5	15.6	15.6	15.3	16.1	15.4	16.3	16.3	15.3	16.5	14.9	**15.8**
Nov	12.7	12.7	12.7	12.6	12.7	12.6	12.7	13.4	12.2	13.0	12.2	12.6	11.9	12.9	12.4	13.4	13.0	11.9	13.6	12.3	**12.7**
Dec	9.9	9.9	9.9	9.6	9.7	9.8	9.7	10.5	9.8	10.4	9.8	9.8	9.5	10.0	9.3	10.8	10.5	9.5	12.1	9.9	**10.0**
Year	**14.4**	**14.4**	**14.4**	**14.0**	**12.9**	**13.4**	**14.8**	**14.1**	**14.0**	**13.9**	**14.0**	**13.4**	**14.0**	**14.8**	**13.8**	**14.6**	**14.5**	**14.0**	**15.0**	**13.7**	**14.1**

ESTIMATED AVERAGE MONTHLY MINIMUM TEMPERATURES

	Ulster Province					Connaught Province						Leinster Province				Munster Province					
	Aldergrove	Hillsborough	Armagh	Lough Navar	Malin Head	Sligo	Ardtarmon	Belmullet	Drumsna	Galway	Roscommon	Markree Castle	Edenderry	Kilkenny	Dublin	Shannon	Newport	Tralee	Killarney	Cork	All
Jan	4.6	4.1	4.1	3.0	5.0	5.3	5.3	5.6	4.0	3.8	4.1	5.3	3.7	4.3	4.3	5.5	5.6	7.5	7.5	5.1	**4.9**
Feb	2.6	0.9	0.9	-0.5	4.2	2.6	2.6	3.4	0.1	0.7	0.9	2.6	0.3	0.8	1.6	1.8	2.9	4.3	4.3	2.6	**2.0**
Mar	4.3	3.6	3.6	1.9	5.0	4.9	4.9	5.7	3.3	2.6	3.2	4.9	3.1	2.7	3.5	4.6	5.3	6.0	6.0	3.3	**4.1**
Apr	4.9	3.8	3.8	0.5	6.0	5.5	5.5	5.6	2.8	2.8	3.8	5.5	3.2	3.7	4.0	4.7	5.5	5.3	5.3	5.6	**4.4**
May	8.0	7.6	7.6	5.0	8.3	8.5	8.5	8.5	7.2	6.6	7.3	8.5	7.2	7.5	7.2	8.5	8.7	9.0	9.0	8.2	**7.8**
Jun	9.7	9.1	9.1	6.9	10.3	10.3	10.3	9.6	8.8	9.5	9.3	10.3	8.6	8.9	7.9	10.2	10.7	10.3	10.3	9.5	**9.5**
Jul	11.1	11.0	11.0	9.2	11.2	11.3	11.3	11.9	11.1	10.7	10.8	11.3	10.6	10.4	10.2	11.8	12.2	12.4	12.4	10.5	**11.1**
Aug	11.3	10.8	10.8	8.3	11.7	11.5	11.5	11.7	10.6	9.4	10.7	11.5	10.4	10.4	9.8	11.8	12.1	12.7	12.7	11.1	**11.0**
Sep	11.9	11.5	11.5	9.3	12.2	12.3	12.3	12.1	11.0	9.7	11.3	12.3	10.6	10.6	10.2	11.6	12.7	12.1	12.1	11.8	**11.5**
Oct	9.6	8.7	8.7	6.5	10.5	9.8	9.8	10.5	7.4	7.5	8.2	9.8	7.4	7.4	7.3	9.0	10.1	10.3	10.3	9.1	**8.9**
Nov	7.4	6.3	6.3	5.0	8.5	8.5	8.5	8.8	6.3	5.9	6.8	8.5	5.4	5.2	5.8	7.6	8.5	9.4	9.4	7.1	**7.3**
Dec	4.7	3.7	3.7	2.4	5.8	5.7	5.7	5.9	3.4	3.1	3.9	5.7	2.0	3.8	3.5	5.5	5.8	7.4	7.4	4.9	**4.7**
Year	**7.5**	**6.8**	**6.8**	**4.8**	**8.2**	**8.0**	**8.0**	**8.3**	**6.3**	**6.0**	**6.7**	**8.0**	**6.0**	**6.3**	**6.3**	**7.7**	**8.3**	**8.9**	**8.9**	**7.4**	**7.3**

ESTIMATED MAXIMUM AVERAGE TEMPERATURES – QUICK REFERENCE

MONTHLY

	Ulster Province						Connaught Province							Leinster Province					Munster Province						
	Aldergrove	Hillsborough	Armagh	Lough Navar	Malin Head	Ulster Overall	Sligo	Ardtarmon	Belmullet	Drumsna	Galway	Roscommon	Connaught Overall	Markree Castle	Edenderry	Kilkenny	Dublin	Leinster Overall	Shannon	Newport	Tralee	Killarney	Cork	Munster Overall	Overall
Jan	wmr	wmr	wmr	wmr	wmr	wmr	wmr	wmr	wmr	wmr	wmr	wmr	wmr	wmr	wmr	wmr	wmr	wmr	wmr	wmr	av	wmr	wmr	wmr	wmr
Feb	av	wmr	av	av	av	av	a-w	av	av	a-w	av	av	av	av	av	av	av	av	a-c	a-w	clr	a-w	av	av	av
Mar	a-w	wmr	av	av	a-w	a-w	wmr	wmr	av	wmr	av	wmr	a-w	a-w	a-w	av	av	a-w	av	a-w	clr	wmr	av	av	a-w
Apr	wmr	wmr	wmr	wmr	wmr	wmr	wmr	wmr	wmr	wmr	wmr	wmr	wmr	wmr	wmr	wmr	wmr	wmr	wmr	wmr	wmr	wmr	wmr	wmr	wmr
May	wmr	wmr	a-w	a-w	a-w	wmr	a-w	wmr	a-w	wmr	av	wmr	wmr	a-w	wmr	wmr	a-w	a-w	av	wmr	av	wmr	a-w	wmr	wmr
Jun	a-w	wmr	av	a-w	av	a-w	av	wmr	wmr	wmr	a-c	a-w	wmr	av	av	av	a-c	av	av	av	a-c	wmr	av	av	av
Jul	a-c	a-w	clr	clr	clr	a-c	a-c	wmr	av	av	av	clr	av	a-c	clr	clr	clr	clr	clr	clr	clr	av	clr	clr	a-c
Aug	av	wmr	a-c	a-c	a-c	av	av	wmr	av	wmr	av	av	a-w	av	av	a-c	a-c	av	a-c	av	av	wmr	a-c	av	av
Sep	wmr	wmr	wmr	wmr	a-w	wmr	wmr	wmr	a-w	wmr	wmr	wmr	wmr	wmr	wmr	wmr	wmr	wmr	wmr	wmr	wmr	wmr	wmr	wmr	wmr
Oct	wmr	wmr	wmr	wmr	wmr	wmr	wmr	wmr	wmr	wmr	wmr	wmr	wmr	wmr	wmr	wmr	wmr	wmr	wmr	wmr	a-w	wmr	wmr	wmr	wmr
Nov	wmr	wmr	wmr	wmr	wmr	wmr	wmr	wmr	wmr	wmr	wmr	wmr	wmr	wmr	wmr	wmr	wmr	wmr	wmr	wmr	av	wmr	wmr	wmr	wmr
Dec	wmr	wmr	wmr	wmr	wmr	wmr	wmr	wmr	wmr	wmr	wmr	wmr	wmr	wmr	wmr	wmr	a-w	wmr	wmr	wmr	a-c	wmr	wmr	wmr	wmr
Year	wmr	wmr	a-w	a-w	a-w	wmr	wmr	wmr	a-w	wmr	a-w	wmr	wmr	wmr	wmr	a-w	av	a-w	a-w	wmr	av	wmr	a-w	a-w	wmr

wmr = 1.0> degrees above average | a-w = 0.6-1.0 degrees above average | av = 0.0-0.5 degrees from average | a-c = 0.6-1.0 degrees below average | clr = 1.0 > degrees below average

SEASONAL

	Ulster Province						Connaught Province							Leinster Province					Munster Province						
	Aldergrove	Hillsborough	Armagh	Lough Navar	Malin Head	Ulster Overall	Sligo	Ardtarmon	Belmullet	Drumsna	Galway	Roscommon	Connaught Overall	Markree Castle	Edenderry	Kilkenny	Dublin	Leinster Overall	Shannon	Newport	Tralee	Killarney	Cork	Munster Overall	Overall
Winter	wmr	wmr	wmr	a-w	wmr	wmr	wmr	wmr	a-w	wmr	wmr	wmr	wmr	wmr	wmr	wmr	wmr	wmr	a-w	wmr	clr	wmr	a-w	wmr	wmr
Spring	wmr	wmr	a-w	a-w	a-w	wmr	wmr	wmr	a-w	wmr	a-w	wmr	wmr	a-w	wmr	wmr	a-w	wmr	a-w	wmr	av	wmr	wmr	wmr	wmr
Summer	av	wmr	a-c	av	a-c	av	av	wmr	av	a-w	av	av	av	av	av	a-c	clr	a-c	a-c	av	a-c	a-w	a-c	av	av
Autumn	wmr	wmr	wmr	wmr	wmr	wmr	wmr	wmr	wmr	wmr	wmr	wmr	wmr	wmr	wmr	wmr	wmr	wmr	wmr	wmr	a-w	wmr	wmr	wmr	wmr
December	wmr	wmr	wmr	wmr	wmr	wmr	wmr	wmr	wmr	wmr	wmr	wmr	wmr	wmr	wmr	wmr	a-w	wmr	wmr	wmr	a-c	wmr	wmr	wmr	wmr
Year	wmr	wmr	a-w	a-w	a-w	wmr	wmr	wmr	a-w	wmr	a-w	wmr	wmr	wmr	wmr	a-w	av	a-w	a-w	wmr	av	wmr	a-w	a-w	wmr

wmr = 1.0> degrees above average | a-w = 0.6-1.0 degrees above average | av = 0.0-0.5 degrees from average | a-c = 0.6-1.0 degrees below average | clr = 1.0 > degrees below average

ESTIMATED MINIMUM AVERAGE TEMPERATURES – QUICK REFERENCE

MONTHLY

	Ulster Province						Connaught Province							Leinster Province					Munster Province						
	Aldergrove	Hillsborough	Armagh	Lough Navar	Malin Head	Ulster Overall	Sligo	Ardtarmon	Belmullet	Drumsna	Galway	Roscommon	Connaught Overall	Markree Castle	Edenderry	Kilkenny	Dublin	Leinster Overall	Shannon	Newport	Tralee	Killarney	Cork	Munster Overall	Overall
Jan	wmr	wmr	wmr	wmr	wmr	wmr	wmr	wmr	wmr	wmr	wmr	wmr	wmr	wmr	wmr	wmr	wmr	wmr	wmr	wmr	wmr	wmr	wmr	wmr	wmr
Feb	wmr	av	av	a-c	a-w	av	av	av	av	clr	clr	a-c	av	a-w	a-c	clr	a-c	av	clr	av	wmr	wmr	av	a-w	av
Mar	wmr	wmr	wmr	a-c	a-w	a-w	wmr	wmr	a-w	a-w	a-c	a-w	a-w	wmr	a-w	av	av	a-w	av	wmr	wmr	wmr	a-c	wmr	a-w
Apr	wmr	av	av	clr	av	av	wmr	wmr	av	a-c	clr	av	av	wmr	av	av	a-c	av	a-c	wmr	wmr	av	a-w	a-w	av
May	wmr	wmr	wmr	clr	av	a-w	wmr	wmr	a-w	wmr	av	wmr	wmr	wmr	wmr	a-w	av	wmr	av	wmr	wmr	wmr	a-w	wmr	wmr
Jun	av	av	av	clr	av	av	wmr	wmr	a-c	av	av	a-w	av	wmr	av	av	clr	av	a-c	wmr	wmr	av	av	av	av
Jul	av	av	av	clr	a-c	av	a-w	a-w	av	av	a-c	av	av	a-w	av	a-c	clr	av	clr	a-w	wmr	a-w	clr	av	av
Aug	a-w	av	av	clr	a-c	av	wmr	wmr	av	av	clr	av	av	wmr	av	a-c	clr	av	a-c	a-w	wmr	wmr	a-c	a-w	av
Sep	wmr	wmr	wmr	av	wmr	wmr	wmr	wmr	wmr	wmr	av	wmr	wmr	wmr	wmr	wmr	av	wmr	a-w	wmr	wmr	wmr	wmr	wmr	wmr
Oct	wmr	wmr	wmr	av	wmr	wmr	wmr	wmr	wmr	a-w	av	wmr	wmr	wmr	a-w	a-w	av	wmr	a-w	wmr	wmr	wmr	wmr	wmr	wmr
Nov	wmr	wmr	wmr	wmr	wmr	wmr	wmr	wmr	wmr	wmr	wmr	wmr	wmr	wmr	wmr	wmr	wmr	wmr	wmr	wmr	wmr	wmr	wmr	wmr	wmr
Dec	wmr	wmr	wmr	a-w	wmr	wmr	wmr	wmr	wmr	a-w	av	wmr	wmr	wmr	av	wmr	a-w	wmr	wmr	wmr	wmr	wmr	wmr	wmr	wmr
Year	wmr	wmr	wmr	a-c	a-w	a-w	wmr	wmr	a-w	a-w	av	wmr	wmr	wmr	a-w	av	av	a-w	av	wmr	wmr	wmr	av	wmr	wmr

wmr = 1.0> degrees above average | a-w = 0.6-1.0 degrees above average | av = 0.0-0.5 degrees from average | a-c = 0.6-1.0 degrees below average | clr = 1.0 > degrees below average

SEASONAL

	Ulster Province						Connaught Province							Leinster Province					Munster Province						
	Aldergrove	Hillsborough	Armagh	Lough Navar	Malin Head	Ulster Overall	Sligo	Ardtarmon	Belmullet	Drumsna	Galway	Roscommon	Connaught Overall	Markree Castle	Edenderry	Kilkenny	Dublin	Leinster Overall	Shannon	Newport	Tralee	Killarney	Cork	Munster Overall	Overall
Winter	wmr	wmr	wmr	a-w	wmr	wmr	wmr	wmr	wmr	a-w	av	wmr	wmr	wmr	wmr	wmr	a-w	wmr	a-w	wmr	wmr	wmr	a-w	wmr	wmr
Spring	wmr	wmr	a-w	clr	av	av	wmr	wmr	av	av	clr	a-w	a-w	wmr	av	av	av	a-w	av	wmr	wmr	wmr	av	wmr	a-w
Summer	av	av	av	clr	av	av	a-w	a-w	av	av	a-c	av	av	wmr	av	a-c	clr	av	a-c	a-w	wmr	a-w	a-c	av	av
Autumn	wmr	wmr	wmr	av	wmr	wmr	wmr	wmr	wmr	wmr	av	wmr	wmr	wmr	wmr	wmr	a-w	wmr	wmr	wmr	wmr	wmr	wmr	wmr	wmr
December	wmr	wmr	wmr	a-w	wmr	wmr	wmr	wmr	wmr	a-w	av	wmr	wmr	wmr	av	wmr	a-w	wmr	wmr	wmr	wmr	wmr	wmr	wmr	wmr
Year	wmr	wmr	wmr	a-c	a-w	a-w	wmr	wmr	a-w	a-w	av	wmr	wmr	wmr	a-w	av	av	a-w	av	wmr	wmr	wmr	av	wmr	wmr

wmr = 1.0> degrees above average | a-w = 0.6-1.0 degrees above average | av = 0.0-0.5 degrees from average | a-c = 0.6-1.0 degrees below average | clr = 1.0 > degrees below average

GARDENING GUIDE 2025

REST PERIOD:
The Rest Period in the garden is as important as the activity times. It's a time when the Moon is Void of Course, that is, traveling between one sign and another and during this time the Moon is deemed to have "no energizing power", also Rest Periods on the day of the Full Moon, New Moon and when the moon is at Southern and Northern Declinations. Hence it is a good time to sit and enjoy your garden or to attend to non-plant activities such as tidy your shed, repair or maintain outdoor structures or even to sit and plan your next "to do" list when the Moon is once more energized. Eclipse days and the day after the eclipse are also deemed days to rest in the garden.

ALL ABOVE-GROUND ACTIVITY SHOULD TAKE PLACE when the Moon is Waxing – i.e. building from New Moon through the 1st Quarter and up to the Full Moon. This is a time of increasing light and with the Moon pulling the sap upwards into the leaves and flowers.

ALL BELOW-GROUND ACTIVITY SHOULD TAKE PLACE when the Moon is Waning – i.e. from the Full Moon through the 3rd Quarter and up to the New Moon. This is a phase of decreasing light and the general trend for sap to be pulled downwards and into the root and tuber systems.

THE NEW MOON PHASE:
Activities can begin in the garden once the crescent of the New Moon begins to be visible.

At this time:

DO start sowing, transplanting, grafting or strike cuttings of all ABOVE-GROUND producers. Flowering annuals, brassicas (cabbage, broccoli etc) cover crops (mustard, lupins etc) and annual fruiting crops (cucumber, pumpkin, tomatoes etc) and house-plants all benefit from attention during this phase. Also, mow lawns should you wish to encourage growth or sow any large areas such as lawns.

DON'T plant root crops (as they will only produce plenty of leaves and little roots, or go to seed too early), nor prune at this time – your plants, and trees in particular, will be prone to excessive sap loss and suffer from die-back.

THE 1ST QUARTER PHASE:
Continue on with the New Moon phase activities and especially sow any seeds two days before the Full Moon to yield optimum results. During drought times, sow during this phase when the Moon is in Capricorn or Taurus for the strongest of survival qualities.

At this time:

DO sow and transplant all legume, grain, flowering plants (especially roses) and pot plants and start harvesting any above-ground produce, with 2-3 days before the Full Moon being the most beneficial time (especially for herbs and grape harvesting).

DO give plants a food boost if required, but do water it down and apply as close to the Full Moon as possible, particularly for those plants that require phosphorus.

DON'T harvest with the Moon in a water sign (Cancer, Scorpio or Pisces) unless for immediate use, nor prune or plant out root crops.

THE FULL MOON PHASE:
The Moon will have reached its full potency by the time of the Full Moon, and from this point to the New Moon, there is a gradual decrease in its vitality and the sap flow is now downwards into the roots – it is a time when the plants build up their root systems and store their goodness. However, you can continue to harvest above-ground producers for a few more days after the Full Moon. Leeks, onions, garlic etc are at their best potency when harvested with the Full Moon in Leo.

At this time:

DO sow or plant out any root crops (potatoes, kumaras, carrots etc) especially during the two days in the middle of the Full Moon phase. Also sow or plant out fruting plants (e.g. strawberries, passion fruit etc) as well as perennial flowering plants, bulbs and tubers (e.g. daffodils, dahlias, crocus etc) and this is also a good time to spray, prune and trim hedges.

DO perform drying activities of herbs, flowers and fruits.

DO dig herb roots or harvest leaves and bark intended for medicinal teas.

DO spray fruit trees, mulch your garden and kill weeds.

THE 3RD QUARTER PHASE:
This is considered to be the resting period of the Lunar cycle, the time when the sap in the plants runs at its slowest.

DO harvest all root crops that you wish to keep for long term storage (e.g. potatoes, garlic, onions and root herbs etc), especially if the Moon is in Taurus or Capricorn. Also lift your bulbs and tubers for their seasonal storage. You can also use this time to build up your compost, especially when the Moon is in Scorpio and attend to maintenance chores such as spreading compost, mulching, staking taller plants etc, as well as attend to all pruning and trimming of trees, vines and bushes, as there is less chance of die back and excessive sap bleeding. Cut your lawns during this phase to discourage growth.

DO plant trees and saplings during this phase for good strong foundations of root growth and bark development as well as plant strawberries or their runners so their growth is well developed before fruiting.

DO fertilize your crops for potassium if needed, as potassium absorption is at its peak at this phase.

When the Moon is in Cancer, Scorpio or Pisces:
This is a beneficial time for attending to vegetables such as lettuce, cabbage, herbs, broccoli, leafy pot plants, etc. Avoid any harvesting except for immediate use during these Moon signs as anything harvested for long term storage will deteriorate faster and be prone to mildew, except of course, unless you are attempting to build compost – choose when the Moon is in Scorpio for best results, as vegetation decomposes fastest under this influence.

When the Moon is in Taurus, Virgo or Capricorn:
This is a beneficial time for attending to vegetables such as potatoes, carrots, leeks etc. as well as to non-flowering shrubs and trees and ground covers such as lawns all respond best attended to under these signs. A Virgo Moon (particularly in the 3rd Quarter phase) is best for harvesting any medicinal leafy herbs due to the usual desire for long term storage.

When the Moon is in Gemini, Libra, Aquarius:
This is a beneficial time to attend to vegetables such as globe artichokes, flowers for medicinal purposes, and flowers in general. (Brocolli and cauliflower etc are considered "leafy" as if planted during a Flower phase they are more likely to bolt straight to seed). The Moon's focus is more directed towards developing perfume, fragrance, flower vibrancy and beauty as well as medicinal properties with the Full Moon in one of these signs.

When the Moon is in Aries, Leo or Sagittarius:
This is a beneficial time to attend to any fruit, berry, seed or grain crops such as tomatoes, peas, beans, peppers, wheat or corn, pumpkin, apples, strawberries, raspberries etc. Any seeds you wish to store for next year's planting, or for long term storage are best attended to during these Moon signs, as well as harvesting crops such as onions and garlic for their most potent flavour.

Above-ground - Leafy, Fruits and Seeds examples:

Vegetables	beans, broccoli, brussels sprouts, cabbages, cauliflower, chicory, celery, cress, cucumber, endive, lettuces, marrow, mustard, parsley, rhubarb, sage, silverbeet, corn, etc.
Flowers	cornflowers, forget-me-nots, marigolds, pansies, poppies, roses etc.
Fruits and seeds	apples, blue-berries, grapes, passion-fruit, pears, peas, peppers, raspberries, strawberries, tomatoes, wheat

Below-ground - Roots, Bulbs and Non-Flowering shrubs examples:

Vegetables	beetroot, carrots, garlic, leeks, onions, parsnips, radish, salsify, shallots, swedes, turnips etc.
Flowers	agapanthus, anemone, crocus, daffodils, dahlias, freesias, gladioli, hyacinth, ranunculus, red-hot pokers, tuberose, tulips
Fruits and Seeds	garlic, onions, peanuts

BASIC MOON IN SIGN GUIDE

Moon in Aries – Harvest root and fruit for storage. Good for destroying weeds and pests.

Moon in Taurus – Excellent time for planting and transplanting, particularly root crops. Also good for leaf crops.

Moon in Gemini – Good for harvesting root and fruit for storage. Good for destroying weeds and pests. Melon seeds respond well.

Moon in Cancer – Best time for all planting and transplanting. Good for grafting and irrigation.

Moon in Leo – Good for harvesting root and fruit for storage. Excellent for destroying weeds and pest.

Moon in Virgo – Good for destroying weeds and pests. Vines and some flowers respond well.

Moon in Libra – Best time for planting for fragrance and beauty of flowers, vines and herbs.

Moon in Scorpio – Best time for planting sturdy plants and vines. Especially favoured by tomatoes, corn and squash. Good for irrigation and transplanting.

Moon in Sagittarius – Good for harvesting roots. Especially favoured by onions and fruit trees. Good time to cultivate the soil.

Moon in Capricorn – Plant potatoes and root crops and encouraging hardy growth. Good for grafting and pruning to promote healing and applying organic fertilizers.

Moon in Aquarius – Harvest root and fruit for storage. Good for destroying weeds and pest. Good for planting onions.

Moon in Pisces – Good for planting and transplanting, especially for root growth. Good for irrigation.

JANUARY	2025			Hour: 1am 2am 3am 4am 5am 6am 7am 8am 9am 10am 11am Noon 1pm 2pm 3pm 4pm 5pm 6pm 7pm 8pm 9pm 10pm 11pm Midnight
	1st	Wed	Above	Capricorn / REST / Aquarius
	2nd	Thur	Above	Aquarius
	3rd	Fri	Above	Aquarius / REST / Pisces
	4th	Sat	Above	Pisces
	5th	Sun	Above	Pisces / REST / Aries
1Q	6th	Mon	Above	Aries
P13 XhN	7th	Tue	Above	Aries / REST / Taurus
	8th	Wed	Above	Taurus
	9th	Thur	Above	Taurus / REST
	10th	Fri	Above	REST / Gemini
	11th	Sat	Above	Gemini
^	12th	Sun	Above	REST DAY - Northern Declination
F	13th	Mon	Below	REST DAY - Full Moon
	14th	Tue	Below	Cancer / REST / Leo
	15th	Wed	Below	Leo
	16th	Thur	Below	Leo / REST / Virgo
	17th	Fri	Below	Virgo
	18th	Sat	Below	Virgo
	19th	Sun	Below	Virgo / REST / Libra
XhS	20th	Mon	Below	Libra
3Q A	21st	Tue	Below	Libra / REST / Scorpio
	22nd	Wed	Below	Scorpio
	23rd	Thur	Below	Scorpio
	24th	Fri	Below	REST / Sagittarius
	25th	Sat	Below	Sagittarius
V	26th	Sun	Below	REST DAY - Southern Declination
	27th	Mon	Below	Capricorn
	28th	Tue	Below	Capricorn / REST / Aquarius
N	29th	Wed	Above	REST DAY - New Moon
	30th	Thur	Above	Aquarius / REST / Pisces
	31st	Fri	Above	Pisces

FEBRUARY	2025			Hour: 1am 2am 3am 4am 5am 6am 7am 8am 9am 10am 11am Noon 1pm 2pm 3pm 4pm 5pm 6pm 7pm 8pm 9pm 10pm 11pm Midnight
	1st	Sat	Above	Pisces / REST
P10	2nd	Sun	Above	REST / Aries
XhN	3rd	Mon	Above	Aries / REST
	4th	Tue	Above	REST / Taurus
1Q	5th	Wed	Above	Taurus
	6th	Thur	Above	Taurus / REST / Gemini
	7th	Fri	Above	Gemini
^	8th	Sat	Above	REST DAY - Northern Declination
	9th	Sun	Above	Cancer
	10th	Mon	Above	Cancer / REST / Leo
	11th	Tue	Above	Leo
F	12th	Wed	Below	REST DAY - Full Moon
	13th	Thur	Below	REST / Virgo
	14th	Fri	Below	Virgo
	15th	Sat	Below	Virgo / REST / Libra
XhS	16th	Sun	Below	Libra

FEBRUARY		2025	Hour:	1am – Midnight
	17th	Mon	Below	Libra / REST
A	18th	Tue	Below	RE / Scorpio
	19th	Wed	Below	Scorpio
3Q	20th	Thur	Below	Scorpio / REST / Sagittarius
	21st	Fri	Below	Sagittarius
V	22nd	Sat	Below	REST DAY - Southern Declination
	23rd	Sun	Below	Capricorn
	24th	Mon	Below	Capricorn
	25th	Tue	Below	Capricorn / REST / Aquarius
	26th	Wed	Below	Aquarius / REST
	27th	Thur	Below	REST / Pisces
N	28th	Fri	A	REST DAY - New Moon

MARCH		2025	Hour:	1am – Midnight
P7	1st	Sat	Above	Pisces / REST / Aries
XhN	2nd	Sun	Above	Aries / REST
	3rd	Mon	Above	REST / Taurus
	4th	Tue	Above	Taurus
	5th	Wed	Above	Taurus / REST / Gemini
1Q	6th	Thur	Above	Gemini
^	7th	Fri	Above	REST DAY - Northern Declination
	8th	Sat	Above	Cancer
	9th	Sun	Above	Cancer / REST / Leo
	10th	Mon	Above	Leo
	11th	Tue	Above	Leo / REST
	12th	Wed	Above	REST / Virgo
	13th	Thur	Above	Virgo
F	14th	Fri	Below	REST DAY - Full Moon - Lunar Total Eclipse
XhS	15th	Sat	Below	REST DAY - Day after Eclipse Day
	16th	Sun	Below	Libra / REST
A	17th	Mon	Below	REST / Scorpio
	18th	Tue	Below	Scorpio
	19th	Wed	Below	Scorpio / REST / Sagittarius
	20th	Thur	Below	Sagittarius
	21st	Fri	Below	Sagittarius
3Q V	22nd	Sat	Below	REST DAY - Southern Declination
	23rd	Sun	Below	Capricorn
	24th	Mon	Below	Capricorn / RE / Aquarius
	25th	Tue	Below	Aquarius
	26th	Wed	Below	Aquarius / REST / Pisces
	27th	Thur	Below	Pisces
	28th	Fri	Below	Pisces / RE / Aries
N XhN	29th	Sat	Above	REST DAY - New Moon - Partial Solar Eclipse
P4	30th	Sun	Above	REST DAY - Day after Eclipse Day (Daylight Savings Starts)
	31st	Mon	Above	Taurus

APRIL		2025	Hour:	1am – Midnight
	1st	Tue	Above	Taurus · REST · Gemini
	2nd	Wed		Gemini
^	3rd	Thur		REST DAY - Northern Declination
	4th	Fri		Cancer
1Q	5th	Sat		Cancer · REST
	6th	Sun		REST · Leo
	7th	Mon		Leo
	8th	Tue		Leo · REST · Virgo
	9th	Wed		Virgo
	10th	Thur		Virgo · REST
XhS	11th	Fri		REST · Libra
	12th	Sat		Libra
F A	13th	Sun	Below	REST DAY - Full Moon
	14th	Mon		Scorpio
	15th	Tue		Scorpio
	16th	Wed		Scorpio · RE · Sagittarius
	17th	Thur		Sagittarius
V	18th	Fri		REST DAY - Southern Declination
	19th	Sat		Capricorn
	20th	Sun		Capricorn · REST · Aquar
3Q	21st	Mon		Aquarius
	22nd	Tue		Aquarius · REST
	23rd	Wed		REST · Pisces
	24th	Thur		Pisces
	25th	Fri		Pisces · REST · Aries
XhN	26th	Sat		Aries · REST
N P3	27th	Sun	Above	REST DAY - New Moon
	28th	Mon		Taurus
	29th	Tue		Taurus · REST · Gemini
	30th	Wed		Gemini

MAY		2025	Hour:	1am – Midnight
^	1st	Thur	Above	REST DAY - Northern Declination
	2nd	Fri		Cancer
	3rd	Sat		Cancer · REST · Leo
1Q	4th	Sun		Leo
	5th	Mon		Leo · REST · Virgo
	6th	Tue		Virgo
	7th	Wed		Virgo
	8th	Thur		Virgo · REST · Libra
XhS	9th	Fri		Libra
	10th	Sat		Libra · REST · Scorpio
A	11th	Sun		Scorpio
F	12th	Mon	Below	REST DAY - Full Moon
	13th	Tue		Scorpio · REST · Sagittarius
	14th	Wed		Sagittarius
V	15th	Thur		REST DAY - Southern Declination
	16th	Fri		Capricorn
	17th	Sat		Capricorn

MAY		2025	Hour:	1am	2am	3am	4am	5am	6am	7am	8am	9am	10am	11am	Noon	1pm	2pm	3pm	4pm	5pm	6pm	7pm	8pm	9pm	10pm	11pm	Midnight
	18th	Sun	Below			Capricorn		REST										Aquarius									
	19th	Mon													Aquarius												
3Q	20th	Tue							Aquarius						REST							Pisces					
	21st	Wed													Pisces												
	22nd	Thur									Pisces									RE				Aries			
XhN	23rd	Fri													Aries												
	24th	Sat							Aries									REST						Taurus			
	25th	Sun													Taurus												
P5	26th	Mon								Taurus									REST					Gemini			
N	27th	Tue	Above												REST DAY - New Moon												
^	28th	Wed													REST DAY - Northern Declination												
	29th	Thur													Cancer												
	30th	Fri										Cancer										REST			Leo		
	31st	Sat													Leo												

JUNE		2025	Hour:	1am	2am	3am	4am	5am	6am	7am	8am	9am	10am	11am	Noon	1pm	2pm	3pm	4pm	5pm	6pm	7pm	8pm	9pm	10pm	11pm	Midnight
	1st	Sun	Above												Leo												REST
	2nd	Mon			REST												Virgo										
1Q	3rd	Tue													Virgo												
	4th	Wed							Virgo							REST						Libra					
XhS	5th	Thur													Libra												
	6th	Fri													Libra												
A	7th	Sat		Libra	REST												Scorpio										
	8th	Sun													Scorpio												
	9th	Mon							Scorpio								REST						Sagittarius				
	10th	Tue													Sagittarius												
F V	11th	Wed	Below												REST DAY - Full Moon / Southern Declination												
	12th	Thur		REST												Capricorn											
	13th	Fri													Capricorn												
	14th	Sat						Capricorn					REST								Aquarius						
	15th	Sun													Aquarius												
	16th	Mon										Aquarius									RE			Pisces			
	17th	Tue													Pisces												
3Q	18th	Wed												Pisces											REST		Aries
XhN	19th	Thur													Aries												
	20th	Fri													Aries												
	21st	Sat		Aries	RE											Taurus											
	22nd	Sun													Taurus												
P8	23rd	Mon		Taurus	REST												Gemini										
	24th	Tue													Gemini												
N ^	25th	Wed	Above												REST DAY - New Moon / Northern Declination												
	26th	Thur													Cancer												
	27th	Fri				Cancer			REST									Leo									
	28th	Sat													Leo												
	29th	Sun							Leo						REST						Virgo						
	30th	Mon													Virgo												

JULY		2025	Hour:	1am – Midnight
	1st	Tue	Above	Virgo / REST / Libra
1Q XhS	2nd	Wed	Above	Libra / REST
	3rd	Thur	Above	REST DAY - Moon Void of Course all day
	4th	Fri	Above	REST / Scorpio
A	5th	Sat	Above	Scorpio
	6th	Sun	Above	Scorpio / RE / Sagittarius
	7th	Mon	Above	Sagittarius / REST
	8th	Tue	Above	REST DAY - Moon Void of Course all day
V	9th	Wed	Above	REST DAY - Southern Declination
F	10th	Thur	Below	REST DAY - Full Moon
	11th	Fri	Below	REST / Aquarius
	12th	Sat	Below	Aquarius / REST
	13th	Sun	Below	REST DAY - Moon Void of Course all day
	14th	Mon	Below	Pisces
	15th	Tue	Below	Pisces / REST
	16th	Wed	Below	REST / Aries
XhN	17th	Thur	Below	Aries
3Q	18th	Fri	Below	Ar / REST / Taurus
	19th	Sat	Below	Taurus
P11	20th	Sun	Below	Taurus / REST / Gemini
	21st	Mon	Below	Gemini / REST
^	22nd	Tue	Below	REST DAY - Northern Declination
	23rd	Wed	Below	Cancer
N	24th	Thur	Above	REST DAY - New Moon
	25th	Fri	Above	Leo
	26th	Sat	Above	Leo / REST / Virgo
	27th	Sun	Above	Virgo
	28th	Mon	Above	Virgo
	29th	Tue	Above	Virgo / REST / Libra
XhS	30th	Wed	Above	Libra / REST
	31st	Thur	Above	REST / Scorpio

AUGUST		2025	Hour:	1am – Midnight
1Q A	1st	Fri	Above	Scorpio
	2nd	Sat	Above	Scorpio
	3rd	Sun	Above	Scorp / REST / Sagittarius
	4th	Mon	Above	Sagittarius
V	5th	Tue	Above	REST DAY - Southern Declination
	6th	Wed	Above	Capricorn / REST
	7th	Thur	Above	REST DAY - Moon Void of Course all day
	8th	Fri	Above	REST / Aquarius
F	9th	Sat	Below	REST DAY - Full Moon
	10th	Sun	Below	REST / Pisces
	11th	Mon	Below	Pisces / REST
	12th	Tue	Below	REST / Aries
XhN	13th	Wed	Below	Aries / REST
P12	14th	Thur	Below	REST / Taurus
	15th	Fri	Below	Taurus
3Q	16th	Sat	Below	Taurus / REST / Gemini
	17th	Sun	Below	Gemini

AUGUST		2025	Hour:	1am – Midnight
^	18th	Mon	Below	REST DAY - Northern Declination
	19th	Tue	Below	Cancer
	20th	Wed	Below	Cancer; REST; Leo
	21st	Thur	Below	Leo; REST
	22nd	Fri	Below	REST DAY - Moon Void of Course all day
N	23rd	Sat	Above	REST DAY - New Moon
	24th	Sun	Above	Virgo
	25th	Mon	Above	Virgo; RE; Libra
XhS	26th	Tue	Above	Libra
	27th	Wed	Above	Libra; REST
	28th	Thur	Above	REST; Scorpio
A	29th	Fri	Above	Scorpio
	30th	Sat	Above	Sco; REST; Sagittarius
1Q	31st	Sun	Above	Sagittarius

SEPTEMBER		2025	Hour:	1am – Midnight
V	1st	Mon	Above	REST DAY - Southern Declination
	2nd	Tue	Above	REST; Capricorn
	3rd	Wed	Above	Capricorn
	4th	Thur	Above	Capricorn; RE; Aquarius
	5th	Fri	Above	Aquarius; REST
	6th	Sat	Above	REST; Pisces
F	7th	Sun	Below	REST DAY - Full Moon - Lunar Total Eclipse
	8th	Mon	Below	REST DAY - Day after Eclipse Day
XhN	9th	Tue	Below	Aries
P9	10th	Wed	Below	Aries; REST; Taurus
	11th	Thur	Below	Taurus
	12th	Fri	Below	Taurus; REST; Gemini
	13th	Sat	Below	Gemini
3Q ^	14th	Sun	Below	REST DAY - Northern Declination
	15th	Mon	Below	RE; Cancer
	16th	Tue	Below	Cancer
	17th	Wed	Below	Cancer; REST; Leo
	18th	Thur	Below	Leo
	19th	Fri	Below	Leo; RE; Virgo
	20th	Sat	Below	Virgo
N	21st	Sun	Above	REST DAY - New Moon / Partial Solar Eclipse
XhS	22nd	Mon	Above	REST DAY - Day after Eclipse Day
	23rd	Tue	Above	Libra; REST
	24th	Wed	Above	REST; Scorpio
	25th	Thur	Above	Scorpio
A	26th	Fri	Above	Scorpio; REST; Sagittarius
	27th	Sat	Above	Sagittarius
	28th	Sun	Above	Sagittarius
1Q V	29th	Mon	Above	REST DAY - Southern Declination
	30th	Tue	Above	Capricorn

OCTOBER 2025

Hour: 1am 2am 3am 4am 5am 6am 7am 8am 9am 10am 11am Noon 1pm 2pm 3pm 4pm 5pm 6pm 7pm 8pm 9pm 10pm 11pm Midnight

	1st	Wed	Above	Capricorn / REST / Aquarius
	2nd	Thur	Above	Aquarius
	3rd	Fri	Above	Aquarius / REST
	4th	Sat	Above	REST / Pisces
	5th	Sun	Above	Pisces
	6th	Mon	Above	Pis / REST / Aries
F XhN	7th	Tue	Below	REST DAY - Full Moon
P6	8th	Wed	Below	REST / Taurus
	9th	Thur	Below	Taurus
	10th	Fri	Below	Tau / REST / Gemini
	11th	Sat	Below	Gemini
^	12th	Sun	Below	REST DAY - Northern Declination
3Q	13th	Mon	Below	Cancer
	14th	Tue	Below	Cancer / REST / Leo
	15th	Wed	Below	Leo
	16th	Thur	Below	Leo / REST / Virgo
	17th	Fri	Below	Virgo
	18th	Sat	Below	Virgo / REST
	19th	Sun	Below	REST / Libra
XhS	20th	Mon	Below	Libra
N	21st	Tue	Above	REST DAY - New Moon
	22nd	Wed	Above	Scorpio
A	23rd	Thur	Above	Scorpio
	24th	Fri	Above	Scorpio / RE / Sagittarius
	25th	Sat	Above	Sagittarius
V	26th	Sun	Above	REST DAY - Southern Declination / Daylight Savings Ends
	27th	Mon	Above	Capricorn
	28th	Tue	Above	Capricorn
1Q	29th	Wed	Above	Capricorn / RE / Aquarius
	30th	Thur	Above	Aquarius
	31st	Fri	Above	Aquarius / REST / Pisces

NOVEMBER 2025

Hour: 1am 2am 3am 4am 5am 6am 7am 8am 9am 10am 11am Noon 1pm 2pm 3pm 4pm 5pm 6pm 7pm 8pm 9pm 10pm 11pm Midnight

	1st	Sat	Above	Pisces
	2nd	Sun	Above	Pisces / REST / Aries
XhN	3rd	Mon	Above	Aries
	4th	Tue	Above	Aries / REST / Taurus
F P1	5th	Wed	Below	REST DAY - Full Moon
	6th	Thur	Below	Taurus / REST / Gemini
	7th	Fri	Below	Gemini
^	8th	Sat	Below	REST DAY - Northern Declination
	9th	Sun	Below	Cancer
	10th	Mon	Below	Cancer / RE / Leo
	11th	Tue	Below	Leo
3Q	12th	Wed	Below	Leo / RE / Vir
	13th	Thur	Below	Virgo
	14th	Fri	Below	Virgo
	15th	Sat	Below	Virgo / REST / Libra
XhS	16th	Sun	Below	Libra
	17th	Mon	Below	Libra / REST / Scorpio

NOVEMBER		2025	Hour:	Schedule (1am – Midnight)
	18th	Tue	Belo	Scorpio
	19th	Wed		Scorpio
N A	20th	Thur	Above	REST DAY - New Moon
	21st	Fri		Sagittarius
V	22nd	Sat		REST DAY - Southern Declination
	23rd	Sun		Capricorn
	24th	Mon		Capricorn
	25th	Tue		Capricorn / REST / Aquarius
	26th	Wed		Aquarius
	27th	Thur		Aquarius / REST / Pisces
1Q	28th	Fri		Pisces
	29th	Sat		Pisces
XhN	30th	Sun		REST / Aries

DECEMBER		2025	Hour:	Schedule (1am – Midnight)
	1st	Mon	Above	Aries / REST
	2nd	Tue		REST / Taurus
	3rd	Wed		Taurus
F P2	4th	Thur	Below	REST DAY - Full Moon
^	5th	Fri		REST DAY - Northern Declination
	6th	Sat		Gemin / REST / Cancer
	7th	Sun		Cancer
	8th	Mon		Cancer / REST / Leo
	9th	Tue		Leo
	10th	Wed		Leo / REST / Virgo
3Q	11th	Thur		Virgo
	12th	Fri		Virgo / REST / Libı
XhS	13th	Sat		Libra
	14th	Sun		Libra
	15th	Mon		Libra / RE / Scorpio
	16th	Tue		Scorpio
A	17th	Wed		Scorpio / REST / Sagittarius
	18th	Thur		Sagittarius
V	19th	Fri		REST DAY - Southern Declination
N	20th	Sat	Above	REST DAY - New Moon
	21st	Sun		Capricorn
	22nd	Mon		Capricorn / REST / Aquarius
	23rd	Tue		Aquarius
	24th	Wed		Aquarius / REST
	25th	Thur		REST / Pisces
	26th	Fri		Pisces
1Q XhN	27th	Sat		Pisces / REST / Aries
	28th	Sun		Aries
	29th	Mon		Aries / REST / Taurus
	30th	Tue		Taurus
	31st	Wed		Taurus / REST / Gemini

FISHING CALENDAR 2025

AV	VG	EX
1 fish	2 fish	3 fish

JANUARY	1st	2nd	3rd	4th	5th	6th	7th	8th	9th	10th	11th	12th	13th	14th	15th	16th	17th	18th	19th	20th	21st	22nd	23rd	24th	25th	26th	27th	28th	29th	30th	31st
MOON						First Q	P13						Full Moon								Third Q, Ap								New Moon		
Best Bite Time am/pm	1.00	2.00	3.00	4.00	5.00	6.00	6.30	7.00	8.00	9.00	10.00	11.00	12.00	12.30	1.00	2.00	3.00	4.00	4.30	5.00	6.00	6.30	7.00	8.00	9.00	10.00	10.30	11.00	12.00	12.30	1.00
Next Best Bite Time am/pm	7.00	8.00	9.00	10.00	11.00	12.00	12.30	1.00	2.00	3.00	4.00	5.00	6.00	6.30	7.00	8.00	9.00	10.00	10.30	11.00	12.00	12.30	1.00	2.00	3.00	4.00	4.30	5.00	6.00	6.30	7.00
RATING	EX	AV	AV	AV	VG	VG	VG	AV	AV	AV	AV	EX	EX	EX	AV	AV	AV	AV	AV	VG	VG	VG	AV	AV	AV	AV	AV	EX	EX	EX	AV

FEBRUARY	1st	2nd	3rd	4th	5th	6th	7th	8th	9th	10th	11th	12th	13th	14th	15th	16th	17th	18th	19th	20th	21st	22nd	23rd	24th	25th	26th	27th	28th
MOON		P10			First Q							Full Moon						Ap			Third Q							New Moon
Best Bite Time am/pm	2.00	3.00	4.00	5.00	6.00	7.00	8.00	9.00	10.00	10.30	11.00	12.00	12.30	1.00	2.00	3.00	3.30	4.00	4.30	5.00	6.00	7.00	8.00	9.00	10.00	11.00	11.30	12.00
Next Best Bite Time am/pm	8.00	9.00	10.00	11.00	12.00	1.00	2.00	3.00	4.00	4.30	5.00	6.00	6.30	7.00	8.00	9.00	9.30	10.00	10.30	11.00	12.00	1.00	2.00	3.00	4.00	5.00	5.30	6.00
RATING	AV	AV	AV	VG	VG	VG	AV	AV	AV	AV	EX	EX	EX	AV	AV	AV	AV	AV	AV	VG	VG	VG	AV	AV	AV	AV	EX	EX

MARCH	1st	2nd	3rd	4th	5th	6th	7th	8th	9th	10th	11th	12th	13th	14th	15th	16th	17th	18th	19th	20th	21st	22nd	23rd	24th	25th	26th	27th	28th	29th	30th	31st
MOON	P7						First Q							Full Moon				Ap					Third Q					New Moon		P4	
Best Bite Time am/pm	1.00	2.00	3.00	4.00	5.00	5.30	6.00	7.00	8.00	9.00	10.00	11.00	11.30	12.00	1.00	2.00	3.00	3.30	4.00	4.30	5.00	6.00	7.00	8.00	9.00	10.00	11.00	12.00	1.00	2.00	3.00
Next Best Bite Time am/pm	7.00	8.00	9.00	10.00	11.00	11.30	12.00	1.00	2.00	3.00	4.00	5.00	5.30	6.00	7.00	8.00	9.00	9.30	10.00	10.30	11.00	12.00	1.00	2.00	3.00	4.00	5.00	6.00	7.00	8.00	9.00
RATING	EX	AV	AV	AV	AV	VG	VG	VG	AV	AV	AV	AV	EX	EX	EX	AV	AV	AV	AV	AV	VG	VG	VG	AV	AV	AV	EX	EX	EX	AV	AV

APRIL	1st	2nd	3rd	4th	5th	6th	7th	8th	9th	10th	11th	12th	13th	14th	15th	16th	17th	18th	19th	20th	21st	22nd	23rd	24th	25th	26th	27th	28th	29th	30th
MOON					First Q								Full Moon	Ap						Third Q							P3 New Moon			
Best Bite Time am/pm	3.30	4.00	5.00	5.30	6.00	7.00	8.00	8.30	9.00	10.00	11.00	11.30	12.00	12.30	1.00	2.00	3.00	4.00	5.00	6.00	6.30	7.00	8.00	9.00	10.00	11.00	12.00	12.00	1.00	2.00
Next Best Bite Time am/pm	9.30	10.00	11.00	11.30	12.00	1.00	2.00	2.30	3.00	4.00	5.00	5.30	6.00	6.30	7.00	8.00	9.00	10.00	11.00	12.00	12.30	1.00	2.00	3.00	4.00	5.00	6.00	6.00	7.00	8.00
RATING	AV	AV	AV	VG	VG	VG	AV	AV	AV	AV	AV	EX	EX	EX	AV	AV	AV	AV	VG	VG	VG	AV	AV	AV	AV	EX	EX	EX	AV	AV

MAY	1st	2nd	3rd	4th	5th	6th	7th	8th	9th	10th	11th	12th	13th	14th	15th	16th	17th	18th	19th	20th	21st	22nd	23rd	24th	25th	26th	27th	28th	29th	30th	31st
MOON					First Q						Ap	Full Moon								Third Q						P5	New Moon				
Best Bite Time am/pm	2.30	3.00	4.00	5.00	6.00	7.00	8.00	8.30	9.00	10.00	11.00	12.00	12.30	1.00	2.00	3.00	4.00	4.30	5.00	6.00	7.00	8.00	9.00	9.30	10.00	11.00	12.00	1.00	2.00	3.00	3.30
Next Best Bite Time am/pm	8.30	9.00	10.00	11.00	12.00	1.00	2.00	2.30	3.00	4.00	5.00	6.00	6.30	7.00	8.00	9.00	10.00	10.30	11.00	12.00	1.00	2.00	3.00	3.30	4.00	5.00	6.00	7.00	8.00	9.00	9.30
RATING	AV	AV	AV	VG	VG	VG	AV	AV	AV	AV	EX	EX	EX	AV	AV	AV	AV	AV	VG	VG	VG	AV	AV	AV	AV	EX	EX	EX	AV	AV	AV

JUNE	1st	2nd	3rd	4th	5th	6th	7th	8th	9th	10th	11th	12th	13th	14th	15th	16th	17th	18th	19th	20th	21st	22nd	23rd	24th	25th	26th	27th	28th	29th	30th
MOON			First Q				Ap				Full Moon								Third Q				P8		New Moon					
Best Bite Time am/pm	4.30	5.00	6.00	6.30	7.00	8.00	9.00	10.00	11.00	11.30	12.00	1.00	2.00	3.00	3.30	4.30	5.00	5.30	6.00	7.00	8.00	9.00	10.00	11.00	12.00	1.00	2.00	3.00	3.30	4.30
Next Best Bite Time am/pm	10.30	11.00	12.00	12.30	1.00	2.00	3.00	4.00	5.00	5.30	6.00	7.00	8.00	9.00	9.30	10.30	11.00	11.30	12.00	1.00	2.00	3.00	4.00	5.00	6.00	7.00	8.00	9.00	9.30	10.30
RATING	AV	VG	VG	VG	AV	AV	AV	AV	AV	EX	EX	EX	AV	AV	AV	AV	AV	VG	VG	VG	AV	AV	AV	EX	EX	EX	AV	AV	AV	AV

AV	VG	EX

JULY	1st	2nd	3rd	4th	5th	6th	7th	8th	9th	10th	11th	12th	13th	14th	15th	16th	17th	18th	19th	20th	21st	22nd	23rd	24th	25th	26th	27th	28th	29th	30th	31st
MOON			First Q		Ap					Full Moon							Third Q			P11				New Moon							
Best Bite Time am/pm	5.00	5.30	6.00	6.30	7.00	8.00	9.00	10.00	11.00	12.00	1.00	2.00	2.30	3.00	4.00	5.00	6.00	6.30	7.00	8.00	9.00	10.00	11.00	12.00	1.00	2.00	2.30	3.00	3.30	4.00	5.00
Next Best Bite Time am/pm	11.00	11.30	12.00	12.30	1.00	2.00	3.00	4.00	5.30	7.00	7.00	8.00	8.30	9.00	10.00	11.00	12.00	12.30	1.00	2.00	3.00	4.00	5.30	7.00	7.00	8.00	8.30	9.00	9.30	10.00	11.00
RATING	AV	VG	VG	VG	AV	AV	AV	AV	EX	EX	EX	AV	AV	AV	AV	VG	VG	VG	AV	AV	AV	AV	EX	EX	EX	AV	AV	AV	AV	AV	AV

AUGUST	1st	2nd	3rd	4th	5th	6th	7th	8th	9th	10th	11th	12th	13th	14th	15th	16th	17th	18th	19th	20th	21st	22nd	23rd	24th	25th	26th	27th	28th	29th	30th	31st
MOON		First Q Ap							Full Moon					P12		Third Q							New Moon							Ap	First Q
Best Bite Time am/pm	5.30	6.00	7.00	7.30	8.00	9.00	10.00	11.00	12.00	1.00	2.00	3.00	4.00	4.30	5.00	6.00	6.30	7.00	8.00	9.00	10.00	11.00	12.00	1.00	2.00	3.00	4.00	4.30	5.00	5.30	6.00
Next Best Bite Time am/pm	11.30	12.00	1.00	1.30	2.00	3.00	4.00	5.00	6.00	7.00	8.00	9.00	10.00	10.30	11.00	12.00	12.30	1.00	2.00	3.00	4.00	5.00	6.00	7.00	8.00	9.00	10.00	10.30	11.00	11.30	12.00
RATING	VG	VG	VG	AV	AV	AV	AV	EX	EX	EX	AV	AV	AV	AV	VG	VG	VG	AV	AV	AV	AV	EX	EX	EX	AV	AV	AV	AV	AV	VG	VG

SEPTEMBER	1st	2nd	3rd	4th	5th	6th	7th	8th	9th	10th	11th	12th	13th	14th	15th	16th	17th	18th	19th	20th	21st	22nd	23rd	24th	25th	26th	27th	28th	29th	30th
MOON							Full Moon			P9				Third Q							New Moon					Ap				First Q
Best Bite Time am/pm	6.30	7.00	8.00	9.00	10.00	11.00	12.00	1.00	2.00	3.00	3.30	4.00	5.00	6.00	6.30	7.00	8.00	9.00	10.00	11.00	12.00	1.00	2.00	3.00	3.30	4.00	4.30	5.00	5.30	6.00
Next Best Bite Time am/pm	12.30	1.00	2.00	3.00	4.00	5.00	6.00	7.00	8.00	9.00	9.30	10.00	11.00	12.00	12.30	1.00	2.00	3.00	4.00	5.00	6.00	7.00	8.00	9.00	9.30	10.00	10.30	11.00	11.30	12.00
RATING	VG	AV	AV	AV	AV	EX	EX	EX	AV	AV	AV	AV	VG	VG	VG	AV	AV	AV	AV	EX	EX	EX	AV	AV	AV	AV	AV	AV	VG	VG

OCTOBER	1st	2nd	3rd	4th	5th	6th	7th	8th	9th	10th	11th	12th	13th	14th	15th	16th	17th	18th	19th	20th	21st	22nd	23rd	24th	25th	26th	27th	28th	29th	30th	31st
MOON							Full Moon	P6						Third Q							New Moon			Ap						First Q	
Best Bite Time am/pm	7.00	7.30	8.00	9.00	10.00	11.00	12.00	1.00	2.00	2.30	3.00	4.00	5.00	6.00	6.30	7.00	8.00	9.00	10.00	11.00	12.00	1.00	1.00	2.00	3.00	4.00	4.30	5.00	6.00	6.30	7.00
Next Best Bite Time am/pm	1.00	1.30	2.00	3.00	4.00	5.00	6.00	7.00	8.00	8.30	9.00	10.00	11.00	12.00	12.30	1.00	2.00	3.00	4.00	5.00	6.00	7.00	7.00	8.00	9.00	10.00	10.30	11.00	12.00	12.30	1.00
RATING	VG	AV	AV	AV	AV	EX	EX	EX	AV	AV	AV	AV	VG	VG	VG	AV	AV	AV	AV	EX	EX	EX	AV	AV	AV	AV	AV	VG	VG	VG	AV

NOVEMBER	1st	2nd	3rd	4th	5th	6th	7th	8th	9th	10th	11th	12th	13th	14th	15th	16th	17th	18th	19th	20th	21st	22nd	23rd	24th	25th	26th	27th	28th	29th	30th
MOON					P1 Full Moon							Third Q								New Moon Ap								First Q		
Best Bite Time am/pm	8.00	9.00	10.00	11.00	12.00	1.00	1.30	2.00	3.00	4.00	5.00	6.00	6.30	7.00	7.30	8.00	9.00	10.00	11.00	12.00	1.00	2.00	2.30	3.00	4.00	5.00	6.00	7.00	8.00	9.00
Next Best Bite Time am/pm	2.00	3.00	4.00	5.00	6.00	7.00	7.00	8.00	9.00	10.00	11.00	12.00	12.30	1.00	1.30	2.00	3.00	4.00	5.00	6.00	7.00	8.00	8.30	9.00	10.00	11.00	12.00	1.00	2.00	3.00
RATING	AV	AV	AV	EX	EX	EX	AV	AV	AV	AV	VG	VG	VG	AV	AV	AV	AV	AV	EX	EX	EX	AV	AV	AV	AV	VG	VG	VG	AV	AV

DECEMBER	1st	2nd	3rd	4th	5th	6th	7th	8th	9th	10th	11th	12th	13th	14th	15th	16th	17th	18th	19th	20th	21st	22nd	23rd	24th	25th	26th	27th	28th	29th	30th	31st
MOON				Full moon, P2								Third Q					Ap			New Moon							First Q				
Best Bite Time am/pm	10.00	10.30	11.00	12.00	12.30	1.00	2.00	3.00	4.00	4.30	5.00	6.00	6.30	7.00	8.00	9.00	10.00	10.30	11.00	12.00	1.00	2.00	3.00	3.30	4.00	5.30	6.00	7.00	7.30	8.00	9.00
Next Best Bite Time am/pm	4.00	4.30	5.00	6.00	6.30	7.00	8.00	9.00	10.00	10.30	11.00	12.00	12.30	1.00	2.00	3.00	4.00	4.30	5.00	6.00	7.00	8.00	9.00	9.30	10.00	11.30	12.00	1.00	1.30	2.00	3.00
RATING	AV	AV	EX	EX	EX	AV	AV	AV	AV	AV	VG	VG	VG	AV	AV	AV	AV	AV	EX	EX	EX	AV	AV	AV	AV	VG	VG	VG	AV	AV	AV

www.ingramcontent.com/pod-product-compliance
Lightning Source LLC
LaVergne TN
LVHW070850210926
848482LV00016B/904